Solved Problems in Quantum and Statistical Mechanics

Michele Cini
Francesco Fucito
Mauro Sbragaglia

Solved Problems
in Quantum
and Statistical Mechanics

 Springer

Michele Cini
Department of Physics
University of Rome
Tor Vergata,
INFN
Laboratori Nazionali Frascati

Francesco Fucito
Department of Physics
University of Rome
Tor Vergata and INFN

Mauro Sbragaglia
Department of Physics
University of Rome
Tor Vergata and INFN

UNITEXT- Collana di Fisica e Astronomia
ISSN print edition: 2038-5714   ISSN electronic edition: 2038-5765

ISBN 978-88-470-2314-7   ISBN 978-88-470-2315-4 (eBook)
DOI 10.1007/978-88-470-2315-4

Library of Congress Control Number: 2011940537

Springer Milan Dordrecht Heidelberg London New York

Cover-Design: Simona Colombo, Milano
Typesetting with LaTeX: CompoMat S.r.l., Configni (RI)

Springer-Verlag Italia S.r.l., Via Decembrio 28, I-20137 Milano
Springer fa parte di Springer Science + Business Media (www.springer.com)

Preface

Italian students start studying Quantum and Statistical Mechanics in the last year of their undergraduate studies. Many physicists think these subjects are the core of an education in physics. At the same time, these two subjects are not easily learnt by the average student. In Italy the final exam is divided in two separate parts: there is, in fact, a written and an oral one. Most textbooks concentrate on the principles of the theory and the applications are dealt with at the end of each chapter under the headings 'problems' or 'exercises'. Most of the times the latter consist only of the text of the problem with some vague indications on how to proceed with the solution. Some other times the solution is completely left to the student. The authors of the present book think this is didactically wrong: these applications are crucial for a correct understanding of the subject and help the students get acquainted with the mathematical tools they have learnt in other classes. Many times we have noticed that a simple change in the denominations of the letters was enough to throw a student in disarray: a function whose behaviour was familiar when the independent variable was called x, became unfathomable when the very same variable was the energy E. In reality, during the elementary study of classical physics, the exercises are mostly straightforward applications of the general formulae deduced from experience or at the most require the simplest notions of differential and integral calculus. Things change with Quantum and Statistical Mechanics whose mathematical formalism is more complex. Also the problems and exercises reflect this point and the required solutions often need longer and more elaborate manipulations.

It is for these reasons that in our teaching we have always dedicated a large amount of time to the discussion of the applications, to the correction of the problems, and have tried to elaborate written solutions with lengthy discussions to help the students get ready for the written exams. From this point to the publication of our notes it has been a natural step.

Acknowledgements

Francesco Fucito wishes to specially thank M. Guagnelli with whom he has shared the teaching of Statistical Mechanics at the Università di Roma *Tor Vergata* for many years and with whom he started this project which, for various reasons, could not be ended together. He also wants to thank M.G. De Divitiis for collaboration in an early stage of this work. Francesco Fucito and Mauro Sbragaglia are also grateful to their students in the classes of Statistical Mechanics in the years 2007 − 2010 and in particular we wish to thank Cristina Bertulli, Andrea Bussone, Roberta De Angelis, Giulio De Magistris, Davide Liberati, Sebastian Grothans, Valerio Latini, Francesca Mancini, Stefano Marchesani, Claudia Narcisi, Francesco Nazzaro, Claudia Violante. Their questions and support have been of valuable help for us.

Rome, September 2011   *Michele Cini*
                                                              Francesco Fucito
                                                             Mauro Sbragaglia

Contents

Part I Theoretical Background

1   Summary of Quantum and Statistical Mechanics 3
1.1   One Dimensional Schrödinger Equation 3
1.2   One Dimensional Harmonic Oscillator 5
1.3   Variational Method 6
1.4   Angular Momentum 8
1.5   Spin 10
1.6   Hydrogen Atom 10
1.7   Solutions of the Three Dimensional Schrödinger Equation 12
1.8   WKB Method 19
1.9   Perturbation Theory 21
1.10 Thermodynamic Potentials 23
1.11 Fundamentals of Ensemble Theory 26
      1.11.1 Microcanonical Ensemble 28
      1.11.2 Canonical Ensemble 28
      1.11.3 Grand Canonical Ensemble 29
      1.11.4 Quantum Statistical Mechanics 29
1.12 Kinetic Approach 30
1.13 Fluctuations 31
1.14 Mathematical Formulae 32
References 36

Part II Quantum Mechanics – Problems

2   Formalism of Quantum Mechanics and One Dimensional Problems .. 39

3   Angular Momentum and Spin 113

4 Central Force Field 145

5 Perturbation Theory and WKB Method 163

Part III Statistical Mechanics – Problems

6 Thermodynamics and Microcanonical Ensemble 193

7 Canonical Ensemble 227

8 Grand Canonical Ensemble 289

9 Kinetic Physics 301

10 Bose-Einstein Gases 315

11 Fermi-Dirac Gases 337

12 Fluctuations and Complements 363

Index 393

Part I

Theoretical Background

Summary of Quantum and Statistical Mechanics

1.1 One Dimensional Schrödinger Equation

In Quantum Mechanics, the state of a particle in one dimension and in presence of a potential $U(x,t)$, is entirely described by a complex wave function $\psi(x,t)$ obeying the time dependent Schrödinger equation

$$i\hbar\frac{\partial \psi(x,t)}{\partial t} = -\frac{\hbar^2}{2m}\frac{\partial^2 \psi(x,t)}{\partial x^2} + U(x,t)\psi(x,t)$$

where m is the mass of the particle and $\hbar$ is the Planck constant, h, divided by 2π. If we multiply the Schrödinger equation by the complex conjugate wave function $\psi^*(x,t)$, take the complex conjugate of the Schrödinger equation and multiply by $\psi(x,t)$, and finally subtract both expressions, we find the so-called *continuity equation*

$$\frac{\partial |\psi(x,t)|^2}{\partial t} + \frac{\partial}{\partial x}\left[\frac{i\hbar}{2m}\left(\psi(x,t)\frac{\partial \psi^*(x,t)}{\partial x} - \psi^*(x,t)\frac{\partial \psi(x,t)}{\partial x}\right)\right] = 0.$$

This equation represents the conservation law for the quantity $\int |\psi(x,t)|^2 dx$ and allows us to interpret $|\psi(x,t)|^2$ as the probability density function to find the particle in the point x at a time t. The quantity

$$J(x,t) = \frac{i\hbar}{2m}\left(\psi(x,t)\frac{\partial \psi^*(x,t)}{\partial x} - \psi^*(x,t)\frac{\partial \psi(x,t)}{\partial x}\right)$$

is the density flux for such probability. The physical interpretation of $|\psi(x,t)|^2$ sets some conditions on $\psi(x,t)$ that has to be chosen as a continuous not multivalued function without singularities. Also, the derivatives of $\psi(x,t)$ have to be continuous, with the exception of a moving particle in a potential field possessing some discontinuities, as we will see explicitly in the exercises. If the potential does not depend explicitly on time, $U(x,t) = U(x)$, the time dependence can be separated out

Cini M., Fucito F., Sbragaglia M.: Solved Problems in Quantum and Statistical Mechanics.
DOI 10.1007/978-88-470-2315-4_1, © Springer-Verlag Italia 2012

from the Schrödinger equation and the solutions, named *stationary*, satisfy

$$\psi(x,t) = e^{-\frac{i}{\hbar}Et}\psi(x).$$

In such a case, the functions $\rho = |\psi|^2$ and J are independent of time. Using the form of $\psi(x,t)$ in the original equation, we end up with the stationary Schrödinger equation

$$\left[-\frac{\hbar^2}{2m}\frac{d^2}{dx^2} + U(x)\right]\psi(x) = \hat{H}\psi(x) = E\psi(x).$$

The operator $\hat{H}$ is known as the Hamiltonian of the system. Continuous, non multivalued and finite functions which are solutions of this equation exist only for particular values of the parameter E, which has to be identified with the energy of the particle. The energy values may be continuous (the case of a continuous spectrum for the Hamiltonian $\hat{H}$), discrete (discrete spectrum), or even present a discrete and continuous part together. For a discrete spectrum, the associated ψ may be normalized to unity

$$\int |\psi(x)|^2\,dx = 1.$$

All the functions ψ corresponding to precise values of the energy are called eigenfunctions and are orthogonal. With a continuous spectrum, the condition of orthonormality may be written using the Dirac delta function

$$\int \psi_E^*(x)\psi_{E'}(x)\,dx = \delta(E - E').$$

The condition of continuity for the wave function and its derivatives is valid even in the case when the potential energy $U(x)$ is discontinuous. Nevertheless, such conditions are not valid when the potential energy becomes infinite outside the domain where we solve our differential equations. The particle cannot penetrate a region of the space where $U = +\infty$ (you can imagine electrons inside a box), and in such a region we must have $\psi = 0$. The condition of continuity imposes a vanishing wave function where the potential energy barrier is infinite and, consequently, the derivatives may present discontinuities.

Let $\tilde{U}$ be the minimum of the potential. Since the average value of the energy is $\bar{E} = \bar{T} + \bar{U}$, and since $\bar{U} > \tilde{U}$, we conclude that

$$\bar{E} > \tilde{U}$$

due to the positive value of $\bar{T}$, that is the average kinetic energy of the particle. This relation is true for a generic state and, in particular, is still valid for an eigenfunction of the discrete spectrum. It follows that $E_n > \tilde{U}$, with E_n any of the eigenvalues of the discrete spectrum. If we now define the potential energy in such a way that it vanishes at infinity ($U(\pm\infty) = 0$), the discrete spectrum is characterized by all those energy levels $E < 0$ which represent bound states. In fact, if the particle is in a bound state, its motion takes place between two points (say x_1, x_2) so that $\psi(\pm\infty) = 0$.

This constraints the normalization condition for the states. In Classical Mechanics, the inaccessible regions where $E < U$ have an imaginary velocity. In Quantum Mechanics, instead, particle motion can also take place in those regions where $E < U$, although the probability density function is going rapidly to zero there.

The continuous spectrum is described by positive values of the energy. In such a case, the region of motion is not bounded ($\psi(\pm\infty) \neq 0$) and the resulting wave function cannot be normalized.

We finally end this section giving some general properties of the solution to the one dimensional Schrödinger equation:

- all the energy levels of the discrete spectrum are non degenerate;
- the eigenfunction $\psi_n(x)$, corresponding to the eigenvalue E_n ($n = 0, 1, 2, ...$; $E_i < E_j$ if $i < j$), vanishes n times for finite values of x (oscillation theorem);
- if the potential energy is symmetrical, $U(x) = U(-x)$, all the eigenfunctions of the discrete spectrum must be either even or odd.

1.2 One Dimensional Harmonic Oscillator

When treating the one dimensional harmonic oscillator from the point of view of Quantum Mechanics, we need to replace the usual classical variables x (position) and p (momentum) with the associated operators satisfying the commutation rule $[\hat{x}, \hat{p}] = i\hbar\mathbb{1}$, where $\mathbb{1}$ is the identity operator. A stationary state with energy E satisfies the following differential equation

$$-\frac{\hbar^2}{2m}\frac{d^2\psi(x)}{dx^2} + \frac{1}{2}m\omega^2 x^2\psi(x) = E\psi(x)$$

with m the mass of the oscillator and ω the angular frequency of the oscillations. Eigenvalues and eigenstates of the Hamiltonian are given by

$$E_n = \hbar\omega\left(n + \frac{1}{2}\right)$$

$$\psi_n(x) = \langle x|n\rangle = C_n H_n(\xi)e^{-\xi^2/2} \quad C_n = \frac{1}{2^{n/2}\sqrt{n!}}\left(\frac{m\omega}{\hbar\pi}\right)^{1/4}$$

where we have used a rescaled variable ξ defined by

$$\xi = \sqrt{\frac{m\omega}{\hbar}}x$$

and where

$$H_n(\xi) = (-1)^n e^{\xi^2}\frac{d^n}{d\xi^n}e^{-\xi^2}$$

represents the n-th order Hermite polynomial. The first Hermite polynomials are

$$H_0(\xi) = 1 \qquad H_1(\xi) = 2\xi$$

$$H_2(\xi) = 4\xi^2 - 2 \qquad H_3(\xi) = 8\xi^3 - 12\xi.$$

Equivalently, we can describe the properties of the harmonic oscillator with the *creation and annihilation operators*, $\hat{a}^\dagger$ and $\hat{a}$, such that $[\hat{a}, \hat{a}^\dagger] = \mathbb{1}$. In this case, the Hamiltonian becomes

$$\hat{H} = \hbar\omega \left(\hat{a}^\dagger \hat{a} + \frac{1}{2}\mathbb{1} \right) = \hbar\omega \left(\hat{n} + \frac{1}{2}\mathbb{1} \right)$$

where $\hat{n} = \hat{a}^\dagger \hat{a}$ is the number operator with the property $\hat{n}|n\rangle = n|n\rangle$. The relations connecting the creation and annihilation operators with the position and momentum operators are

$$\begin{cases} \hat{a} = \sqrt{\frac{m\omega}{2\hbar}}\hat{x} + \frac{i}{\sqrt{2m\hbar\omega}}\hat{p} \\ \hat{a}^\dagger = \sqrt{\frac{m\omega}{2\hbar}}\hat{x} - \frac{i}{\sqrt{2m\hbar\omega}}\hat{p} \end{cases}$$

$$\begin{cases} \hat{x} = \sqrt{\frac{\hbar}{2m\omega}}(\hat{a}^\dagger + \hat{a}) \\ \hat{p} = i\sqrt{\frac{m\hbar\omega}{2}}(\hat{a}^\dagger - \hat{a}). \end{cases}$$

The creation and annihilation operators act on a generic eigenstate as *step up* and *step down* operators

$$\hat{a}^\dagger |n\rangle = \sqrt{n+1}\,|n+1\rangle \qquad \hat{a}|n\rangle = \sqrt{n}\,|n-1\rangle$$

so that

$$|n\rangle = \frac{(\hat{a}^\dagger)^n}{\sqrt{n!}}|0\rangle \qquad \langle n|m\rangle = \delta_{nm}.$$

1.3 Variational Method

The *variational method* is an approximation method used to find approximate ground and excited states. The basis for this method is the *variational principle* which we briefly describe now. Let $|\phi\rangle$ be a state of an arbitrary quantum system with one or many particles normalized such that

$$N = \langle \phi|\phi \rangle = 1.$$

The energy of the quantum system is a quadratic functional of $|\phi\rangle$

$$E = \langle \phi|\hat{H}|\phi \rangle$$

and cannot be lower than the ground state ε_0; in fact, let us expand the state $|\phi\rangle$ in the eigenfunction basis $|\psi_n\rangle$ (each one corresponding to the eigenvalue ε_n) of $\hat{H}$

$$|\phi\rangle = \sum_n \langle \psi_n|\phi\rangle |\psi_n\rangle \qquad \sum_n |\langle \psi_n|\phi\rangle|^2 = 1.$$

Since $\varepsilon_n \geq \varepsilon_0$, we find that

$$E = \langle \phi|\hat{H}|\phi\rangle = \sum_n |\langle \psi_n|\phi\rangle|^2 \varepsilon_n \geq \sum_n |\langle \psi_n|\phi\rangle|^2 \varepsilon_0 = \varepsilon_0.$$

From this we infer that the energy of the ground state can be found by minimization. Let us start by variating the state $|\phi\rangle$

$$|\phi\rangle \rightarrow |\phi\rangle + |\delta\phi\rangle.$$

The variation of the energy which follows is

$$\delta E = \langle \delta\phi|\hat{H}|\phi\rangle + \langle \phi|\hat{H}|\delta\phi\rangle + \mathcal{O}(\delta^2).$$

At the same time, the normalization changes as

$$\delta N = \langle \delta\phi|\phi\rangle + \langle \phi|\delta\phi\rangle + \mathcal{O}(\delta^2).$$

The extremum we are looking for must be represented by functions whose norm is 1. This condition can be efficiently imposed by introducing a Lagrangian multiplier and minimizing the quantity

$$Q_\lambda(\phi) = \langle \phi|\hat{H} - \lambda \, \mathbb{1}|\phi\rangle = (E - \lambda N).$$

The variation of $Q_\lambda(\phi)$ must be zero for variations in both ϕ and λ

$$\delta_\phi Q_\lambda(\phi) = \langle \delta\phi|\hat{H} - \lambda \, \mathbb{1}|\phi\rangle = 0 \qquad \delta_\lambda Q_\lambda(\phi) = \delta\lambda(\langle \phi|\phi\rangle - 1) = 0.$$

The second equation is the constraint, the first must be valid for arbitrary variations, leading to

$$\hat{H}|\phi\rangle = \lambda|\phi\rangle$$

which is the stationary Schrödinger equation. Multiplying by $\langle \phi|$, we get the value of the multiplier, i.e. the energy of the ground state. We remark that the condition of constrained minimum, follows from that of unconstrained minimum $\delta(\langle \phi|\hat{H}|\phi\rangle) = 0$ substituting $\hat{H}$ with $(\hat{H} - \lambda \, \mathbb{1})$. This can also be done by making $|\phi\rangle$ depend on the multiplier λ, whose value is fixed imposing

$$N = \langle \phi(\lambda)|\phi(\lambda)\rangle = 1.$$

1.4 Angular Momentum

From the definition of the angular momentum in Classical Mechanics

$$L = r \wedge p$$

with $r \equiv (x, y, z)$, $p \equiv (p_x, p_y, p_z)$, we get its quantum mechanical expression, once the vectors r, p are substituted by their correspondent operators. Once the commutation rule between r, p is known, it is immediate to deduce the commutation rules of the different components of the angular momentum

$$[\hat{L}_x, \hat{L}_y] = i\hbar \hat{L}_z \qquad [\hat{L}_y, \hat{L}_z] = i\hbar \hat{L}_x \qquad [\hat{L}_x, \hat{L}_z] = -i\hbar \hat{L}_y.$$

From the theory of Lie Algebras, we know that a complete set of states is determined from a set of quantum numbers whose number is that of the maximum number of commuting operator we can build starting from the generators (in our case $\hat{L}_x, \hat{L}_y, \hat{L}_z$). One of these operators is the *Casimir operator*

$$\hat{L}^2 = \hat{L}_x^2 + \hat{L}_y^2 + \hat{L}_z^2$$

which is commuting with all the generators of the group, i.e. $[\hat{L}^2, \hat{L}_i] = 0$, $i = x, y, z$. The other element of this sub algebra is one of the three generators $\hat{L}_x, \hat{L}_y, \hat{L}_z$: the convention is to choose $\hat{L}_z$. The quantum states are thus labelled by the quantum numbers l, m such that

$$\hat{L}^2 |l, m\rangle = \hbar^2 l(l+1) |l, m\rangle \qquad \hat{L}_z |l, m\rangle = \hbar m |l, m\rangle \qquad -m \leq l \leq m$$

$$\hat{L}_+ |l, m\rangle = \hbar \sqrt{(l-m)(l+m+1)} |l, m+1\rangle$$

$$\hat{L}_- |l, m\rangle = \hbar \sqrt{(l+m)(l-m+1)} |l, m-1\rangle$$

where $\hat{L}_\pm = (\hat{L}_x \pm i\hat{L}_y)$ are known as *raising and lowering operators* for the z component of the angular momentum. Other useful relations are

$$[\hat{L}_z, \hat{L}_\pm] = \pm\hbar L_\pm \qquad [\hat{L}^2, \hat{L}_\pm] = 0 \qquad [\hat{L}_+, \hat{L}_-] = 2\hbar \hat{L}_z$$

$$\hat{L}_+ \hat{L}_- = \hat{L}^2 - \hat{L}_z^2 + \hbar \hat{L}_z$$

$$\hat{L}_- \hat{L}_+ = \hat{L}^2 - \hat{L}_z^2 - \hbar \hat{L}_z.$$

When acting on functions of the spherical polar coordinates, the generators and the Casimir operator take the form

$$\hat{L}_x = i\hbar \left(\sin\phi \frac{\partial}{\partial\theta} + \cot\theta \cos\phi \frac{\partial}{\partial\phi} \right)$$

$$\hat{L}_y = -i\hbar \left(\cos\phi \frac{\partial}{\partial\theta} - \cot\theta \sin\phi \frac{\partial}{\partial\phi} \right)$$

$$\hat{L}_z = -i\hbar \frac{\partial}{\partial \phi}$$

$$\hat{L}^2 = -\hbar^2 \left[\frac{1}{\sin \theta} \frac{\partial}{\partial \theta} \left(\sin \theta \frac{\partial}{\partial \theta} \right) + \frac{1}{\sin^2 \theta} \frac{\partial^2}{\partial \phi^2} \right].$$

To write the state $|l,m\rangle$ in spherical coordinates it is useful to introduce the *spherical harmonics*

$$Y_{l,m}(\theta, \phi) = \langle \theta, \phi | l, m \rangle$$

which enjoy the property

$$Y_{l,m}(\theta, \phi) = (-1)^m \sqrt{\frac{(2l+1)}{4\pi} \frac{(l-m)!}{(l+m)!}} e^{im\phi} P_l^m(\cos \theta) \qquad m \geq 0$$

$$Y_{l,m}(\theta, \phi) = Y_{l,-|m|}(\theta, \phi) = (-1)^{|m|} Y^*_{l,|m|}(\theta, \phi) \qquad m < 0$$

where $P_l^m(\cos \theta)$ is the associated Legendre polynomial defined by

$$P_l^m(u) = (1-u^2)^{m/2} \frac{d^m}{du^m} P_l(u) \qquad 0 \leq m \leq l$$

$$P_l(u) = \frac{1}{2^l l!} \frac{d^l}{du^l} \left[(u^2-1)^l \right]$$

where $P_l(u)$ is the Legendre polynomial of order l. Some explicit expressions for $l = 0, 1, 2$ are

$$Y_{0,0}(\theta, \phi) = \frac{1}{\sqrt{4\pi}}$$

$$Y_{1,0}(\theta, \phi) = \sqrt{\frac{3}{4\pi}} \cos \theta \qquad Y_{1,\pm1}(\theta, \phi) = \mp \sqrt{\frac{3}{8\pi}} \sin \theta e^{\pm i\phi}$$

$$Y_{2,0}(\theta, \phi) = \sqrt{\frac{5}{16\pi}} (3\cos^2 \theta - 1) \qquad Y_{2,\pm1}(\theta, \phi) = \mp \sqrt{\frac{15}{8\pi}} \sin \theta \cos \theta e^{\pm i\phi}$$

$$Y_{2,\pm2}(\theta, \phi) = \sqrt{\frac{15}{32\pi}} \sin^2 \theta e^{\pm 2i\phi}.$$

Given the two angular momentum operators $\hat{L}_1, \hat{L}_2$, we now want to deduce the states of the angular momentum operator sum of the two, $\hat{L} = \hat{L}_1 + \hat{L}_2$. This is possible using the states $|l_1, l_2, m_1, m_2\rangle$ or $|l_1, l_2, l, m\rangle$, where $\hbar^2 l(l+1)$, $\hbar m$ are the eigenvalues of the operators $\hat{L}^2 = \hat{L}_1^2 + \hat{L}_2^2$ and $\hat{L}_z = \hat{L}_{1z} + \hat{L}_{2z}$ respectively. The relation between these two sets of states is

$$|l_1, l_2, l, m\rangle = \sum_{m_1, m_2} \langle l_1, l_2, m_1, m_2 | l_1, l_2, l, m \rangle |l_1, l_2, m_1, m_2\rangle.$$

The total angular momentum gets the values $l = l_2 + l_1, l_2 + l_1 - 1, ..., |l_2 - l_1|$, $m = m_1 + m_2$ and the coefficients $\langle l_1, l_2, m_1, m_2 | l_1, l_2, l, m \rangle$ are known as *Clebsh-Gordan* or *Wigner* coefficients.

1.5 Spin

From the experiment of Stern and Gerlach or from the splitting of the electron energy levels in an atom it follows that, besides an angular momentum, a moving electron has a *spin*. Such quantity has no classical correspondence and can get the two values $\pm \hbar/2$. The generators of this physical quantity are

$$\hat{S}_x = \frac{\hbar}{2} \hat{\sigma}_x \qquad \hat{S}_y = \frac{\hbar}{2} \hat{\sigma}_y \qquad \hat{S}_z = \frac{\hbar}{2} \hat{\sigma}_z$$

where $\hat{\sigma}_x$, $\hat{\sigma}_y$, $\hat{\sigma}_z$ are the *Pauli matrices* . A possible representation of these matrices (the one where $\hat{\sigma}_z$ is diagonal) is

$$\hat{\sigma}_x = \begin{pmatrix} 0 & 1 \\ 1 & 0 \end{pmatrix} \qquad \hat{\sigma}_y = \begin{pmatrix} 0 & -i \\ i & 0 \end{pmatrix} \qquad \hat{\sigma}_z = \begin{pmatrix} 1 & 0 \\ 0 & -1 \end{pmatrix}.$$

The states which describe the spin are two dimensional vectors

$$\chi = \begin{pmatrix} \chi_1 \\ \chi_2 \end{pmatrix}$$

and $|\chi_1|^2$, $|\chi_2|^2$ is the probability to get $\pm \hbar/2$ out of a spin measurement.

1.6 Hydrogen Atom

The problem of the motion of two interacting particles with coordinates r_1, r_2 can be reduced, in analogy with Classical Mechanics, to the motion of a single particle at distance r from a fixed centre. The Hamiltonian of the two particles of masses m_1 and m_2 and interacting with a centrally symmetric potential $U(r)$ is given by

$$\hat{H} = -\frac{\hbar^2}{2m_1} \nabla_1^2 - \frac{\hbar^2}{2m_2} \nabla_2^2 + \hat{U}(\hat{r})$$

where ∇_1^2, ∇_2^2 are the Laplacian operators in the coordinates r_1, r_2 and $r = r_1 - r_2$. Using the coordinate of the center of mass

$$R = \frac{m_1 r_1 + m_2 r_2}{m_1 + m_2}$$

the Hamiltonian becomes

$$\hat{H} = -\frac{\hbar^2}{2(m_1 + m_2)} \nabla_R^2 - \frac{\hbar^2}{2m_r} \nabla_r^2 + \hat{U}(\hat{r})$$

where $m_1 + m_2$ is the total mass and $m_r = m_1 m_2/(m_1 + m_2)$ the reduced mass. Then, we seek the solution in the form $\Psi(r_1, r_2) = \phi(R)\psi(r)$, where $\phi(R)$ describes the motion of a free particle and $\psi(r)$ describes the motion of a particle subject to the centrally symmetric potential $U(r)$

$$\nabla^2 \psi + \frac{2m_r}{\hbar^2}[E - U(r)]\psi = 0.$$

With the use of spherical polar coordinates, the equation becomes

$$\frac{1}{r^2}\frac{\partial}{\partial r}\left(r^2 \frac{\partial \psi}{\partial r}\right) + \frac{1}{r^2}\left[\frac{1}{\sin\theta}\frac{\partial}{\partial\theta}\left(\sin\theta\frac{\partial\psi}{\partial\theta}\right) + \frac{1}{\sin^2\theta}\frac{\partial^2\psi}{\partial\phi^2}\right] + \frac{2m_r}{\hbar^2}[E - U(r)]\psi = 0.$$

The differential operator dependent on the angular variables coincides with the Casimir operator $\hat{L}^2$, so that

$$\frac{\hbar^2}{2m_r}\left[-\frac{1}{r^2}\frac{\partial}{\partial r}\left(r^2\frac{\partial\psi}{\partial r}\right) + \frac{\hat{L}^2}{\hbar^2 r^2}\psi\right] + U(r)\psi = E\psi.$$

In the motion in a central field, the angular momentum is conserved: let us take two arbitrary values l, m for the angular momentum and its projection on the z axis. Given that the differential equation is separable, we look for a solution using the ansatz

$$\psi(r) = R(r)Y_{l,m}(\theta, \phi)$$

where $Y_{l,m}$ are the spherical harmonics, which are eigenfunctions of both $\hat{L}_z$ and $\hat{L}^2$. The above equation becomes

$$\frac{1}{r^2}\frac{d}{dr}\left(r^2\frac{dR}{dr}\right) - \frac{l(l+1)}{r^2}R + \frac{2m_r}{\hbar^2}[E - U(r)]R = 0.$$

Since the quantum number m is not appearing in this equation, the solutions will be $2l + 1$ degenerate with respect to the angular momentum. The dependence on $Y_{l,m}$ has been removed by multiplying by $Y_{l,m}^*$ and integrating over the angular part of the volume. Let us focus now on the radial part of the wave function and let us perform a further change of variables, by setting $R(r) = \frac{\Theta(r)}{r}$, to get

$$\frac{d^2\Theta}{dr^2} + \left[\frac{2m_r}{\hbar^2}(E - U(r)) - \frac{l(l+1)}{r^2}\right]\Theta = 0.$$

The domain of variation of r is now $[0, +\infty]$, and at the boundary of this region the wave function must vanish to guarantee that it can be normalized, thus leading to a discrete spectrum. The equation we got after these manipulations looks like a one dimensional Schrödinger equation with potential

$$U_{eff}(r) = U(r) + \frac{\hbar^2}{2m_r}\frac{l(l+1)}{r^2}.$$

For a fixed l, the radial part is determined by the quantum number labelling the energy, since in a one dimensional motion the eigenvalues are not degenerate. The angular part with quantum numbers l, m, and the energy spectrum, E_n, determine the particle motion without ambiguities. To label states with different angular momenta, we use the notation

$$l = \quad 0\,(s) \quad 1\,(p) \quad 2\,(d) \quad 3\,(f) \quad 4\,(g)\dots$$

Using the theorem of oscillations we see that the ground state is always an s wave since the wave function cannot have zeros for the lowest level, while the $Y_{l,m}$ for $l \neq 0$ are always oscillating functions with positive and negative values.

To close this section on the theory of the angular momentum, we report its application to the case of hydrogen-like atoms , i.e. those atoms with one electron with charge $-e$ (and mass m_e) and nuclear charge Ze. After the center of mass is separated out, the stationary Schrödinger equation describing the wave function of the electron becomes

$$\frac{\hbar^2}{2m_e}\left[-\frac{1}{r^2}\frac{\partial}{\partial r}\left(r^2\frac{\partial \psi_{n,l,m}}{\partial r}\right) + \frac{l(l+1)}{r^2}\psi_{n,l,m}\right] - \frac{Ze^2}{r}\psi_{n,l,m} = E_n\psi_{n,l,m}$$

where we have used the wave function for the orbital motion

$$\psi_{n,l,m}(r,\theta,\phi) = R_{n,l}(r)Y_{l,m}(\theta,\phi)$$

with n the principal quantum number giving the energy

$$E_n = -\frac{Z^2e^2}{2n^2a} \qquad a = \frac{\hbar^2}{m_e e^2} \qquad \text{(Bohr radius)}$$

and $R_{n,l}(r)$ the radial part of the wave function. Some of the expressions of $R_{n,l}(r)$ are here reported

$$R_{1,0}(r) = \left(\frac{Z}{a}\right)^{3/2} 2e^{-Zr/a}$$

$$R_{2,0}(r) = \left(\frac{Z}{2a}\right)^{3/2}\left(2 - \frac{Zr}{a}\right)e^{-Zr/2a}$$

$$R_{2,1}(r) = \left(\frac{Z}{2a}\right)^{3/2}\frac{Zr}{\sqrt{3}a}e^{-Zr/2a}.$$

1.7 Solutions of the Three Dimensional Schrödinger Equation

Let us now discuss the three dimensional solutions of the Schrödinger equation in full generality. We start from

$$\left(\frac{\hbar^2}{2m}\nabla^2 - U + E\right)\psi = 0$$

and we will choose an appropriate coordinate system according to the symmetry of the problem. This choice is crucial to get a separable differential equation and a simple form for the potential. The standard example is the hydrogen atom where the choice of spherical coordinates and a potential which is dependent only on the radius naturally lead to variable separation. Once this is done, the left over problem is to solve a one dimensional differential equation. The coefficients of these equations have pole singularities. There is only a finite number of coordinate systems leading to separable equations and their singularities can be classified according to their number and type. The solutions are usually classified in mathematics manuals, but to access these results some preliminary work is needed. This paragraph is meant to be a guide on how to use these results, neglecting all mathematical rigor and giving only the results of the theorems we will need (not many indeed) without demonstrations. The starting point is the study of the singularities which leads to the classification. Let us start with the one dimensional differential equation

$$y'' + p(x)y' + q(x)y = 0$$

where $y = y(x)$ is the unknown function, and $p(x)$, $q(x)$ some coefficients. Then, we define the different kinds of singularities associated with $p(x)$ and $q(x)$.

A point $x_0 (x_0 \neq +\infty)$ is *ordinary* if $p(x)$ and $q(x)$ are analytic functions (without singularities) in a neighborhood of x_0. As an example, take the two equations

$$y'' - e^x y = 0$$

$$x^5 y'' - y = 0.$$

In the first, all the points x_0 ($x_0 \neq +\infty$) are ordinary. In the second, all the points x_0, except $x_0 = 0, +\infty$ are ordinary. As for the behaviour close to an ordinary point x_0, the Fuchs theorem guarantees that the solution may be expanded in Taylor series, and that the radius of convergence of this series is at least equal to the distance between x_0 and the nearest singularity in the complex plane . For example, take the equation

$$(x^2 + 1)y' + 2xy = 0$$

that is of the first order for simplicity. The solution is $y = 1/(1 + x^2)$ that can be expanded in Taylor series with radius of convergence equal to 1, that is the distance between $x_0 = 0$ and i in the complex plane.

A point $x_0 (x_0 \neq +\infty)$ is a *regular singularity* if $p(x)$ has at most a single pole and $q(x)$ at most a pole of order two, i.e. they are of the form

$$p(x) = \frac{p_0}{(x - x_0)} \qquad q(x) = \frac{q_0}{(x - x_0)^2}$$

with p_0 and q_0 non singular coefficients . To give some examples, consider the three equations

$$(x - 2)^2 y'' - x^3 y = 0$$

$$x^3 y'' + \frac{x^2}{x-1} y' + xy = 0$$

$$x^3 y'' + \frac{x}{x-1} y' + xy = 0.$$

The first equation has a regular singularity in $x_0 = 2$, the second has regular singular points in $x_0 = 0$ and $x_0 = 1$, the third has a regular singularity in $x_0 = 1$ and a singularity that is *non regular* in $x_0 = 0$.

In the case of regular singularities, a well developed theory exists and, in particular, Fuchs proved that these equations always possess a solution of the form

$$y(x) = (x - x_0)^\alpha F(x)$$

where α is called *indicial exponent*, and $F(x)$ is an analytic function in a neighborhood of x_0. $F(x)$ can be expanded in Taylor series with radius of convergence at least equal to the distance between x_0 and the nearest singularity. To make an example, let us consider a second order differential equation with constant coefficients

$$y'' - \frac{4}{x} y' + \frac{4}{x^2} y = 0$$

that, with the substitution $x = e^t$, becomes

$$y''(t) - 5y'(t) + 4y(t) = 0.$$

To find the solution we use the *ansatz* $y = e^{\alpha t}$. Substituting, we get the equation for the index: $\alpha^2 - 5\alpha + 4 = 0$ which is solved by $\alpha = 1, 4$. With respect to x, the general solution is $y = ax + bx^4$ where a, b are integration constants. This makes the Fuchs solution look more familiar. The Fuchs solution is more general than what we just saw in this example since it can have singular points. If two solutions of the indicial equation are coincident, the second solution in the neighborhood of x_0 looks like

$$y(x) = (x - x_0)^\alpha F(x) \ln(x - x_0).$$

A point $x_0 \neq +\infty$ is an *irregular singularity* if it is not an ordinary point or a regular singularity .

To control the singularity at $x_0 = +\infty$, the strategy is the following: we first change variable as

$$x = \frac{1}{t} \qquad \frac{d}{dx} = -t^2 \frac{d}{dt} \qquad \frac{d^2}{dx^2} = t^4 \frac{d^2}{dt^2} + 2t^3 \frac{d}{dt}$$

and then study the equation close to $t = 0$. For example, consider

$$y' - \frac{y}{2x} = 0.$$

This equation has a regular singularity in $x_0 = 0$. With the change of variable $x = 1/t$, we get

$$y'(t) + \frac{y(t)}{2t} = 0$$

and we see that $t_0 = 0$ (that means $x_0 = +\infty$) is a regular singularity.

We are now ready to classify the solutions in terms of the singularities of the equation. We will explicitly treat the cases with one, two, and three regular singularities. We will also treat a case with both regular and irregular singularities. This will allow us to discuss the relevant properties of the *hypergeometric series* and *confluent hypergeometric functions* that we will encounter in the exercises proposed in this book.

- **1 regular singularity.** The equation takes the form

$$y'' + \frac{2}{x-a}y' = 0$$

  where the coefficient of the first derivative must be 2 not to have a singularity at $+\infty$. The solution is

$$y = C_1 + \frac{C_2}{(x-a)}$$

  where C_1, C_2 are the integration constants.

- **2 regular singularities.** When we have two regular singularities (say in a and b), a typical example is provided by the equation

$$y'' = \left(\frac{\lambda + \mu - 1}{x - a} - \frac{\lambda + \mu + 1}{x - b} \right) y' - \frac{\lambda \mu (a - b)^2}{(x-a)^2 (x-b)^2} y.$$

  Let us first change variables with $z = (x - a)/(x - b)$, so that $x = a, b$ implies $z = 0, +\infty$. The differential equation for $y(z)$ with regular singularities in $z = 0, +\infty$ has the form

$$y'' - \frac{\tilde{p}}{z}y' + \frac{\tilde{q}}{z^2}y = 0$$

  with $\tilde{p} = \lambda + \mu - 1, \tilde{q} = \lambda \mu$. With this notation, λ and μ are the solutions of the indicial equation close to $z = 0$. We also remark that there must be a relation between the indicial exponents in order to have a regular singularity at infinity. The solution is

$$y = C_1 z^\lambda + C_2 z^\mu = C_1 \left(\frac{x-a}{x-b} \right)^\lambda + C_2 \left(\frac{x-a}{x-b} \right)^\mu$$

  where C_1, C_2 are constants of integration. If $\lambda = \mu$ the solution becomes

$$y = z^\lambda (C_1 + C_2 \ln z) = \left(\frac{x-a}{x-b} \right)^\lambda \left(C_1 + C_2 \ln \left(\frac{x-a}{x-b} \right) \right).$$

- **3 regular singularities.** This is the most interesting case, since it leads to the *hypergeometric series*. We have already seen how a solution looks like around regular singularities. Given an equation with three regular singularities, we first send these points to $0, 1, +\infty$. Then, we divide the solution by $x^\alpha, (x-1)^\beta$, where α, β are solutions of the indicial equation. The resulting differential equation has a solution, $F(x)$, which is called hypergeometric series or hypergeometric function; we must put our differential equation in this form to use the known formulae in the literature. We then start from the following general form of the equation with three regular singularities in the points a, b, c

$$y'' = \left[\frac{\lambda + \lambda' - 1}{x-a} + \frac{\mu + \mu' - 1}{x-b} + \frac{\nu + \nu' - 1}{x-c}\right] y' +$$
$$\left[\frac{\lambda\lambda'(a-b)(c-a)}{(x-a)^2(x-b)(x-c)} + \frac{\mu\mu'(b-c)(a-b)}{(x-a)(x-b)^2(x-c)} + \frac{\nu\nu'(c-a)(b-c)}{(x-a)(x-b)(x-c)^2}\right] y$$

known as *Papperitz-Riemann equation*. $\lambda, \lambda', \mu, \mu', \nu, \nu'$ are the solutions of the indicial equation for the singularities a, b, c. To send a, b, c to pre-assigned values, we must change variables according to

$$z = \gamma \frac{(x-a)}{(x-c)} \qquad x = \frac{(\gamma a - cz)}{(\gamma - z)} \qquad \gamma = \frac{b-c}{b-a}.$$

To have $+\infty$ as a regular point we must, furthermore, satisfy

$$\lambda + \lambda' + \mu + \mu' + \nu + \nu' = 1.$$

The differential equation becomes

$$y'' = \left[\frac{\lambda + \lambda' - 1}{z} + \frac{\mu + \mu' - 1}{z-1}\right] y' +$$
$$\left[\frac{\lambda\lambda'}{z} - \frac{\mu\mu'}{z-1} + \nu(\lambda + \lambda' + \mu + \mu' + \nu - 1)\right] \frac{y}{z(z-1)}.$$

To get to our final form, we must first make the indices as simple as possible. Let us then take a new solution in the form $y(z) = z^\lambda (z-1)^\mu F(z)$. Substituting, we get a differential equation for $F(z)$. This equation is again of Fuchsian type with singularities at $0, 1, +\infty$. Due to our substitution, it is easy to verify that the indices around $0, 1, +\infty$ are $(0, \lambda' - \lambda), (0, \mu' - \mu), (\nu + \lambda + \mu, 1 - \lambda' - \mu' - \nu)$ respectively. Since two indices have been set to zero, we are left with another four which must satisfy the constraint that their sum must be one. Therefore, the final number of independent indices in the differential equation is three. We then call A and B the indicial exponents in $z = +\infty$, and $1 - C, C - A - B$ the two non zero indicial exponents for the points $z = 0$ and $z = 1$, respectively. In the new variables, the differential equation becomes

$$z(z-1)F'' + [(A+B+1)z - C]F' + ABF = 0$$

that is the second order differential equation defining the hypergeometric series $F(A,B,C|z)$. Before giving information on this function, let us recapitulate the steps which brought us to this point: starting from the Papperitz-Riemann equation, which is the most general form of a differential equation with three singularities, we sent the latter to three standard positions $0, 1, +\infty$. Then we made the indices as simple as possible, scaling the solution for the factors which give the singularities. The final result is the standard form of the differential equation which defines the hypergeometric series. Let us now analyze the solution. The series expansion for $F(A,B,C|z)$ is (note that $F(A,B,C|0) = 1$)

$$F(A,B,C|z) = 1 + \frac{AB}{C}z + \frac{A(A+1)B(B+1)}{2!C(C+1)}z^2 + \cdots .$$

The general solution for the second order differential equation is given by

$$F(z) = C_1 F(A,B,C|z) + C_2 z^{1-C} F(1+B-C, 1+A-C, 2-C|z)$$

where C_1, C_2 are constants of integration.

The last argument we discuss is the analytical continuation of the hypergeometric series. The hypergeometric series close to the origin is the one discussed above: it possesses a radius of convergence equal to the distance between $z = 0$ and the nearest singularity in the complex plane , i.e. $z = 1$. How to connect the behaviour of the hypergeometric series close to the origin to that close to the other points of singularity? The formula giving the relation between hypergeometric series with variables z and $1 - z$ is

$$F(A,B,C|z) = \frac{\Gamma(C)\Gamma(C-A-B)}{\Gamma(C-A)\Gamma(C-B)} F(A,B,A+B-C+1 \mid 1-z) +$$
$$\frac{\Gamma(C)\Gamma(A+B-C)}{\Gamma(A)\Gamma(B)} (1-z)^{C-A-B} F(C-A,C-B,C-B-A+1|1-z).$$

A similar formula holds for the relation between hypergeometric series with variables z and $1/z$

$$F(A,B,C|z) = \frac{\Gamma(C)\Gamma(B-A)}{\Gamma(B)\Gamma(C-A)} (-z)^{-A} F(A, 1-C+A, 1-B+A|\frac{1}{z}) +$$
$$\frac{\Gamma(C)\Gamma(A-B)}{\Gamma(A)\Gamma(C-B)} (-z)^{-B} F(B, 1-C+B, 1+B-A|\frac{1}{z}).$$

- **Regular and irregular singularities.** In this case it is standard to put the regular singularity in zero and the irregular one at $+\infty$. This type of equation is found after having separated the variables for the kinetic operator in the case of the hydrogen atom. Let us start with

$$y'' + p(x)y' + q(x)y = 0$$

and

$$p(x) = \frac{(1 - \lambda - \lambda')}{x} \qquad q(x) = -k^2 + \frac{2\alpha}{x} + \frac{\lambda\lambda'}{x^2}$$

with k, α, λ and λ' constants. This is not the most general choice for $p(x)$, $q(x)$, but is the most popular for applications of interest for physics. Looking for a solution of the form $y(x) = x^{\lambda} f(x)$, with λ a solution of the indicial equation around the singularity in zero, we get

$$f'' + \frac{1 + \lambda - \lambda'}{x} f' + \left(\frac{2\alpha}{x} - k^2 \right) f = 0.$$

To study the point at infinity we change variables according to $x = 1/t$. With respect to this variable, the differential equation becomes

$$f'' + \frac{1 + \lambda' - \lambda}{t} f' + \left(\frac{2\alpha}{t^3} - \frac{k^2}{t^4} \right) f = 0.$$

The singularity is irregular due to the term $2\alpha/t^3 - k^2/t^4$ on the l.h.s. To remove the singularity in k^2/t^4, we have to seek the solution in the form $f(t) = e^{-k/t} F(t)$, and the equation becomes

$$F'' + \left(\frac{2k}{t^2} + \frac{1 + \lambda' - \lambda}{t} \right) F' - \frac{k(1 - \lambda' + \lambda) - 2\alpha}{t^3} F = 0$$

which has an indicial equation with a solution $F(t) \approx t^{\beta}$. After having neglected all the less divergent terms, this equation is

$$2k\beta - k(1 - \lambda' + \lambda) + 2\alpha = 0$$

with solution $\beta = (1 - \lambda' + \lambda)/2 - \alpha/k$. A solution for $F(t)$ is $t^{\beta} v_1(t)$, where $v_1(t)$ is an analytic function in $t = 0$. We can now go back to the original equation and use $y(x) = x^{\lambda} e^{-kx} F(x)$ to obtain

$$F'' + \left(\frac{1 + \lambda - \lambda'}{x} - 2k \right) F' - \frac{k(1 + \lambda - \lambda') - 2\alpha}{x} F = 0.$$

If we set $z = 2kx$, $C = 1 + \lambda - \lambda'$, $A = (1 + \lambda - \lambda')/2 - \alpha/k$, we get

$$zF'' + (C - z)F' - AF = 0$$

that is the equation defining the *confluent hypergeometric function* with solution $F(A, C|z)$. This name is due to the fact that this equation can be obtained from the case with three regular singularities, by making the singularity around 1 merge with the singularity at $+\infty$. The confluent hypergeometric function $F(A, C|z)$ is defined by the series

$$F(A,C|z) = 1 + \frac{A}{C}\frac{z}{1!} + \frac{A(A+1)}{C(C+1)}\frac{z^2}{2!} + \dots$$

and the series reduces to a polynomial of degree $|A|$ when $A = -n$, with n a non negative integer. A general solution for the differential equation is given by

$$y(x) = C_1 e^{-kx} x^\lambda F\left(\frac{1+\lambda-\lambda'}{2} - \frac{\alpha}{k}, 1+\lambda-\lambda' | 2kx\right) +$$
$$C_2 e^{-kx} x^{\lambda'} F\left(\frac{1-\lambda+\lambda'}{2} - \frac{\alpha}{k}, 1-\lambda+\lambda' | 2kx\right)$$

where C_1, C_2 are constants of integrations. When treating problems with spherical symmetry, it will be useful to connect the confluent hypergeometric functions to the spherical Bessel functions. The formula of interest is the following

$$J_{n+\frac{1}{2}}(x) = \frac{1}{\Gamma(n+\frac{3}{2})} \left(\frac{1}{2}x\right)^{n+\frac{1}{2}} e^{ix} F(n+1, 2n+2| -2ix)$$

where $J_{n+1/2}$ are the Bessel functions of half-integral order, with the property

$$j_n(x) = \sqrt{\frac{\pi}{2x}} J_{n+1/2}(x)$$

where $j_n(x)$ are the spherical Bessel functions. For the first values of n we get

$$j_0(x) = \frac{\sin x}{x} \qquad j_1(x) = \frac{\sin x}{x^2} - \frac{\cos x}{x} \qquad j_2(x) = \sin x\left(\frac{3}{x^3} - \frac{1}{x}\right) - \frac{3\cos x}{x^2}.$$

1.8 WKB Method

It is a distinctive feature of Quantum Mechanics that particles exhibit wave-like properties. In particular, the *De Broglie equation* relates the wavelength λ to the momentum p of a free material particle

$$p = \frac{h}{\lambda}.$$

When the De Broglie wavelength of a particle becomes small with respect to the typical dimensional scale of our problem, our system is said to be *quasi-classical*. In this limit, using an analogy with the case in which geometric optics is derived starting from the equation of the electromagnetic waves, the wave function can be sought for in the form (we take the one dimensional case)

$$\psi(x) = e^{\frac{i}{\hbar}S(x)}.$$

Substituting this expression in the stationary Schrödinger equation, we get

$$\left(\frac{dS}{dx}\right)^2 - i\hbar\left(\frac{d^2S}{dx^2}\right) - 2m(E - U(x)) = 0.$$

The so-called quasi-classical approximation consists of the expansion of $S(x)$ in powers of $\hbar$

$$S(x) = S_0(x) + \frac{\hbar}{i}S_1(x) + \left(\frac{\hbar}{i}\right)^2 S_2(x) + \ldots$$

Using this expansion in the original equation and imposing the consistency order by order in $\hbar$, we get

$$\mathcal{O}(\hbar): \quad \left(\frac{dS_0}{dx}\right)^2 = 2m(E - U(x))$$

$$\mathcal{O}(\hbar^2): \quad 2\left(\frac{dS_1}{dx}\right) = -\frac{\left(\frac{d^2S_0}{dx^2}\right)}{\left(\frac{dS_0}{dx}\right)}$$

$$\mathcal{O}(\hbar^3): \quad 2\left(\frac{dS_2}{dx}\right) = -\frac{\left(\frac{d^2S_1}{dx^2}\right) + \left(\frac{dS_1}{dx}\right)^2}{\left(\frac{dS_0}{dx}\right)}.$$

The first equation sets the zeroth order approximation: the wave function is a linear combination of the exponential functions $e^{\pm i/\hbar \int p(x)dx}$, where $p(x) = \sqrt{2m(E - U(x))}$ is the classical momentum of the particle. The zeroth order is obtained by neglecting the second order derivative with respect to the square of the first order derivative

$$\hbar\left|\frac{\left(\frac{d^2S_0}{dx^2}\right)}{\left(\frac{dS_0}{dx}\right)^2}\right| = \hbar\left|\frac{1}{p^2}\frac{dp}{dx}\right| = \frac{1}{2\pi}\left|\frac{d\lambda}{dx}\right| \ll 1.$$

The zeroth order approximation will then be valid in the limit in which the oscillations of the wave function are small with respect to the typical scale of our problem or when the momentum is large. This approximation will not be valid in the points in which the classical motion gets inverted since in the inversion point $p = 0$. The WKB approximation (after G. Wentzel, H.A. Kramers and L.Brillouin who put forward the proposal for the first time in 1926) consists in solving the first two equations in the series for $\hbar$. After having solved the first equation, we can solve the second getting $S_1 = -\frac{1}{2}\ln p + \text{const}$. The wave function becomes

$$\psi = \frac{A}{\sqrt{p}}e^{\frac{i}{\hbar}\int p\,dx} + \frac{B}{\sqrt{p}}e^{-\frac{i}{\hbar}\int p\,dx}.$$

The approximation is now valid for $\hbar|S_2| \ll 1$. The third equation in $\hbar$ becomes (after having substituted the $S_0(x)$ and $S_1(x)$ we have just found)

$$\hbar\frac{dS_2}{dx} = -\frac{1}{8\pi}\left(\frac{d^2\lambda}{dx^2}\right) + \frac{1}{16\pi\lambda}\left(\frac{d\lambda}{dx}\right)^2$$

and, integrating once

$$\hbar S_2 = -\frac{1}{8\pi}\left(\frac{d\lambda}{dx}\right) + \int\frac{1}{16\pi\lambda}\left(\frac{d\lambda}{dx}\right)^2 dx.$$

We see that the condition $\hbar|S_2| \ll 1$ is satisfied when $|d\lambda/dx|$ is very small.

Within the WKB approach, the requirement that the wave function is not multi-valued leads to

$$\oint p\,dx = \left(n+\frac{1}{2}\right)h \quad n = 0,1,2,3,\ldots$$

This is known as the *Bohr-Sommerfeld quantization rule*. In the above expression, $\oint p\,dx = 2\int_{x_1}^{x_2} p\,dx$, where $x_{1,2}$ are the turning points of the classical motion.

1.9 Perturbation Theory

When an eigenvalue problem is too complicated to be solved exactly, one can use static *perturbation theory*. The theory of perturbations is an extremely important computational tool in modern physics. In fact, it allows to describe real quantum systems whose eigenvalues equations are, in general, not amenable to an exact treatment. The method is based on the introduction of a "small" perturbation in the Hamiltonian which allows for a series expansion. Let us suppose to have exactly solved the eigenvalue problem for the Hamiltonian $\hat{H}_0$

$$\hat{H}_0|\psi_k^{(0)}\rangle = E_k^{(0)}|\psi_k^{(0)}\rangle.$$

Let us then consider the potential energy $\varepsilon\hat{U}$, with $\varepsilon \ll 1$. The eigenvalues and eigenfunctions of the Hamiltonian $\hat{H} = \hat{H}_0 + \varepsilon\hat{U}$

$$\hat{H}|\psi_k\rangle = E_k|\psi_k\rangle$$

may be found with a power series in the parameter ε. Let us then start with the case where the eigenvalues of the Hamiltonian $\hat{H}_0$ are not degenerate. The eigenstates are expanded as

$$|\psi_k\rangle = |\psi_k^{(0)}\rangle + \sum_{n\neq k} c_{nk}|\psi_n^{(0)}\rangle$$

and, substituting this expansion back in the original equation, we find

$$\left(\hat{H}_0 + \varepsilon\hat{U}\right)\left(|\psi_k^{(0)}\rangle + \sum_{n\neq k} c_{nk}|\psi_n^{(0)}\rangle\right) = E_k\left(|\psi_k^{(0)}\rangle + \sum_{n\neq k} c_{nk}|\psi_n^{(0)}\rangle\right).$$

The coefficients c_{nk} and the energy E_k are expanded as

$$E_k = E_k^{(0)} + \varepsilon\Delta E_k^{(1)} + \varepsilon^2\Delta E_k^{(2)} + \cdots \qquad c_{nk} = \varepsilon c_{nk}^{(1)} + \varepsilon^2 c_{nk}^{(2)} + \cdots.$$

Projecting the Schrödinger equation on the eigenstate $\langle\psi_k^0|$, and retaining only the first order in ε, we get

$$\Delta E_k^{(1)} = U_{kk}$$

$$c_{nk}^{(1)} = \frac{U_{nk}}{E_k^{(0)} - E_n^{(0)}} \qquad k \neq n$$

where the matrix element U_{nk} is defined as

$$U_{nk} = \langle\psi_n^{(0)}|\hat{U}|\psi_k^{(0)}\rangle.$$

This procedure can be extended to the second order in ε, and we find the correction

$$\Delta E_k^{(2)} = \sum_{n\neq k} \frac{|U_{kn}|^2}{E_k^{(0)} - E_n^{(0)}}.$$

When the eigenvalues of the Hamiltonian $\hat{H}_0$ are degenerate, let $\psi_{k_1}^{(0)}, \psi_{k_2}^{(0)}, \ldots$ be the eigenstates of $\hat{H}_0$ corresponding to the same energy E_k. At the first order in ε, the correction to the unperturbed eigenvalue is found by diagonalizing the perturbation matrix $U_{ps} = \langle\psi_p^{(0)}|\hat{U}|\psi_s^{(0)}\rangle$ with $p,s = k_1, k_2, \ldots$.

Let us now proceed with the properties of time dependent perturbations. To this end, we consider a quantum system described by a time independent Hamiltonian $\hat{H}_0$. Let us then assume that at time $t = 0$ we act on such system with time dependent forces until a later time $t = \tau$. Let us further suppose that the contribution of these forces to the Hamiltonian is given by a perturbation $\hat{H}'$ such that

$$\begin{cases} \hat{H}' = \hat{U}(t) & 0 \leq t \leq \tau \\ \hat{H}' = 0 & t < 0, t > \tau. \end{cases}$$

The probability that this perturbation generates a transition from the state $|n\rangle$ to the state $|m\rangle$ (both of them eigenstates of $\hat{H}_0$) is given by the formula

$$P_{m,n} = \frac{1}{\hbar^2}\left|\int_0^\tau \langle m|\hat{U}(t)|n\rangle e^{i\frac{(E_m - E_n)}{\hbar}t}dt\right|^2.$$

1.10 Thermodynamic Potentials

Thermodynamic systems are described by measurable parameters, such as energy E, volume V, temperature T, pressure P, etc. The transfer of heat and energy in the various processes involved is regulated by the laws of thermodynamics. For infinitesimal changes towards another state, the conservation of the energy for a thermodynamic system can be stated as follows

$$dE = \delta Q + dW$$

where dE is the infinitesimal change in the internal energy, δQ the amount of heat exchanged and dW the infinitesimal work done on the system. This is known as *the first law of thermodynamics*. For example, in the case of a fluid, we have $dW = -PdV$ (this is a case frequently considered hereafter), where P is the hydrostatic pressure and V the volume. It is obvious that a positive work (compression, $dV < 0$) done on the system reduces its volume, in agreement with experimental observations.

The *second law of thermodynamics* states that no thermodynamic process is possible whose only result is the transfer of heat from a body of lower temperature to a body of higher temperature. Since the quantity δQ is not an exact differential, one introduces the entropy S as the *thermodynamic potential* whose change for an infinitesimal and reversible transformation between two states at an absolute temperature T is

$$dS = \frac{\delta Q}{T}.$$

Moreover, if the number of particles changes, the first law becomes $TdS = dE + PdV - \mu dN$, where μ is the chemical potential. The differential of the entropy S is therefore written as

$$dS = \frac{1}{T}dE + \frac{P}{T}dV - \frac{\mu}{T}dN$$

so that E, V, N are the *natural* variables for the entropy S. The derivatives of S with respect to the natural variables lead to specific thermodynamic quantities

$$\frac{1}{T} = \left(\frac{\partial S}{\partial E}\right)_{V,N} \qquad \frac{P}{T} = \left(\frac{\partial S}{\partial V}\right)_{E,N} \qquad -\frac{\mu}{T} = \left(\frac{\partial S}{\partial N}\right)_{E,V}.$$

Consistently with these constraints, the state of equilibrium is the state with maximum entropy. When the control variables of a system are different from E, V, N, other thermodynamic potentials are used. These are the enthalpy $H(S,V,N)$, the free energy $F(T,V,N)$, the Gibbs potential $\Phi(T,P,N)$ and the grand potential $\Omega(T,V,\mu)$. Their definitions are given below, together with their natural variables and the resulting control variables obtained after differentiation:

- Enthalpy: $H = E + PV$

$$dH = TdS + VdP + \mu dN \Rightarrow H(S,P,N)$$

$$T = \left(\frac{\partial H}{\partial S}\right)_{P,N} \qquad V = \left(\frac{\partial H}{\partial P}\right)_{S,N} \qquad \mu = \left(\frac{\partial H}{\partial N}\right)_{S,P}.$$

- Free Energy: $F = E - TS$

$$dF = -SdT - PdV + \mu dN \Rightarrow F(T,V,N)$$

$$-S = \left(\frac{\partial F}{\partial T}\right)_{V,N} \qquad -P = \left(\frac{\partial F}{\partial V}\right)_{T,N} \qquad \mu = \left(\frac{\partial F}{\partial N}\right)_{T,V}.$$

- Gibbs Potential: $\Phi = F + PV = H - TS$

$$d\Phi = -SdT + VdP + \mu dN \Rightarrow \Phi(T,P,N)$$

$$-S = \left(\frac{\partial \Phi}{\partial T}\right)_{P,N} \qquad V = \left(\frac{\partial \Phi}{\partial P}\right)_{T,N} \qquad \mu = \left(\frac{\partial \Phi}{\partial N}\right)_{T,P}.$$

- Grand Potential: $\Omega = F - \mu N$

$$d\Omega = -SdT - PdV - Nd\mu \Rightarrow \Omega(T,V,\mu)$$

$$S = -\left(\frac{\partial \Omega}{\partial T}\right)_{V,\mu} \qquad P = -\left(\frac{\partial \Omega}{\partial V}\right)_{T,\mu} \qquad N = -\left(\frac{\partial \Omega}{\partial \mu}\right)_{T,V}.$$

From all these relations, a variety of useful constraints between second order derivatives of the potentials may be obtained. For example, considering the enthalpy, we have

$$T = \left(\frac{\partial H}{\partial S}\right)_{P,N} \qquad V = \left(\frac{\partial H}{\partial P}\right)_{S,N}.$$

We derive the first equation with respect to P and the second with respect to S

$$\left(\frac{\partial}{\partial P}\left(\frac{\partial H}{\partial S}\right)_{P,N}\right)_{S,N} = \left(\frac{\partial}{\partial S}\left(\frac{\partial H}{\partial P}\right)_{S,N}\right)_{P,N}$$

from which we find (using Schwartz lemma for mixed partial derivatives) the following identity

$$\left(\frac{\partial T}{\partial P}\right)_{S,N} = \left(\frac{\partial V}{\partial S}\right)_{P,N}$$

that is the mathematical condition for dH to be an exact differential. The associated relation is called *Maxwell relation* . Proceeding in a similar way for the other potentials, one may prove other Maxwell relations

$$\left(\frac{\partial S}{\partial V}\right)_{T,N} = \left(\frac{\partial P}{\partial T}\right)_{V,N}$$

$$\left(\frac{\partial S}{\partial P}\right)_{T,N} = -\left(\frac{\partial V}{\partial T}\right)_{P,N}$$

$$\left(\frac{\partial S}{\partial V}\right)_{T,\mu} = \left(\frac{\partial P}{\partial T}\right)_{V,\mu}.$$

Another set of relations is obtained by considering that the thermodynamic potentials are *extensive*. For example, take the free energy $F = F(T,V,N)$: if we rescale the volume V and the number of particles N with the same rescaling factor ℓ, the free energy rescales accordingly

$$F(T,\ell V,\ell N) = \ell F(T,V,N)$$

that implies $F = Nf(T,\frac{V}{N})$, where f is the free energy density.

In order to quantify the change of temperature as a function of the absorbed heat, *specific heats* are frequently introduced. By definition, the specific heat of a system characterizes the heat required to change the temperature by a given amount. One usually defines a particular heating process by keeping fixed some thermodynamic variable. For a fluid, we will frequently use the specific heats at constant volume and pressure

$$C_V = T\left(\frac{\partial S}{\partial T}\right)_{V,N} = \left(\frac{\partial U}{\partial T}\right)_{V,N}$$

$$C_P = T\left(\frac{\partial S}{\partial T}\right)_{P,N} = \left(\frac{\partial H}{\partial T}\right)_{P,N}.$$

When working with partial derivatives of thermodynamic variables, it is sometimes convenient to use the method of Jacobians. Let us consider two generic functions $u(x,y), v(x,y)$, where x and y are independent variables. The Jacobian is defined as the determinant

$$J(u,v) = \frac{\partial(u,v)}{\partial(x,y)} = \det\begin{pmatrix} \left(\frac{\partial u}{\partial x}\right)_y & \left(\frac{\partial v}{\partial x}\right)_y \\ \left(\frac{\partial u}{\partial y}\right)_x & \left(\frac{\partial v}{\partial y}\right)_x \end{pmatrix} = \left(\frac{\partial u}{\partial x}\right)_y\left(\frac{\partial v}{\partial y}\right)_x - \left(\frac{\partial u}{\partial y}\right)_x\left(\frac{\partial v}{\partial x}\right)_y.$$

The Jacobian has the following properties

$$\frac{\partial(u,v)}{\partial(x,y)} = -\frac{\partial(v,u)}{\partial(x,y)} = \frac{\partial(v,u)}{\partial(y,x)}$$

$$\frac{\partial(u,y)}{\partial(x,y)} = \left(\frac{\partial u}{\partial x}\right)_y$$

$$\frac{\partial(u,v)}{\partial(x,y)} = \frac{\partial(u,v)}{\partial(t,s)}\frac{\partial(t,s)}{\partial(x,y)}$$

with t and s two other generic variables.

When dealing with thermodynamic states, it is also common to find an *equation of state*, i.e. a relation between three thermodynamic variables, say x, y, z. In the most general case, such equation of state may be formulated as

$$f(x,y,z) = 0$$

with $f(x,y,z)$ some given function. Such relation clearly reduces the number of independent variables from three to two. On the manifold where the three variables x,y,z are still consistent with the equation of state, we get

$$df = 0 = \left(\frac{\partial f}{\partial x}\right)_{y,z} dx + \left(\frac{\partial f}{\partial y}\right)_{x,z} dy + \left(\frac{\partial f}{\partial z}\right)_{x,y} dz$$

from which

$$\left(\frac{\partial x}{\partial y}\right)_{z} = -\left(\frac{\partial f}{\partial y}\right)_{x,z} \Big/ \left(\frac{\partial f}{\partial x}\right)_{y,z}$$

$$\left(\frac{\partial x}{\partial z}\right)_{y} = -\left(\frac{\partial f}{\partial z}\right)_{x,y} \Big/ \left(\frac{\partial f}{\partial x}\right)_{y,z}$$

$$\left(\frac{\partial y}{\partial z}\right)_{x} = -\left(\frac{\partial f}{\partial z}\right)_{x,y} \Big/ \left(\frac{\partial f}{\partial y}\right)_{x,z}.$$

It follows that

$$\left(\frac{\partial x}{\partial y}\right)_{z} \left(\frac{\partial y}{\partial z}\right)_{x} \left(\frac{\partial z}{\partial x}\right)_{y} = -1$$

that is a chain rule for x, y, z which can be used as an equation relating the variables entering the equation of state.

1.11 Fundamentals of Ensemble Theory

Let us denote by $q = (q_1, q_2, \ldots, q_n)$ the generalized coordinates of a system with n degrees of freedom, and $p = (p_1, p_2, \ldots, p_n)$ the associated momenta. For example, in the case of a fluid with N particles in three dimensions, we have $n = 3N$. A microscopic state is defined by specifying the values of the $2n$ variables (q, p), and the corresponding $2n$ dimensional space is called the *phase space*. A given microscopic state evolves in time along a trajectory given by the solution of the following $2n$ differential equations

$$\frac{dp_i}{dt} = -\left(\frac{\partial H}{\partial q_i}\right) \qquad \frac{dq_i}{dt} = \left(\frac{\partial H}{\partial p_i}\right) \qquad i = 1, 2, \ldots, n$$

where $H = H(p,q)$ is the Hamiltonian of the system and where the derivative with respect to q_i (p_i) is performed by keeping fixed all the other variables. For a conservative system, this trajectory lies on a surface of constant energy

$$H(p,q) = E$$

sometimes called *ergodic* surface. During the finite time of a measurement, microscopic fluctuations are so rapid that the system explores many microstates. Therefore, given some observable $O(p,q)$, its time average is equivalent to the average over an ensemble of infinite copies of the system

$$\bar{O} = \lim_{T \to +\infty} \frac{1}{T} \int_0^T O(p(s), q(s)) \, ds = \langle O \rangle$$

where the average $\langle ... \rangle$ is computed with some probability density function of the phase space variables $f(p,q)$. The resulting classical average is

$$\langle O \rangle = \int f(p,q) O(p,q) \, d^n p \, d^n q.$$

It has to be noted that in Quantum Mechanics, since p and q cannot be measured simultaneously, the concept of the phase space is somehow meaningless. Nevertheless, a quantum stationary state, say $|\psi_i\rangle$, with a well defined energy E_i, can be defined from the stationary Schrödinger equation

$$\hat{H}|\psi_i\rangle = E_i|\psi_i\rangle \qquad\qquad i = 1, 2, \ldots$$

and a microscopic state is defined as a superposition of a set of states $|\psi_i\rangle$, chosen to be consistent with some macroscopic requirements. Let us call $\mathcal{M}$ this set (or ensemble). Consistently, we have some expectation value for the observable (an operator) $\hat{O}$ on the i-th state

$$O_i = \langle \psi_i|\hat{O}|\psi_i\rangle \qquad\qquad i \in \mathcal{M}$$

and the average over the above mentioned set is

$$\langle \hat{O} \rangle = \sum_{i \in \mathcal{M}} O_i f_i$$

with f_i the probability associated with O_i. Usually, one defines the density matrix corresponding to a given ensemble as

$$\hat{\rho} = \sum_{i \in \mathcal{M}} w_i |\psi_i\rangle\langle\psi_i|$$

with w_i the weight (characteristic of the ensemble) associated with the state $|\psi_i\rangle$. The corresponding ensemble average of an observable $\hat{O}$ is

$$\langle \hat{O} \rangle = \frac{\sum_{i \in \mathcal{M}} w_i \langle \psi_i|\hat{O}|\psi_i\rangle}{\sum_{i \in \mathcal{M}} w_i} = \frac{\sum_{i \in \mathcal{M}} w_i O_i}{\sum_{i \in \mathcal{M}} w_i} = \frac{Tr(\hat{\rho}\hat{O})}{Tr(\hat{\rho})}.$$

Therefore, the probability previously mentioned becomes $f_i = \frac{w_i}{\sum_{i \in \mathcal{M}} w_i}$.

1.11.1 Microcanonical Ensemble

The microcanonical ensemble describes a *closed* physical system. For example, in the case of a thermodynamic fluid, the associated probability distribution function assigns equal probabilities to each microstate consistent with a fixed energy E, fixed volume V, and fixed number of particles N. The number of states $\Omega(E,V,N)$ is connected to the thermodynamic entropy

$$S(E,V,N) = k \ln \Omega(E,V,N).$$

This equation is known as the *Boltzmann formula*, $k = 1.38 \times 10^{-23} JK^{-1}$ is the Boltzmann constant. For a system with a discrete set of microstates, the number of states is just a discrete sum. When dealing with a continuous set of microstates, say a classical fluid with N particles and Hamiltonian $H_N(p,q)$ in three dimensions, the phase space volume occupied by the microcanonical ensemble is then

$$\Omega(E,V,N) = \int_{H_N(p,q)=E} \frac{d^{3N}p\, d^{3N}q}{h^{3N}}$$

where $p = (p_1, p_2, ..., p_{3N})$ and $q = (q_1, q_2, ..., q_{3N})$ are the momenta and positions of the particles. The quantity h is a constant with dimension of action ($\sim [pq]$), useful to make $\Omega(E,V,N)$ dimensionless, and it is appropriate to identify it with the Planck constant. It represents the minimal volume that can be measured according to the Heisenberg indetermination principle. When the number of degrees of freedom is very large ($N \gg 1$) one can define the phase space volume $\Sigma(E,V,N)$ enclosed by the energy surface $H_N(p,q) = E$

$$\Sigma(E,V,N) = \int_{H_N(p,q)\leq E} \frac{d^{3N}p\, d^{3N}q}{h^{3N}}$$

and show that, apart from corrections of order $\ln N$, the following definition of entropy

$$S(E,V,N) = k \ln \Sigma(E,V,N)$$

is equivalent to the previous one.

1.11.2 Canonical Ensemble

The canonical ensemble describes a system with a fixed volume V, fixed number of particles N, and in thermal equilibrium with a reservoir at temperature T. The system can exchange energy with the reservoir. A state is specified by the energy E (we use the notation E for the energy of the microstate and U for the averaged one) and the associated statistical weight is proportional to $e^{-\beta E}$, where $\beta = 1/kT$. The normalization factor, called *canonical partition function*, takes the form (still for the classical fluid considered in section 1.11.1)

$$Q_N(T,V,N) = \frac{1}{N! h^{3N}} \int e^{-\beta H_N(p,q)} d^{3N}p\, d^{3N}q.$$

The factor $N!$, called *Gibbs factor*, accounts for the indistinguishability of the particles. The Thermodynamics of the system is obtained from the relation

$$F(T,V,N) = -kT \ln Q_N(T,V,N)$$

where $F(T,V,N)$ is the thermodynamic free energy.

1.11.3 Grand Canonical Ensemble

The grand canonical ensemble is used to describe a system inside a volume V and in equilibrium with a reservoir at temperature T and with a chemical potential μ. Both the energy and particles exchanges are allowed in this case. A state is specified by the energy E and the number N of particles, and the associated statistical weight is proportional to $z^N e^{-\beta E}$, where $z = e^{\beta \mu}$ is called *fugacity*. The normalization factor, called *grand canonical partition function*, takes the form (still for the classical fluid considered in section 1.11.1)

$$\mathcal{Q}(T,V,\mu) = \sum_N z^N Q_N(T,V,N) = \sum_N e^{\beta \mu N} Q_N(T,V,N)$$

where $Q_N(T,V,N)$ is the canonical partition function seen in section 1.11.2. The thermodynamic interpretation of this ensemble is given by

$$\frac{PV}{kT} = \ln \mathcal{Q}(T,V,\mu)$$

where P is the pressure of the system.

1.11.4 Quantum Statistical Mechanics

When we deal with quantum mechanical problems where *indistinguishable* particles are present, we need to distinguish two cases: particles with integer spin obeying the Bose-Einstein statistics and called *bosons*; particle with half-odd-integer spin obeying Fermi-Dirac statistics and called *fermions*. A generic energy state can be occupied by an arbitrary number of bosons; for fermions, because of the Pauli exclusion principle, it can be occupied by at most one particle. Quantum mechanical effects usually emerge at high density and low temperatures while, at high temperatures and low densities, the classical limit (i.e. the Maxwell-Boltzmann statistics) is recovered. A system of not interacting quantum particles is easily treated in the grand canonical ensemble, where

$$PV = kT \ln \mathscr{Q}(T,V,\mu) = kT \sum_{\varepsilon} \frac{1}{a} \ln(1 + ae^{-\beta(\varepsilon-\mu)}).$$

In the above expression, ε stands for the single particle energy and a discrete spectrum has been assumed (the continuous case is recovered by properly replacing the summation with an integral). The Bose-Einstein statistics corresponds to $a = -1$, the Fermi-Dirac statistics to $a = 1$, and the limit $a \to 0$ (or $z = e^{\beta\mu} \ll 1$) corresponds to Maxwell-Boltzmann particles. The mean occupation number $\langle n_\varepsilon \rangle$ associated to the energy level ε is given by

$$\langle n_\varepsilon \rangle = -\frac{1}{\beta}\left(\frac{\partial \ln \mathscr{Q}}{\partial \varepsilon}\right)_{\beta\mu} = \frac{1}{e^{\beta(\varepsilon-\mu)}+a}.$$

1.12 Kinetic Approach

When working with very large volumes, the single particle energy levels would be so close, that a summation over them may be replaced by an integral. Let us indicate with $\varepsilon = \varepsilon(p)$ the single particle energy, solely dependent on the absolute value of the momentum p. Using the grand canonical ensemble, one gets the following results for the average number of particles and pressure of an ideal quantum gas in three dimensions

$$N = \int \langle n_p \rangle \frac{Vd^3p}{h^3} = \frac{4\pi V}{h^3} \int_0^{+\infty} \frac{1}{z^{-1}e^{\beta\varepsilon}+a} p^2 dp$$

$$P = \frac{4\pi}{3h^3} \int_0^{+\infty} \frac{1}{z^{-1}e^{\beta\varepsilon}+a}\left(p\frac{d\varepsilon}{dp}\right)p^2 dp = \frac{n}{3}\left\langle p\frac{d\varepsilon}{dp}\right\rangle = \frac{n}{3}\langle pv \rangle$$

where v is the absolute value of the speed of each particle, $n = N/V$ the particles density, and where we have used

$$\left\langle p\frac{d\varepsilon}{dp}\right\rangle = \frac{\int_0^{+\infty} \langle n_p \rangle \left(p\frac{d\varepsilon}{dp}\right)d^3p}{\int_0^{+\infty} \langle n_p \rangle d^3p} = \frac{\int_0^{+\infty} \frac{1}{z^{-1}e^{\beta\varepsilon}+a}\left(p\frac{d\varepsilon}{dp}\right)p^2 dp}{\int_0^{+\infty} \frac{1}{z^{-1}e^{\beta\varepsilon}+a}p^2 dp}.$$

The above pressure arises from the microscopic motion of the particles and can be deduced from purely kinetic considerations. To show this point, let us take an infinitesimal element of area dA perpendicular to the z axis and located on the wall of the container where the gas is placed. If we focus our attention on the particles with velocity between v and $v + dv$, with $f(v)$ the probability density function, the relevant number of particles that in the time interval dt are able to hit the surface wall within the area dA and with velocities between v and $v + dv$ is

$$dN_{hit} = n(dA \cdot v) dt \times f(v) dv.$$

Due to the reflection from the wall, the normal component of the momentum undergoes a change from p_z to $-p_z$; as a result, the normal momentum transfered by these particles per unit time to a unit area of the wall is $2p_z v_z n f(v) dv$. By definition, the kinetic pressure of the gas is

$$P = 2n \int_{-\infty}^{+\infty} dv_x \int_{-\infty}^{+\infty} dv_y \int_{0}^{+\infty} f(v) p_z v_z dv_z = n \langle pv \cos^2 \theta \rangle = \frac{n}{3} \langle pv \rangle$$

with θ the angle that the velocity is forming with the z axis. The previous equation is indeed the very same equation obtained with the grand canonical ensemble with the specification that $f(v)$ is just the Bose-Einstein or Fermi-Dirac probability distribution function. In a similar way, we can determine the rate of *effusion* of the gas through the hole (of unit area) in the wall

$$R = n \int_{-\infty}^{+\infty} dv_x \int_{-\infty}^{+\infty} dv_y \int_{0}^{+\infty} f(v) v_z dv_z = \frac{n}{4} \langle v \rangle.$$

1.13 Fluctuations

In the previous sections we have treated thermodynamic systems in equilibrium. Nevertheless, *fluctuations* occur around the equilibrium states, and a precise probability distribution law may be derived in the framework of statistical mechanics. If we look at a system (s) in contact with a reservoir (r), the total variation of the entropy $\Delta S = \Delta S_s + \Delta S_r = S - S_0$ with respect to its equilibrium value S_0 is

$$\Delta S = S - S_0 = k \ln \Omega_f - k \ln \Omega_0$$

where Ω_f (Ω_0) denotes the number of distinct microstates in the presence (or in the absence) of the fluctuations. The probability that the fluctuation may occur is then

$$p \propto \frac{\Omega_f}{\Omega_0} = e^{\Delta S / k}.$$

If the exchange of particles between the system and the reservoir (whose temperature is T) is not allowed, the total variation of the entropy ΔS can be expressed only in terms of the variation of the system's temperature (ΔT_s), entropy (ΔS_s), pressure (ΔP_s), and volume (ΔV_s) to yield

$$\Delta S = -\frac{1}{2T} (\Delta S_s \Delta T_s - \Delta P_s \Delta V_s).$$

We may now drop the subscript s knowing that each quantity refers to the properties of the system and write down the probability as

$$p \propto e^{-\frac{1}{2kT}(\Delta S \Delta T - \Delta P \Delta V)}.$$

We note, however, that only two of the four Δ appearing in this probability can be chosen independently. For instance, if we choose ΔT and ΔV as independent variables, then ΔS and ΔP may be expanded with the help of Maxwell relations for S and P

$$\Delta S = \left(\frac{\partial S}{\partial T}\right)_V \Delta T + \left(\frac{\partial S}{\partial V}\right)_T \Delta V = \frac{C_V}{T} \Delta T + \left(\frac{\partial P}{\partial T}\right)_V \Delta V$$

$$\Delta P = \left(\frac{\partial P}{\partial T}\right)_V \Delta T + \left(\frac{\partial P}{\partial V}\right)_T \Delta V = \left(\frac{\partial P}{\partial T}\right)_V \Delta T - \frac{1}{\kappa_T V} \Delta V$$

where we have used the isothermal compressibility $\kappa_T = -\frac{1}{V}\left(\frac{\partial V}{\partial P}\right)_T$. The associated probability is then

$$p \propto e^{-\frac{C_V}{2kT^2}(\Delta T)^2 - \frac{1}{2kT\kappa_T V}(\Delta V)^2}$$

which shows that the fluctuations in T and V are statistically independent *Gaussian* variables with variances related to the specific heat and the isothermal compressibility.

1.14 Mathematical Formulae

In this section, useful formulae are given and briefly commented.

- **Gamma Function.** The Gamma function is defined by the integral

$$\Gamma(v) = \int_0^{+\infty} e^{-x} x^{v-1} dx \quad v > 0.$$

After the integration by parts, we can prove that

$$\Gamma(v) = \frac{1}{v}\Gamma(v+1)$$

which can be iterated to give

$$\Gamma(v+1) = v \times (v-1) \times (v-2) \times ... \times (1+p) \times p \times \Gamma(p) \qquad 0 < p \leq 1.$$

Therefore, for integer values of v (say $v = m$), we have the factorial representation

$$\Gamma(v) = \Gamma(m) = (m-1)! = (m-1) \times (m-2) \times (m-3)... \times 2 \times 1$$

while, when v is half-odd integer (say $v = m + \frac{1}{2}$), we have

$$\Gamma(v) = \Gamma\left(m+\frac{1}{2}\right) = \left(m-\frac{1}{2}\right)! = \left(m-\frac{1}{2}\right) \times \left(m-\frac{3}{2}\right)...\frac{3}{2} \times \frac{1}{2} \times \sqrt{\pi}$$

where we have used

$$\Gamma\left(\frac{1}{2}\right) = \sqrt{\pi}.$$

- **Stirling approximation.** We start from the integral representation for the factorial

$$\Gamma(v+1) = v! = \int_0^{+\infty} e^{-x} x^v \, dx \quad v > 0$$

and we derive an asymptotic expression for it. It is not difficult to see that when $v \gg 1$, the main contribution to this integral comes from the region around $x \approx v$, with a width of order $\sqrt{v}$. In view of this, if we write

$$x = v + \sqrt{v}\xi$$

and plug it back into the integral, we get

$$v! = \sqrt{v} \left(\frac{v}{e}\right)^v \int_{-\sqrt{v}}^{+\infty} e^{-\sqrt{v}\xi} \left(1 + \frac{\xi}{\sqrt{v}}\right)^v d\xi.$$

The integrand has a maximum in $\xi = 0$ and goes very fast to zero on both sides of it. We therefore expand the logarithm of the integrand around $\xi = 0$, and take the exponential of the resulting expression

$$v! = \sqrt{v} \left(\frac{v}{e}\right)^v \int_{-\sqrt{v}}^{+\infty} e^{-\frac{\xi^2}{2} + \frac{\xi^3}{3\sqrt{v}} - \cdots} d\xi.$$

When v is large, we can approximate the integrand with a Gaussian and send the lower limit of integration to infinity. The resulting expression is known as the *Stirling* formula

$$v! \approx \left(\frac{v}{e}\right)^v \sqrt{2\pi v} \quad v \gg 1.$$

- **Multidimensional sphere.** Consider a d dimensional space with coordinates x_i ($i = 1, 2, ..., d$). The infinitesimal volume element of this space is

$$dV_d = \prod_{i=1}^d (dx_i)$$

and the volume of a d dimensional sphere with radius R may be written as

$$V_d = C_d R^d$$

with its infinitesimal variation connected to the d dimensional surface S_d

$$dV_d = S_d dR = C_d dR^{d-1} dR.$$

In the above expression, the constant C_d has to be determined. To do that, we make use of the formula

$$\sqrt{\pi} = \int_{-\infty}^{+\infty} e^{-x^2} dx.$$

Using an integral of this type for each of the x_i involved, we obtain

$$\pi^{d/2} = \int_{-\infty}^{+\infty} dx_1 \int_{-\infty}^{+\infty} dx_2 ... \int_{-\infty}^{+\infty} dx_d \, e^{-\Sigma_{i=1}^{d} x_i^2} = \int_{0}^{+\infty} e^{-R^2} C_d dR^{d-1} dR$$

where we have used polar coordinates with the radius given by

$$R = \sqrt{x_1^2 + x_2^2 + ... + x_d^2}.$$

We now use the definition of Gamma function to get

$$\pi^{d/2} = \frac{1}{2} dC_d \Gamma\left(\frac{d}{2}\right) = C_d(d/2)!$$

so that the volume and surface for the d dimensional sphere with radius R are

$$V_d = \frac{\pi^{d/2}}{(d/2)!} R^d \qquad S_d = \frac{2\pi^{d/2}}{(d/2-1)!} R^{d-1}.$$

- **Bose-Einstein functions.** In the theory of the Bose-Einstein gas, we will use the following integrals

$$g_\nu(z) = \frac{1}{\Gamma(\nu)} \int_{0}^{+\infty} \frac{x^{\nu-1} dx}{z^{-1} e^x - 1}.$$

When $\nu = 1$, the integral can be solved exactly

$$g_1(z) = \frac{1}{\Gamma(1)} \int_{0}^{+\infty} \frac{dx}{z^{-1} e^x - 1} = \ln(1 - ze^{-x})\big|_{0}^{+\infty} = -\ln(1 - z).$$

A simple differentiation of $g_\nu(z)$ leads to the following recurrence formula

$$z \frac{dg_\nu(z)}{dz} = g_{\nu-1}(z).$$

When z is small, the integrand may be expanded in powers of z

$$g_\nu(z) = \frac{1}{\Gamma(\nu)} \int_{0}^{+\infty} x^{\nu-1} \sum_{l=1}^{+\infty} (ze^{-x})^l \, dx = \sum_{l=1}^{+\infty} \frac{z^l}{l^\nu} = z + \frac{z^2}{2^\nu} + \frac{z^3}{3^\nu} + ...$$

When $z \to 1$ and $\nu > 1$, the function $g_\nu(z)$ approaches the Riemann *zeta function* $\zeta(\nu)$

$$g_\nu(1) = \frac{1}{\Gamma(\nu)} \int_{0}^{+\infty} x^{\nu-1} \sum_{l=1}^{+\infty} (e^{-x})^l \, dx = \sum_{l=1}^{+\infty} \frac{1}{l^\nu} = \zeta(\nu).$$

Some of the useful values of the Riemann zeta function are here reported

$$\zeta(2) = \frac{\pi^2}{6} \quad \zeta(4) = \frac{\pi^4}{90} \quad \zeta(6) = \frac{\pi^6}{945}.$$

The behaviour of $g_v(z)$ for z close to 1 and $0 < v < 1$ is given by the following approximate formula

$$g_v(e^{-\alpha}) \approx \frac{\Gamma(1-v)}{\alpha^{1-v}} \qquad z = e^{-\alpha} \approx 1.$$

- **Fermi-Dirac functions.** In the theory of the Fermi-Dirac gas, we will use the following integrals

$$f_v(z) = \frac{1}{\Gamma(v)} \int_0^{+\infty} \frac{x^{v-1}dx}{z^{-1}e^x + 1}.$$

When $v = 1$, the integral can be solved exactly

$$f_1(z) = \frac{1}{\Gamma(1)} \int_0^{+\infty} \frac{dx}{z^{-1}e^x + 1} = -\ln(1 + ze^{-x})\big|_0^{+\infty} = \ln(1+z).$$

As for the case of the Bose-Einstein functions, we have a recurrence relation

$$z\frac{df_v(z)}{dz} = f_{v-1}(z)$$

and an expansion for small z

$$f_v(z) = \frac{1}{\Gamma(v)} \int_0^{+\infty} x^{v-1} \sum_{l=1}^{+\infty} (-1)^{l-1} (ze^{-x})^l dx = \sum_{l=1}^{+\infty} (-1)^{l-1} \frac{z^l}{l^v} = z - \frac{z^2}{2^v} + \frac{z^3}{3^v} - \dots$$

The limit $v \to 1$ is connected to the Riemann zeta function $\zeta(v)$

$$f_v(1) = \frac{1}{\Gamma(v)} \int_0^{+\infty} \frac{x^{v-1}dx}{e^x + 1} = \left(1 - \frac{1}{2^{v-1}}\right)\zeta(v).$$

References

1. G.F. Carrier, M. Krook and C.E. Pearson, *Functions of a Complex Variable: Theory and Technique* (SIAM, New York, 2005)
2. K. Huang, *Statistical Mechanics* (John Wiley, New York, 1987)
3. J.W. Gibbs, *Elementary Principles in Statistical Mechanics* (Yale University Press, New haven , 1902); Reprinted by Dover Publications, New York (1960)
4. L.D. Landau and E.M. Lifshitz, *Statistical Physics* (Pergamon Press, Oxford, 1958)
5. L.D. Landau and E.M. Lifshitz, *Quantum Mechanics: Non relativistic theory* (Pergamon Press, Oxford, 1958)
6. J.J. Sakurai, *Modern Quantum Mechanics* (Addison-Wesley, Redwood City, CA, 1985)
7. R. K. Pathria, *Statistical Mechanics* (Pergamon Press, Oxford, 1972)
8. I. Romanovic Prigogine, *Introduction to Thermodynamics of Irreversible Processes* (John Wiley, New York, 1967)
9. R. Kubo, *Statistical Mechanics* (Interscience Publishers, New York, 1965)

Part II

Quantum Mechanics – Problems

Formalism of Quantum Mechanics and One Dimensional Problems

Problem 2.1.

Let $\hat{A} = \hat{A}^\dagger$ be an observable operator with a complete set of eigenstates $|\phi_n\rangle$ with eigenvalues α_n ($n = 0, 1, 2, ...$). A generic state is given by

$$|\psi\rangle = N\left(3|\psi_1\rangle - 4i|\psi_2\rangle\right)$$

where $|\psi_1\rangle, |\psi_2\rangle$ are orthonormal. Find N and the probability P_4 that a measurement of $\hat{A}$ yields α_4. What does it happen in case of degeneracy? Specialize these calculations to the case $|\phi_n\rangle = |\psi_n\rangle$.

Solution

The principles of Quantum Mechanics are encoded in four postulates:

- the first postulate states that all the information for a physical system is contained in a state vector $|\psi(t)\rangle$ properly defined in Hilbert space;
- the second postulate fixes the properties of the Hermitian operators that represent the classical variables like x (position) and p (momentum);
- the third postulate says that regardless the state $|\psi(t)\rangle$ of a particle, the measurement of an observable (with $\hat{O}$ the associated self-adjoint operator) produces as result one of the eigenvalues O of $\hat{O}$ with probability $P(O) = |\langle O|\psi(t)\rangle|^2$. Soon after the measurement, the system is projected into the eigenstate $|O\rangle$ corresponding to the eigenvalue O;
- the fourth postulate gives the time evolution according to the Schrödinger equation

$$i\hbar \frac{\partial}{\partial t}|\psi(t)\rangle = \hat{H}|\psi(t)\rangle$$

where $\hat{H}$ is the Hermitian operator known as Hamiltonian of the system.

It is the third postulate that applies here. However, preliminarily, we need to fix N in order to normalize the wave function. A direct calculation shows that

$$1 = \langle\psi|\psi\rangle = N^2\left(9 + 16\right)$$

Cini M., Fucito F., Sbragaglia M.: Solved Problems in Quantum and Statistical Mechanics.
DOI 10.1007/978-88-470-2315-4_2, © Springer-Verlag Italia 2012

from which $N = \frac{1}{5}$ and

$$|\psi\rangle = \frac{1}{5}\left(3|\psi_1\rangle - 4i|\psi_2\rangle\right).$$

The probability that a measurement of the observable $\hat{A}$ gives a particular eigenvalue α_n of the associated Hermitian operator, is given by the square modulus of the overlap between the state and the eigenstate $|\phi_n\rangle$. In our case, the desired probability is

$$P_4 = \frac{1}{25}|3\langle\phi_4|\psi_1\rangle - 4i\langle\phi_4|\psi_2\rangle|^2.$$

In case of degeneracy, we have to sum over all the eigenstates corresponding to the same eigenvalue. When $|\phi_n\rangle = |\psi_n\rangle$, we only have two possible outcomes for the measurement, i.e. α_1 and α_2. The associated probabilities can be calculated explicitly

$$P_1 = \frac{9}{25} \qquad P_2 = \frac{16}{25}$$

that correctly satisfy $P_1 + P_2 = 1$.

Problem 2.2.
Consider the operators $\hat{A}$, $\hat{B}$, $\hat{C}$, $\hat{D}$, $\hat{E}$, $\hat{F}$ and simplify the commutator $[\hat{A}\hat{B}\hat{C}, \hat{D}\hat{E}\hat{F}]$ so as to show only commutators of type $[\hat{X}, \hat{Y}]$, with $\hat{X}, \hat{Y}$ chosen among the above mentioned operators.

Solution
We start by analyzing the simple commutator $[\hat{A}\hat{B}, \hat{C}]$. Expanding it and introducing the term $\hat{A}\hat{C}\hat{B}$, we get

$$[\hat{A}\hat{B}, \hat{C}] = \hat{A}\hat{B}\hat{C} - \hat{C}\hat{A}\hat{B} = \hat{A}\hat{B}\hat{C} - \hat{C}\hat{A}\hat{B} \pm \hat{A}\hat{C}\hat{B} =$$
$$\hat{A}\hat{B}\hat{C} - \hat{A}\hat{C}\hat{B} - \hat{C}\hat{A}\hat{B} + \hat{A}\hat{C}\hat{B} = \hat{A}[\hat{B}, \hat{C}] + [\hat{A}, \hat{C}]\hat{B}$$

which is a bilinear relation also known as the *Jacobi identity*. By a systematic use of this property, we see that

$$[\hat{A}\hat{B}\hat{C}, \hat{D}\hat{E}\hat{F}] = \hat{A}[\hat{B}\hat{C}, \hat{D}\hat{E}\hat{F}] + [\hat{A}, \hat{D}\hat{E}\hat{F}]\hat{B}\hat{C}$$

and, iterating this procedure, we obtain

$$[\hat{A}\hat{B}\hat{C}, \hat{D}\hat{E}\hat{F}] = \hat{A}\hat{B}\{[\hat{C}, \hat{D}]\hat{E}\hat{F} + \hat{D}[\hat{C}, \hat{E}]\hat{F} + \hat{D}\hat{E}[\hat{C}, \hat{F}]\} +$$
$$\hat{A}\{[\hat{B}, \hat{D}]\hat{E}\hat{F} + \hat{D}[\hat{B}, \hat{E}]\hat{F} + \hat{D}\hat{E}[\hat{B}, \hat{F}]\}\hat{C} +$$
$$\{[\hat{A}, \hat{D}]\hat{E}\hat{F} + \hat{D}[\hat{A}, \hat{E}]\hat{F} + \hat{D}\hat{E}[\hat{A}, \hat{F}]\}\hat{B}\hat{C}.$$

Problem 2.3.
When electrons impinge on a double slit, a diffraction pattern is obtained on a screen located at distance l ($a \ll l$, with a the distance between the slits) from the slits: a sketch is in Fig. 2.1, where O denotes the central maximum and x is a point seen at angle θ_1 from a slit and θ_2 from the other. Let $2\Delta x$ be the position of the second maximum. Determine Δx. For a given wavelength λ, determine the momenta (p_1

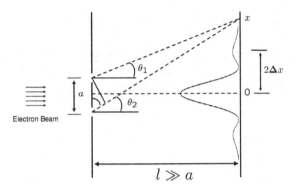

Fig. 2.1 Sketch of the diffraction pattern due to an electron beam passing through a double slit. The slits are separated by a distance a. The diffraction pattern emerges on a screen located at distance l from the slits. The resulting momentum transferred to the slits is connected to the position of the maxima of intensity in the diffraction pattern through the Heisenberg uncertainty relation. Details are reported in Problem 2.3

and p_2) transferred to the slits and verify that the product $\Delta x \Delta p$ ($\Delta p = |p_1 - p_2|$) agrees with Heisenberg uncertainty relation.

Solution
Electrons and any other kind of microscopic particles have wave properties in agreement with the De Broglie relation

$$p = \hbar k = \frac{2\pi\hbar}{\lambda} = \frac{h}{\lambda}$$

involving momentum p and wavelength λ. If a particle has such a high momentum that the wavelength is smaller than all the characteristic lengths in the experiment, the typical wave phenomena like interference and diffraction may be hard to see. The electron mass is light, and if the energy is in the Electronvolt (eV) range, p is such that λ is comparable with the typical lattice spacing in many crystals. Anyhow one can see wave phenomena in experiments such as electron diffraction through thin metal films. The present problem illustrates this situation in the simple special case of the double slit. Electrons hitting x on the second screen are deflected by the first screen while passing it. Suppose the screen is set in motion by the electron and we can calculate the momentum transfer from its recoil speed. We need to consider only the vertical components of the momenta. The electron trajectories arriving at point x through slit 1 or 2 have different deflections, and the momentum transfer is different, namely

$$p_1 = \frac{h}{\lambda}\sin\theta_1 \approx \frac{h}{\lambda}\theta_1 \qquad p_2 = \frac{h}{\lambda}\sin\theta_2 \approx \frac{h}{\lambda}\theta_2$$

where we have assumed that θ_1, θ_2 are small, corresponding to the condition that the distance between the slits is very small respect to the distance of the screen from the slits, i.e. $a \ll l$. Due to the geometry of the problem, we get

$$\tan\theta_1 = \frac{x - \frac{1}{2}a}{l} \approx \theta_1$$

$$\tan\theta_2 = \frac{x + \frac{1}{2}a}{l} \approx \theta_2$$

from which we obtain

$$\Delta p = |p_1 - p_2| = \frac{h}{\lambda}|\sin\theta_1 - \sin\theta_2| \approx \frac{h}{\lambda}|\theta_1 - \theta_2| = \frac{ha}{\lambda l}.$$

Let us consider two waves with the same frequency and at the same time t: $A_1 e^{i(\omega t + \phi_1)}, A_2 e^{i(\omega t + \phi_2)}$, with A_1, A_2 real numbers. Let I be the intensity of the wave resulting from the superposition of the two

$$I = A_1^2 + A_2^2 + 2A_1 A_2 \cos(\phi_1 - \phi_2).$$

The term $\cos(\phi_1 - \phi_2)$ is responsible for the phenomenon of interference, which is constructive if

$$\phi_1 - \phi_2 = kx_{12} = ka\sin\theta = \frac{2\pi}{\lambda}a\sin\theta \approx \frac{2\pi}{\lambda}a\theta = 2\pi m$$

with m an integer number and x_{12} the difference in the distances travelled by the two beams arriving in x. As for the angle θ, we can use the average value of θ_1 and θ_2 given above

$$\theta = \frac{1}{2}(\theta_1 + \theta_2) \approx \frac{(x + \frac{1}{2}a) + (x - \frac{1}{2}a)}{2l} = \frac{x}{l}.$$

Let $2\Delta x$ be the distance between the second maximum and the origin O. We may see Δx as the maximum error allowed on the distance, i.e. if the error is above Δx we are unable to distinguish the location of the maxima. The relation between Δx, a and λ is

$$\theta = \frac{2\Delta x}{l} = \frac{\lambda}{a}$$

where we have used the previous result with $m = 1$ and $\theta = \frac{2\Delta x}{l}$. The uncertainty principle in this experiment takes the form: any determination of the two possible alternatives for the electron destroys the interference between them. In other terms, any measurement that allows one to know which slit the electron went through destroys the interference. Putting together all the results for Δx and Δp, we obtain

$$\Delta x \Delta p = \frac{1}{2}\hbar$$

that is Heisenberg uncertainty relation. A possible interpretation is that the first screen recoils due to momentum transfer and this causes a shift of order Δx on the second screen.

Problem 2.4.
Consider a one dimensional quantum harmonic oscillator with frequency ω and mass m. Using the creation and annihilation operators to represent the position operator $\hat{x}$, determine the matrix elements $(\hat{x}^2)_{0,0}$, $(\hat{x}^2)_{1,1}$, $(\hat{x}^2)_{2,2}$, $(\hat{x}^2)_{0,2}$, $(\hat{x}^5)_{10,5}$. The matrix element $(\hat{x})_{m,n} = \langle m|\hat{x}|n\rangle$ is defined on the eigenstates $|n\rangle$ of the Hamiltonian with eigenvalues $E_n = (n + \frac{1}{2})\hbar\omega$ with $n = 0,1,2,3,\ldots$.

Solution
The relation between the position and the creation and annihilation operators is

$$\hat{x} = \frac{x_0}{\sqrt{2}}(\hat{a} + \hat{a}^\dagger)$$

where $x_0 = \sqrt{\frac{\hbar}{m\omega}}$ is the characteristic length scale of the oscillator. The momentum operator $\hat{p}$ is written as

$$\hat{p} = -i\frac{\hbar}{x_0\sqrt{2}}(\hat{a} - \hat{a}^\dagger)$$

and the Hamiltonian becomes

$$\hat{H} = \frac{\hat{p}^2}{2m} + \frac{1}{2}m\omega\hat{x}^2 = \left(\hat{a}^\dagger\hat{a} + \frac{1}{2}\mathbf{1}\right)\hbar\omega = \left(\hat{n} + \frac{1}{2}\mathbf{1}\right)\hbar\omega$$

where we have defined the number operator $\hat{n} = \hat{a}^\dagger\hat{a}$ such that $\hat{n}|n\rangle = n|n\rangle$. The creation and annihilation operators satisfy the commutation rule

$$[\hat{a}, \hat{a}^\dagger] = \hat{a}\hat{a}^\dagger - \hat{a}^\dagger\hat{a} = \hat{a}\hat{a}^\dagger - \hat{n} = 1.$$

Moreover, we know that $\hat{a}$ and $\hat{a}^\dagger$ act as step down and step up operators on the eigenstates $|n\rangle$

$$\hat{a}|n\rangle = \sqrt{n}|n-1\rangle \qquad \hat{a}^\dagger|n\rangle = \sqrt{n+1}|n+1\rangle$$

from which we see that

$$|n\rangle = \frac{1}{\sqrt{n!}}(\hat{a}^\dagger)^n|0\rangle.$$

Since $\hat{a}^\dagger$ increases n by 1 while $\hat{a}$ decreases n by 1, the only terms in the expansion of $(\hat{a} + \hat{a}^\dagger)^n$ that contribute are those with an equal number of $\hat{a}$ and $\hat{a}^\dagger$. Thus, squaring

$\hat{x}$ one finds the operator identity

$$\hat{x}^2 = \frac{x_0^2}{2}(\hat{a}^2 + (\hat{a}^\dagger)^2 + \hat{a}\hat{a}^\dagger + \hat{a}^\dagger\hat{a}) = \frac{x_0^2}{2}(\hat{a}^2 + (\hat{a}^\dagger)^2 + 1 + 2\hat{n})$$

and

$$(\hat{x}^2)_{n,n} = \frac{x_0^2}{2}(1 + 2n)$$

and we find

$$(\hat{x}^2)_{0,0} = \frac{1}{2}x_0^2 \quad (\hat{x}^2)_{1,1} = \frac{3}{2}x_0^2 \quad (\hat{x}^2)_{2,2} = \frac{5}{2}x_0^2.$$

By the same token, only products of annihilation operators count in the calculation of matrix elements like $(\hat{x}^2)_{0,2}$ where one must go down by one step at each occurrence of $\hat{x}$. Since $\frac{\hat{a}}{\sqrt{2}}|2\rangle = |1\rangle$ and $\hat{a}|1\rangle = |0\rangle$, we find

$$(\hat{x}^2)_{0,2} = \frac{1}{2}x_0^2\langle 0|\hat{a}^2|2\rangle = \frac{1}{\sqrt{2}}x_0^2.$$

A similar reasoning helps one to obtain $(\hat{x}^5)_{10,5}$. In such a case, we consider $\hat{x}^5 = \frac{x_0^5}{2^{5/2}}(\hat{a} + \hat{a}^\dagger)^5$, and select the only term allowing for 5 steps up, i.e. the term $(\hat{a}^\dagger)^5$. The result is

$$(\hat{x}^5)_{10,5} = \frac{x_0^5}{\sqrt{2^5}}\langle 10|(\hat{a}^\dagger)^5|5\rangle = \frac{x_0^5}{\sqrt{5!2^5}}\langle 10|(\hat{a}^\dagger)^{10}|0\rangle = \frac{x_0^5}{4\sqrt{2}}\sqrt{\frac{10!}{5!}}.$$

Problem 2.5.

Consider the wave packet

$$\psi(x, t = 0) = Ae^{-x^2/4a^2}e^{ik_0x}$$

with a, k_0 constants and A a normalization factor. Show that this wave packet minimizes Heisenberg uncertainty relations for the position and momentum operators. Finally, determine the time evolution of the wave packet at a generic time t.

Solution

We start by a statement of the relation between the wave packet at time $t = 0$ and the energy eigenfunctions. These are solutions of the Schrödinger equation with the energy eigenvalue falling in the free-particle continuum. Such solutions may be labeled e.g. by the momentum. This set is complete. This is tantamount to say that any reasonable function $\psi(x)$ can be expanded in this basis

$$\psi(x) = \int c_F \psi_F(x)dF$$

where F is an appropriate set of quantum numbers. Since a generic quantum number can take continuous values, as is the case when F stands for the momentum components, we are using a notation involving the integral sign rather than the summation sign, which is typical of the discrete spectrum. In general, one has to consider continuous *and* discrete summations, as is the case if the continuum states bear angular momentum quantum numbers. For definiteness, here we develop the case when F is a continuous set, like the momentum components. The continuum eigenfunctions ψ_F cannot be normalized like those of the discrete spectrum, by setting the integral of the square modulus equal to 1. Unbound particles have a comparable probability to be at any distance from the origin, so the wave function does not vanish at infinity, and the integral blows up. We can normalize differently. We impose that $|c_F|^2 dF$ is the probability that a measurement of $\hat{F}$ is found to be between F and $F + dF$. By completeness, we have

$$\int \psi^*(x)\psi(x)\,dx = \int c_F^* c_F\,dF = 1.$$

The coefficients c_F are found, in complete analogy with the discrete spectrum, by projecting the function $\psi(x)$ on the $\psi_F(x)$. From the last equation, using the expansion of $\psi(x)$ in terms of the $\psi_F(x)$, we deduce that

$$\int c_F^* \left(\int \psi(x)\psi_F^*(x)\,dx - c_F \right) dF = 0.$$

For an arbitrary value of the coefficient c_F, the above equation is satisfied only if the integrand is zero, so that

$$c_F = \int \psi(x)\psi_F^*(x)\,dx.$$

Again, using the expansion of ψ in terms of the ψ_F, we find

$$c_F = \int \psi(x)\psi_F^*(x)\,dx = \int c_{F'}\psi_{F'}(x)\psi_F^*(x)\,dx\,dF'.$$

For an arbitrary value of c_F, the last equation is satisfied only if

$$\int \psi_F(x)\psi_{F'}^*(x)\,dx = \delta(F - F')$$

and the functions $\psi_F(x)$ are orthogonal for $F \neq F'$.

By this formalism we are now in position to deal with the wave packet. The ψ_F eigenfunctions cannot be realized physically. For example, if they have a well-defined momentum, the particle cannot be localized in any spatial domain however large. The wave packet given by the problem is still a free particle wave function, but it is localized in a region of size a. It can be normalized by taking A such that

$$1 = |A|^2 \int_{-\infty}^{+\infty} |\psi(x)|^2\,dx = |A|^2 \int_{-\infty}^{+\infty} e^{-\frac{x^2}{2a^2}}\,dx = |A|^2 a\sqrt{2\pi}$$

from which $A = 1/(2\pi a^2)^{1/4}$. We can also expand $\psi(x)$ in terms of plane waves, corresponding to a Fourier integral of the wave packet

$$\psi(x) = \frac{1}{\sqrt{2\pi}} \int_{-\infty}^{+\infty} \psi(k) e^{ikx} dk$$

where $\psi(k)$, e^{ikx} play the role of the c_F, ψ_F we have previously introduced. The factor $\sqrt{2\pi}$ is required for the correct normalization of the plane waves

$$\langle x|k \rangle = \frac{1}{\sqrt{2\pi}} e^{ikx} \qquad \langle k|k' \rangle = \frac{1}{2\pi} \int_{-\infty}^{+\infty} e^{i(k-k')x} dx = \delta(k-k').$$

The integral determining $\psi(k)$ can be done by completing the square, and it is found to be proportional to an exponential function

$$\psi(k) \propto \int_{-\infty}^{+\infty} e^{-ikx} e^{-\frac{x^2}{4a^2}} e^{ik_0 x} dx \propto e^{-a^2(k-k_0)^2}.$$

We can fix the normalization constant in front of it by imposing that $|\psi(k)|^2$ is normalized to unity. We then get

$$\psi(k) = \left(\frac{2a^2}{\pi} \right)^{\frac{1}{4}} e^{-a^2(k-k_0)^2}$$

that is the Gaussian function centered in k_0, i.e. the expectation value for $p/\hbar$. By the properties of the Gaussian distributions, we find that the uncertainties are

$$\Delta x = \sqrt{\langle \hat{x}^2 \rangle - \langle \hat{x} \rangle^2} = a \qquad \Delta k = \sqrt{\langle \hat{k}^2 \rangle - \langle \hat{k} \rangle^2} = \frac{1}{2a}.$$

We can verify explicitly this result in the case of the position operator $\hat{x}$. The quantity $\langle \hat{x} \rangle$ is zero due to the symmetries in the integral. At the same time, the average squared position is

$$\langle \hat{x}^2 \rangle = \frac{1}{\sqrt{2\pi}a} \int_{-\infty}^{+\infty} x^2 e^{-\frac{x^2}{2a^2}} dx = \frac{1}{\sqrt{2\pi}a} \lim_{\beta \to 1} (-2a^2) \frac{d}{d\beta} \int_{-\infty}^{+\infty} e^{-\frac{x^2 \beta}{2a^2}} dx =$$

$$\frac{1}{\sqrt{2\pi}a} (-2a^2) \sqrt{2\pi}a \lim_{\beta \to 1} \frac{d}{d\beta} \beta^{-\frac{1}{2}} = a^2$$

from which we prove that $\Delta x = a$. Given the relation between the wave vector and the momentum, $p = \hbar k$, we find

$$\Delta x \Delta p = \frac{\hbar}{2}$$

that is the Heisenberg uncertainty relation for momentum and position.

We can now discuss the time evolution. We need to solve the Schrödinger equation

$$i\hbar \frac{\partial |\psi(t)\rangle}{\partial t} = \hat{H}|\psi(t)\rangle$$

when the Hamiltonian does not depend explicitly on time. The general *method of separation of variables* applies and we seek a solution of the form $|\psi(t)\rangle = A(t)|\psi\rangle$, with the result

$$\begin{cases} i\hbar \frac{dA(t)}{dt} = EA(t) \\ \hat{H}|\psi_E\rangle = E|\psi_E\rangle \end{cases}$$

that is a couple of ordinary differential equations with constant coefficients. The second is the stationary Schrödinger equation and yields the Hamiltonian eigenfunctions with energy E. Once E is known, we can insert it into the solution of the first equation

$$A(t) = e^{-\frac{iEt}{\hbar}}.$$

We see that the states with a well defined value of the energy evolve like

$$\psi_E(x,t) = e^{-\frac{iEt}{\hbar}}\psi_E(x).$$

Going back to our wave packet, all the plane waves evolve in time with a well defined phase

$$\psi(x,t) = \frac{1}{\sqrt{2\pi}} \int_{-\infty}^{+\infty} \psi(k)e^{i(kx-\omega t)}\,dk$$

where, for the plane wave, we know that $E = \hbar\omega = p^2/2m = \hbar^2 k^2/2m$. The above integral can be done exactly, once we use the $\psi(k)$ previously determined

$$\psi(x,t) = \frac{1}{(2\pi)^{1/2}}\left(\frac{2a^2}{\pi}\right)^{\frac{1}{4}} \int_{-\infty}^{+\infty} e^{-a^2(k-k_0)^2 + ikx - \frac{i\hbar k^2 t}{2m}}\,dk =$$

$$\frac{1}{(2\pi)^{1/2}}\left(\frac{2a^2}{\pi}\right)^{\frac{1}{4}} e^{-a^2 k_0^2} \int_{-\infty}^{+\infty} e^{-k^2\left(a^2 + \frac{i\hbar t}{2m}\right) + k(2k_0 a^2 + ix)}\,dk =$$

$$\frac{1}{(2\pi)^{1/2}}\left(\frac{2a^2}{\pi}\right)^{\frac{1}{4}} \sqrt{\frac{\pi}{\left(a^2 + \frac{i\hbar t}{2m}\right)}}\, e^{-a^2 k_0^2}\, e^{\frac{\left(2k_0 a^2 + ix\right)^2}{4\left(a^2 + \frac{i\hbar t}{2m}\right)}}$$

where we have completed the square in the exponential function. The resulting probability density function is

$$P(x,t) = |\psi(x,t)|^2 = \frac{1}{\sqrt{2\pi a^2}\sqrt{1 + \left(\frac{\hbar t}{2ma^2}\right)^2}} e^{-\frac{\left(x - \frac{\hbar k_0 t}{m}\right)^2}{2a^2\left(1 + \left(\frac{\hbar t}{2ma^2}\right)^2\right)}}.$$

The width (the variance of the Gaussian distribution) of the wave packet is proportional to $a^2\left(1 + \left(\frac{\hbar t}{2ma^2}\right)^2\right)$, i.e. it increases as a function of time. The maximum of

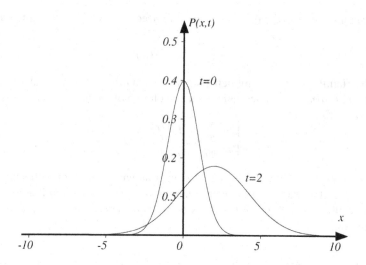

Fig. 2.2 The probability distribution function $P(x,t)$ obtained in Problem 2.5, starting from the initial wave packet $\psi(x,t=0) = \frac{1}{(2\pi a^2)^{1/4}} e^{-x^2/4a^2} e^{ik_0 x}$. We plot the case with $a=1$, $\frac{\hbar}{2ma^2}=1$ and $\frac{\hbar k_0}{m}=1$ for two characteristic times $t=0$ and $t=2.0$

$P(x,t)$ is not in $x=0$ any longer, and it has moved to $x_0 = \hbar k_0 t/m = p_0 t/m$. We see that the average value of the position evolves in time as the position of a point-like particle with mass m and constant velocity p_0/m

$$\frac{d\langle \hat{x}\rangle_t}{dt} = \frac{p_0}{m}.$$

The evolution of $P(x,t)$ is sketched in Fig. 2.2.

Problem 2.6.

Let us consider a quantum system with two states. The matrix representation of the Hamiltonian in a given vector basis (assume $\hbar = 1$ for simplicity) is

$$\hat{H} = \begin{pmatrix} 0 & 1 \\ 1 & 0 \end{pmatrix};$$

- determine the eigenstates and eigenvalues of $\hat{H}$;
- determine the time evolution operator $e^{-i\hat{H}t}$;
- let

$$|\psi(0)\rangle = \begin{pmatrix} 1 \\ 0 \end{pmatrix} \qquad \hat{O} = \begin{pmatrix} 1 & 0 \\ 0 & 2 \end{pmatrix}$$

be the wave function at a time $t=0$ and an observable $\hat{O}$. Using the Schrödinger representation, find the probability that a measurement of the observable $\hat{O}$ at a time $t>0$ gives 2. Repeat the calculation with the Heisenberg representation.

Solution
The Hamiltonian matrix $\hat{H}$ coincides with the famous Pauli $\hat{\sigma}_x$ matrix and its eigen-vectors and eigenvalues are particularly simple. To determine the eigenvalues, we have to solve

$$\det\begin{pmatrix} -\lambda & 1 \\ 1 & -\lambda \end{pmatrix} = \lambda^2 - 1 = 0$$

from which we get $\lambda = \pm 1$. To determine the eigenvector $|\psi_1\rangle$, corresponding to $\lambda = 1$, we set

$$\begin{pmatrix} 0 & 1 \\ 1 & 0 \end{pmatrix}\begin{pmatrix} a \\ b \end{pmatrix} = \begin{pmatrix} a \\ b \end{pmatrix}$$

yielding $a = b$. The normalization condition ($a^2 + b^2 = 1$), completely fixes $|\psi_1\rangle$ up to an unessential constant phase factor. The procedure must be repeated to find the eigenvector corresponding to the eigenvalue -1. Eventually, we get

$$|\psi_1\rangle = \frac{1}{\sqrt{2}}\begin{pmatrix} 1 \\ 1 \end{pmatrix} \qquad |\psi_2\rangle = \frac{1}{\sqrt{2}}\begin{pmatrix} 1 \\ -1 \end{pmatrix}.$$

We evaluate $e^{-i\hat{H}t}$ using three independent (but of course equivalent) approaches. First, we rely on the definition of an analytic function of a matrix through the Taylor series expansion

$$e^{-i\hat{H}t} = \sum_{n=0}^{+\infty}\frac{1}{n!}(-i\hat{H}t)^n = \sum_{n=0}^{+\infty}\frac{(-1)^n}{2n!}t^{2n} - i\hat{H}\sum_{n=0}^{+\infty}\frac{(-1)^n}{(2n+1)!}t^{(2n+1)} =$$

$$1\cos t - i\hat{H}\sin t = \begin{pmatrix} \cos t & -i\sin t \\ -i\sin t & \cos t \end{pmatrix}$$

where we have used the property $\hat{\sigma}_x^{2n} = \mathbb{1}$ and $\hat{\sigma}_x^{2n+1} = \hat{\sigma}_x$, where n is an integer number. Alternatively, we can use the Cauchy integral

$$e^{-i\hat{H}t} = \frac{1}{2\pi i}\oint\frac{e^{-itz}}{(z\mathbb{1}-\hat{H})}dz = \frac{1}{2\pi i}\oint\frac{e^{-itz}}{(z^2-1)}\begin{pmatrix} z & 1 \\ 1 & z \end{pmatrix}dz =$$

$$\begin{pmatrix} \frac{1}{2\pi i}\oint\frac{e^{-itz}}{(z^2-1)}z\,dz & \frac{1}{2\pi i}\oint\frac{e^{-itz}}{(z^2-1)}dz \\ \frac{1}{2\pi i}\oint\frac{e^{-itz}}{(z^2-1)}dz & \frac{1}{2\pi i}\oint\frac{e^{-itz}}{(z^2-1)}z\,dz \end{pmatrix} = \begin{pmatrix} \cos t & -i\sin t \\ -i\sin t & \cos t \end{pmatrix}.$$

As a third possibility, one starts defining a function $\hat{F}(\hat{A})$ of a *diagonal* matrix $\hat{A}$ in the most obvious way, as the diagonal matrix obtained by applying F to the elements on the diagonal. The definition extends naturally to all matrices that can be diagonalized through a similarity transformation $\hat{C}$. In other terms, $\hat{C}\hat{A}\hat{C}^{-1}$ is diagonal, so we apply $\hat{F}$, go back to the original basis, and define

$$\hat{F}(\hat{A}) = \hat{C}\hat{F}(\hat{C}\hat{A}\hat{C}^{-1})\hat{C}^{-1}.$$

The required matrix is

$$\hat{C} = \hat{C}^{-1} = \frac{1}{\sqrt{2}} \begin{pmatrix} 1 & 1 \\ 1 & -1 \end{pmatrix}$$

and its columns are $|\psi_1\rangle, |\psi_2\rangle$. By these formulae, in our case we get again

$$\hat{U}(t) = e^{-i\hat{H}t} = \frac{1}{2} \begin{pmatrix} 1 & 1 \\ 1 & -1 \end{pmatrix} \begin{pmatrix} e^{-it} & 0 \\ 0 & e^{it} \end{pmatrix} \begin{pmatrix} 1 & 1 \\ 1 & -1 \end{pmatrix} = \begin{pmatrix} \cos t & -i\sin t \\ -i\sin t & \cos t \end{pmatrix}.$$

The operator $\hat{U}(t)$ enables us to proceed at once with the time evolution of $|\psi(0)\rangle$ according to $|\psi(t)\rangle = e^{-i\hat{H}t}|\psi(0)\rangle$

$$|\psi(t)\rangle = e^{-i\hat{H}t}|\psi(0)\rangle = \begin{pmatrix} \cos t - i\sin t \\ -i\sin t & \cos t \end{pmatrix} \begin{pmatrix} 1 \\ 0 \end{pmatrix} = \begin{pmatrix} \cos t \\ -i\sin t \end{pmatrix}.$$

This is a toy example, but generally the explicit calculation of $e^{-i\hat{H}t}$ in closed form is prohibitively difficult. It may be easier to project the wave function on the basis of the eigenfunctions of the Hamiltonian ($|\psi_1\rangle, |\psi_2\rangle$). Then $e^{-i\hat{H}t}$ is diagonal with eigenvalues $e^{\mp it}$. We first expand $|\psi(0)\rangle$

$$|\psi(0)\rangle = \frac{1}{\sqrt{2}}(|\psi_1\rangle + |\psi_2\rangle).$$

Then, the associated time evolution is

$$|\psi(t)\rangle = e^{-i\hat{H}t}|\psi(0)\rangle = \frac{1}{\sqrt{2}}(e^{-it}|\psi_1\rangle + e^{it}|\psi_2\rangle) = \begin{pmatrix} \cos t \\ -i\sin t \end{pmatrix}.$$

Let us now discuss the properties of the operator $\hat{O}$. The eigenvectors of $\hat{O}$ are

$$|\tilde{\psi}_1\rangle = \begin{pmatrix} 1 \\ 0 \end{pmatrix} \qquad |\tilde{\psi}_2\rangle = \begin{pmatrix} 0 \\ 1 \end{pmatrix}.$$

They differ from the $|\psi_1\rangle, |\psi_2\rangle$ we found before. A measurement of $\hat{O}$ can give the eigenvalue 2, only if the system is described by $|\tilde{\psi}_2\rangle$. Writing

$$|\psi(t)\rangle = \begin{pmatrix} \cos t \\ -i\sin t \end{pmatrix} = \cos t \begin{pmatrix} 1 \\ 0 \end{pmatrix} - i\sin t \begin{pmatrix} 0 \\ 1 \end{pmatrix} = c_1(t)|\tilde{\psi}_1\rangle + c_2(t)|\tilde{\psi}_2\rangle$$

the sought probability is

$$P(2)_{t>0} = |\langle\psi(t)|\tilde{\psi}_2\rangle|^2 = \sin^2 t.$$

Note that the initial wave function, $|\psi(0)\rangle = |\tilde{\psi}_1\rangle$, has no $|\tilde{\psi}_2\rangle$ component and the probability of a measurement giving 2 is $P(2)_{t=0} = \sin^2(0) = 0$. It is the time evolution that produces this probability. In summary: the probability that a measurement of an observable $\hat{O}$ at time $t = 0$ yields O_k (eigenvalue of $\hat{O}$, which belongs to the eigenvector $|\tilde{\psi}_k\rangle$) is the square modulus of the coefficient c_k of $|\tilde{\psi}_k\rangle$ in the expansion of the wave function in the basis of eigenvectors of $\hat{O}$. If $\hat{O}$ commutes with $\hat{H}$ the coefficient c_k is constant for $t > 0$. Otherwise, if $\hat{O}$ fails to commute with $\hat{H}$, its time dependence must be computed as we have seen.

In what we have seen above, the time dependence is entirely in the quantum state $|\psi(t)\rangle$ and the operators do not depend on time (this is known as *Schrödinger (S) representation*). Quantum Mechanics, however, can also be formulated in a different but equivalent form, in which the time dependence is passed from the quantum states to the operators (this is known as *Heisenberg (H) representation*), i.e. the operators evolve in time, while the wave function is kept the same as the initial time. The Heisenberg operator is

$$\hat{O}_H(t) = \hat{U}^{-1}(t)\hat{O}\hat{U}(t) = e^{i\hat{H}t}\hat{O}e^{-i\hat{H}t} = \begin{pmatrix} 1+\sin^2 t & i\sin t\cos t \\ -i\sin t\cos t & 1+\cos^2 t \end{pmatrix}.$$

Since $\hat{O}_H(t) = \hat{U}^{-1}(t)\hat{O}\hat{U}(t)$ is a unitary transformation, the eigenvalues of $\hat{O}_H$ are still $1, 2$ and the eigenvectors are the columns of

$$\hat{U}^{-1}(t) = e^{i\hat{H}t} = \begin{pmatrix} \cos t & i\sin t \\ i\sin t & \cos t \end{pmatrix}.$$

In this way, we find the relation between the Heisenberg and Schrödinger representations

$$|\psi_S(t)\rangle = \hat{U}(t)|\psi_H\rangle$$
$$\hat{O}_H(t) = \hat{U}^{-1}(t)\hat{O}_S\hat{U}(t).$$

The initial state is $|\tilde{\psi}_1\rangle$, and we use it to calculate the expectation value in the Heisenberg picture

$$\langle\tilde{\psi}_1|\hat{O}_H(t)|\tilde{\psi}_1\rangle = 1+\sin^2 t = \lambda_1|c_1|^2 + \lambda_2|c_2|^2$$

where $\lambda_1, \lambda_2 = 1, 2$ are the eigenvalues of $\hat{O}_H(t)$ and c_1, c_2 the coefficients of the expansion in the vector basis where $\hat{O}_H(t)$ is diagonal. Since $|c_1|^2 + |c_2|^2 = 1$, we get

$$P(2)_{t>0} = |c_2|^2 = \sin^2 t$$

in agreement with the previous result.

Problem 2.7.
A quantum harmonic oscillator has the Hamiltonian

$$\hat{H} = \left(\hat{a}^\dagger\hat{a} + \frac{1}{2}1\right) \qquad \hbar = \omega = m = 1$$

where $\hat{a}^\dagger$ and $\hat{a}$ are the creation/annihilation operators. The oscillator is such that, at a time $t = 0$, measurements of the energy never give results above $E > 2$. Give a matrix representation for the operators $\hat{H}$, $\quad \hat{x} = (\hat{a}+\hat{a}^\dagger)/\sqrt{2}$, $\hat{p} = -i(\hat{a}-\hat{a}^\dagger)/\sqrt{2}$

using the formalism of the creation/annihilation operators. Verify the results in the Schrödinger formalism. Then, for $t > 0$, determine the matrices $\hat{x}(t)$, $\hat{p}(t)$ in the Heisenberg representation. Finally, verify the matrix elements of $\hat{x}(t)$, $\hat{p}(t)$ with the time evolution of the wave function and the Schrödinger representation.

Solution

The eigenvalues of the Hamiltonian of the quantum harmonic oscillator are

$$E_n = \left(n + \frac{1}{2}\right)$$

because $\hbar = \omega = 1$. The only eigenvalues lower than 2 are $E_0 = \frac{1}{2}$ and $E_1 = \frac{3}{2}$. Therefore, the wave function is a superposition of the first two states $|0\rangle$ and $|1\rangle$. The matrix representations correspond to 2×2 matrices whose elements are the scalar product with these states. For a generic operator $\hat{A}$ we have

$$\hat{A} = \begin{pmatrix} \langle 0|\hat{A}|0\rangle & \langle 0|\hat{A}|1\rangle \\ \langle 1|\hat{A}|0\rangle & \langle 1|\hat{A}|1\rangle \end{pmatrix}.$$

Using the relation between $\hat{a}$, $\hat{a}^\dagger$, $\hat{x}$, $\hat{p}$ given in the text and recalling the step up and step down action of the creation and annihilation operators on the generic eigenstate

$$\hat{a}^\dagger |n\rangle = \sqrt{n+1}\,|n+1\rangle \qquad \hat{a}|n\rangle = \sqrt{n}\,|n-1\rangle$$

we easily obtain all the operators at time $t = 0$

$$\hat{H} = \frac{1}{2}\,1 + \begin{pmatrix} 0 & 0 \\ 0 & 1 \end{pmatrix} \qquad \hat{x}(0) = \frac{1}{\sqrt{2}} \begin{pmatrix} 0 & 1 \\ 1 & 0 \end{pmatrix} \qquad \hat{p}(0) = \frac{i}{\sqrt{2}} \begin{pmatrix} 0 & -1 \\ 1 & 0 \end{pmatrix}$$

and the eigenstates

$$|0\rangle = \begin{pmatrix} 1 \\ 0 \end{pmatrix} \qquad |1\rangle = \begin{pmatrix} 0 \\ 1 \end{pmatrix}.$$

We can now verify these results in the Schrödinger formalism. The normalized eigenfunctions of the ground state and the first excited state are

$$\psi_0(x) = \langle x|0\rangle = \frac{1}{\pi^{\frac{1}{4}}} e^{-\frac{1}{2}x^2} \qquad \psi_1(x) = \langle x|1\rangle = \frac{\sqrt{2}}{\pi^{\frac{1}{4}}} x e^{-\frac{1}{2}x^2}$$

and we want to calculate the matrix elements $\langle 0|\hat{x}|0\rangle$, $\langle 1|\hat{x}|1\rangle$, $\langle 0|\hat{x}|1\rangle$, $\langle 1|\hat{x}|0\rangle$ and those with $\hat{p}$ in place of $\hat{x}$. The first two of these elements are always zero because the integrand function is even while the operator is odd. The other elements are

$$\langle 0|\hat{x}|1\rangle = \langle 1|\hat{x}|0\rangle = \int_{-\infty}^{+\infty} \langle 1|\hat{x}|x\rangle\langle x|0\rangle\,dx =$$

$$\int_{-\infty}^{+\infty} x\langle 1|x\rangle\langle x|0\rangle\,dx = \sqrt{\frac{2}{\pi}} \int_{-\infty}^{+\infty} x^2 e^{-x^2}\,dx = \frac{1}{\sqrt{2}}.$$

Similarly, for the matrix elements $\langle 0|\hat{p}|1\rangle$ and $\langle 1|\hat{p}|0\rangle$, we find

$$\langle 1|\hat{p}|0\rangle = -\langle 0|\hat{p}|1\rangle = \int_{-\infty}^{+\infty}\langle 1|x\rangle\langle x|\hat{p}|0\rangle\,dx = -i\int_{-\infty}^{+\infty}\langle 1|x\rangle\frac{d}{dx}\langle x|0\rangle\,dx =$$

$$-i\sqrt{\frac{2}{\pi}}\int_{-\infty}^{+\infty}xe^{-\frac{1}{2}x^2}\frac{d}{dx}\left(e^{-\frac{1}{2}x^2}\right)dx = i\sqrt{\frac{2}{\pi}}\int_{-\infty}^{+\infty}x^2 e^{-x^2}\,dx = \frac{i}{\sqrt{2}}.$$

The expressions of $\hat{x}(t)$ and $\hat{p}(t)$ in the Heisenberg representation (see also Problem 2.6) are

$$\hat{x}(t) = e^{i\hat{H}t}\hat{x}(0)e^{-i\hat{H}t} = \frac{1}{\sqrt{2}}\begin{pmatrix} e^{\frac{it}{2}} & 0 \\ 0 & e^{\frac{3it}{2}} \end{pmatrix}\begin{pmatrix} 0 & 1 \\ 1 & 0 \end{pmatrix}\begin{pmatrix} e^{-\frac{it}{2}} & 0 \\ 0 & e^{-\frac{3it}{2}} \end{pmatrix} =$$

$$\frac{1}{\sqrt{2}}\begin{pmatrix} 0 & e^{-it} \\ e^{it} & 0 \end{pmatrix}$$

$$\hat{p}(t) = e^{i\hat{H}t}\hat{p}(0)e^{-i\hat{H}t} = \frac{i}{\sqrt{2}}\begin{pmatrix} e^{\frac{it}{2}} & 0 \\ 0 & e^{\frac{3it}{2}} \end{pmatrix}\begin{pmatrix} 0 & -1 \\ 1 & 0 \end{pmatrix}\begin{pmatrix} e^{-\frac{it}{2}} & 0 \\ 0 & e^{-\frac{3it}{2}} \end{pmatrix} =$$

$$\frac{i}{\sqrt{2}}\begin{pmatrix} 0 & -e^{-it} \\ e^{it} & 0 \end{pmatrix}.$$

As for the Schrödinger representation, the time evolution of the eigenstates is

$$|0(t)\rangle = e^{-\frac{it}{2}}\begin{pmatrix} 1 \\ 0 \end{pmatrix} \qquad |1(t)\rangle = e^{-\frac{3it}{2}}\begin{pmatrix} 0 \\ 1 \end{pmatrix}$$

and the operators are kept the same as those at time $t = 0$. The relevant matrix elements are

$$\langle 0(t)|\hat{x}(0)|1(t)\rangle = \frac{1}{\sqrt{2}}\begin{pmatrix} e^{\frac{it}{2}} & 0 \end{pmatrix}\begin{pmatrix} 0 & 1 \\ 1 & 0 \end{pmatrix}\begin{pmatrix} 0 \\ e^{-\frac{3it}{2}} \end{pmatrix} = \frac{1}{\sqrt{2}}e^{-it}$$

$$\langle 1(t)|\hat{x}(0)|0(t)\rangle = \frac{1}{\sqrt{2}}\begin{pmatrix} 0 & e^{\frac{3it}{2}} \end{pmatrix}\begin{pmatrix} 0 & 1 \\ 1 & 0 \end{pmatrix}\begin{pmatrix} e^{-\frac{it}{2}} \\ 0 \end{pmatrix} = \frac{1}{\sqrt{2}}e^{it}$$

$$\langle 0(t)|\hat{p}(0)|1(t)\rangle = \frac{i}{\sqrt{2}}\begin{pmatrix} e^{\frac{it}{2}} & 0 \end{pmatrix}\begin{pmatrix} 0 & -1 \\ 1 & 0 \end{pmatrix}\begin{pmatrix} 0 \\ e^{-\frac{3it}{2}} \end{pmatrix} = -\frac{i}{\sqrt{2}}e^{-it}$$

$$\langle 1(t)|\hat{p}(0)|0(t)\rangle = \frac{i}{\sqrt{2}}\begin{pmatrix} 0 & e^{\frac{3it}{2}} \end{pmatrix}\begin{pmatrix} 0 & -1 \\ 1 & 0 \end{pmatrix}\begin{pmatrix} e^{-\frac{it}{2}} \\ 0 \end{pmatrix} = \frac{i}{\sqrt{2}}e^{it}$$

and coincide with those of the Heisenberg representation.

Problem 2.8.
A quantum system with two orthonormal states (say $|1\rangle$ and $|2\rangle$) is described by the following Hamiltonian

$$\hat{H} = |1\rangle\langle 2| + |2\rangle\langle 1|.$$

At time $t = 0$, the average value of the observable

$$\hat{O} = 3|1\rangle\langle 1| - |2\rangle\langle 2|$$

is $\langle\hat{O}\rangle = -1$. Determine the state $|\psi(0)\rangle$ at $t = 0$ and the smallest time $t > 0$ such that $|\psi(t)\rangle = |1\rangle$.

Solution

A comment about the notation is in order. The text does not contain matrices, yet this problem is an exercise on the Heisenberg formulation of Quantum Mechanics, which is equivalent in principle to the Schrödinger continuous formulation but is suitable for systems with a finite number of states. The matrix representation of the Hamiltonian in the vector basis $|1\rangle$, $|2\rangle$ is given by $\langle i|\hat{H}|j\rangle$ where $|i\rangle, |j\rangle = |1\rangle, |2\rangle$

$$\hat{H} = \begin{pmatrix} 0 & 1 \\ 1 & 0 \end{pmatrix}.$$

By diagonalisation and normalisation, one finds the well known eigenvectors of $\hat{\sigma}_x$

$$|+\rangle = \frac{1}{\sqrt{2}}\begin{pmatrix} 1 \\ 1 \end{pmatrix} \qquad |-\rangle = \frac{1}{\sqrt{2}}\begin{pmatrix} 1 \\ -1 \end{pmatrix}$$

with eigenvalues ± 1. The inverse relation is given by

$$|1\rangle = \frac{|+\rangle + |-\rangle}{\sqrt{2}} \qquad |2\rangle = \frac{|+\rangle - |-\rangle}{\sqrt{2}}.$$

The matrix representation of the operator $\hat{O}$ is

$$\hat{O} = \begin{pmatrix} 3 & 0 \\ 0 & -1 \end{pmatrix}.$$

The fact that at time $t = 0$ any measurement of $\hat{O}$ yields the eigenvalue -1, implies that the initial state coincides with $|2\rangle$

$$|\psi(0)\rangle = |2\rangle.$$

To determine the time evolution of this state, we need to express it in terms of the eigenstates of $\hat{H}$, which evolve by simple phase factors: from

$$|\psi(0)\rangle = \frac{|+\rangle - |-\rangle}{\sqrt{2}}$$

we immediately obtain

$$|\psi(t)\rangle = \frac{e^{-it}|+\rangle - e^{it}|-\rangle}{\sqrt{2}}$$

and

$$|\langle 1|\psi(t)\rangle|^2 = \left| \left(\frac{\langle +| + \langle -|}{\sqrt{2}} \right) \left(\frac{e^{-it}|+\rangle - e^{it}|-\rangle}{\sqrt{2}} \right) \right|^2 = \sin^2 t$$

is the probability that at time t the system is in $|1\rangle$. The first time that $\sin^2 t = 1$ is at $t = \frac{\pi}{2}$.

Problem 2.9.
A harmonic oscillator has angular frequency ω and Hamiltonian (in standard notation)

$$\hat{H} = \left(\hat{a}^\dagger \hat{a} + \frac{1}{2} 1 \right) \hbar\omega = \left(\hat{n} + \frac{1}{2} 1 \right) \hbar\omega.$$

We denote by $|n\rangle$ the n-th eigenstate of the Hamiltonian. At time $t = 0$ the oscillator is prepared in the state

$$|\psi(0)\rangle = \frac{|2\rangle + |3\rangle}{\sqrt{2}}.$$

Write the wave function and compute the expectation values of the energy, position, and momentum operators for $t > 0$.

Solution
The eigenstates of the Hamiltonian evolve in time with well defined phase factors given by $e^{-i\frac{Et}{\hbar}}$, with E the energy of the eigenstate. In the case of the harmonic oscillator, the generic eigenstate $|n\rangle$ takes the phase factor $e^{-i\left(n+\frac{1}{2}\right)\omega t}$ in the time evolution. Therefore, we find

$$|\psi(t)\rangle = \frac{|2\rangle e^{-i\frac{5}{2}\omega t} + |3\rangle e^{-i\frac{7}{2}\omega t}}{\sqrt{2}}.$$

The average energy on $|\psi(t)\rangle$ is

$$\langle \hat{H} \rangle = \left(\frac{\langle 2|e^{i\frac{5}{2}\omega t} + \langle 3|e^{i\frac{7}{2}\omega t}}{\sqrt{2}} \right) \hat{H} \left(\frac{|2\rangle e^{-i\frac{5}{2}\omega t} + |3\rangle e^{-i\frac{7}{2}\omega t}}{\sqrt{2}} \right) = \frac{1}{2} \left(\frac{5}{2} + \frac{7}{2} \right) \hbar\omega = 3\hbar\omega$$

where we have used $\hat{H}|n\rangle = \left(\hat{n} + \frac{1}{2} 1\right)\hbar\omega|n\rangle = \left(n + \frac{1}{2}\right)\hbar\omega|n\rangle$. The simplest way to calculate the expectation values of $\hat{x}$ and $\hat{p}$ takes advantage of the relations to the creation and annihilation operators $\hat{a}^\dagger$ and $\hat{a}$

$$\begin{cases} \hat{x} = \frac{x_0}{\sqrt{2}}(\hat{a} + \hat{a}^\dagger) \\ \hat{p} = -\frac{i\hbar}{x_0\sqrt{2}}(\hat{a} - \hat{a}^\dagger) \end{cases}$$

where $x_0 = \sqrt{\frac{\hbar}{m\omega}}$ is a characteristic length scale of the oscillator. We recall the step up and step down action of the creation and annihilation operators on the generic eigenstate

$$\hat{a}^\dagger|n\rangle = \sqrt{n+1}|n+1\rangle \qquad \hat{a}|n\rangle = \sqrt{n}|n-1\rangle$$

and, using the orthogonality of the eigenstates ($\langle n|m \rangle = \delta_{nm}$), one concludes that

$$\langle \hat{x} \rangle_t = \frac{x_0}{2\sqrt{2}} \left(\langle 2|e^{i\frac{5}{2}\omega t} + \langle 3|e^{i\frac{7}{2}\omega t} \right) (\hat{a} + \hat{a}^\dagger) \left(|2\rangle e^{-i\frac{5}{2}\omega t} + |3\rangle e^{-i\frac{7}{2}\omega t} \right) =$$

$$\frac{x_0}{2\sqrt{2}} \left(\langle 2|e^{i\frac{5}{2}\omega t} + \langle 3|e^{i\frac{7}{2}\omega t} \right) \left((\sqrt{2}|1\rangle + \sqrt{3}|3\rangle) e^{-i\frac{5}{2}\omega t} + \right.$$

$$\left. (\sqrt{3}|2\rangle + 2|4\rangle) e^{-i\frac{7}{2}\omega t} \right) = \sqrt{\frac{3}{2}} x_0 \cos(\omega t)$$

$$\langle \hat{p} \rangle_t = -i\frac{\hbar}{2\sqrt{2}x_0} \left(\langle 2|e^{i\frac{5}{2}\omega t} + \langle 3|e^{i\frac{7}{2}\omega t} \right) (\hat{a} - \hat{a}^\dagger) \left(|2\rangle e^{-i\frac{5}{2}\omega t} + |3\rangle e^{-i\frac{7}{2}\omega t} \right) =$$

$$-i\frac{\hbar}{2\sqrt{2}x_0} \left(\langle 2|e^{i\frac{5}{2}\omega t} + \langle 3|e^{i\frac{7}{2}\omega t} \right) \left((\sqrt{2}|1\rangle - \sqrt{3}|3\rangle) e^{-i\frac{5}{2}\omega t} + \right.$$

$$\left. (\sqrt{3}|2\rangle - 2|4\rangle) e^{-i\frac{7}{2}\omega t} \right) = -\frac{\hbar}{x_0}\sqrt{\frac{3}{2}}\sin(\omega t) = -m\omega x_0\sqrt{\frac{3}{2}}\sin(\omega t)$$

corresponding to the dynamical evolution of a classical harmonic oscillator with zero initial momentum

$$\begin{cases} \langle \hat{x} \rangle_t = \langle \hat{x} \rangle_0 \cos(\omega t) \\ \langle \hat{p} \rangle_t = -m\omega \langle \hat{x} \rangle_0 \sin(\omega t) \end{cases}$$

where $\langle \hat{x} \rangle_0 = \sqrt{\frac{3}{2}}x_0$ (see also Problem 2.32).

Problem 2.10.
Calculate the energy levels of a Schrödinger particle in a tree dimensional potential well with the shape of a parallelepiped of edges a, b, c and infinite walls. This could be a rough model for an electron in a quantum dot or a metal particle of this shape, with a size of a few tens of atomic units, such that many properties depend on the discrete energy levels.

Solution
The three dimensional stationary Schrödinger equation reads

$$-\frac{\hbar^2}{2m}\nabla^2\psi + (U - E)\psi = 0$$

where $\psi = \psi(x,y,z)$ and where the Laplacian operator is acting on all the three components x,y,z

$$-\frac{\hbar^2}{2m}\left(\frac{\partial^2\psi}{\partial x^2} + \frac{\partial^2\psi}{\partial y^2} + \frac{\partial^2\psi}{\partial z^2}\right) + (U - E)\psi = 0.$$

The potential $U(x,y,z)$ vanishes for $0 \le x \le a, 0 \le y \le b, 0 \le z \le c$. The wave function cannot penetrate where U diverges: $\psi(a,y,z) = \psi(0,y,z) = \psi(x,b,z) = \psi(x,0,z) = \psi(x,y,c) = \psi(x,y,0) = 0$. Since $\psi = 0$ outside the parallelepiped, the

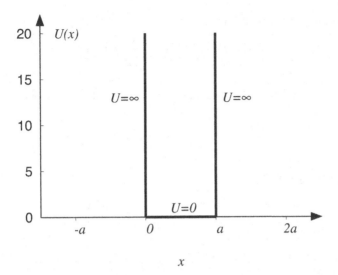

Fig. 2.3 An infinite one dimensional potential well with a very strong repulsion in $x = 0$ and $x = a$. In Problem 2.10 we present the solution of the Schrödinger equation with such a kind of potential

derivatives also vanish there, but not inside, so they must be discontinuous at the boundary. Thanks to the symmetry, the partial differential equation is separable

$$\psi(x,y,z) = \psi_x(x)\,\psi_y(y)\,\psi_z(z)$$

and in the parallelepiped with edges a, b, c

$$\frac{\hbar^2}{2m}\left(\frac{1}{\psi_x}\frac{d^2\psi_x}{dx^2} + \frac{1}{\psi_y}\frac{d^2\psi_y}{dy^2} + \frac{1}{\psi_z}\frac{d^2\psi_z}{dz^2}\right) + E = 0.$$

We now set $k^2 = k_x^2 + k_y^2 + k_z^2 = 2mE/\hbar^2 = 2m(E_x + E_y + E_z)/\hbar^2$ and solve, separately for each variable, three one dimensional identical problems. We choose (arbitrarily) to work in the x direction and set $\psi_x \equiv \psi$. The resulting potential is plotted in Fig. 2.3. The solution of the one dimensional Schrödinger equation gives the following wave function

$$\psi(x) = A\sin(kx + \delta).$$

The boundary conditions $\psi(0) = \psi(a) = 0$, if used separately, yield $\delta = 0$ and $ka = n_x\pi$, with n_x a positive integer number ($n_x = 1, 2, 3, ...$). The associated eigenvalues are

$$E_{n_x} = \frac{\hbar^2 \pi^2 n_x^2}{2ma^2}$$

and the wave function is

$$\psi_{n_x}(x) = A\sin\left(\frac{n_x \pi x}{a}\right)$$

with A a normalization constant

$$A^2 \int_0^a \sin^2\left(\frac{n_x \pi x}{a}\right) dx = \frac{A^2 a}{n_x \pi} \int_0^{n_x \pi} \sin^2 x\, dx = \frac{A^2 a}{n_x \pi} \left.\left(\frac{1 - \cos(2x)}{2}\right)\right|_0^{n_x \pi} = \frac{A^2 a}{2} = 1$$

so that $A = \sqrt{2/a}$. The normalized eigenstates for the three dimensional problem are

$$\psi(x,y,z) = \psi_{n_x}(x)\,\psi_{n_y}(y)\,\psi_{n_z}(z) = \sqrt{\frac{8}{abc}} \sin\left(\frac{n_x \pi x}{a}\right) \sin\left(\frac{n_y \pi y}{b}\right) \sin\left(\frac{n_z \pi z}{c}\right)$$

and the discrete spectrum is given by

$$E_{n_x, n_y, n_z} = \frac{\hbar^2 \pi^2}{2m} \left(\frac{n_x^2}{a^2} + \frac{n_y^2}{b^2} + \frac{n_z^2}{c^2}\right)$$

with n_x, n_y, n_z positive integer numbers.

Problem 2.11.
A Schrödinger particle with Hamiltonian $\hat{H} = \frac{\hat{p}^2}{2m} + \hat{V}$ is confined in a one dimensional potential well with infinite walls ($V = 0$ for $0 \le x \le a$; $V = +\infty$ otherwise) and its mass m is such that

$$\frac{\hbar^2 \pi^2}{2ma^2} = \varepsilon$$

with ε a given constant. The eigenfunctions of the Hamiltonian are (see Problem 2.10)

$$\psi_n(x) = \sqrt{\frac{2}{a}} \sin\left(\frac{n\pi x}{a}\right)$$

with $n = 1, 2, 3, \ldots$ a positive integer number. At time $t = 0$ the particle is prepared in the state described by

$$\phi(x,0) = \frac{\psi_1(x) + i\psi_2(x)}{\sqrt{2}}.$$

Compute $\phi(x,t)$ and the probability $P_{left}(t)$ of finding the particle at time t in the region $0 \le x \le \frac{a}{2}$. Compute also the probability current density $J(x,t)$. Verify the continuity equation. How does $\phi(x,t)$ transform under parity $\hat{P}$ (with respect to the center of the well) and time reversal $\hat{T}$?

Solution
Let us write for short

$$s_n(x) = \sin\left(\frac{n\pi x}{a}\right) \qquad c_n(x) = \cos\left(\frac{n\pi x}{a}\right).$$

The time evolution driven by $\hat{H}$ is such that each eigenstate takes a phase factor $e^{-iE_n t/\hbar}$, where the energies can be written as $E_n = n^2 \varepsilon$. Therefore, we find

$$\phi(x,t) = \frac{s_1(x)e^{-i\frac{\varepsilon t}{\hbar}} + i s_2(x)e^{-4i\frac{\varepsilon t}{\hbar}}}{\sqrt{a}}.$$

The probability density at time t of finding the particle at x is the square modulus

$$|\phi(x,t)|^2 = \rho(x,t) = \frac{1}{a}\left(s_1^2(x) + s_2^2(x) + 2s_1(x)s_2(x)\sin(\omega_{12}t)\right)$$

where $\omega_{12} = \frac{3\varepsilon}{\hbar}$. The probability $P_{left}(t)$ is obtained with the integral of $\rho(x,t)$ in the interval $0 \le x \le \frac{a}{2}$

$$P_{left}(t) = \int_0^{\frac{a}{2}} \rho(x,t)\,dx = \frac{1}{2} + \frac{4}{3\pi}\sin(\omega_{12}t)$$

where we have used the following indefinite integrals

$$\int_0^{\pi/2} \sin^2 x\,dx = \frac{1}{2}\int_0^{\pi/2}(1 - 2\cos(2x))\,dx = \frac{\pi}{4}$$

$$\int_0^{\pi/2} \sin x\sin(2x)\,dx = 2\int_0^{\pi/2}\sin^2 x\cos x\,dx = 2\int_0^1 y^2\,dy = \frac{2}{3}.$$

The probability density flux $J(x,t)$ is

$$J(x,t) = \frac{i\hbar}{2m}\left(\phi\frac{\partial \phi^*}{\partial x} - \phi^*\frac{\partial \phi}{\partial x}\right) = \frac{\hbar}{m}\Im\left(\phi^*\frac{\partial \phi}{\partial x}\right) = \frac{\hbar\pi}{ma^2}[2s_1c_2 - s_2c_1]\cos(\omega_{12}t).$$

Moreover, when we calculate the time derivative of the probability density function, we get

$$\frac{\partial \rho}{\partial t} = \frac{3\pi^2\hbar}{ma^3}s_2s_1\cos(\omega_{12}t).$$

We see that

$$\frac{\partial J(x,t)}{\partial x} = \frac{\hbar\pi}{ma^2}\cos(\omega_{12}t)\frac{d}{dx}[2s_1c_2 - s_2c_1] =$$

$$\frac{\hbar\pi^2}{ma^3}\cos(\omega_{12}t)[2c_1c_2 - 4s_1s_2 - 2c_2c_1 + s_2s_1] = -\frac{3\pi^2\hbar}{ma^3}s_2s_1\cos(\omega_{12}t)$$

from which we can check the continuity equation

$$\frac{\partial \rho(x,t)}{\partial t} + \frac{\partial J(x,t)}{\partial x} = \frac{\partial |\phi(x,t)|^2}{\partial t} + \frac{\partial J(x,t)}{\partial x} = 0.$$

The above equation is a direct consequence of the Schrödinger equation and is fundamental for the Copenhagen interpretation of Quantum Mechanics.

A parity transformation $\hat{P}$, with respect to the center of the well, is such that $\tilde{x} \to -\tilde{x}$, where $\tilde{x} = x - \frac{a}{2}$.

$$\hat{P}s_1 = s_1 \qquad \hat{P}s_2 = -s_2.$$

In the Schrödinger theory, the time reversal operator is $\hat{T} = \hat{K}$, where $\hat{K}$ is the Kramers operator which takes the complex conjugate; this means that if $\phi(t)$ is

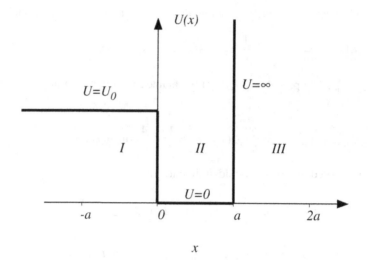

Fig. 2.4 We plot the asymmetric potential well with a depth U_0 and an infinite barrier on one side. The width of the well is a. The solution of the Schrödinger equation with this potential is discussed in Problem 2.12

the solution of the Schrödinger equation with $\hat{H}(t)$, then $\phi'(t) = \phi^*(-t)$ solves the Schrödinger equation with $\hat{H}(-t)$ in place of $\hat{H}(t)$; therefore $\phi^*(-t)$ is called the time reversed wave function. Putting all these results together, we find

$$\hat{P}\phi(x,t) = \frac{s_1 e^{-i\frac{\mathcal{E}t}{\hbar}} - is_2 e^{-4i\frac{\mathcal{E}t}{\hbar}}}{\sqrt{a}}$$

$$\hat{T}\hat{P}\phi(x,t) = \frac{s_1 e^{-i\frac{\mathcal{E}t}{\hbar}} + is_2 e^{-4i\frac{\mathcal{E}t}{\hbar}}}{\sqrt{a}}.$$

Consequently, the problem is invariant under $\hat{T}\hat{P}$.

Problem 2.12.
A particle with mass $m = 1/2$ moves in the one dimensional potential well $U(x)$ (see Fig. 2.4) such that

$$\begin{cases} U(x) = U_0 & x \leq 0 \\ U(x) = 0 & 0 < x < a \\ U(x) = +\infty & x \geq a. \end{cases}$$

Determine the values of U_0 and a for which we find bound states.

Solution
The potential is characterized by three distinct regions and we need to solve the stationary Schrödinger equation separately in $x \leq 0$ (region *I*), $0 < x < a$ (region

II), and $x \geq a$ (region *III*). In these regions, the potential is constant and the general integral of the resulting equation with constant coefficients is well known. We solve the Schrödinger equation in each region and then impose the continuity conditions. The particle cannot be in the region *III* where the potential is infinite; therefore, for $x > a$ we set $\psi(x) = 0$. In particular, we impose $\psi(a) = 0$. However, $\psi'(x)$ cannot also vanish for $x \to a$ from the left, because then $\psi(x) = 0$ everywhere; it follows that $\psi'(x)$ has a jump at $x = a$, which is due to the divergence in the potential. On the other hand, for $x = 0$ the potential is discontinuous but finite and we can assume the continuity of both ψ and ψ'. In the region *I* (where $U_0 > E$, with E the bound state energy) we must solve

$$(\hat{H} - E)\psi_I = 0 = -\frac{\hbar^2}{2m}\frac{d^2\psi_I}{dx^2} + (U_0 - E)\psi_I.$$

Moreover, in the region *II*, we find

$$(\hat{H} - E)\psi_{II} = 0 = -\frac{\hbar^2}{2m}\frac{d^2\psi_{II}}{dx^2} - E\psi_{II}.$$

The solutions of the above differential equations are

$$\begin{cases} \psi_I(x) = Ae^{\frac{\sqrt{2m(U_0 - E)}}{\hbar}x} & x \leq 0 \\ \psi_{II}(x) = C\sin\left(\frac{\sqrt{2mE}}{\hbar}x + \delta\right) & 0 < x < a \end{cases}$$

where A, C, δ are integration constants that we shall find by the boundary conditions. We note that the region *I* is classically forbidden, because the particle momentum becomes imaginary. As a consequence of the Heisenberg uncertainty principle, we cannot say that the particle is in a precise position of the region *I* without mixing many momenta. Therefore, for a given momentum, this region is allowed and we find a non zero probability (decaying exponentially to zero at infinity) to find the particle there. The boundary condition $\psi_{II}(a) = 0$ leads to

$$\sin\left(\frac{\sqrt{2mE}}{\hbar}a + \delta\right) = 0$$

from which

$$\delta = -\frac{\sqrt{2mE}}{\hbar^2}a + n\pi$$

with n an integer number. As for the continuity of the first derivative in $x = 0$, it is convenient to combine it with the continuity of the function, that is

$$\frac{\psi_I'(0)}{\psi_I(0)} = \frac{\psi_{II}'(0)}{\psi_{II}(0)}$$

so that

$$\frac{\sqrt{2m(U_0 - E)}}{\hbar} = \frac{\sqrt{2mE}}{\hbar} \cot\left(-\frac{\sqrt{2mE}}{\hbar}a + n\pi\right) = -\frac{\sqrt{2mE}}{\hbar} \cot\left(\frac{\sqrt{2mE}}{\hbar}a\right)$$

which is a transcendental equation. The solution can be worked out graphically. If we define $y = a\frac{\sqrt{2m(U_0-E)}}{\hbar}$ and $x = a\frac{\sqrt{2mE}}{\hbar}$ (not to be confused with the position in the beginning of the problem), the solutions are found from the intersection of the two curves

$$y = -x\cot x$$

and

$$y^2 + x^2 = a^2 \bar{U}_0 = R^2 \quad \bar{U}_0 = \frac{2mU_0}{\hbar^2}$$

given in Fig. 2.5. We note that $\cot x = \cos x / \sin x \approx 1/x$ when x is small. Therefore, $y = -x\cot x$ is negative close to the origin, while $y^2 + x^2 = a^2 \bar{U}_0$ is a circle with radius $R = a\sqrt{\bar{U}_0}$ that is positive: near the origin there is no crossing. The function $y = -x\cot x$ is zero when $x = \pi/2$. If $R = \pi/2$ we have an intersection and a bound state. There are no bound states if

$$a^2 \bar{U}_0 = a^2 \frac{2mU_0}{\hbar^2} < \frac{\pi^2}{4}.$$

If a and U_0 are such that this inequality is not satisfied, there are as many energy levels as intersections, $k = 1, 2, \ldots$, a finite number at any rate. The intersections occur in the first quadrant and the arc is a decreasing function; therefore the discrete eigenvalues are ordered by the quantum number k.

Problem 2.13.
Find the energy spectrum for a particle with mass m in the symmetric potential well (see Fig. 2.6)

$$\begin{cases} U(x) = U_0 & x \leq 0 \\ U(x) = 0 & 0 < x < a \\ U(x) = U_0 & x \geq a \end{cases}$$

with $U_0 > 0$.

Solution
We first find the general integral in the three regions where $U(x)$ is constant, then we discard the solutions that blow up for $x \to +\infty$ in the region *III* or for $x \to -\infty$ in the region *I*. Following the same procedures of other problems (see also Problems 2.10

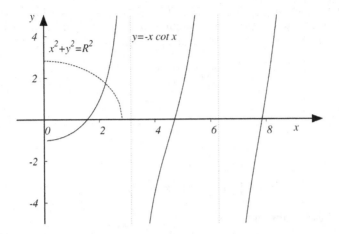

Fig. 2.5 We look for the intersection between the two curves $y = -x \cot x$ and $y^2 + x^2 = R^2$, where $R^2 = a^2 \bar{U}_0 = a^2 \frac{2mU_0}{\hbar^2}$. The case considered here refers to $R^2 = 8$. These intersections are important to characterize the bound states of a particle with mass m in the asymmetric potential well with depth U_0, width a, and an infinite barrier on one side (see Fig. 2.4)

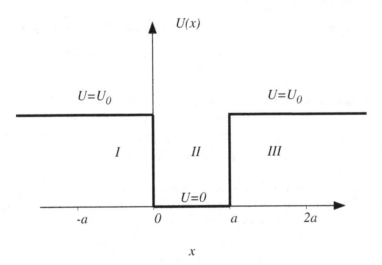

Fig. 2.6 A symmetric potential well with depth U_0 and width a. In Problem 2.13 we study the solution of the Schrödinger equation for a particle with mass m in this potential field

and 2.12), we arrive at

$$
\begin{cases}
\psi_I(x) = A e^{k_1 x} & x \leq 0 \\
\psi_{II}(x) = C \sin(kx + \delta) & 0 < x < a \\
\psi_{III}(x) = B e^{-k_1 x} & x \geq a
\end{cases}
$$

where A, B, C, δ are integration constants, $k^2 = 2mE/\hbar^2$, $k_1^2 = 2m(U_0 - E)/\hbar^2$ and $0 < E < U_0$. We need to impose the continuity conditions for the wave function and its derivative in $x = 0$ and $x = a$. These are equivalent to

$$\begin{cases} \dfrac{\psi_I'(0)}{\psi_I(0)} = \dfrac{\psi_{II}'(0)}{\psi_{II}(0)} \\ \dfrac{\psi_{II}'(a)}{\psi_{II}(a)} = \dfrac{\psi_{III}'(a)}{\psi_{III}(a)} \end{cases}$$

obtaining

$$\begin{cases} k_1 = k \cot \delta \\ -k_1 = k \cot(ka + \delta). \end{cases}$$

Combining these results, one finds

$$\cot \delta = -\cot(ka + \delta)$$

that implies $\delta = -ka - \delta + n\pi$ and

$$\delta = \frac{n\pi}{2} - \frac{ka}{2}$$

with n an integer number. Plugging this value of δ in the above equations, we get

$$\cot\left(\frac{n\pi - ka}{2}\right) = \frac{k_1}{k}$$

that is an equation whose solution depends on whether n is even or odd

$$\begin{cases} \cot\left(\frac{ka}{2}\right) = -\frac{k_1}{k} & n \text{ even} \\ \tan\left(\frac{ka}{2}\right) = \frac{k_1}{k} & n \text{ odd}. \end{cases}$$

If we define $x = ka/2$ (not to be confused with the position in the beginning of the problem), $y = k_1 a/2$, the solutions of the transcendental equations are given by the intersection of the curves (see Fig. 2.7)

$$\begin{cases} x^2 + y^2 = R^2 = \frac{mU_0 a^2}{2\hbar^2} \\ y = -x \cot x \\ y = x \tan x. \end{cases}$$

We note that the discrete spectrum is always present because the curves always intersect. It is also instructive to compare this result with the one of Problem 2.12, where the potential well has an infinite barrier on one side. In this case, the curve $y = x \tan x$ would be missing and there is a range of R without a discrete spectrum.

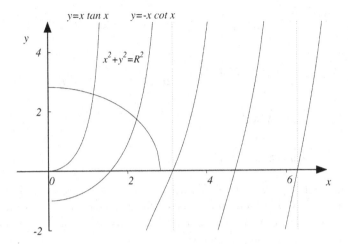

Fig. 2.7 We look for the intersection between the curve $x^2 + y^2 = R^2 = \frac{mU_0 a^2}{2\hbar^2}$ and $y = x\tan x$ or $y = -x\cot x$. The case considered here refers to $R^2 = 8$. The variable x is related to the energy of the bound states for a particle with mass m in the symmetric potential well given in Fig. 2.6. The choice of the function $\tan x$ of $\cot x$ depends on the symmetry of the solution (odd or even)

Problem 2.14.
A Schrödinger particle in one dimension is in the state whose amplitude is described by the wave function

$$\psi(x) = N e^{-\frac{x^2}{a^2}}$$

where N is a normalization constant. Determine the average values of the kinetic energy and position and the density flux associated with the probability density function $|\psi(x)|^2$.

Solution
First of all, we need to determine the normalization constant N from the condition

$$\int_{-\infty}^{+\infty} |\psi(x)|^2 \, dx = 1.$$

Physically, this says that the probability of spotting the particle in a finite region of the x axis approximates 1 as closely as we wish, provided that the region is chosen large enough. Such a scheme fails for 'free' states like plane waves: no finite region however large has any appreciable probability of containing a plane wave state. If we use the Gaussian integral $\int_{-\infty}^{+\infty} e^{-y^2} dy = \sqrt{\pi}$, we find

$$N^2 \int_{-\infty}^{+\infty} e^{-\frac{2x^2}{a^2}} \, dx = N^2 \frac{a}{\sqrt{2}} \int_{-\infty}^{+\infty} e^{-y^2} \, dy = 1$$

so that $N = \sqrt{\frac{1}{a}\sqrt{\frac{2}{\pi}}}$. The average value of the kinetic energy $\hat{T}$ is

$$\langle \psi|\hat{T}|\psi \rangle = \int_{-\infty}^{+\infty} \langle \psi|x\rangle\langle x|\hat{T}|\psi\rangle\, dx = -\frac{\hbar^2}{2ma}\sqrt{\frac{2}{\pi}}\int_{-\infty}^{+\infty} e^{-\frac{x^2}{a^2}}\frac{d^2}{dx^2}e^{-\frac{x^2}{a^2}}\, dx =$$

$$-\frac{\hbar^2}{2ma}\sqrt{\frac{2}{\pi}}\int_{-\infty}^{+\infty} e^{-\frac{x^2}{a^2}}\left(-\frac{2}{a^2}+\frac{4x^2}{a^4}\right)e^{-\frac{x^2}{a^2}}\, dx =$$

$$\frac{\hbar^2}{2ma}\frac{a}{\sqrt{\pi}}\int_{-\infty}^{+\infty}\left(\frac{2}{a^2}-\frac{2y^2}{a^2}\right)e^{-y^2}\, dy.$$

Using the integral $\int_{-\infty}^{+\infty} y^2 e^{-y^2}\, dy = \frac{\sqrt{\pi}}{2}$, we get $\langle \hat{T} \rangle = \frac{\hbar^2}{2ma^2}$. The average value of the position operator $\langle \hat{x} \rangle$ is zero: this is due to the symmetry of the wave function and can be explicitly verified because the calculation of $\langle \hat{x} \rangle$ leads to the integral $\int_{-\infty}^{+\infty} y e^{-y^2}\, dy = 0$. As for the density flux

$$J = \frac{i\hbar}{2m}\left(\psi\frac{d\psi^*}{dx}-\psi^*\frac{d\psi}{dx}\right) = \frac{\hbar}{m}\Im\left(\psi^*\frac{d\psi}{dx}\right)$$

it vanishes (it vanishes identically for $\psi = \psi^*$, that is, for any real ψ).

Problem 2.15.
Determine the discrete energy spectrum for a particle with mass m in the asymmetric potential well

$$\begin{cases} U(x) = U_1 & x \le 0 \\ U(x) = 0 & 0 < x < a \\ U(x) = U_2 & x \ge a \end{cases}$$

with $0 < U_1 < U_2$.

Solution
We need to solve the Schrödinger equation in the three different regions: $x \le 0$ (region *I*), $0 < x < a$ (region *II*) and $x \ge a$ (region *III*). Following the same procedures of other problems (see Problems 2.10, 2.12, 2.13) we find the solutions in the three regions

$$\begin{cases} \psi_I(x) = Ae^{k_1 x} & x < 0 \\ \psi_{II}(x) = C\sin(kx+\delta) & 0 \le x \le a \\ \psi_{III}(x) = Be^{-k_2 x} & x > a \end{cases}$$

where A, B, C, δ are integration constants and

$$k = \sqrt{\frac{2mE}{\hbar^2}} \qquad k_1 = \sqrt{\frac{2m(U_1-E)}{\hbar^2}} \qquad k_2 = \sqrt{\frac{2m(U_2-E)}{\hbar^2}}$$

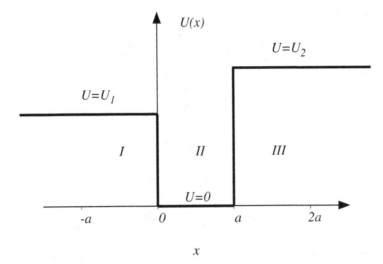

Fig. 2.8 We plot the asymmetric potential well with depths U_1, U_2 and width a. In Problem 2.15 we determine the energy spectrum for this potential

with $0 < E < U_1 < U_2$. The wave functions for $x > 0$ and $x < 0$ go to zero when $x \to \pm\infty$. We need to impose the continuity conditions for the wave function and its derivative in $x = 0$ and $x = a$. These are equivalent to

$$\begin{cases} \dfrac{\psi_I'(0)}{\psi_I(0)} = \dfrac{\psi_{II}'(0)}{\psi_{II}(0)} \\ \dfrac{\psi_{II}'(a)}{\psi_{II}(a)} = \dfrac{\psi_{III}'(a)}{\psi_{III}(a)} \end{cases}$$

and we get

$$\begin{cases} k_1 \tan\delta = k \\ -k_2 \tan(ka + \delta) = k \end{cases}$$

that implies

$$\delta = \arctan\left(\frac{k}{k_1}\right) + n_1\pi$$

and

$$ka + \delta = -\arctan\left(\frac{k}{k_2}\right) + n_2\pi$$

with n_1, n_2 integers. Using the relation

$$\arctan x = \arcsin\left(\frac{x}{\sqrt{1 + x^2}}\right)$$

and the property

$$\frac{(k/k_{1,2})}{\sqrt{1+(k/k_{1,2})^2}} = \sqrt{\frac{E}{U_{1,2}}} = \frac{k\hbar}{\sqrt{2mU_{1,2}}}$$

we obtain

$$\delta = \arcsin\left(\frac{k\hbar}{\sqrt{2mU_1}}\right) + n_1\pi$$

and

$$ka + \delta = -\arcsin\left(\frac{k\hbar}{\sqrt{2mU_2}}\right) + n_2\pi.$$

The above equations can be subtracted with the result $(n = n_2 - n_1)$

$$n\pi - ka = \arcsin\left(\frac{k\hbar}{\sqrt{2mU_1}}\right) + \arcsin\left(\frac{k\hbar}{\sqrt{2mU_2}}\right).$$

If we define $x = k/b = k\hbar/\sqrt{2mU_1} = \sqrt{E/U_1}$ (not to be confused with the position in the beginning of the problem) we rewrite the previous equation as

$$n\pi - abx = \arcsin x + \arcsin\left(x\sqrt{\frac{U_1}{U_2}}\right) = \arcsin x + \arcsin(x\sin\gamma).$$

Since we know that $U_1 < U_2$ and $0 \leq \sqrt{U_1/U_2} \leq 1$, we have set $\sin\gamma = \sqrt{U_1/U_2}$ with $0 \leq \gamma \leq \pi/2$. Moreover, we define

$$y_n(x) = n\pi - abx$$

identifying different functions, each of them with a different value in $x = 0$, $y_n(x = 0) = n\pi$. Finally, we define

$$y(x) = \arcsin x + \arcsin(x\sin\gamma).$$

We need $0 \leq E \leq U_1$ for the discrete spectrum. When E varies in this interval, $0 \leq x \leq 1$ and $0 \leq \arcsin x \leq \pi/2$. Also, when $0 \leq x \leq 1$, we see that $y(x)$ is a monotonically increasing function with the property

$$0 \leq y(x) \leq \pi/2 + \arcsin(\sin\gamma) = \pi/2 + \gamma$$

while the $y_n(x)$, which are monotonically decreasing functions, have the property $n\pi \geq y_n(x) \geq n\pi - ab$. The solutions are given by the intersection of the two curves $y_n(x)$ and $y(x)$, i.e. the condition $y_n(x) = y(x)$ (see Fig. 2.9). For a fixed (positive) n, due to the property of $y(x)$, the condition that there is at least one intersection is

$$n\pi - ab \leq \frac{\pi}{2} + \gamma$$

that implies the minimum value of $y_n(x)$ is below the maximum value of $y(x)$.

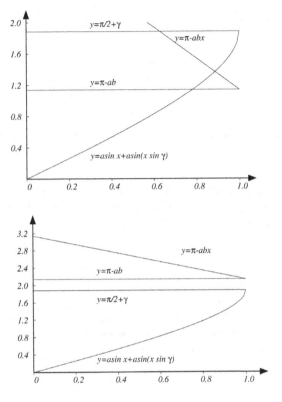

Fig. 2.9 We plot the functions $y(x) = \arcsin x + \arcsin(x \sin \gamma)$ and $y_n(x) = n\pi - abx$ (n is an integer) for the case $n = 1$. In the upper panel we report the case with $\gamma = \frac{\pi}{10}$ and $ab = 2.0$ while, in the lower panel, we report the case with $\gamma = \frac{\pi}{10}$ and $ab = 1.0$. For a generic integer value n, the condition that $y_n(x)$ intersects $y(x)$ in the interval $0 \le x \le 1$ is $n\pi - ab \le \frac{\pi}{2} + \gamma$. This condition corresponds to the existence of a bound state for a particle with mass m in the asymmetric potential well of Fig. 2.8

Problem 2.16.

A particle with mass m is subject to the following one dimensional potential

$$\begin{cases} U(x) = +\infty & x \le 0, x \ge a \\ U(x) = 0 & 0 < x < a. \end{cases}$$

At time $t = 0$ the wave function $\psi(x)$ is such that any measurement of the energy cannot give results larger than $3\hbar^2 \pi^2 / ma^2$. Besides, the mean energy is $7\pi^2\hbar^2/8ma^2$ and the average value of the momentum operator $\hat{p}$ is $\frac{8\hbar}{3a}$. Find as much as you can about $\psi(x)$ and the mean value of $\hat{p}^4$ at time $t \ge 0$.

Solution

We see that the particle is inside a well with infinite potential barriers in $x = 0$ and $x = a$. The normalized eigenstates and eigenvalues of the Hamiltonian have been determined in Problem 2.10, and they are

$$\psi_n(x) = \sqrt{\frac{2}{a}} \sin\left(\frac{\pi n x}{a}\right)$$

$$E_n = \frac{\pi^2 \hbar^2 n^2}{2ma^2}$$

with n a positive integer number. The condition $E \leq 3\hbar^2\pi^2/ma^2$ requires $n \leq 2$ and we are left with

$$|\psi\rangle = c_1|\psi_1\rangle + c_2|\psi_2\rangle$$

where c_1, c_2 are complex numbers whose moduli are easily computed from the condition of normalization plus the condition that the average value of the energy is equal to $\frac{7\pi^2\hbar^2}{8ma^2}$, namely

$$\begin{cases} |c_1|^2 + |c_2|^2 = 1 \\ E = \frac{7\pi^2\hbar^2}{8ma^2} = E_1|c_1|^2 + E_2|c_2|^2 \end{cases}$$

yielding $|c_1|^2 = 3/4, |c_2|^2 = 1/4$. We wish to find ψ up to a constant phase factor which has no physical meaning. To this end, we need the phase difference between c_1 and c_2. This can be deduced from the mean value of the momentum operator, since

$$\langle\psi|\hat{p}|\psi\rangle = \frac{8\hbar}{3a} = \int_0^a \psi^*(x)\left(-i\hbar\frac{d}{dx}\right)\psi(x)\,dx =$$

$$-i\hbar\int_0^a \left(|c_1|^2\psi_1^*\frac{d\psi_1}{dx} + |c_2|^2\psi_2^*\frac{d\psi_2}{dx} + c_1^*c_2\psi_1^*\frac{d\psi_2}{dx} + c_2^*c_1\psi_2^*\frac{d\psi_1}{dx}\right)dx$$

where, however, the first two terms can be dropped since the wave functions vanish in $x = 0$ and $x = a$ and so

$$\int_0^a \psi_1^*\frac{d\psi_1}{dx}dx = \int_0^a \psi_2^*\frac{d\psi_2}{dx}dx = 0.$$

If we set $c_1 = |c_1|$, $c_2 = e^{i\alpha}|c_2|$ and we use the indefinite integral

$$\int \sin(m_1 x)\cos(m_2 x)\,dx = -\frac{\cos(m_1 - m_2)x}{2(m_1 - m_2)} - \frac{\cos(m_1 + m_2)x}{2(m_1 + m_2)} + \text{const.}$$

for m_1, m_2 integer numbers, we get the following result

$$\int_0^a \psi_1^*\frac{d\psi_2}{dx}dx = -\frac{8}{3a}$$

$$\int_0^a \psi_2^*\frac{d\psi_1}{dx}dx = \frac{8}{3a}$$

and we find $\sin \alpha = -2/\sqrt{3}$. We cannot find α uniquely. We can do the calculation of $\langle \psi | \hat{p}^4 | \psi \rangle$ without this information, since

$$\langle x | \hat{p}^4 | \psi_n \rangle = (-i\hbar)^4 \frac{d^4 \psi_n}{dx^4} = \left(\frac{-i\hbar\pi n}{a} \right)^4 \psi_n(x)$$

and so

$$\langle \psi_1 | \hat{p}^4 | \psi_2 \rangle = 0 \quad \langle \psi_n | \hat{p}^4 | \psi_n \rangle = \left(\frac{-i\hbar\pi n}{a} \right)^4$$

with the result

$$\langle \psi | \hat{p}^4 | \psi \rangle = |c_1|^2 \frac{\pi^4 \hbar^4}{a^4} + |c_2|^2 \frac{16\pi^4 \hbar^4}{a^4} = \frac{19\pi^4 \hbar^4}{4a^4}.$$

Note that this does not evolve at all in time.

Problem 2.17.
Let $|\psi_A\rangle, |\psi_B\rangle$ be the eigenvectors of the Hamiltonian $\hat{H}$ of a two-level system

$$\hat{H} |\psi_{A,B}\rangle = E_{A,B} |\psi_{A,B}\rangle \quad E_A > E_B.$$

Another basis $|\psi_1\rangle, |\psi_2\rangle$, with

$$\langle \psi_i | \psi_j \rangle = \delta_{ij} \quad i, j = 1, 2$$

is related to $|\psi_A\rangle, |\psi_B\rangle$ by

$$|\psi_{A,B}\rangle = \frac{1}{\sqrt{2}} (|\psi_1\rangle \pm |\psi_2\rangle).$$

Find the matrix elements of the Hamiltonian $\hat{H}'$ in the basis $|\psi_1\rangle, |\psi_2\rangle$ using the dyadic notation and the matrix notation. Then, assuming that at time $t = 0$ the system is in $|\psi_1\rangle$, find the time evolved state using the time dependence of the $\hat{H}$ eigenstates, and calculate the time t such that for the first time the system has probability 1 to be in $|\psi_2\rangle$. Show how one obtains the same result by the time evolution operator. Assume $\hbar = 1$ for simplicity.

Solution
The Hamiltonian $\hat{H}$ has the matrix representation

$$\hat{H} = \begin{pmatrix} E_A & 0 \\ 0 & E_B \end{pmatrix}$$

on the basis $|\psi_A\rangle, |\psi_B\rangle$. Writing the Hamiltonian in the equivalent dyadic form, we get

$$\hat{H} = E_A |\psi_A\rangle\langle\psi_A| + E_B |\psi_B\rangle\langle\psi_B|.$$

Making the substitutions

$$\hat{H} \to \hat{H}' \quad |\psi_A\rangle \to \frac{1}{\sqrt{2}}(|\psi_1\rangle + |\psi_2\rangle) \quad |\psi_B\rangle \to \frac{1}{\sqrt{2}}(|\psi_1\rangle - |\psi_2\rangle)$$

and collecting terms, one readily arrives at

$$\hat{H}' = \frac{1}{2}[(E_A + E_B)(|\psi_1\rangle\langle\psi_1| + |\psi_2\rangle\langle\psi_2|) + (E_A - E_B)(|\psi_1\rangle\langle\psi_2| + |\psi_2\rangle\langle\psi_1|)].$$

In matrix notation this is

$$\hat{H}' = \begin{pmatrix} \frac{E_A + E_B}{2} & \frac{E_A - E_B}{2} \\ \frac{E_A - E_B}{2} & \frac{E_A + E_B}{2} \end{pmatrix}.$$

One can do the same using the matrix notation by introducing the matrix $\hat{C}$ whose columns are the vectors representing $|\psi_A\rangle$ and $|\psi_B\rangle$ in the new basis

$$\hat{C} = \hat{C}^{-1} = \frac{1}{\sqrt{2}} \begin{pmatrix} 1 & 1 \\ 1 & -1 \end{pmatrix}.$$

This is the transformation matrix that yields directly the result $\hat{H}' = \hat{C}\hat{H}\hat{C}$. The evolved state is

$$|\psi_1(t)\rangle = \frac{e^{-E_A t}|\psi_A\rangle + e^{-E_B t}|\psi_B\rangle}{\sqrt{2}}.$$

The amplitude to find the system in $|\psi_2\rangle = \frac{|\psi_A\rangle - |\psi_B\rangle}{\sqrt{2}}$ is

$$A_{12} = \langle\psi_2|\psi_1(t)\rangle = \frac{e^{-iE_A t} - e^{-iE_B t}}{2}$$

and the probability is therefore

$$P_{12} = A_{12}^2 = \frac{1 - \cos((E_A - E_B)t)}{2}$$

that first attains unity at time $t = \frac{\pi}{(E_A - E_B)}$. Alternatively, one can use the time evolution operator. Since $\hat{H}$ is diagonal, trivially

$$e^{-i\hat{H}t} = \begin{pmatrix} e^{-iE_A t} & 0 \\ 0 & e^{-iE_B t} \end{pmatrix}$$

which leads to

$$e^{-i\hat{H}'t} = \hat{C}e^{-i\hat{H}t}\hat{C} = \frac{1}{2}\begin{pmatrix} e^{-iE_A t} + e^{-iE_B t} & e^{-iE_A t} - e^{-iE_B t} \\ e^{-iE_A t} - e^{-iE_B t} & e^{-iE_A t} + e^{-iE_B t} \end{pmatrix}.$$

One obtains back the above results by computing

$$A_{12} = \langle\psi_2|e^{-i\hat{H}'t}|\psi_1\rangle = \frac{1}{2}\begin{pmatrix} 0 & 1 \end{pmatrix}\begin{pmatrix} e^{-iE_A t} + e^{-iE_B t} & e^{-iE_A t} - e^{-iE_B t} \\ e^{-iE_A t} - e^{-iE_B t} & e^{-iE_A t} + e^{-iE_B t} \end{pmatrix}\begin{pmatrix} 1 \\ 0 \end{pmatrix}.$$

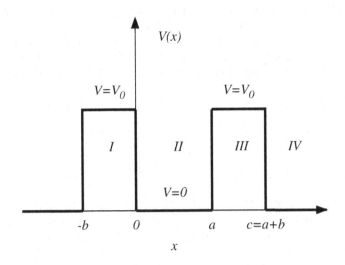

Fig. 2.10 A one dimensional periodic potential, which is a crude sketch of the effective potential for an electron in a crystal. The period is $c = a + b$. For $0 \leq x \leq a$ the potential vanishes, while for $a < x < c$ its value is a constant V_0. In Problem 2.18 we characterize the band structure emerging from the solution of the Schrödinger equation with this potential

Problem 2.18.
Show that the energy spectrum of a particle with mass m in the periodic potential $V(x)$ shown in Fig. 2.10 has a band structure (with allowed and forbidden bands) for energies $0 \leq E \leq V_0$. Analyze in detail the case when $b \to 0$, $V_0 \to +\infty$, with finite $bV_0 \approx bk_3^2$ while $k_3 \to +\infty$, with $k_3^2 = 2m(V_0 - E)/\hbar^2$. This is a one dimensional model of problems that commonly arise when studying the electronic states in solids in the one body approximation; the effective periodic potential stands for the interactions with the ion cores, assumed to be frozen, and all the many-body direct and exchange interactions with the other electrons. The band theory is a useful approximation to many solid state problems, including the conduction of electricity in simple metals like Al at low temperatures.

Solution
The potential $V(x)$ (see Fig. 2.10) is periodic ($V(x + a + b) = V(x)$) with period $c = a + b$. The eigenfunctions of a symmetric Hamiltonian do not generally have the full symmetry of the potential; for instance, the solutions in a central potential are not spherically symmetric, except those with vanishing angular momentum. The solution that we seek will be periodic up to a phase factor

$$\psi(x + n(a + b)) = e^{in\theta}\psi(x)$$

being simultaneous eigenstates of the Hamiltonian and of the (unitary) translation operator that commutes with $\hat{H}$. While a constant phase factor has no physical meaning, a space-time dependent might have; for instance the current and the kinetic energy of a plane wave state are encoded in its phase. The solution of the Schrödinger equations in the four different regions (region I-region IV) given in Fig. 2.10 is

$$\begin{cases} \psi_I(x) = Ae^{ik_1x} + Be^{-ik_1x} & -b \le x < 0 \\ \psi_{II}(x) = Ce^{ik_2x} + De^{-ik_2x} & 0 \le x \le a \\ \psi_{III}(x) = e^{i\theta}\left(Ae^{ik_1(x-a-b)} + Be^{-ik_1(x-a-b)}\right) & a < x < a+b \\ \psi_{IV}(x) = e^{i\theta}\left(Ce^{ik_2(x-a-b)} + De^{-ik_2(x-a-b)}\right) & a+b \le x < 2a+b \end{cases}$$

with $k_1^2 = 2m(E - V_0)/\hbar^2, k_2^2 = 2mE/\hbar^2$. As for the solution in the regions III and IV, we remark that the condition $\psi(x+a+b) = e^{i\theta}\psi(x)$ is equivalent to $\psi(x) = e^{i\theta}\psi(x-a-b)$. Next, we impose the continuity conditions in 0 and a as follows

$$\begin{cases} \psi_I(0) = \psi_{II}(0) \Rightarrow A + B - C - D = 0 \\ \psi_I'(0) = \psi_{II}'(0) \Rightarrow k_1(A - B) - k_2(C - D) = 0 \\ \psi_{II}(a) = \psi_{III}(a) \Rightarrow -e^{i\theta}\left(Ae^{-ik_1b} + Be^{ik_1b}\right) + Ce^{ik_2a} + De^{-ik_2a} = 0 \\ \psi_{II}'(a) = \psi_{III}'(a) \Rightarrow -k_1 e^{i\theta}\left(Ae^{-ik_1b} - Be^{ik_1b}\right) + k_2(Ce^{ik_2a} - De^{-ik_2a}) = 0. \end{cases}$$

To find a non zero solution for this problem, the determinant of the associated matrix must be zero. If we define $\alpha = e^{ik_1b}, \beta = e^{ik_2a}$, the matrix associated to this system is

$$\hat{M} = \begin{pmatrix} 1 & 1 & -1 & -1 \\ -\frac{e^{i\theta}}{\alpha} & -e^{i\theta}\alpha & \beta & \frac{1}{\beta} \\ k_1 & -k_1 & -k_2 & k_2 \\ -k_1\frac{e^{i\theta}}{\alpha} & k_1 e^{i\theta}\alpha & k_2\beta & -\frac{k_2}{\beta} \end{pmatrix}.$$

Only the trivial solution of the system exists, unless $\det\hat{M} = 0$ which requires

$$\det\hat{M} = -4k_1k_2\left(1+e^{2i\theta}\right) - e^{i\theta}\left(k_1^2+k_2^2\right)\left(\frac{\alpha}{\beta} - \frac{1}{\alpha\beta} - \alpha\beta + \frac{\beta}{\alpha}\right) +$$

$$2k_1k_2e^{i\theta}\left(\frac{\alpha}{\beta} + \frac{1}{\alpha\beta} + \alpha\beta + \frac{\beta}{\alpha}\right) =$$

$$e^{i\theta}\{-8k_1k_2\cos\theta - 2(k_1^2+k_2^2)[\cos(k_1b-k_2a) - \cos(k_1b+k_2a)]+$$

$$4k_1k_2[\cos(k_1b-k_2a) + \cos(k_1b+k_2a)]\} = 0.$$

Using the identity

$$\cos(\alpha + \beta) = \cos\alpha\cos\beta - \sin\alpha\sin\beta$$

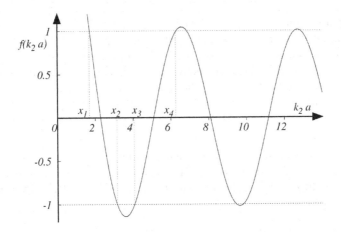

Fig. 2.11 The graphical solution to the problem of a particle in the periodic potential of Fig. 2.10. The intersections between the function $f(k_2a) = \frac{A}{k_2a}\sin(k_2a) + \cos(k_2a)$ and the ± 1 lines are the band edges (since k_2 determines the energy). For the case in this figure we took $A = 2.0$

and setting $k_3^2 = 2m(V_0 - E)/\hbar^2$, with $i\sinh x = \sin ix$ and $\cosh x = \cos ix$, we obtain an equation for $\cos\theta$

$$\cos\theta = \cos(k_2a)\cosh(k_3b) - \frac{k_2^2 - k_3^2}{2k_2k_3}\sin(k_2a)\sinh(k_3b) \qquad 0 \le E \le V_0.$$

The condition for the energy comes from the requirement that θ be real, that is, $-1 \le \cos\theta \le 1$. We concentrate on the case $b \to 0$, $V_0 \to +\infty$, with finite $bV_0 \approx bk_3^2$ while $k_3 \to +\infty$. The argument of $\sinh, \cosh$ is k_3b, which goes to zero like $1/k_3$ enabling us to develop in Taylor series. Approximating $\cosh(k_3b) \approx 1$, $\sinh(k_3b) \approx k_3b$, $k_2 \ll k_3$ and setting $A = abk_3^2/2$, the above condition reads

$$-1 \le A\frac{\sin(k_2a)}{k_2a} + \cos(k_2a) \le 1.$$

The function $f(k_2a) = \frac{A}{k_2a}\sin(k_2a) + \cos(k_2a)$ is plotted in Fig. 2.11. The condition is satisfied for all the continuous values of k_2a in the intervals $[x_1, x_2], [x_3, x_4], \ldots$. This is the band structure requested by the problem.

Problem 2.19.
Let us consider a particle with mass m subject to the one dimensional potential $U(x) = -\alpha x$ with $\alpha > 0$:

- determine the time evolution for $\Delta p = \sqrt{\langle \hat{p}^2 \rangle - \langle \hat{p} \rangle^2}$, where $\hat{p}$ is the momentum operator;
- determine the time evolution of the wave function $\psi(x,t)$ knowing that $\psi(x,0) = e^{ip_0x/\hbar - i\phi_0}$, with p_0 and ϕ_0 constants.

Solution

The potential given in the text is not leading to an eigenvalue problem that can be easily solved: if we write the stationary Schrödinger equation we obtain a continuous spectrum with the Airy functions as eigenfunctions. Thus, instead of applying the definition of average value based on the knowledge of the eigenfunctions of the Hamiltonian, we use the formula for the time evolution of the average value of a generic operator $\hat{A}$

$$\frac{d\langle\hat{A}\rangle_t}{dt} = \frac{i}{\hbar}\langle[\hat{H},\hat{A}]\rangle.$$

This formula is the quantum mechanical analogue of the classical formula giving the time evolution of a quantity in terms of its Poisson bracket with the Hamiltonian. In our case, both $\hat{p}$ and $\hat{p}^2$ commute with the kinetic part of $\hat{H}$, and we find the differential equation

$$\frac{d\langle\hat{p}\rangle_t}{dt} = \frac{i}{\hbar}\langle[\hat{U},\hat{p}]\rangle = -\frac{i\alpha}{\hbar}\langle[\hat{x},\hat{p}]\rangle = \alpha$$

which can be easily integrated

$$\langle\hat{p}\rangle_t = \alpha t + p_0$$

where p_0 (the constant of integration) is the average value of $\hat{p}$ at time $t = 0$. Similarly, for the average of the squared momentum $\langle\hat{p}^2\rangle$, we get

$$\frac{d\langle\hat{p}^2\rangle_t}{dt} = \frac{i}{\hbar}\langle[\hat{U},\hat{p}^2]\rangle = -\frac{i\alpha}{\hbar}\langle[\hat{x},\hat{p}^2]\rangle.$$

We can make use of the Jacobi identity (see also Problem 2.2) to find

$$[\hat{x},\hat{p}^2] = [\hat{x},\hat{p}\hat{p}] = \hat{p}[\hat{x},\hat{p}] + [\hat{x},\hat{p}]\hat{p} = 2i\hbar\hat{p}$$

and, hence

$$\frac{d\langle\hat{p}^2\rangle_t}{dt} = \frac{i}{\hbar}\langle[\hat{U},\hat{p}^2]\rangle = -\frac{i\alpha}{\hbar}\langle[\hat{x},\hat{p}^2]\rangle = 2\alpha\langle\hat{p}\rangle.$$

Integrating the above differential equation, we get

$$\langle\hat{p}^2\rangle_t = \alpha^2 t^2 + 2\alpha p_0 t + \text{const.}$$

Therefore, the final result is

$$\Delta p = \sqrt{\langle\hat{p}^2\rangle_t - \langle\hat{p}\rangle_t^2} = \sqrt{\alpha^2 t^2 + 2\alpha p_0 t + \text{const.} - (\alpha t + p_0)^2} = \sqrt{\text{const.} - p_0^2}.$$

Let us now face the second point. At time $t = 0$ the wave function is a plane wave with an additional phase factor

$$\psi(x,0) = e^{ip_0 x/\hbar - i\phi_0}.$$

When evolving in time the function ψ, both p_0 and ϕ_0 acquire a non trivial time dependence. We therefore consider $\langle \hat{p} \rangle_t$ and $\phi(t)$, whose values at time $t = 0$ are

$$\langle \hat{p} \rangle_{t=0} = p_0 \qquad \phi(0) = \phi_0.$$

The time evolution for $\langle \hat{p} \rangle_t$ has already been determined before. As for ϕ, we request that

$$\psi(x,t) = e^{i\langle \hat{p} \rangle_t x / \hbar - i\phi(t)} = e^{i\alpha t x / \hbar + i p_0 x / \hbar - i\phi(t)}$$

is a solution of the Schrödinger equation

$$i\hbar \frac{\partial \psi(x,t)}{\partial t} = \hat{H} \psi(x,t) = \left(-\frac{\hbar^2}{2m} \frac{\partial^2}{\partial x^2} - \alpha x \right) \psi(x,t).$$

If we perform the derivatives, we get

$$\hbar \frac{d\phi(t)}{dt} = \frac{1}{2m} (\alpha t + p_0)^2$$

leading to

$$\phi(t) = \frac{(\alpha t + p_0)^3}{6m\alpha\hbar} - \frac{(p_0)^3}{6m\alpha\hbar} + \phi_0.$$

Problem 2.20.

A quantum particle is in the ground state of a one dimensional harmonic oscillator with Hamiltonian

$$\hat{H}_1 = \frac{\hat{p}^2}{2m} + \frac{1}{2} m\omega_0^2 \hat{x}^2.$$

At a given time, the frequency ω_0 changes abruptly and the Hamiltonian becomes

$$\hat{H}_\eta = \frac{\hat{p}^2}{2m} + \frac{1}{2} m(\eta\omega_0)^2 \hat{x}^2.$$

Determine the probability $P(\eta)$ that the particle is in the ground state of the new harmonic oscillator with frequency $\eta\omega_0$. Prove that the result is symmetric in η and $\frac{1}{\eta}$ and verify the limit $\eta \to 1$.

Solution

The wave function for the ground state of the harmonic oscillator can be written as

$$\psi_0^{(1)}(x) = \frac{1}{\sqrt{x_1 \sqrt{\pi}}} e^{-\frac{x^2}{2x_1^2}}$$

where $x_1^2 = \frac{\hbar}{m\omega_0}$. We also introduce the variable $x_\eta^2 = \frac{\hbar}{m\eta\omega_0}$ to characterize the ground state of the new harmonic oscillator with frequency $\eta\omega_0$

$$\psi_0^{(\eta)}(x) = \frac{1}{\sqrt{x_\eta \sqrt{\pi}}} e^{-\frac{x^2}{2x_\eta^2}}.$$

In the approximation given by the text, the change from $\hat{H}_1$ to $\hat{H}_\eta$ is abrupt and the requested probability is $P = |A|^2$, where A is the overlap integral (see also Problem 5.13) between the two wave functions $\psi_0^{(1)}(x)$ and $\psi_0^{(\eta)}(x)$

$$A = \int_{-\infty}^{+\infty} \psi_0^{(1)}(x)\psi_0^{(\eta)}(x)dx = \frac{1}{\sqrt{\pi x_1 x_\eta}} \int_{-\infty}^{+\infty} e^{-\frac{x^2}{2x_1^2}} e^{-\frac{x^2}{2x_\eta}} dx = \frac{1}{\sqrt{\pi x_1 x_\eta}} \int_{-\infty}^{+\infty} e^{-\frac{x^2}{x_a^2}} dx$$

where x_a is such that

$$\frac{1}{x_a^2} = \frac{1}{2}\left(\frac{1}{x_1^2} + \frac{1}{x_\eta^2}\right).$$

Using the known result $\int_{-\infty}^{+\infty} e^{-\frac{x^2}{a^2}} dx = a\sqrt{\pi}$, we get

$$A = \frac{\sqrt{\pi}x_a}{\sqrt{\pi x_1 x_\eta}} = \sqrt{2}\frac{\eta^{\frac{1}{4}}}{\sqrt{1+\eta}}$$

and, hence

$$P(\eta) = 2\frac{\sqrt{\eta}}{1+\eta} = \frac{2}{\sqrt{\eta} + \frac{1}{\sqrt{\eta}}}.$$

The result is indeed symmetric in η and $\frac{1}{\eta}$. Moreover, in the limit $\eta \to 1$ we see that the probability becomes 1, as it should be expected because both Hamiltonians become the same in that limit.

Problem 2.21.
A quantum system has two energy states and is characterized by the following Hamiltonian

$$\hat{H} = -\frac{\hbar\omega}{2}(|0\rangle\langle 0| - |1\rangle\langle 1|)$$

where $|0\rangle$ and $|1\rangle$ are normalized orthogonal eigenstates with eigenvalues $-\frac{\hbar\omega}{2}$ and $+\frac{\hbar\omega}{2}$, respectively. Let us consider the operator $\hat{a}$ with the property

$$\hat{a} = |0\rangle\langle 1|$$

and let $\hat{a}^\dagger$ be the adjoint of $\hat{a}$:
- determine $[\hat{a}, \hat{a}^\dagger]_+ = \hat{a}\hat{a}^\dagger + \hat{a}^\dagger\hat{a}$;
- determine $(\hat{a}^\dagger)^2$ and $\hat{a}^2$;
- determine the commutators $[\hat{H}, \hat{a}]_-$ and $[\hat{H}, \hat{a}^\dagger]_-$, with $[\hat{A}, \hat{B}]_- = \hat{A}\hat{B} - \hat{B}\hat{A}$;
- determine the eigenvalues of $\hat{n}$, with $\hat{n} = \hat{a}^\dagger\hat{a}$, and express $\hat{H}$ in terms of $\hat{n}$ and the operator $\hat{E} = |0\rangle\langle 0| + |1\rangle\langle 1|$.

Solution
From the definition of $\hat{a}$ and $\hat{a}^\dagger$, we see that

$$\hat{a}^\dagger = |1\rangle\langle 0|$$

and, hence

$$[\hat{a}, \hat{a}^\dagger]_+ = |0\rangle\langle 1|1\rangle\langle 0| + |1\rangle\langle 0|0\rangle\langle 1| = \hat{E}.$$

As for the square of $\hat{a}$ and $\hat{a}^\dagger$, making use of the orthogonality of states, we find

$$\hat{a}^2 = |0\rangle\langle 1|0\rangle\langle 1| = 0$$

$$(\hat{a}^\dagger)^2 = |1\rangle\langle 0|1\rangle\langle 0| = 0.$$

Let us then determine the commutators $[H, \hat{a}]_-$ and $[\hat{H}, \hat{a}^\dagger]_-$

$$[\hat{H}, \hat{a}]_- = -\frac{\hbar\omega}{2}(|0\rangle\langle 0| - |1\rangle\langle 1|)|0\rangle\langle 1| + \frac{\hbar\omega}{2}|0\rangle\langle 1|(|0\rangle\langle 0| - |1\rangle\langle 1|) = -\hbar\omega\hat{a}$$

$$[\hat{H}, \hat{a}^\dagger]_- = -\frac{\hbar\omega}{2}(|0\rangle\langle 0| - |1\rangle\langle 1|)|1\rangle\langle 0| + \frac{\hbar\omega}{2}|1\rangle\langle 0|(|0\rangle\langle 0| - |1\rangle\langle 1|) = \hbar\omega\hat{a}^\dagger.$$

As for the last point, we note that

$$\hat{n} = \hat{a}^\dagger\hat{a} = |1\rangle\langle 0|0\rangle\langle 1| = |1\rangle\langle 1|$$

that implies $|0\rangle$ and $|1\rangle$ are eigenstates of both $\hat{H}$ and $\hat{n}$. We therefore conclude that

$$\hat{H} = -\frac{\hbar\omega}{2}(|0\rangle\langle 0| - |1\rangle\langle 1| \pm |1\rangle\langle 1|) = \hbar\omega|1\rangle\langle 1| - \frac{\hbar\omega}{2}(|0\rangle\langle 0| + |1\rangle\langle 1|) = \hbar\omega\left(\hat{n} - \frac{\hat{E}}{2}\right).$$

Problem 2.22.
Let us consider the motion of a particle with energy $E > U_0$ ($U_0 > 0$) in the one dimensional potential field $U(x) = \frac{U_0}{(1+e^{-x})}$. Determine the resulting reflection and transmission coefficients.

Solution
To solve the problem we need to determine the properties of the solution of the Schrödinger equation for $x \to \pm\infty$. In the case of Classical Mechanics, a particle with energy $E > U_0$ moving from left to right, is not reflected from the potential, due to the fact that $E > \lim_{x\to+\infty} U(x)$. In Quantum Mechanics, instead, such particle continues to move from left to right but a portion of the associated wave function is reflected from the potential $U(x)$. Let

$$\psi(x) \approx C_T e^{i\frac{\sqrt{2m(E-U_0)}}{\hbar}x}$$

be the solution of the Schrödinger equation for $x \to +\infty$, and

$$\psi(x) \approx e^{i\frac{\sqrt{2mE}}{\hbar}x} + C_R e^{-i\frac{\sqrt{2mE}}{\hbar}x}$$

the form of the wave function for the free particle when $x \to -\infty$. The reflection and transmission coefficients are defined through the density flux

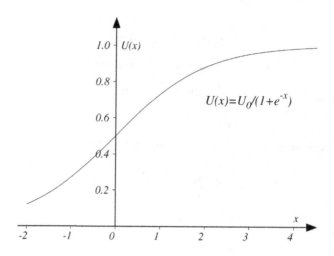

Fig. 2.12 The potential field $U(x) = \frac{U_0}{(1+e^{-x})}$ with $U_0 = 1$. In Problem 2.22 we determine the associated reflection and transmission coefficients

$$J = \frac{i\hbar}{2m} \left(\psi \frac{d\psi^*}{dx} - \psi^* \frac{d\psi}{dx} \right)$$

of the incident wave and the density flux of the reflected/transmitted wave. For simplicity, we have normalized to unity the coefficient of the incident wave travelling from left to right. This is not important because the reflection/transmission coefficient is defined as the ratio of the density flux in the reflected/transmitted $(J_R \propto \sqrt{2mE}|C_R|^2, J_T \propto \sqrt{2m(E-U_0)}|C_T|^2)$ wave to that of the incident wave $(J_I \propto \sqrt{2mE})$. These densities depend on the derivatives of the solutions previously discussed and, consequently, on the associated exponents. Therefore, the transmission coefficient is

$$T = \frac{\sqrt{2m(E-U_0)}}{\sqrt{2mE}} |C_T|^2$$

and the reflection coefficient is given by

$$R = 1 - T = 1 - \frac{\sqrt{2m(E-U_0)}}{\sqrt{2mE}} |C_T|^2 = |C_R|^2$$

where we have used the condition of continuity in the density flux $(J_I + J_R = J_T)$. Let us then characterize the solution of the Schrödinger equation with the potential $U(x)$

$$\psi'' + \left(\alpha - \frac{\beta}{1+e^{-x}} \right) \psi = 0$$

with $\alpha = 2mE/\hbar^2$, $\beta = 2mU_0/\hbar^2$. If we set $z = -e^{-x}$, we can write the Schrödinger equation for $\psi(z)$

$$\frac{d^2\psi}{dz^2} + \frac{1}{z}\frac{d\psi}{dz} + \left(\frac{\alpha}{z^2} + \frac{\beta}{z^2(z-1)}\right)\psi = 0$$

where we have used that

$$\frac{d^2\psi}{dx^2} = z^2\frac{d^2\psi}{dz^2} + z\frac{d\psi}{dz}.$$

When $-\infty < x < +\infty$, we get $-\infty < z < 0$. There are two regular singularities in the points $z = 0$ and $z = 1$ (see also Problems 2.25, 4.1, 4.2, 4.5 for a discussion of the singularities associated with the Schrödinger equation). If we set $z = 1/t$, we get an equation for $\psi(t)$

$$\frac{d^2\psi}{dt^2} + \frac{1}{t}\frac{d\psi}{dt} + \frac{1}{t^2}\left(\alpha - \frac{\beta t}{(t-1)}\right)\psi = 0$$

where we see that also $z = +\infty$ ($t = 0$) is a regular singularity. The whole solution is the product of the singular behaviour and a suitable hypergeometric function. To determine the singular part, we need to calculate the indicial exponents with the substitution $\psi(z) = z^\lambda ((z-1)^\mu)$ in the original equation. In the limit $z \to 0$ ($z \to 1$), we solve the resulting second order algebraic equation and we find the following solutions

$$\lambda = \pm i\sqrt{\alpha - \beta} = \pm\frac{i}{\hbar}\sqrt{2m(E - U_0)} \qquad \mu = 0,\ 1.$$

We then seek the solution in the form

$$\psi(z) = z^\lambda(z-1)^\mu y(z) = z^\lambda y(z)$$

since one of the solutions for μ is zero. As for the value of λ, we choose the minus sign, since for $x \to +\infty$ ($z \to 0$) the solution must represent a transmitted wave travelling from left to right, that is

$$\psi \approx e^{\frac{i}{\hbar}\sqrt{2m(E-U_0)}x} = (e^{-x})^{-\frac{i}{\hbar}\sqrt{2m(E-U_0)}} = z^{-\frac{i}{\hbar}\sqrt{2m(E-U_0)}}.$$

Substituting $\psi(z) = z^\lambda y(z)$ in the original equation, we get

$$z(z-1)y'' + (z-1)(2\lambda + 1)y' + \beta y = 0.$$

If we want to match the general equation for the hypergeometric series

$$z(z-1)y'' + [(A+B+1)z - C]y' + ABy = 0$$

we find

$$A = i(\sqrt{\alpha} - \sqrt{\alpha - \beta})$$
$$B = -i(\sqrt{\alpha} + \sqrt{\alpha - \beta})$$
$$C = 2\lambda + 1 = -2i\sqrt{\alpha - \beta} + 1$$

from which we get the solution

$$y = F(A, B, C|z) = F(i(\sqrt{\alpha} - \sqrt{\alpha - \beta}), -i(\sqrt{\alpha} + \sqrt{\alpha - \beta}), -2i\sqrt{\alpha - \beta} + 1|z).$$

When $z \to 0$ ($x \to +\infty$) the hypergeometric series becomes 1 and ψ represents the plane wave travelling from left to right. To determine the reflection coefficient, we need to determine the solution for $z \to -\infty$ ($x \to -\infty$). The problem is that of the analytical continuation of the hypergeometric series from 0 to $-\infty$. Therefore, we need the formula relating the hypergeometric series with variables z and $1/z$

$$F(A, B, C|z) = G(-z)^{-A} F(A, 1 - C + A, 1 - B + A|\frac{1}{z}) +$$

$$S(-z)^{-B} F(B, 1 - C + B, 1 + B - A|\frac{1}{z})$$

with the constants G and S given by

$$G = \frac{\Gamma(C)\Gamma(B - A)}{\Gamma(B)\Gamma(C - A)} = \frac{\Gamma(-2i\sqrt{\alpha - \beta} + 1)\Gamma(-2i\sqrt{\alpha})}{\Gamma(-i(\sqrt{\alpha} + \sqrt{\alpha - \beta}))\Gamma(-i(\sqrt{\alpha} + \sqrt{\alpha - \beta}) + 1)}$$

$$S = \frac{\Gamma(C)\Gamma(A - B)}{\Gamma(A)\Gamma(C - B)} = \frac{\Gamma(-2i\sqrt{\alpha - \beta} + 1)\Gamma(2i\sqrt{\alpha})}{\Gamma(i(\sqrt{\alpha} - \sqrt{\alpha - \beta}))\Gamma(i(\sqrt{\alpha} - \sqrt{\alpha - \beta}) + 1)}.$$

In the limit $z \to -\infty$, the hypergeometric series becomes 1 and we are left with

$$\psi(z) \approx z^{-i\sqrt{\alpha - \beta}} \left[G(-z)^{i(\sqrt{\alpha - \beta} - \sqrt{\alpha})} + S(-z)^{i(\sqrt{\alpha - \beta} + \sqrt{\alpha})} \right] =$$

$$(-1)^{-i\sqrt{\alpha - \beta}} G \left[e^{i\sqrt{\alpha}x} + \frac{S}{G} e^{-i\sqrt{\alpha}x} \right].$$

Therefore, the reflection coefficient is $R = |S/G|^2$, with S and G given above.

Problem 2.23.
Consider the one dimensional rectangular potential wall (V_0)

$$\begin{cases} V(x) = 0 & x < 0 \\ V(x) = V_0 & x \geq 0. \end{cases}$$

In the case of a particle travelling from left to right, determine the energy $E > V_0$ such that there is the same probability to find the particle in the regions $x \geq 0$ and $x < 0$. Using the same initial condition, repeat the calculation for the rectangular potential barrier (see also Fig. 2.13)

$$\begin{cases} V(x) = 0 & x < 0 \\ V(x) = V_0 & 0 \leq x \leq a \\ V(x) = 0 & x > a. \end{cases}$$

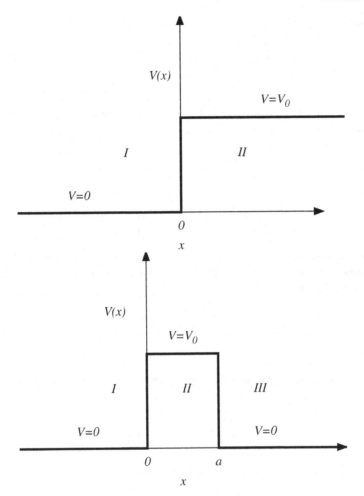

Fig. 2.13 A one dimensional rectangular potential wall (top) and a rectangular potential barrier (bottom). In Problem 2.23 we study the solution of the one dimensional Schrödinger equation with the property that the reflection and transmission coefficients are equal

Solution

The wave functions entering this problem cannot be normalized because $E > V_0$. Consequently, the condition that there is the same probability to find the particle in the regions $x < 0, x \geq 0$ implies that the reflection (R) and transmission (T) coefficients are the same. Furthermore, the condition $R + T = 1$ imposes $R = T = \frac{1}{2}$.

Let us then start with the first case. The Schrödinger equation for $x < 0$ (region *I*) and $x \geq 0$ (region *II*) is satisfied (apart from an overall constant) by the wave functions

$$\begin{cases} \psi_I(x) = e^{ik_1x} + C_R e^{-ik_1x} & x < 0 \\ \psi_{II}(x) = C_T e^{ik_2x} & x \geq 0 \end{cases}$$

with $k_1 = \sqrt{2mE}/\hbar$, $k_2 = \sqrt{2m(E-V_0)}/\hbar$. The constants C_R, C_T are determined from the condition of continuity of the wave function and its first derivative in $x = 0$. In formulae, we get

$$1 + C_R = C_T \quad\quad k_1(1 - C_R) = k_2 C_T$$

leading to

$$C_T = \frac{2k_1}{(k_1 + k_2)} \quad\quad C_R = \frac{(k_1 - k_2)}{(k_1 + k_2)}.$$

The transmission/reflection coefficient is the ratio of the density flux in the transmitted/reflected wave to that of the incident wave (see also Problem 2.22). The definition of the density flux is

$$J = \frac{i\hbar}{2m}\left(\psi \frac{d\psi^*}{dx} - \psi^* \frac{d\psi}{dx} \right).$$

The incident wave is given by e^{ik_1x}. e^{-ik_1x} in ψ_I represents the reflected wave and e^{ik_2x} in ψ_{II} represents the transmitted wave. Therefore, we find $J_I \propto k_1$, $J_R \propto k_1|C_R|^2$, $J_T \propto k_2|C_T|^2$ and the transmission/reflection coefficients

$$T = |C_T|^2 \frac{k_2}{k_1} \quad\quad R = |C_R|^2.$$

The desired condition ($T = R = \frac{1}{2}$) is translated into

$$\frac{1}{2} = |C_R|^2 = \left(\frac{k_1 - k_2}{k_1 + k_2} \right)^2$$

leading to the following equation for E

$$\frac{1}{2} = \left(\frac{\sqrt{E} - \sqrt{E - V_0}}{\sqrt{E} + \sqrt{E - V_0}} \right)^2$$

yielding $E = \frac{(3\sqrt{2}+4)}{8} V_0$.

As for the rectangular potential barrier of the second point, things are more complicated. The Schrödinger equation has to be solved in the three different regions

($x < 0, 0 \leq x \leq a$ and $x > a$) and the solutions are

$$\begin{cases} \psi_I(x) = e^{ik_1 x} + A e^{-ik_1 x} & x < 0 \\ \psi_{II}(x) = B e^{ik_2 x} + B' e^{-ik_2 x} & 0 \leq x \leq a \\ \psi_{III}(x) = C_T e^{ik_1 x} & x > a. \end{cases}$$

The conditions of continuity for the ψ and its first derivative in $x = 0$ and $x = a$ are translated into

$$\begin{cases} 1 + A = B + B' \\ 1 - A = \frac{k_2}{k_1}(B - B') \\ B e^{ik_2 a} + B' e^{-ik_2 a} = C_T e^{ik_1 a} \\ B e^{ik_2 a} - B' e^{-ik_2 a} = \frac{k_1}{k_2} C_T e^{ik_1 a}. \end{cases}$$

If we sum the first and second of these equations and divide the third by the forth, we get

$$\begin{cases} B' = \frac{2k_1}{k_1 - k_2} - B\left(\frac{k_1 + k_2}{k_1 - k_2}\right) \\ \frac{B e^{ik_2 a} + B' e^{-ik_2 a}}{B e^{ik_2 a} - B' e^{-ik_2 a}} = \frac{k_2}{k_1}. \end{cases}$$

Substituting B' in the second equation, we find

$$B = \frac{k_1(k_1 + k_2)e^{-ik_2 a}}{2k_1 k_2 \cos(k_2 a) - i(k_1^2 + k_2^2)\sin(k_2 a)}.$$

Also, in the equations for the coefficients, we can sum the third and forth equations and we get

$$C_T = \frac{2B k_2 e^{i(k_2 - k_1)a}}{k_1 + k_2}.$$

Substituting the value of B in this equation, we find

$$C_T = \frac{2k_1 k_2 e^{-ik_1 a}}{2k_1 k_2 \cos(k_2 a) - i(k_1^2 + k_2^2)\sin(k_2 a)}.$$

Moreover, for this rectangular potential barrier, we have $J_I \propto k_1, J_T \propto k_1 |C_T|^2$ and the resulting transmission coefficient is

$$T = |C_T|^2 = \frac{4k_1^2 k_2^2}{(k_1^2 - k_2^2)^2 \sin^2(k_2 a) + 4k_1^2 k_2^2}.$$

Consequently, the reflection coefficient is

$$R = 1 - T = \frac{(k_1^2 - k_2^2)^2 \sin^2(k_2 a)}{(k_1^2 - k_2^2)^2 \sin^2(k_2 a) + 4k_1^2 k_2^2}.$$

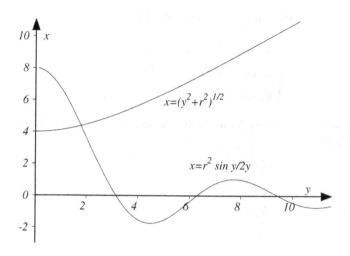

Fig. 2.14 The intersection of the two curves $x = (r^2 \sin y)/2y$ and $x = (y^2 + r^2)^{1/2}$ (with $r = 4$) allows us to determine the energy at which the transmission coefficient equals the reflection coefficient for the rectangular potential barrier of Fig. 2.13

When $\sin^2(k_2 a) = 0$, i.e. $k_2 a = n\pi, n = 1, 2, \ldots$, the reflection and transmission coefficients become $R = 0$ and $T = 1$, i.e. the potential barrier is perfectly transmitting the wave function. This happens when the width of the barrier satisfies $a = n\pi/k_2 = n\lambda/2$, with $\lambda = 2\pi/k_2$. Imposing $T = 1/2$ we find the desired condition for the energy E

$$\frac{4k_1^2 k_2^2}{(k_1^2 - k_2^2)^2} = \sin^2(k_2 a).$$

If we set $x = ak_1$ (not to be confused with the position in the beginning of the problem) and $y = ak_2$, we realize that the desired value for E is found from the intersection of the two curves

$$\begin{cases} x^2 - y^2 = \dfrac{2mV_0 a^2}{\hbar^2} = r^2 \\ x^2 y^2 = \dfrac{r^4}{4} \sin^2 y. \end{cases}$$

In Fig. 2.14 we plot $x = (r^2 \sin y)/2y$, $x = (y^2 + r^2)^{1/2}$. Since $x \propto \sqrt{E}$, we need to look for positive solutions corresponding to a real energy. The first curve is a branch of a hyperbola while the second is a decreasing oscillating function with maximum value in $y = 0$. The intersection is possible when $r^2/2 \geq r$, i.e. $r \geq 2$ and $\frac{\sqrt{mV_0 a^2}}{\hbar} \geq 2$.

Problem 2.24.

Let us consider a one dimensional quantum harmonic oscillator. Using the Heisenberg representation, determine the time evolution of the operators $\hat{x}$, $\hat{p}$, $\hat{a}$, $\hat{a}^\dagger$. Determine the average value of $\hat{x}$, $\hat{p}$, $\hat{x}^2$, $\hat{p}^2$ on the generic eigenstate $|n\rangle$ of the Hamiltonian $\hat{H} = \hbar\omega\left(\hat{a}^\dagger\hat{a} + \frac{1}{2}\,1\right)$, $\hat{H}|n\rangle = \frac{\hbar\omega}{2}\left(n + \frac{1}{2}\right)|n\rangle$, at time $t = 0$ and at the generic time $t > 0$. If at time $t = 0$ the system is in the state

$$|\alpha\rangle = c_0|0\rangle + c_1|1\rangle$$

with c_0 and c_1 real constants, determine the average value of $\hat{x}$, $\hat{p}$, $\hat{x}^2$, $\hat{p}^2$ on such a state at time $t = 0$ and at the generic time $t > 0$.

Solution
To determine the time evolution of the operators, we make use of the following identity

$$e^{\hat{B}}\hat{A}e^{-\hat{B}} = \hat{A} + [\hat{B},\hat{A}] + \frac{1}{2!}[\hat{B},[\hat{B},\hat{A}]] + \frac{1}{3!}[\hat{B},[\hat{B},[\hat{B},\hat{A}]]] + \cdots$$

valid for two generic operators $\hat{A}$ and $\hat{B}$. Therefore, let us start by demonstrating such formula. Let us define $\hat{\phi}(t) = e^{t\hat{B}}\hat{A}e^{-t\hat{B}}$ with the property $\hat{\phi}(1) = e^{\hat{B}}\hat{A}e^{-\hat{B}}$. The derivatives of $\hat{\phi}(t)$ are

$$\frac{d\hat{\phi}(t)}{dt} = \hat{B}e^{t\hat{B}}\hat{A}e^{-t\hat{B}} - e^{t\hat{B}}\hat{A}\hat{B}e^{-t\hat{B}} = [\hat{B},\hat{\phi}(t)]$$

$$\frac{d^2\hat{\phi}(t)}{dt^2} = \frac{d}{dt}[\hat{B},\hat{\phi}(t)] = \left[\hat{B},\frac{d\hat{\phi}(t)}{dt}\right] = [\hat{B},[\hat{B},\hat{\phi}(t)]]$$

$$\frac{d^3\hat{\phi}(t)}{dt^3} = \frac{d}{dt}[\hat{B},[\hat{B},\hat{\phi}(t)]] = \left[\hat{B},\left[\hat{B},\frac{d\hat{\phi}(t)}{dt}\right]\right] = [\hat{B},[\hat{B},[\hat{B},\hat{\phi}(t)]]].$$

A possible way to define $\hat{\phi}(1) = e^{\hat{B}}\hat{A}e^{-\hat{B}}$ takes advantage of the Taylor expansion of $\phi(t)$ around $t = 0$

$$e^{\hat{B}}\hat{A}e^{-\hat{B}} = \hat{\phi}(1) = \hat{\phi}(0) + \left.\frac{d\hat{\phi}(t)}{dt}\right|_{t=0} + \frac{1}{2!}\left.\frac{d^2\hat{\phi}(t)}{dt^2}\right|_{t=0} + \frac{1}{3!}\left.\frac{d^3\hat{\phi}(t)}{dt^3}\right|_{t=0} + \cdots.$$

Since $\hat{\phi}(0) = \hat{A}$, we can use the above expressions for the derivatives evaluated at $t = 0$ and immediately obtain the desired result for $e^{\hat{B}}\hat{A}e^{-\hat{B}}$. Setting $\hat{B} = i\hat{H}t/\hbar$, we are now ready to determine the time evolution for a generic operator $\hat{A}(t)$ in the Heisenberg representation

$$\hat{A}(t) = e^{i\frac{\hat{H}t}{\hbar}}\hat{A}(0)e^{-i\frac{\hat{H}t}{\hbar}} = \hat{A}(0) + \frac{it}{\hbar}[\hat{H},\hat{A}(0)] -$$

$$\frac{1}{2!}\frac{t^2}{\hbar^2}[\hat{H},[\hat{H},\hat{A}(0)]] - i\frac{t^3}{\hbar^3}\frac{1}{3!}[\hat{H},[\hat{H},[\hat{H},\hat{A}(0)]]] + \cdots.$$

When $\hat{A} = \hat{a}$ (the annihilation operator), we obtain

$$\hat{a}(t) = e^{i\omega(\hat{a}^\dagger \hat{a} + \frac{1}{2}1)t} \hat{a} e^{-i\omega(\hat{a}^\dagger \hat{a} + \frac{1}{2}1)t} = e^{i\omega \hat{a}^\dagger \hat{a}t} \hat{a} e^{-i\omega \hat{a}^\dagger \hat{a}t} = \hat{a} + it\omega[\hat{a}^\dagger \hat{a}, \hat{a}]$$
$$- \frac{1}{2!}(\omega t)^2 [\hat{a}^\dagger \hat{a}, [\hat{a}^\dagger \hat{a}, \hat{a}]] - i(\omega t)^3 \frac{1}{3!}[\hat{a}^\dagger \hat{a}, [\hat{a}^\dagger \hat{a}, [\hat{a}^\dagger \hat{a}, \hat{a}]]] + \cdots =$$
$$\hat{a}\left(1 - it\omega - \frac{1}{2!}(\omega t)^2 + i(\omega t)^3 \frac{1}{3!} + \cdots\right) = \hat{a}e^{-i\omega t}.$$

Since $\hat{a}^\dagger$ is the adjoint of $\hat{a}$, we easily obtain $\hat{a}^\dagger(t) = \hat{a}^\dagger e^{i\omega t}$. From the relation between the creation/annihilation operators $(\hat{a}^\dagger, \hat{a})$ and the position/momentum operators $(\hat{x}, \hat{p})$, we know that

$$\begin{cases} \hat{x}(t) = \sqrt{\frac{\hbar}{2m\omega}}(\hat{a}e^{-i\omega t} + \hat{a}^\dagger e^{i\omega t}) \\ \hat{p}(t) = i\sqrt{\frac{m\hbar\omega}{2}}(\hat{a}^\dagger e^{i\omega t} - \hat{a}e^{-i\omega t}). \end{cases}$$

We then consider the matrix elements of $\hat{x}, \hat{p}, \hat{x}^2, \hat{p}^2$ between the states $|m\rangle$ and $|n\rangle$ at time $t = 0$. Such elements evolve in time as

$$A_{nm}(t) = \langle n,t|\hat{A}|m,t\rangle = e^{-i\frac{(E_m - E_n)t}{\hbar}} \langle n|\hat{A}|m\rangle = e^{-i\frac{(E_m - E_n)t}{\hbar}} A_{nm}(t = 0)$$

where $\hat{A}$ is chosen among the operators $\hat{x}, \hat{p}, \hat{x}^2, \hat{p}^2$. We are interested in the diagonal elements $(n = m)$ where the phase $e^{-i\frac{(E_m - E_n)t}{\hbar}}$ is zero and we conclude that those elements do not evolve in time. Therefore, it is necessary to calculate those matrix elements only at time $t = 0$. To determine the average value of the position and momentum operators, we first rewrite $\hat{x}$ and $\hat{p}$ in terms of the creation and annihilation operators. Then, knowing that $\hat{a}$ and $\hat{a}^\dagger$ act as step down and step up operators on the eigenstates $|n\rangle$, we get

$$\begin{cases} \hat{x}|n\rangle = \sqrt{\frac{n\hbar}{2m\omega}}|n-1\rangle + \sqrt{\frac{(n+1)\hbar}{2m\omega}}|n+1\rangle \\ \hat{p}|n\rangle = -i\sqrt{\frac{nm\omega\hbar}{2}}|n-1\rangle + i\sqrt{\frac{(n+1)m\omega\hbar}{2}}|n+1\rangle. \end{cases}$$

Using the orthogonality of the eigenstates, we find

$$\langle n|\hat{x}|n\rangle = \langle n|\hat{p}|n\rangle = 0$$

and, for the squared position and momentum operators, the result is

$$\langle n|\hat{x}^2|n\rangle = \frac{\hbar}{2m\omega}\langle n|(\hat{a} + \hat{a}^\dagger)^2|n\rangle = \frac{\hbar}{2m\omega}\langle n|(\hat{a}\hat{a}^\dagger + \hat{a}^\dagger\hat{a})|n\rangle = \frac{\hbar}{m\omega}\left(n + \frac{1}{2}\right)$$
$$\langle n|\hat{p}^2|n\rangle = -\frac{m\omega\hbar}{2}\langle n|(-\hat{a} + \hat{a}^\dagger)^2|n\rangle = \frac{m\omega\hbar}{2}\langle n|(\hat{a}\hat{a}^\dagger + \hat{a}^\dagger\hat{a})|n\rangle = \hbar m\omega\left(n + \frac{1}{2}\right).$$

To verify the above results, we can compute the average energy

$$E = \langle\hat{H}\rangle = \frac{\langle\hat{p}^2\rangle}{2m} + \frac{m\omega^2}{2}\langle\hat{x}^2\rangle = \hbar\omega\left(n + \frac{1}{2}\right)$$

that is the expected result. Let us now answer the last question. At time $t = 0$ we have

$$\langle \alpha | \hat{x} | \alpha \rangle = (c_0 \langle 0| + c_1 \langle 1|) \hat{x} (c_0 |0\rangle + c_1 |1\rangle) = \sqrt{\frac{2\hbar}{m\omega}} c_0 c_1$$

$$\langle \alpha | \hat{p} | \alpha \rangle = (c_0 \langle 0| + c_1 \langle 1|) \hat{p} (c_0 |0\rangle + c_1 |1\rangle) = 0$$

$$\langle \alpha | \hat{x}^2 | \alpha \rangle = (c_0 \langle 0| + c_1 \langle 1|) \hat{x}^2 (c_0 |0\rangle + c_1 |1\rangle) = \frac{\hbar}{2m\omega} \left(|c_0|^2 + 3|c_1|^2 \right)$$

$$\langle \alpha | \hat{p}^2 | \alpha \rangle = (c_0 \langle 0| + c_1 \langle 1|) \hat{p}^2 (c_0 |0\rangle + c_1 |1\rangle) = \frac{m\hbar\omega}{2} \left(|c_0|^2 + 3|c_1|^2 \right).$$

To compute the average values on the state $|\alpha\rangle$ at time $t > 0$, we need to evolve the state with the action of the time evolution operator $e^{-i\hat{H}t/\hbar}$. Such operator is diagonal with respect to the eigenstates of $\hat{H}$ and, therefore, the eigenstates $|0\rangle$ and $|1\rangle$ acquire the phase factors $e^{-iE_0 t/\hbar}$ and $e^{-iE_1 t/\hbar}$ respectively, with $E_0 = \frac{\hbar\omega}{2}$ and $E_1 = \frac{3\hbar\omega}{2}$

$$|\alpha, t\rangle = c_0 e^{-i\frac{\omega t}{2}} |0\rangle + c_1 e^{-i\frac{3\omega t}{2}} |1\rangle.$$

The average values become

$$\langle \alpha, t | \hat{x} | \alpha, t \rangle = (c_0 e^{i\frac{\omega t}{2}} \langle 0| + c_1 e^{i\frac{3\omega t}{2}} \langle 1|) \hat{x} (c_0 e^{-i\frac{\omega t}{2}} |0\rangle + c_1 e^{-i\frac{3\omega t}{2}} |1\rangle) =$$
$$\sqrt{\frac{2\hbar}{m\omega}} c_0 c_1 \cos(\omega t)$$

$$\langle \alpha, t | \hat{p} | \alpha, t \rangle = (c_0 e^{i\frac{\omega t}{2}} \langle 0| + c_1 e^{i\frac{3\omega t}{2}} \langle 1|) \hat{p} (c_0 e^{-i\frac{\omega t}{2}} |0\rangle + c_1 e^{-i\frac{3\omega t}{2}} |1\rangle) =$$
$$- \sqrt{2m\hbar\omega} c_0 c_1 \sin(\omega t)$$

$$\langle \alpha, t | \hat{x}^2 | \alpha, t \rangle = (c_0 \langle 0| e^{i\frac{\omega t}{2}} + c_1 e^{i\frac{3\omega t}{2}} \langle 1|) \hat{x}^2 (c_0 e^{-i\frac{\omega t}{2}} |0\rangle + c_1 e^{-i\frac{3\omega t}{2}} |1\rangle) =$$
$$\frac{\hbar}{2m\omega} \left(|c_0|^2 + 3|c_1|^2 \right)$$

$$\langle \alpha, t | \hat{p}^2 | \alpha, t \rangle = (c_0 e^{i\frac{\omega t}{2}} \langle 0| + c_1 e^{i\frac{3\omega t}{2}} \langle 1|) \hat{p}^2 (c_0 e^{-i\frac{\omega t}{2}} |0\rangle + c_1 e^{-i\frac{3\omega t}{2}} |1\rangle) =$$
$$\frac{m\hbar\omega}{2} \left(|c_0|^2 + 3|c_1|^2 \right).$$

Problem 2.25.

Determine the energy spectrum for the bound states of a particle with mass m subject to the Morse potential (see Fig. 2.15) defined by $U(x) = M(e^{-2ax} - 2e^{-ax})$, $M, a > 0$. When $\frac{\sqrt{2mM}}{\hbar a} = S$, with S a positive integer, compute the number of bound states.

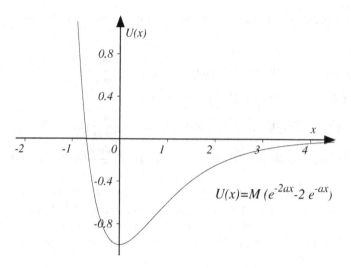

Fig. 2.15 We plot the Morse potential defined by $U(x) = M(e^{-2ax} - 2e^{-ax})$ with $M = a = 1$. The associated energy spectrum for the bound states is characterized in Problem 2.25

Solution

The stationary Schrödinger equation with the Morse potential is

$$\frac{\hbar^2}{2m}\frac{d^2\psi}{dx^2} + (E - M(e^{-2ax} - 2e^{-ax}))\psi = 0$$

where E is the energy. Since the derivative of an exponential function is still an exponential function, it is convenient to define $z = e^{-ax}$ ($z = 0$ is zero when $x \to +\infty$ and $z = +\infty$ when $x \to -\infty$). Therefore, we obtain

$$\frac{d^2\psi}{dx^2} = a^2 z^2 \frac{d^2\psi}{dz^2} + a^2 z \frac{d\psi}{dz}$$

and the equation

$$\frac{d^2\psi}{dz^2} + \frac{1}{z}\frac{d\psi}{dz} + \left(\frac{2\beta}{z} + \frac{\gamma}{z^2} - \beta\right)\psi = 0$$

where $\beta = 2mM/(\hbar^2 a^2)$, $\gamma = 2mE/(\hbar^2 a^2)$. The complete solution of this equation can be found in the class of the confluent hypergeometric functions (see also Problems 2.22, 4.1, 4.2, 4.5 for a discussion of the singularities associated with the Schrödinger equation). First of all, we need to examine the singularities of the equation: the point $z = 0$ is a regular singularity and the indicial exponent is found by plugging the approximate form $\psi(z) \approx z^\lambda$ in the original equation

$$\lambda(\lambda - 1)z^{\lambda-2} + \lambda z^{\lambda-2} + 2\beta z^{\lambda-1} + \gamma z^{\lambda-2} - \beta z^\lambda = 0.$$

When $z \to 0$, the terms proportional to $z^{\lambda-2}$ dominate and we find $\lambda = \pm\sqrt{-\gamma}$. We note that γ must be negative; otherwise the wave function would present oscillations and would be impossible to describe bound states: we therefore conclude that the discrete spectrum is given by the negative energies. Since γ is negative, the correct non divergent solution for $z \approx 0$ is given by $z^{+\sqrt{-\gamma}}$. We next separate out the behaviour close to $z = 0$ and seek the solution in the form $\psi(z) = z^\lambda f(z)$. Plugging this form in the Schrödinger equation, we find

$$f'' + \frac{(2\sqrt{-\gamma}+1)}{z} f' + \left(\frac{2\beta}{z} - \beta\right) f = 0$$

and we observe that the singular behaviour z^{-2} has been removed from the equation. With the substitution $z = 1/t$, we can study the behaviour close to $z = +\infty$ $(t = 0)$

$$f'' - \left(\frac{2\sqrt{-\gamma}-1}{t}\right) f' + \left(\frac{2\beta}{t^3} - \frac{\beta}{t^4}\right) f = 0$$

and we see that there is a singularity in $t = 0$. The terms t^{-3}, t^{-4} make the singularity irregular: with this form of the equation it is not possible to find an indicial exponent and a solution of the form $f(t) \approx t^a$ when $t \approx 0$. To find an indicial equation with a solution, we need to use the substitution $f(t) = e^{-\sqrt{\beta}/t}F(t)$, leading to

$$F'' + \left(\frac{2\sqrt{\beta}}{t^2} - \frac{2\sqrt{-\gamma}-1}{t}\right) F' - \frac{(2\sqrt{-\gamma}+1)\sqrt{\beta}-2\beta}{t^3} F = 0.$$

The effect of this substitution is to cancel out the terms proportional to t^{-4}: it is now possible to find a solution of the form $F(t) \approx t^\delta$ and the Fuchs theorem guarantees that the solution close to $t = 0$ behaves like $t^\delta u(t)$, with $u(t)$ some analytic function. Going back to the original equation for $f(z)$, with the substitution $f(z) = e^{-\sqrt{\beta}z}Y(z)$, we obtain the equation for $Y(z)$

$$Y'' + \left(\frac{2\sqrt{-\gamma}+1}{z} - 2\sqrt{\beta}\right) Y' + \frac{2\beta - 2\sqrt{\beta}(\sqrt{-\gamma}+\frac{1}{2})}{z} Y = 0.$$

With the definition $z = \xi/(2\sqrt{\beta})$, the equation becomes

$$\xi Y'' + (2\sqrt{-\gamma}+1-\xi)Y' - \left(\sqrt{-\gamma}+\frac{1}{2}-\sqrt{\beta}\right) Y = 0$$

that is the confluent hypergeometric equation with solution

$$Y(A,C|\xi) = Y\left(\frac{1}{2}+\sqrt{-\gamma}-\sqrt{\beta}, 2\sqrt{-\gamma}+1, \xi\right)$$

with

$$A = \frac{1}{2}+\sqrt{-\gamma}-\sqrt{\beta} \quad C = 2\sqrt{-\gamma}+1.$$

Going back to the original wave function, this has the form

$$\psi(z) = z^{\sqrt{-\gamma}} e^{-\sqrt{\beta}z} Y\left(\frac{1}{2} + \sqrt{-\gamma} - \sqrt{\beta}, 2\sqrt{-\gamma} + 1 | 2\sqrt{\beta}z\right).$$

To have $\psi(+\infty) = 0$, the dominant contribution in ψ when $z \to +\infty$ must be exponential, i.e. the hypergeometric series must reduce to a polynomial. An infinite number of terms would in fact spoil the exponential behaviour and the convergence of the wave function when $z \to +\infty$. The n-th term in the series of the confluent hypergeometric function is

$$\frac{A(A+1)\cdots(A+n-1)}{(n!C(C+1)\cdots(C+n-1))}.$$

If A is zero or a negative integer number ($A = -n, n = 0, 1, 2, 3, ...$) the series is truncated and reduces to a polynomial: this is the condition of quantization for the discrete spectrum. Expressing β and γ in terms of the original parameters, we find

$$E_n = -M\left(1 - \frac{\hbar a \left(n + \frac{1}{2}\right)}{\sqrt{2mM}}\right)^2.$$

We recall that $\sqrt{-\gamma}$ is positive

$$\sqrt{-\gamma} = \frac{\sqrt{-2mE_n}}{\hbar a} = \frac{\sqrt{2mM}}{\hbar a}\left(1 - \frac{\hbar a \left(n + \frac{1}{2}\right)}{\sqrt{2mM}}\right) = \frac{\sqrt{2mM}}{\hbar a} - \left(n + \frac{1}{2}\right) \geq 0$$

and, therefore, $n + \frac{1}{2} \leq \frac{\sqrt{2mM}}{a\hbar}$, i.e. the number of energy levels is finite. When $\frac{\sqrt{2mM}}{\hbar a} = S$, with S a positive integer, we find $S + 1$ bound states.

Problem 2.26.
Discuss the existence of bound states for a particle with mass m subject to the one dimensional Dirac delta function potential $U(x) = -\delta(x)$.

Solution
Our potential is a Dirac delta function for which

$$\int_{-\varepsilon}^{+\varepsilon} \delta(x)\, dx = 1$$

for $\varepsilon > 0$. The associated Schrödinger equation is given by

$$\frac{\hbar^2}{2m}\frac{d^2\psi(x)}{dx^2} + (E + \delta(x))\psi(x) = 0.$$

The wave function has to be continuous, a condition that is necessary to interpret its square modulus as a probability density function. As for the properties of its derivative, care has to be taken. Let us concentrate on the infinitesimal interval $[-\varepsilon, \varepsilon]$, with ε an arbitrarily small parameter. We first integrate the Schrödinger equation between $-\varepsilon$ and ε, and then send ε to zero

$$\lim_{\varepsilon \to 0} \left(\frac{\hbar^2}{2m} \int_{-\varepsilon}^{+\varepsilon} \left(\psi''(x) + (E + \delta(x)) \psi(x) \right) dx \right) = \frac{\hbar^2}{2m} (\psi'(0_+) - \psi'(0_-)) + \psi(0) = 0$$

where the term multiplying E is zero due to the continuity of ψ in zero. Such equation reveals the singularity of the logarithmic derivative of ψ in the origin

$$\frac{\psi'(0_+)}{\psi(0)} - \frac{\psi'(0_-)}{\psi(0)} = -\frac{2m}{\hbar^2}.$$

To find bound states, a negative energy is required, otherwise we would have a continuous spectrum given by plane waves. Since the $\delta(x)$ is centered in zero, for $x \neq 0$ we find the Schrödinger equation for a free particle, whose solution is $\psi(x) \propto e^{\pm \sqrt{-2mEx}/\hbar}$. To ensure a vanishing wave function at infinity, we have to select

$$\psi(x) = A e^{-\sqrt{-2mEx}/\hbar}$$

for $x > 0$ (A is a normalization constant) and

$$\psi(x) = A e^{\sqrt{-2mEx}/\hbar}$$

for $x < 0$. Using the discontinuity condition previously found, we get

$$\frac{\psi'(0_+)}{\psi(0)} - \frac{\psi'(0_-)}{\psi(0)} = \frac{2\sqrt{-2mE}}{\hbar} = \frac{2m}{\hbar^2}.$$

This condition is enough to determine the only value of the energy for which we find a bound state, i.e. $E = -m/2\hbar^2$. For this value of the energy, the normalization condition (see also Problem 2.29) yields

$$A = \sqrt{\frac{m}{\hbar^2}}.$$

Problem 2.27.
Consider the one dimensional quantum harmonic oscillator with Hamiltonian $\hat{H}_2 = \hat{T} + \hat{V}_2$, where $\hat{T}$ is the kinetic energy ($\hat{T} = \frac{\hat{p}^2}{2m}$) and $\hat{V}_2$ the potential energy ($\hat{V}_2 = \frac{\hbar^2}{2mx_0^4} \hat{x}^2$, $x_0 = \sqrt{\frac{\hbar}{m\omega}}$). Then, consider the Hamiltonian $\hat{H}_4 = \hat{T} + \hat{V}_4$, where $\hat{V}_4 = \frac{\hbar^2}{6mx_0^6} \hat{x}^4$. Using the variational method with the Gaussian trial function

$$\phi(x) = A e^{-bx^2}$$

determine the best estimate for the wave functions $\phi_2^{(0)}, \phi_4^{(0)}$ and the corresponding energies $E_2^{(0)}, E_4^{(0)}$ for the ground states of the two Hamiltonians.

Solution

According to the variational principle, the ground state of the Schrödinger equation corresponds to the condition that $E = \langle \phi | \hat{H} | \phi \rangle$ is minimized with respect to the variations of the wave function ϕ. Besides the theoretical importance, the variational principle is also very useful on practical grounds, because it allows to construct approximate wave functions in many body problems that could not be dealt with other elementary methods. One usually chooses a set of trial functions $\phi(x, \{\lambda_1, \lambda_2, ..., \lambda_n\})$ dependent on a given set of parameters $\{\lambda_1, \lambda_2, ..., \lambda_n\}$, and minimizes the energy as a function of the parameters. If the exact state belongs to the class of functions considered, we get the exact solution. If, on the contrary, the exact solution does not belong to such a class, the minimum always overestimates the ground state energy. Obviously, the larger is the set of parameters and the better is the estimate of the state. When the trial function ϕ is not normalized, the normalization condition can also be imposed with a Lagrange multiplier. In our case, the normalization condition, $\int_{-\infty}^{+\infty} |\phi(x)|^2 \, dx = 1$, is particularly simple because we know the Gaussian integral $\int_{-\infty}^{+\infty} e^{-x^2} \, dx = \sqrt{\pi}$. Consequently, the value of A is $A = \left(\frac{2b}{\pi}\right)^{1/4}$.

In the harmonic case (i.e. when we treat $\hat{H}_2$), the trial function is exactly of the same class of functions (it is a Gaussian) describing the ground state of the harmonic oscillator (see Problems 2.20 and 2.32) with $b = \frac{1}{2x_0^2}$ and $E_2^{(0)} = \frac{\hbar\omega}{2}, \omega = \frac{\hbar}{mx_0^2}$.

In the anharmonic case (i.e. when we treat $\hat{H}_4$), we can use the variational method with this trial function to approximate the ground state. We first need to compute the energy $E(b) = \langle \hat{T} + \hat{V}_4 \rangle$. For the kinetic energy $\langle \hat{T} \rangle$, we use the second order moment of a Gaussian, $\int_{-\infty}^{+\infty} x^2 e^{-x^2} \, dx = \frac{\sqrt{\pi}}{2}$, to get

$$\langle \hat{T} \rangle = -\frac{\hbar^2}{2m} A^2 \int_{-\infty}^{+\infty} e^{-bx^2} \frac{d^2 e^{-bx^2}}{dx^2} \, dx = -\frac{\hbar^2}{2m} A^2 \int_{-\infty}^{+\infty} e^{-bx^2} (-2b + 4b^2 x^2) e^{-bx^2} \, dx = \frac{\hbar^2 b}{2m}.$$

Moreover, for the potential energy, we find

$$\langle \hat{V}_4 \rangle = \frac{\hbar^2 A^2}{6mx_0^6} \int_{-\infty}^{+\infty} e^{-2bx^2} x^4 \, dx = \frac{\hbar^2}{32mx_0^6 b^2}$$

where we have used $\int_{-\infty}^{+\infty} x^4 e^{-x^2} \, dx = \frac{3\sqrt{\pi}}{4}$. We now impose the condition $\frac{dE(b)}{db} = 0$ and find

$$1 - \frac{2}{b^3} \frac{1}{16x_0^6} = 0$$

from which we get $b = \frac{1}{2x_0^2}$ and $E = \frac{3}{8} \frac{\hbar^2}{mx_0^2}$. We see that this energy is lower than the energy of the ground state for the harmonic oscillator, i.e. $E_4^{(0)} = \frac{3}{8}\hbar\omega < \frac{1}{2}\hbar\omega$. This is due to the fact that the anharmonic potential is smaller than the harmonic one for small x. We finally remark that the effect of the anharmonic terms can also be studied using a perturbative approach. As for this point, see the Problems 7.22 and 7.25 in the section of Statistical Mechanics.

Problem 2.28.
Let us consider a particle with mass m subject to the harmonic potential $U(x) = \frac{1}{2}mx^2$. If the particle is in the trial state $\phi_0(x) = Ae^{-Bx^2}$, with A a normalization constant and B a free variational parameter, determine:

- the average value of the energy as a function of B;
- the value of B minimizing the energy.

If at time $t = 0$ the particle is in the state $\psi = a\phi_0 + b\phi_1$, with $\phi_0(x)$ given above and $\phi_1(x)$ the eigenstate of the harmonic oscillator with eigenvalue $\frac{3}{2}\hbar$, determine:

- the probability that at time $t = 0$ a measurement of the energy gives $E = \hbar/2$;
- the probability that at time $t > 0$ the particle is in the interval $[-\varepsilon, \varepsilon]$, $\varepsilon > 0$.

Solution
We first need to determine A in terms of B using the normalization condition

$$1 = \int_{-\infty}^{+\infty} |\phi_0(x)|^2 dx = A^2 \int_{-\infty}^{+\infty} e^{-2Bx^2} dx = A^2 \sqrt{\frac{\pi}{2B}}$$

that implies $A = (2B/\pi)^{1/4}$ and, hence, $\phi_0(x) = (2B/\pi)^{1/4}e^{-Bx^2}$. In the above expression, we have used the Gaussian integral

$$I(\beta) = \int_{-\infty}^{+\infty} e^{-2\beta Bx^2} dx = \sqrt{\frac{\pi}{2B\beta}}$$

with $\beta = 1$. Using the property

$$\int_{-\infty}^{+\infty} x^2 e^{-2\beta Bx^2} dx = -\frac{1}{2B}\lim_{\beta \to 1}\frac{dI(\beta)}{d\beta} = \frac{1}{4B}\sqrt{\frac{\pi}{2B}}$$

we find the average value of the energy

$$E(B) = \langle \hat{H} \rangle = \sqrt{\frac{2B}{\pi}} \int_{-\infty}^{+\infty} e^{-Bx^2}\left(-\frac{\hbar^2}{2m}\frac{d^2}{dx^2} + \frac{1}{2}mx^2\right)e^{-Bx^2} dx =$$

$$\sqrt{\frac{2B}{\pi}}\left[-\frac{\hbar^2}{2m}\int_{-\infty}^{+\infty} e^{-Bx^2}\frac{d^2 e^{-Bx^2}}{dx^2} dx + \frac{1}{2}m\left(-\frac{1}{2B}\right)\lim_{\beta \to 1}\frac{dI(\beta)}{d\beta}\right] =$$

$$\sqrt{\frac{2B}{\pi}}\left[-\frac{\hbar^2}{2m}\int_{-\infty}^{+\infty} e^{-2Bx^2}(4B^2x^2 - 2B) dx + \frac{1}{2}m\left(-\frac{1}{2B}\right)\lim_{\beta \to 1}\frac{d}{d\beta}\sqrt{\frac{\pi}{2B\beta}}\right] =$$

$$\sqrt{\frac{2B}{\pi}}\left[-\frac{\hbar^2}{2m}\frac{4B^2}{(-2B)}\lim_{\beta \to 1}\frac{dI(\beta)}{d\beta} + \frac{\hbar^2 B}{m}\sqrt{\frac{\pi}{2B}} + \frac{m}{8B}\sqrt{\frac{\pi}{2B}}\right] =$$

$$-\frac{\hbar^2 B}{2m} + \frac{\hbar^2 B}{m} + \frac{m}{8B} = \frac{\hbar^2 B}{2m} + \frac{m}{8B}.$$

We take the derivative and set it to zero

$$\frac{dE(B)}{dB} = \frac{\hbar}{m} - \frac{m}{4B^2} = 0$$

to find the value of B and the explicit form of the normalized wave function

$$B = \frac{m}{2\hbar} \quad E = \frac{1}{2}\hbar \quad \phi_0(x) = \left(\frac{m}{\pi\hbar}\right)^{\frac{1}{4}} e^{-\frac{mx^2}{2\hbar}}.$$

We note that $\phi_0(x)$ is the wave function of the ground state of the harmonic oscillator. As given in the text, also $\phi_1(x)$ is an eigenstate of the harmonic oscillator with energy $E_1 = 3/2\hbar$, i.e. the first excited state. The probability that a measurement of the energy gives $E = \hbar/2$ is $P(E = \hbar/2) = |a|^2$. The time evolution of ψ is given by the action of $e^{-i\hat{H}t/\hbar}$ on the wave function at time $t = 0$

$$\langle x|\psi(t)\rangle = \langle x|e^{-i\hat{H}t/\hbar}|\psi(0)\rangle = ae^{-\frac{it}{2\hbar}}\phi_0(x) + be^{-\frac{i3t}{2\hbar}}\phi_1(x).$$

The probability to find the particle in the interval $[-\varepsilon, \varepsilon]$ is the integral of $|\psi(x,t)|^2$ between $-\varepsilon$ and ε

$$P = \int_{-\varepsilon}^{\varepsilon} |\psi(x,t)|^2\, dx = |a|^2 \int_{-\varepsilon}^{\varepsilon} |\phi_0(x)|^2\, dx + |b|^2 \int_{-\varepsilon}^{\varepsilon} |\phi_1(x)|^2\, dx$$

which is not dependent on time t, due to the fact that $\int_{-\varepsilon}^{\varepsilon} \phi_0(x)\phi_1(x)dx = 0$ because $\phi_0(x)\phi_1(x)$ is an odd function.

Problem 2.29.

Consider the following Hamiltonian for a particle with mass m in one dimension

$$\hat{H} = -\frac{\hbar^2}{2m}\frac{d^2}{dx^2} - \delta(\hat{x}) + b|\hat{x}| \quad b \geq 0.$$

For the special case $b = 0$, find the energy of the bound state with the variational method using a trial function of the form $\psi(x) = Ne^{-\lambda|x|}$, with λ a variational parameter and N the normalization factor (to be determined). Then, suppose that $b > 0$, and find the variational condition with the same form of trial functions previously used. Solve the case with small b. Is the energy increasing or decreasing? Try to give some qualitative explanation.

Solution

The case with $b = 0$ is a one dimensional problem with an attractive potential energy. Such problems always have at least one bound state in the discrete spectrum, whose squared wave function can be integrated.

First of all, we need to determine the normalization constant N. To do that, we use the integral $\int_0^{+\infty} e^{-2\lambda x}\, dx = \frac{1}{2\lambda}$ in the normalization condition

$$\int_{-\infty}^{+\infty} |\psi(x)|^2\, dx = 2N^2 \int_0^{+\infty} e^{-2\lambda x}\, dx = \frac{N^2}{\lambda} = 1.$$

We therefore have the normalized trial function

$$\psi(x) = \sqrt{\lambda}\, e^{-\lambda |x|}.$$

We have already discussed in Problem 2.26 the properties of the one dimensional Schrödinger equation in presence of a Dirac delta function potential $-\delta(x)$, and we have seen that the derivative of the wave function has a discontinuity in the origin. We now want to analyze the same physical picture from the point of view of the variational method (see also Problem 2.27 for a discussion on the method). We need to compute the average energy on the normalized trial functions, and find its minimum to determine the optimal value of λ. We start by calculating the average kinetic energy on our trial functions. To do this, we note that the function $\psi(x)$ has a discontinuity with the property

$$\frac{\psi'(0_+)}{\psi(0)} - \frac{\psi'(0_-)}{\psi(0)} = -2\lambda.$$

This means that we have to take care in evaluating the second derivative of ψ close to the origin. When averaging on the state ψ, we can divide the domain of integration from $-\infty$ to $+\infty$ in the three regions $[-\infty, -\varepsilon[, [-\varepsilon, +\varepsilon],] + \varepsilon, +\infty]$. We get

$$\langle \psi | \hat{T} | \psi \rangle = -\frac{\hbar^2}{2m} \int_{-\infty}^{+\infty} \psi(x) \psi''(x)\, dx =$$

$$-\frac{\hbar^2}{2m} \int_{-\varepsilon}^{+\varepsilon} \psi(x) \psi''(x)\, dx - \frac{\hbar^2}{m} \int_{+\varepsilon}^{+\infty} \psi(x) \psi''(x)\, dx$$

where we have used the symmetry properties of $\psi(x)$ to write

$$\int_{+\varepsilon}^{+\infty} \psi(x) \psi''(x)\, dx + \int_{-\infty}^{-\varepsilon} \psi(x) \psi''(x)\, dx = 2 \int_{+\varepsilon}^{+\infty} \psi(x) \psi''(x)\, dx.$$

In the limit $\varepsilon \to 0$, we find

$$-\frac{\hbar^2}{m} \int_{+\varepsilon}^{+\infty} \psi(x) \psi''(x)\, dx = -\frac{\hbar^2}{m} \int_{+\varepsilon}^{+\infty} \psi(x) \lambda^2 \psi(x)\, dx \approx -\frac{\lambda^2 \hbar^2}{2m}.$$

Close to the origin, if we want to take into account the above mentioned singularity of $\psi(x)$, we can use

$$\frac{\psi'(x)}{\psi(x)} = -\lambda\, [\theta(x) - \theta(-x)]$$

with $\theta(x)$ the Heaviside function. Therefore, in the limit $\varepsilon \to 0$, we find

$$-\frac{\hbar^2}{2m} \int_{-\varepsilon}^{+\varepsilon} \psi(x) \psi''(x)\, dx \approx \frac{\hbar^2}{m} \lambda \int_{-\varepsilon}^{+\varepsilon} |\psi(x)|^2 \delta(x)\, dx = \frac{\hbar^2}{m} \lambda^2.$$

The average kinetic energy is

$$\langle \psi | \hat{T} | \psi \rangle = -\frac{\lambda^2 \hbar^2}{2m} + \frac{\lambda^2 \hbar^2}{m} = \frac{\lambda^2 \hbar^2}{2m}.$$

We now need to determine the average potential energy

$$\langle \psi | \hat{V} | \psi \rangle = -\int_{-\infty}^{+\infty} |\psi(x)|^2 \delta(x)\, dx = -\psi^2(0) = -\lambda$$

so that the total average energy is

$$E(\lambda) = \langle \psi | \hat{T} | \psi \rangle + \langle \psi | \hat{V} | \psi \rangle = \frac{\hbar^2 \lambda^2}{2m} - \lambda.$$

The variational condition is then imposed

$$\frac{dE(\lambda)}{d\lambda} = \frac{\hbar^2 \lambda}{m} - 1 = 0$$

leading to $\lambda = \frac{m}{\hbar^2}$ and $E = -\frac{m}{2\hbar^2}$. With such a choice for λ we get

$$\psi(x) = \sqrt{\frac{m}{\hbar^2}} e^{-\frac{m}{\hbar^2}|x|}$$

that is the exact solution already studied in Problem 2.26. Let us now switch on the term $b|\hat{x}|$ in the potential energy and compute its average on the state $\psi(x)$

$$b\langle |\hat{x}| \rangle = 2b\lambda \int_0^{+\infty} x e^{-2\lambda x}\, dx = -bx e^{-2\lambda x}\Big|_0^{+\infty} + b \int_0^{+\infty} e^{-2\lambda x}\, dx = \frac{b}{2\lambda}.$$

The new average energy has the form

$$E(\lambda) = \frac{\hbar^2 \lambda^2}{2m} - \lambda + \frac{b}{2\lambda}$$

and the resulting variational condition is

$$\frac{dE(\lambda)}{d\lambda} = \frac{\hbar^2 \lambda}{m} - 1 - \frac{b}{2\lambda^2} = 0.$$

To find λ, we need to solve a cubic equation. The solution is simpler when b is small (it is the case of our problem), because we can use an iterative method to find such solution. A zeroth order approximation delivers the same result as before, i.e. $\lambda = \lambda_0 = \frac{m}{\hbar^2}$. As a first order approximation, we can determine a λ_1 such that

$$\frac{\hbar^2 \lambda_1}{m} - 1 - \frac{b}{2\lambda_0^2} = 0.$$

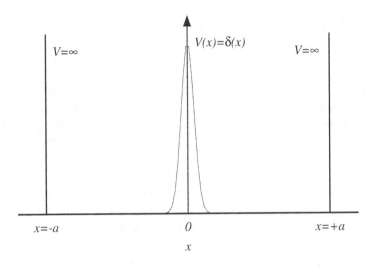

Fig. 2.16 A one dimensional potential well with a Dirac delta function $\delta(x)$ in the middle. In Problem 2.30 we characterize the eigenstates and eigenvalues for a particle with mass m subject to this potential

This produces $\lambda_1 = \frac{m}{\hbar^2} + b\frac{\hbar^2}{2m}$ to be used in the expression of $E(\lambda)$. The result is

$$E \approx -\frac{m}{2\hbar^2} + b\frac{\hbar^2}{2m} + \dots$$

and the energy increases due to the terms proportional to b, that is an expected result because the term $b|\hat{x}|$ is a repulsive potential.

Problem 2.30.
Determine the eigenstates and eigenvalues for a particle with mass m subject to the potential

$$\begin{cases} V(x) = +\infty & x \le -a, x \ge a \\ V(x) = \delta(x) & -a < x < a \end{cases}$$

reported in Fig. 2.16.

Solution
The Schrödinger equation for $-a < x < a$ is

$$\frac{\hbar^2}{2m}\frac{d^2\psi}{dx^2} + (E - \delta(x))\psi = 0.$$

The boundary conditions yield $\psi(\pm a) = 0$, plus the condition of continuity in $x = 0$. The first derivative is not continuous in $x = 0$ (see also Problem 2.26). We also note that the operator

$$\frac{\hbar^2}{2m}\frac{d^2}{dx^2} + (E - \delta(x))$$

is invariant under the spatial inversion $x \to -x$ (we recall that $\delta(x) = \delta(-x)$). This means that if $\psi(x)$ is a solution to our problem, also $\psi(-x)$ is automatically a solution. Moreover, we are dealing with a one dimensional problem, and there cannot be two independent solutions with the same eigenvalue. To prove this statement we suppose the contrary to be true, and consider ψ_1, ψ_2 as two independent solutions with the same potential V and the same eigenvalue E

$$\psi_{1,2}'' = \frac{2m}{\hbar^2}(V - E)\psi_{1,2}.$$

If we divide the two Schrödinger equations by ψ_1, ψ_2, we get

$$\frac{\psi_1''}{\psi_1} = \frac{2m}{\hbar^2}(V - E) = \frac{\psi_2''}{\psi_2}$$

that implies $\psi_1''\psi_2 - \psi_2''\psi_1 = 0$. One integration leads to

$$\psi_1'\psi_2 - \psi_2'\psi_1 = \text{const.}$$

where the constant of integration is zero due to the boundary condition $\psi_{1,2}(\pm\infty) = 0$. This happens because $\psi_{1,2}$ is an eigenfunction of the discrete spectrum and must go to zero at $\pm\infty$ to be normalized. Therefore, we obtain

$$\psi_1'\psi_2 = \psi_2'\psi_1$$

and another integration gives $\psi_1 = c\psi_2$, where c is an integration constant: we see that the two functions ψ_1, ψ_2 are linearly dependent and this violates the previous assumption of independence.

Let us then go back to our problem considering the two solutions $\psi(x)$ and $\psi(-x)$. Applying the previous argument, we find $\psi(x) = c\psi(-x)$ and, applying another spatial inversion, we get $\psi(x) = c^2\psi(x)$. This means that $c = \pm 1$ and all the solutions of the Schrödinger equation must be either even or odd. In general, this is true for a generic one dimensional symmetrical potential $V(x) = V(-x)$. When we integrate the Schrödinger equation through the discontinuity (see also Problem 2.26), we get

$$\lim_{\varepsilon \to 0}\left(\frac{\hbar^2}{2m}\int_{-\varepsilon}^{+\varepsilon}\left(\psi''(x) + (E - \delta(x))\psi(x)\right)dx\right) = \frac{\hbar^2}{2m}(\psi'(0_+) - \psi'(0_-)) - \psi(0) = 0$$

from which

$$\frac{\psi'(0_+)}{\psi(0)} - \frac{\psi'(0_-)}{\psi(0)} = \frac{2m}{\hbar^2}.$$

We note that the odd functions have $\psi(0) = 0$ and do not present a discontinuity in the derivative. For these functions, the Dirac delta function in $x = 0$ does not exist, and they possess the energy spectrum of a potential well (see Problem 2.10). For the even functions things change. When $x \neq 0$, the solution of the Schrödinger equation takes the form

$$\psi(x) = A \sin\left[\sqrt{\frac{2mE}{\hbar^2}}(x+\gamma)\right]$$

where A, γ are constants set by the boundary conditions. Imposing $\psi(\pm a) = 0$, we find $\gamma = \pm a$ and

$$\begin{cases} \psi(x) = A_{x<0} \sin\left[\sqrt{\frac{2mE}{\hbar^2}}(x+a)\right] & -a \leq x < 0 \\ \psi(x) = A_{x>0} \sin\left[\sqrt{\frac{2mE}{\hbar^2}}(x-a)\right] & 0 < x \leq a. \end{cases}$$

The condition $\psi_{x>0}(x) = \psi_{x<0}(-x)$ imposes

$$A_{x<0} \sin\left[\sqrt{\frac{2mE}{\hbar^2}}(-x+a)\right] = A_{x>0} \sin\left[\sqrt{\frac{2mE}{\hbar^2}}(x-a)\right]$$

that means $-A_{x<0} = A_{x>0} = \tilde{A}$. The solutions are

$$\begin{cases} \psi(x) = -\tilde{A} \sin\left[\sqrt{\frac{2mE}{\hbar^2}}(x+a)\right] & -a \leq x < 0 \\ \psi(x) = +\tilde{A} \sin\left[\sqrt{\frac{2mE}{\hbar^2}}(x-a)\right] & 0 < x \leq a. \end{cases}$$

The condition of discontinuity of the first derivative in $x = 0$ is

$$\frac{\psi'(0_+)}{\psi(0)} - \frac{\psi'(0_-)}{\psi(0)} = -2\sqrt{\frac{2mE}{\hbar^2}} \cot\left(\sqrt{\frac{2mE}{\hbar^2}}a\right) = \frac{2m}{\hbar^2}.$$

The energy spectrum E is given by the solution of this equation. Setting $A^2 = 2ma^2/\hbar^2$, $x = \sqrt{E}$ (not to be confused with the position in the beginning of the problem), the solutions can be found by looking at the intersection of the two curves

$$\begin{cases} y = x\cot(Ax) \\ y = -\frac{A}{2a} \end{cases}$$

reported in Fig. 2.17.

Problem 2.31.
Determine the transmission and reflection coefficients for a particle with mass m in a potential barrier given by a one dimensional Dirac delta function $V(x) = V_0 \delta(x)$, $V_0 > 0$.

Solution
We need to determine the solution of the Schrödinger equation in the two regions $x \leq$

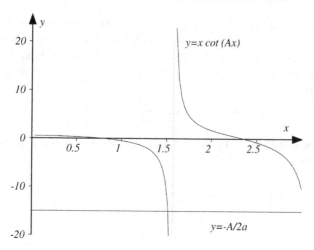

Fig. 2.17 We plot the two curves $y = x\cot(Ax)$ and $y = -\frac{A}{2a}$ with $A = 2$ and $a = \frac{1}{15}$. The intersection is important to characterize the eigenstates and eigenvalues for a particle with mass m subject to the potential shown in Fig. 2.16

0 (region *I*) and $x > 0$ (region *II*). The resulting wave function must be continuous in $x = 0$ while its first derivative must present a discontinuity (see Problem 2.26). The wave functions are

$$\begin{cases} \psi_I(x) = e^{ikx} + C_R e^{-ikx} & x \le 0 \\ \psi_{II}(x) = C_T e^{ikx} & x > 0 \end{cases}$$

with $k = \frac{\sqrt{2mE}}{\hbar}$. The condition of continuity yields $\psi_I(0) = \psi_{II}(0)$ and this is equivalent to

$$1 + C_R = C_T.$$

As for the discontinuity of the first derivative in $x = 0$, we get

$$\frac{\psi'(0_+)}{\psi(0)} - \frac{\psi'(0_-)}{\psi(0)} = \frac{ikC_T}{(1+C_R)} - \frac{ik(1-C_R)}{(1+C_R)} = \frac{2mV_0}{\hbar^2}.$$

Using these two conditions (continuity of the function and discontinuity of the derivative), we get

$$C_R = \frac{mV_0}{\hbar^2} \frac{1}{\left(ik - \frac{mV_0}{\hbar^2}\right)} \qquad C_T = 1 + C_R = \frac{ik}{\left(ik - \frac{mV_0}{\hbar^2}\right)}.$$

The resulting density fluxes are proportional to

$$J_I \propto k \qquad J_R \propto k|C_R|^2 \qquad J_T \propto k|C_T|^2.$$

The transmission and reflection coefficients are

$$R = |C_R|^2 = \frac{1}{1 + \frac{2\hbar^2 E}{mV_0^2}} \qquad T = |C_T|^2 = \frac{1}{1 + \frac{mV_0^2}{2\hbar^2 E}}.$$

With these expressions is immediate to verify that $T + R = 1$.

Problem 2.32.

Characterize the uncertainty relations for two generic self-adjoint operators ($\hat{K}$ and $\hat{F}$) with commutation rule $[\hat{K}, \hat{F}] = i\hat{M}$ (also $\hat{M}$ is a self-adjoint operator), and determine the most general form of the wave packet minimizing such relation when $\hat{K} = \hat{x}, \hat{F} = \hat{p}, \hat{M} = \hbar 1$. Using such wave packet, determine the averages $\langle \hat{x} \rangle$, $\langle \hat{p} \rangle$, $\langle \hat{x}^2 \rangle$ and $\langle \hat{p}^2 \rangle$. Finally, determine the time evolution of the wave packet with the Hamiltonian of the harmonic oscillator and verify that the average values of the position and momentum operators satisfy the classical equations of motion.

Solution
Let us start by defining the average of $\hat{K}$ and $\hat{F}$ on a generic state $\psi(x)$

$$\langle \hat{K} \rangle = \int \psi^*(x) K \psi(x) \, dx \qquad \langle \hat{F} \rangle = \int \psi^*(x) F \psi(x) \, dx$$

and introducing the operators $\Delta \hat{K}$ and $\Delta \hat{F}$

$$\Delta \hat{K} = \hat{K} - \langle \hat{K} \rangle \, 1 \qquad \Delta \hat{F} = \hat{F} - \langle \hat{F} \rangle \, 1.$$

These new operators satisfy the commutation rule

$$[\Delta \hat{K}, \Delta \hat{F}] = i\hat{M}.$$

We next introduce the integral $I(\alpha)$, with α a generic real parameter

$$I(\alpha) = \int |(\alpha \Delta K - i \Delta F) \psi(x)|^2 \, dx \geq 0.$$

The above inequality is surely true, since $I(\alpha)$ is defined as the integral of a squared function. If we use the fact that $\hat{K}$ and $\hat{F}$ are self-adjoint operators, we can write $I(\alpha)$ as

$$I(\alpha) = \int \psi^*(x)(\alpha \Delta K + i \Delta F)(\alpha \Delta K - i \Delta F) \psi(x) \, dx =$$
$$\int \psi^*(x)(\alpha^2 (\Delta K)^2 + \alpha M + (\Delta F)^2) \psi(x) \, dx =$$
$$\left(\alpha \sqrt{\langle (\Delta \hat{K})^2 \rangle} + \frac{\langle \hat{M} \rangle}{2\sqrt{\langle (\Delta \hat{K})^2 \rangle}} \right)^2 - \frac{\langle \hat{M} \rangle^2}{4\langle (\Delta \hat{K})^2 \rangle} + \langle (\Delta \hat{F})^2 \rangle =$$
$$\langle (\Delta \hat{K})^2 \rangle \left(\alpha + \frac{\langle \hat{M} \rangle}{2\langle (\Delta \hat{K})^2 \rangle} \right)^2 + \langle (\Delta \hat{F})^2 \rangle - \frac{\langle \hat{M} \rangle^2}{4\langle (\Delta \hat{K})^2 \rangle} \geq 0.$$

The first term of the inequality is positive because it is the square of a function. The last two terms contribute with a positive term when

$$\langle (\Delta \hat{F})^2 \rangle \langle (\Delta \hat{K})^2 \rangle \geq \frac{1}{4} \langle \hat{M} \rangle^2$$

that is the desired form of the uncertainty relation. If we set

$$\alpha = \bar{\alpha} = -\frac{\langle \hat{M} \rangle}{2 \langle (\Delta \hat{K})^2 \rangle}$$

the integral I becomes

$$I(\bar{\alpha}) = \int \left| \left(\frac{\langle \hat{M} \rangle \Delta K}{2 \langle (\Delta \hat{K})^2 \rangle} + i \Delta F \right) \psi(x) \right|^2 dx = \langle (\Delta \hat{F})^2 \rangle - \frac{\langle \hat{M} \rangle^2}{4 \langle (\Delta \hat{K})^2 \rangle} \geq 0$$

that is another way to produce the uncertainty relation. The condition

$$\langle (\Delta \hat{F})^2 \rangle \langle (\Delta \hat{K})^2 \rangle = \frac{1}{4} \langle \hat{M} \rangle^2$$

is obtained when $I(\bar{\alpha}) = 0$. This means that the wave packet minimizing the uncertainty relation has to satisfy

$$\left(\frac{\langle \hat{M} \rangle \Delta \hat{K}}{2 \langle (\Delta \hat{K})^2 \rangle} + i \Delta \hat{F} \right) |\psi\rangle = 0.$$

When we set

$$\Delta \hat{K} = \Delta \hat{x} = \hat{x} - \bar{x}\,1 \qquad \Delta \hat{F} = \Delta \hat{p} = \hat{p} - \bar{p}\,1 = -i\hbar \frac{d}{dx} - \bar{p}\,1 \qquad \hat{M} = \hbar\,1$$

with $\bar{x} = \langle \hat{x} \rangle$ and $\bar{p} = \langle \hat{p} \rangle$, we obtain the differential equation

$$\left(\frac{x - \bar{x}}{x_0^2} + \frac{d}{dx} - i\frac{\bar{p}}{\hbar} \right) \psi(x) = 0$$

with $x_0^2 = 2\langle (\Delta \hat{K})^2 \rangle = 2\langle (\Delta \hat{x})^2 \rangle$, whose solution is the following normalized wave packet

$$\psi(x) = \frac{1}{\sqrt{\sqrt{\pi} x_0}} e^{-\frac{(x - \bar{x})^2}{2 x_0^2} + i\frac{\bar{p}x}{\hbar}}.$$

We can determine the average values $\langle \hat{x} \rangle, \langle \hat{p} \rangle, \langle \hat{x}^2 \rangle, \langle \hat{p}^2 \rangle$ using some properties of the Gaussian integrals

$$\langle \hat{x} \rangle = \int_{-\infty}^{+\infty} \psi^*(x) x \psi(x)\, dx = \frac{1}{\sqrt{\pi}} \int_{-\infty}^{+\infty} e^{-\left(\frac{x}{x_0} - \frac{\bar{x}}{x_0} \right)^2} \left(\frac{x}{x_0} \right) x_0 d\left(\frac{x}{x_0} \right) = \bar{x}$$

$$\langle \hat{p} \rangle = \int_{-\infty}^{+\infty} \psi^*(x) \left(-i\hbar \frac{d}{dx} \right) \psi(x)\, dx = \bar{p} \int_{-\infty}^{+\infty} \psi^*(x) \psi(x)\, dx + i\hbar \left\langle \frac{\hat{x} - \bar{x}\,1}{x_0^2} \right\rangle = \bar{p}$$

$$\langle \hat{x}^2 \rangle = \langle (\hat{x} \pm \bar{x}\, 1)^2 \rangle = \langle (\hat{x} - \bar{x}\, 1)^2 \rangle + \langle \bar{x}^2\, 1 \rangle + 2\bar{x}\langle (\hat{x} - \bar{x}\, 1) \rangle = \frac{x_0^2}{2} + \bar{x}^2$$

$$\langle \hat{p}^2 \rangle = -\hbar^2 \int_{-\infty}^{+\infty} \psi^*(x) \frac{d^2 \psi(x)}{dx^2}\, dx =$$

$$-\frac{\hbar^2}{\sqrt{\pi}} \int_{-\infty}^{+\infty} \left[-\frac{1}{x_0^2} + \left(\frac{i\bar{p}}{\hbar} - \left(\frac{x}{x_0^2} - \frac{\bar{x}}{x_0^2} \right) \right)^2 \right] e^{-\left(\frac{x}{x_0} - \frac{\bar{x}}{x_0} \right)^2} \frac{dx}{x_0} =$$

$$\frac{\hbar^2}{x_0^2} + \bar{p}^2 - \frac{\hbar^2}{\sqrt{\pi}x_0^2} \int_{-\infty}^{+\infty} \left(\frac{x}{x_0} - \frac{\bar{x}}{x_0} \right)^2 e^{-\left(\frac{x}{x_0} - \frac{\bar{x}}{x_0} \right)^2} \frac{dx}{x_0} +$$

$$\frac{i\hbar\bar{p}}{\sqrt{\pi}x_0^2} \int_{-\infty}^{+\infty} \left(\frac{x}{x_0} - \frac{\bar{x}}{x_0} \right) e^{-\left(\frac{x}{x_0} - \frac{\bar{x}}{x_0} \right)^2} \frac{dx}{x_0} = \frac{\hbar^2}{2x_0^2} + \bar{p}^2.$$

We see that our wave packet has non zero average values for the position and momentum operators. We will now show that the average values evolve in time according to the solutions of the equations of motion of the classical harmonic oscillator

$$\begin{cases} x(t) = x(0)\cos(\omega t) + \frac{p(0)}{m\omega} \sin(\omega t) \\ p(t) = -m\omega x(0)\sin(\omega t) + p(0)\cos(\omega t). \end{cases}$$

To determine the time evolution, we write $|\psi\rangle = \sum_n c_n |\psi_n\rangle$ where $|\psi_n\rangle$ are the eigenstates of the harmonic oscillator. Once we know the generic projection coefficient c_n, the time evolution of the wave packet is given by the phase factors $e^{-iE_n t/\hbar}$ multiplying the c_n, where $E_n = \hbar\omega(n + \frac{1}{2})$ is the eigenvalue of the n-th eigenstate. We first introduce the scalar quantity

$$a = \frac{\bar{x}}{\sqrt{2}x_0} + \frac{ix_0\bar{p}}{\hbar\sqrt{2}}$$

and rewrite the wave packet as

$$\psi(x) = \frac{1}{\sqrt{\sqrt{\pi}x_0}} e^{-\frac{1}{2}\left(\frac{x}{x_0}\right)^2 - \frac{1}{2}\left(\frac{\bar{x}}{x_0}\right)^2 + \sqrt{2}\left(\frac{x}{x_0}\right)a}.$$

Then, we consider the n-th eigenstate of the harmonic oscillator

$$\psi_n(x) = \frac{1}{\sqrt{2^n n! \sqrt{\pi}x_0}} e^{-\frac{1}{2}\left(\frac{x}{x_0}\right)^2} H_n\left(\frac{x}{x_0}\right) = \frac{(-1)^n}{\sqrt{2^n n! \sqrt{\pi}x_0}} e^{\frac{1}{2}\left(\frac{x}{x_0}\right)^2} \frac{d^n e^{-\left(\frac{x}{x_0}\right)^2}}{d\left(\frac{x}{x_0}\right)^n}$$

with H_n the n-th order Hermite polynomial, and compute the projection coefficient c_n as

$$c_n = \int_{-\infty}^{+\infty} \psi(x)\,\psi_n^*(x)\,dx = \frac{(-1)^n}{\sqrt{2^n n!\pi}} e^{-\frac{1}{2}\left(\frac{\bar{x}}{x_0}\right)^2} \int_{-\infty}^{+\infty} e^{\sqrt{2}a\left(\frac{x}{x_0}\right)} \frac{d^n e^{-\left(\frac{x}{x_0}\right)^2}}{d\left(\frac{x}{x_0}\right)^n} \frac{dx}{x_0} =$$

$$\frac{(-1)^n e^{-\frac{1}{2}\left(\frac{\bar{x}}{x_0}\right)^2}}{\sqrt{2^n n!\pi}} \left(\frac{d^{n-1} e^{-\left(\frac{x}{x_0}\right)^2}}{d\left(\frac{x}{x_0}\right)^{n-1}} e^{\sqrt{2}a\left(\frac{x}{x_0}\right)} - \frac{d^{n-2} e^{-\left(\frac{x}{x_0}\right)^2}}{d\left(\frac{x}{x_0}\right)^{n-2}} \frac{d\, e^{\sqrt{2}a\left(\frac{x}{x_0}\right)}}{d\left(\frac{x}{x_0}\right)} + \cdots \right)\Bigg|_{-\infty}^{+\infty} +$$

$$\frac{(-1)^n e^{-\frac{1}{2}\left(\frac{\bar{x}}{x_0}\right)^2}}{\sqrt{2^n n!\pi}} (-1)^n \int_{-\infty}^{+\infty} \frac{d^n e^{\sqrt{2}a\left(\frac{x}{x_0}\right)}}{d\left(\frac{x}{x_0}\right)^n} e^{-\left(\frac{x}{x_0}\right)^2} \frac{dx}{x_0} =$$

$$\frac{a^n e^{-\frac{1}{2}\left(\frac{\bar{x}}{x_0}\right)^2}}{\sqrt{n!\pi}} \int_{-\infty}^{+\infty} e^{\sqrt{2}a\left(\frac{x}{x_0}\right) - \left(\frac{x}{x_0}\right)^2} \frac{dx}{x_0} = \frac{a^n e^{-\frac{1}{2}\left(\frac{\bar{x}}{x_0}\right)^2 + \frac{1}{2}a^2}}{\sqrt{n!}} = \frac{1}{\sqrt{n!}} a^n e^{\frac{i\bar{p}\bar{x}}{2\hbar} - \frac{1}{2}|a|^2}$$

where we have used the integration by parts and set $|a|^2 = aa^*$. The exponential $e^{i\bar{p}\bar{x}/2\hbar}$ is a constant with respect to the variable x and we neglect it, since we can define the wave packet with an additional unimportant phase factor

$$\psi(x) \to e^{-\frac{i\bar{p}\bar{x}}{2\hbar}} \psi(x) = \frac{1}{\sqrt{\sqrt{\pi}x_0}} e^{-\frac{(x-\bar{x})^2}{2x_0^2}}.$$

The time evolution is then given by

$$\psi(x,t) = \sum_n c_n e^{-\frac{iE_n t}{\hbar}} \psi_n(x) = e^{-\frac{1}{2}|a|^2} e^{-\frac{i\omega t}{2}} \sum_n \frac{1}{\sqrt{n!}} (ae^{-i\omega t})^n \psi_n(x)$$

where we see that the quantity a previously defined gets the phase factor $e^{-i\omega t}$ (it is exactly the time evolution of the annihilation operator we have seen in Problem 2.24). The time dependence in a is directly translated in a time dependence of the average values $\bar{x}(t)$ and $\bar{p}(t)$

$$a(t) = ae^{-i\omega t} = \frac{\bar{x}(t)}{\sqrt{2}x_0} + \frac{ix_0 \bar{p}(t)}{\hbar\sqrt{2}}$$

so that

$$a^*(t) = a^* e^{i\omega t} = \frac{\bar{x}(t)}{\sqrt{2}x_0} - \frac{ix_0 \bar{p}(t)}{\hbar\sqrt{2}}.$$

The average values $\bar{x}(t)$ and $\bar{p}(t)$ are then obtained from the two previous relations. If we define $x_0 = \sqrt{\hbar/m\omega}$, we get

$$\bar{x}(t) = \frac{x_0}{\sqrt{2}}(a(t) + a^*(t)) = \sqrt{\frac{\hbar}{2m\omega}}(a(t) + a^*(t)) =$$

$$\sqrt{\frac{2\hbar}{m\omega}}(\Re(a)\cos(\omega t) + \Im(a)\sin(\omega t)) =$$

$$\bar{x}(0)\cos(\omega t) + \frac{1}{m\omega}\bar{p}(0)\sin(\omega t)$$

$$\bar{p}(t) = i\frac{\hbar}{x_0\sqrt{2}}(-a(t) + a^*(t)) = i\sqrt{\frac{1}{2}m\hbar\omega}(-a(t) + a^*(t)) =$$

$$\sqrt{2m\hbar\omega}(-\Re(a)\sin(\omega t) + \Im(a)\cos(\omega t)) =$$

$$-m\omega\bar{x}(0)\sin(\omega t) + \bar{p}(0)\cos(\omega t)$$

that is the solution of the equations of motion for the classical harmonic oscillator previously anticipated.

Problem 2.33.
We consider a particle with charge q and mass $m = 1$ subject to the one dimensional harmonic potential $U(x) = \frac{x^2}{2}$ ($\hbar = \omega = 1$), and placed in a constant electric field, E_x, directed along the positive x direction. Determine the eigenfunctions and eigenvalues of the Hamiltonian and the average values $\langle \hat{x} \rangle, \langle \hat{p} \rangle, \langle \hat{x}^2 \rangle, \langle \hat{p}^2 \rangle$. If at time $t = 0$ the wave function is

$$\psi(x,0) = (\pi)^{-\frac{1}{4}} e^{-\frac{1}{2}x^2 - i\phi(0)}$$

with $\phi(0) = 0$ the initial phase factor, determine the wave function $\psi(x,t)$ at time t.

Solution
The Hamiltonian of the problem is

$$\hat{H} = \frac{\hat{p}^2}{2} + \frac{\hat{x}^2}{2} - qE_x\hat{x} = \frac{\hat{p}^2}{2} + \frac{\hat{x}^2}{2} - 2A\hat{x}$$

where we have introduced the constant $A = \frac{qE_x}{2}$. We can complete the square and rewrite the Hamiltonian as

$$\hat{H} = \frac{\hat{p}^2}{2} + \frac{\hat{x}^2}{2} - 2A\hat{x} + 2A^2\,1 - 2A^2\,1 = \frac{\hat{p}^2}{2} + \frac{(\hat{x} - 2A\,1)^2}{2} - 2A^2\,1 =$$

$$\frac{\hat{p}^2}{2} + \frac{\hat{y}^2}{2} - 2A^2\,1 = \hat{H}_{ho} - 2A^2\,1$$

with $\hat{y} = \hat{x} - 2A\,1$. We also note that the derivative is not changed by the translation, i.e. $d^2/dx^2 = d^2/dy^2$; therefore, the momentum operators in the coordinates x and y are the same and the Hamiltonian $\hat{H}_{ho} = \frac{\hat{p}^2}{2} + \frac{\hat{y}^2}{2}$ is the one of the harmonic oscillator with coordinate y. The eigenvalue problem in the stationary Schrödinger equation can be solved

$$\hat{H}\left|n\right\rangle = (\hat{H}_{ho} - 2A^2\mathbf{1})\left|n\right\rangle = \left[\left(n+\frac{1}{2}\right) - 2A^2\right]\left|n\right\rangle$$

where the eigenstates $\left|n\right\rangle$ are those of the harmonic oscillator in the coordinate y with energies diminished by the quantity $-2A^2$

$$\langle y|n\rangle = \psi_n(y) = \psi_n(x - 2A) = \frac{1}{\sqrt{2^n n!\sqrt{\pi}}}e^{-\frac{1}{2}y^2}H_n(y) = \frac{(-1)^n}{\sqrt{2^n n!\sqrt{\pi}}}e^{\frac{1}{2}y^2}\frac{d^n e^{-y^2}}{dy^n}$$

$$E_n = \left(n+\frac{1}{2}\right) - 2A^2$$

with $H_n(y)$ the n-th order Hermite polynomial. From the relation between $\hat{y}$, $\hat{p}$ and the creation and annihilation operators $\hat{a}^\dagger$, $\hat{a}$

$$\begin{cases} \hat{y} = \frac{1}{\sqrt{2}}(\hat{a} + \hat{a}^\dagger) \\ \hat{p} = -\frac{i}{\sqrt{2}}(\hat{a} - \hat{a}^\dagger) \end{cases}$$

plus the step down and step up action of $\hat{a}$, $\hat{a}^\dagger$ on the eigenstates $\left|n\right\rangle$

$$\hat{a}\left|n\right\rangle = \sqrt{n}\left|n-1\right\rangle \qquad \hat{a}^\dagger\left|n\right\rangle = \sqrt{n+1}\left|n+1\right\rangle$$

we find the following average values

$$\langle n|\hat{x}|n\rangle = \langle n|\hat{y} + 2A\mathbf{1}\,|n\rangle = \frac{1}{\sqrt{2}}\langle n|\,(\hat{a} + \hat{a}^\dagger)\,|n\rangle + 2A = 2A$$

$$\langle n|\hat{p}|n\rangle = -\frac{i}{\sqrt{2}}\langle n|\,(\hat{a} - \hat{a}^\dagger)\,|n\rangle = 0$$

$$\langle n|\hat{x}^2\,|n\rangle = \langle n|\,(\hat{y} + 2A\mathbf{1}\,)^2\,|n\rangle = \frac{1}{2}\langle n|\,(\hat{a} + \hat{a}^\dagger)^2\,|n\rangle + 4A^2 =$$

$$\frac{1}{2}\langle n|\,(\hat{a}\hat{a}^\dagger + \hat{a}^\dagger\hat{a})\,|n\rangle + 4A^2 = \left(n+\frac{1}{2}\right) + 4A^2$$

$$\langle n|\hat{p}^2\,|n\rangle = -\frac{1}{2}\langle n|\,(\hat{a} - \hat{a}^\dagger)^2\,|n\rangle = \frac{1}{2}\langle n|\,(\hat{a}\hat{a}^\dagger + \hat{a}^\dagger\hat{a})\,|n\rangle = n+\frac{1}{2}.$$

We now discuss the time evolution of the wave function

$$\psi(x,0) = (\pi)^{-\frac{1}{4}}e^{-\frac{1}{2}x^2 - i\phi(0)} = (\pi)^{-\frac{1}{4}}e^{-\frac{1}{2}(y+2A)^2 - i\phi(0)} = (\pi)^{-\frac{1}{4}}e^{-\frac{1}{2}(y-\bar{y})^2 - i\phi(0)}$$

where we have used the definition $\bar{y} = -2A$. We expect the time evolution to appear in the average values of the position $(\bar{y}(t))$ and momentum $(\bar{p}(t))$ and in the phase factor (see also Problem 2.17). Therefore, we rewrite

$$\psi(y,t) = (\pi)^{-\frac{1}{4}}e^{-\frac{1}{2}(y-\bar{y}(t))^2 - i\phi(t) + i\bar{p}(t)y}.$$

The average value of the position $\bar{y}(t)$ is found with

$$\frac{d\bar{y}(t)}{dt} = i\langle[\hat{H},\hat{y}]\rangle = \frac{i}{2}\langle[\hat{p}^2,\hat{y}]\rangle = \frac{i}{2}\langle[\hat{p},\hat{y}]\hat{p}+\hat{p}[\hat{p},\hat{y}]\rangle = \langle\hat{p}\rangle = \bar{p}(t).$$

As for the average value of the momentum operator $\bar{p}(t)$, we get

$$\frac{d\bar{p}(t)}{dt} = i\langle[\hat{H},\hat{p}]\rangle = \frac{i}{2}\langle[\hat{y}^2,\hat{p}]\rangle = -\frac{i}{2}\langle[\hat{p},\hat{y}]\hat{y}+\hat{y}[\hat{p},\hat{y}]\rangle = -\langle\hat{y}\rangle = -\bar{y}(t).$$

If we take the derivative of the first expression, we find

$$\frac{d^2\bar{y}(t)}{dt^2} = \frac{d\bar{p}(t)}{dt} = -\bar{y}(t)$$

that is the equation of motion for the classical harmonic oscillator with solution

$$\bar{y}(t) = C_1 \sin t + C_2 \cos t.$$

Imposing the initial conditions $\bar{y}(0) = -2A$, $\bar{p}(0) = 0$, we get $C_1 = 0$, $C_2 = -2A$. To determine the phase factor, we use the time dependent Schrödinger equation

$$i\frac{\partial\psi(y,t)}{\partial t} = \left(-\frac{1}{2}\frac{\partial^2}{\partial y^2}+\frac{y^2}{2}-2A^2\right)\psi(y,t)$$

which leads to

$$i\frac{\partial\psi(y,t)}{\partial t} = \left[-y\frac{d\bar{p}}{dt}+i(y-\bar{y}(t))\frac{d\bar{y}}{dt}+\frac{d\phi}{dt}\right]\psi(y,t)$$

$$\frac{\partial^2\psi(y,t)}{\partial y^2} = \left[-1-\bar{p}(t)^2+(y-\bar{y}(t))^2-2i(y-\bar{y}(t))\bar{p}(t)\right]\psi(y,t).$$

Plugging this back into the Schrödinger equation and using the relations $d\bar{p}/dt = -\bar{y}(t)$, $\bar{p} = d\bar{y}/dt$ we find

$$\frac{d\phi(t)}{dt} = \frac{1}{2}+\frac{1}{2}\left(\left(\frac{d\bar{y}}{dt}\right)^2-\bar{y}^2(t)\right)-2A^2.$$

The above relation is integrated with the initial condition $\phi(t=0)=0$ leading to

$$\phi(t) = \frac{1}{2}t+\frac{1}{2}\int_0^t\left(\left(\frac{d\bar{y}(s)}{ds}\right)^2-\bar{y}^2(s)\right)ds - 2A^2 t =$$

$$\frac{1}{2}t+\frac{1}{2}\left(\bar{y}(t)\frac{d\bar{y}(t)}{dt}-\int_0^t\bar{y}(s)\left(\frac{d^2\bar{y}(s)}{ds^2}+\bar{y}(s)\right)ds\right)-2A^2 t =$$

$$\frac{t}{2}+\frac{\bar{y}(t)\bar{p}(t)}{2}-2A^2 t$$

where we have used $\frac{d^2\bar{y}(s)}{ds^2} = -\bar{y}(s)$, $0 \leq s \leq t$. Summarizing, the wave function is

$$\psi(y,t) = (\pi)^{-\frac{1}{4}} e^{-\frac{1}{2}(y-\bar{y}(t))^2 - \frac{it}{2} - \frac{1}{2}i\bar{y}(t)\bar{p}(t) + i\bar{p}(t)y + 2iA^2 t}$$

where $\bar{y}(t)$, $\bar{p}(t)$ are calculated from the classical equations of motion (see also Problems 2.9 and 2.32).

Problem 2.34.

Let us consider the one dimensional quantum harmonic oscillator with frequency ω, and let $\hat{a}$ and $|n\rangle$ be the annihilation operator and the normalized eigenstate of $\hat{a}^\dagger \hat{a}$ with eigenvalue n. Determine the commutator $[\hat{a}, (\hat{a}^\dagger)^n]$, with n a positive integer number. Then, consider the generic normalized state $|\alpha\rangle = \sum_n c_n |n\rangle$, and determine the coefficients c_n in such a way that $\hat{a}|\alpha\rangle = \alpha|\alpha\rangle$, with α a real number.

Solution

Let us start by determining the commutator $[\hat{a}, (\hat{a}^\dagger)^n]$. The idea is to use the Jacobi identity (see also Problem 2.2)

$$[\hat{A}, \hat{B}\hat{C}] = \hat{B}[\hat{A}, \hat{C}] + [\hat{A}, \hat{B}]\hat{C}$$

which, applied to our case, yields

$$[\hat{a}, (\hat{a}^\dagger)^n] = (\hat{a}^\dagger)^{n-1}[\hat{a}, \hat{a}^\dagger] + [\hat{a}, (\hat{a}^\dagger)^{n-1}]\hat{a}^\dagger = (\hat{a}^\dagger)^{n-1} + [\hat{a}, (\hat{a}^\dagger)^{n-1}]\hat{a}^\dagger$$

where we have used $[\hat{a}, \hat{a}^\dagger] = 1$. We can also determine $[\hat{a}, (\hat{a}^\dagger)^{n-1}]$

$$[\hat{a}, (\hat{a}^\dagger)^{n-1}] = (\hat{a}^\dagger)^{n-2}[\hat{a}, \hat{a}^\dagger] + [\hat{a}, (\hat{a}^\dagger)^{n-2}]\hat{a}^\dagger = (\hat{a}^\dagger)^{n-2} + [\hat{a}, (\hat{a}^\dagger)^{n-2}]\hat{a}^\dagger$$

which, plugged back in the previous expression, leads to

$$[\hat{a}, (\hat{a}^\dagger)^n] = 2(\hat{a}^\dagger)^{n-1} + [\hat{a}, (\hat{a}^\dagger)^{n-2}](\hat{a}^\dagger)^2.$$

By iteration, we obtain

$$[\hat{a}, (\hat{a}^\dagger)^n] = n(\hat{a}^\dagger)^{n-1}.$$

As for the second point, we know that

$$\hat{a}|n\rangle = \sqrt{n}|n-1\rangle$$

and the condition $\hat{a}|\alpha\rangle = \alpha|\alpha\rangle$ is equivalent to

$$c_n\sqrt{n} = \alpha c_{n-1}.$$

Therefore, if we define $c_0 = C$, we find $c_1 = C\alpha$, $c_2 = C\frac{\alpha^2}{\sqrt{2}}$ and, finally, $c_n = C\frac{\alpha^n}{\sqrt{n!}}$. The normalization condition fixes the constant $C = e^{-\frac{\alpha^2}{2}}$. This state is known as *coherent state*

$$|\alpha\rangle = e^{-\frac{\alpha^2}{2}} \sum_n \frac{\alpha^n}{\sqrt{n!}} |n\rangle.$$

In Problem 2.35 we will further characterize the properties of this kind of states and determine the time evolution.

Problem 2.35.
Consider the one dimensional quantum harmonic oscillator with $\hbar = m = \omega = 1$. We define the coherent state as

$$|\alpha\rangle = A e^{\alpha \hat{a}^\dagger} |0\rangle$$

with α a complex number, $|0\rangle$ the ground state, and $\hat{a}^\dagger$ the creation operator:

- determine the normalization constant A;
- show that $|\alpha\rangle$ is an eigenstate of the annihilation operator $\hat{a}$ with eigenvalue α;
- show that the average value of the position operator $\hat{x}$ on $|\alpha\rangle$ is non zero;
- determine the probability to find the n-th eigenvalue of the energy in the state $|\alpha\rangle$;
- determine the time evolution of $|\alpha\rangle$.

Solution
Let us consider the action of $e^{\alpha \hat{a}^\dagger}$ on the state $|0\rangle$

$$e^{\alpha \hat{a}^\dagger} |0\rangle = \sum_{n=0}^{+\infty} \frac{\alpha^n}{n!} (\hat{a}^\dagger)^n |0\rangle = \sum_{n=0}^{+\infty} \frac{\alpha^n}{\sqrt{n!}} |n\rangle$$

where we have used (by iteration) the following relation

$$\hat{a}^\dagger |n\rangle = \sqrt{n+1} |n+1\rangle.$$

If we take the square modulus of the coherent state, we find

$$1 = \langle \alpha | \alpha \rangle = |A|^2 \sum_{n=0}^{+\infty} \sum_{m=0}^{+\infty} \frac{\alpha^n}{\sqrt{n!}} \frac{(\alpha^*)^m}{\sqrt{m!}} \langle m | n \rangle = e^{\alpha \alpha^*} |A|^2.$$

Therefore, the normalized state is

$$|\alpha\rangle = e^{-\frac{1}{2}\alpha\alpha^*} e^{\alpha \hat{a}^\dagger} |0\rangle.$$

As for the second point, we write $e^{\alpha \hat{a}^\dagger}$ with its Taylor expansion

$$\hat{a} |\alpha\rangle = A \hat{a} \sum_{n=0}^{+\infty} \frac{\alpha^n}{n!} (\hat{a}^\dagger)^n |0\rangle.$$

The next step is the calculation (see also Problem 2.34) of the commutator $[\hat{a}, (\hat{a}^\dagger)^n]$

$$[\hat{a}, (\hat{a}^\dagger)^n] = [\hat{a}, \hat{a}^\dagger](\hat{a}^\dagger)^{n-1} + \hat{a}^\dagger [\hat{a}, (\hat{a}^\dagger)^{n-1}] =$$
$$(\hat{a}^\dagger)^{n-1} + (\hat{a}^\dagger)[\hat{a}, \hat{a}^\dagger](\hat{a}^\dagger)^{n-2} + (\hat{a}^\dagger)^2 [\hat{a}, (\hat{a}^\dagger)^{n-2}] =$$
$$2(\hat{a}^\dagger)^{n-1} + (\hat{a}^\dagger)^2 [\hat{a}, (\hat{a}^\dagger)^{n-2}] = \ldots = n(\hat{a}^\dagger)^{n-1}$$

where we have used the Jacobi identity (see also Problem 2.2). This result allows us to rewrite $\hat{a}|\alpha\rangle$ as

$$\hat{a}|\alpha\rangle = A\hat{a}\sum_{n=0}^{+\infty}\frac{\alpha^n}{n!}(\hat{a}^\dagger)^n|0\rangle = A\sum_{n=0}^{+\infty}\frac{\alpha^n}{n!}\left((\hat{a}^\dagger)^n\hat{a}+n(\hat{a}^\dagger)^{n-1}\right)|0\rangle = A\sum_{n=0}^{+\infty}\frac{\alpha^n}{n!}n(\hat{a}^\dagger)^{n-1}|0\rangle.$$

Setting $n' = n - 1$ we can rewrite the right hand side as $\alpha|\alpha\rangle$: the extra term generated by this substitution is indeed zero because $(-1)! = +\infty$. As for the third point, we need to calculate the average value of the position operator on $|\alpha\rangle$. To do this, we use the relation

$$\langle\alpha|\hat{x}|\alpha\rangle = \frac{1}{\sqrt{2}}\langle\alpha|(\hat{a}+\hat{a}^\dagger)|\alpha\rangle = \frac{1}{\sqrt{2}}(\alpha+\alpha^*) = \sqrt{2}\Re(\alpha)$$

and write

$$\langle\alpha|\hat{a}^\dagger = \alpha^*\langle\alpha|$$

with $\hat{a}|\alpha\rangle = \alpha|\alpha\rangle$. The probability to find the energy eigenvalue $E_n = n + \frac{1}{2}$ in the state $|\alpha\rangle$ is

$$|c_n|^2 = e^{-\alpha\alpha^*}|\langle n|\sum_{m=0}^{+\infty}\frac{\alpha^m}{m!}(\hat{a}^\dagger)^m|0\rangle|^2 = e^{-\alpha\alpha^*}|\langle n|\sum_{m=0}^{+\infty}\frac{\alpha^m}{\sqrt{m!}}|m\rangle|^2 = e^{-\alpha\alpha^*}\frac{(\alpha\alpha^*)^n}{n!}$$

i.e. a Poisson distribution. The time evolution of the state is given by

$$|\alpha,t\rangle = e^{-\frac{1}{2}\alpha\alpha^*}\sum_{n=0}^{+\infty}\frac{\alpha^n}{\sqrt{n!}}e^{-iE_nt}|n\rangle = e^{-\frac{1}{2}\alpha\alpha^*}\sum_{n=0}^{+\infty}\frac{\alpha^n}{\sqrt{n!}}e^{-i(n+\frac{1}{2})t}|n\rangle =$$

$$e^{-\frac{1}{2}\alpha\alpha^*}e^{-\frac{1}{2}it}\sum_{n=0}^{+\infty}\frac{(\alpha e^{-it})^n}{\sqrt{n!}}|n\rangle = e^{-\frac{1}{2}\alpha\alpha^*}e^{-\frac{1}{2}it}\sum_{n=0}^{+\infty}\frac{(\alpha\hat{a}^\dagger e^{-it})^n}{n!}|0\rangle$$

where the last series sums to an exponential function (we already met this kind of wave function in Problem 2.32).

3

Angular Momentum and Spin

Problem 3.1.

Determine the uncertainty relations between the orbital angular momentum $\hat{L} = (\hat{L}_x, \hat{L}_y, \hat{L}_z)$ and the components of the position and of the momentum operators $\hat{r} = (\hat{x}, \hat{y}, \hat{z})$, $\hat{p} = (\hat{p}_x, \hat{p}_y, \hat{p}_z)$. Then, find the operator $\hat{L}_z$ in spherical polar coordinates and explain why the operators $\hat{\phi}$ (azimuthal angle) and $\hat{L}_z$ can be measured simultaneously. What are the functions of $\hat{\phi}$ whose commutator with $\hat{L}_z$ has a physical sense?

Solution

We start from the commutation rules involving the position and the momentum operators

$$[\hat{x}_i, \hat{p}_j] = i\hbar \delta_{ij} 1 \, .$$

The orbital angular momentum is given by $L = r \wedge p$. The different components of such vector, using the Einstein convention for summation on repeated indexes, can be written as

$$\hat{L}_i = \varepsilon_{ijk} \hat{r}_j \hat{p}_k$$

where ε_{ijk} is the Levi-Civita tensor and $i, j, k = x, y, z$. Such tensor is totally antisymmetric and conventionally chosen in such a way that $\varepsilon_{xyz} = 1$. Using the commutation rule between position and momentum operators, we find

$$[\hat{L}_i, \hat{r}_j] = \varepsilon_{ikl} \hat{r}_k [\hat{p}_l, \hat{r}_j] = -i\hbar \varepsilon_{ikl} \hat{r}_k \delta_{lj} = i\hbar \varepsilon_{ijk} \hat{r}_k$$

$$[\hat{L}_i, \hat{p}_j] = \varepsilon_{ikl} [\hat{r}_k, \hat{p}_j] \hat{p}_l = i\hbar \varepsilon_{ikl} \hat{p}_l \delta_{kj} = i\hbar \varepsilon_{ijl} \hat{p}_l \, .$$

The operator $\hat{L}_z$, when acting on a function in Cartesian coordinates, is a differential operator with the property

$$\hat{L}_z = -i\hbar \left(x \frac{\partial}{\partial y} - y \frac{\partial}{\partial x} \right) \, .$$

Cini M., Fucito F., Sbragaglia M.: Solved Problems in Quantum and Statistical Mechanics.
DOI 10.1007/978-88-470-2315-4_3, © Springer-Verlag Italia 2012

The useful relations to rewrite it in spherical polar coordinates are

$$\begin{cases} x = r\sin\theta\cos\phi \\ y = r\sin\theta\sin\phi \\ z = r\cos\theta \end{cases}$$

and the inverse

$$\begin{cases} r = \sqrt{x^2 + y^2 + z^2} \\ \theta = \arccos\left(\frac{z}{r}\right) \\ \phi = \arctan\left(\frac{y}{x}\right). \end{cases}$$

Furthermore

$$\begin{cases} \frac{\partial r}{\partial y} = \sin\theta\sin\phi; & \frac{\partial r}{\partial x} = \sin\theta\cos\phi \\ \frac{\partial\theta}{\partial y} = \frac{1}{r}\cos\theta\sin\phi; & \frac{\partial\theta}{\partial x} = \frac{1}{r}\cos\theta\cos\phi \\ \frac{\partial\phi}{\partial y} = \frac{\cos\phi}{r\sin\theta}; & \frac{\partial\phi}{\partial x} = -\frac{\sin\phi}{r\sin\theta} \end{cases}$$

leading to

$$\hat{L}_z = -i\hbar\left(x\frac{\partial}{\partial y} - y\frac{\partial}{\partial x}\right) = -i\hbar r\sin\theta\cos\phi\left(\frac{\partial r}{\partial y}\frac{\partial}{\partial r} + \frac{\partial\theta}{\partial y}\frac{\partial}{\partial\theta} + \frac{\partial\phi}{\partial y}\frac{\partial}{\partial\phi}\right) +$$

$$i\hbar r\sin\theta\sin\phi\left(\frac{\partial r}{\partial x}\frac{\partial}{\partial r} + \frac{\partial\theta}{\partial x}\frac{\partial}{\partial\theta} + \frac{\partial\phi}{\partial x}\frac{\partial}{\partial\phi}\right) = -i\hbar\frac{\partial}{\partial\phi}.$$

We now write the Heisenberg uncertainty relations (see Problem 2.32)

$$\langle(\Delta\hat{F})^2\rangle\langle(\Delta\hat{K})^2\rangle \geq \frac{1}{4}\langle\hat{M}\rangle^2$$

where

$$[\hat{K},\hat{F}] = i\hat{M}$$

with $\hat{K},\hat{F},\hat{M}$ self-adjoint operators. If we consider the operators $\hat{\phi}$ and $\hat{L}_z$, we see that their commutator gives a finite non zero result. Nevertheless, we note that they have different domains. As for the case of $\hat{L}_z$, its eigenfunctions and eigenvalues are obtained from the solutions of the ordinary differential equation

$$-i\hbar\frac{d\psi(\phi)}{d\phi} = L_z\psi(\phi)$$

with $0 \leq \phi \leq 2\pi$. The solutions of such equation are

$$\psi(\phi) = Ae^{iL_z\phi/\hbar}.$$

The need to interpret the square modulus as a probability distribution prevents the functions from being multivalued

$$\psi(\phi + 2\pi) = \psi(\phi)$$

and imposes that the eigenvalues of $\hat{L}_z$ are quantized: $L_z = m\hbar$, with m an integer number. The corresponding normalized eigenfunctions are

$$\psi(\phi) = \frac{1}{\sqrt{2\pi}}e^{im\phi}.$$

We see that the operator $\hat{L}_z$ acts on periodic functions. The action of $\hat{\phi}$ on a periodic function $\psi(\phi)$, leads to $\hat{\phi}\psi(\phi) = \phi\psi(\phi) = f(\phi)$, which is manifestly non periodic

$$f(\phi + 2\pi) = (\phi + 2\pi)\psi(\phi + 2\pi) = (\phi + 2\pi)\psi(\phi) \neq f(\phi).$$

Therefore, we conclude that, before taking the commutator between two operators and ask if they can (or cannot) be measured simultaneously, we need to be sure they act on the same functional space. In the case of $\hat{L}_z$, we need operators acting on periodic functions, as for example $\cos\phi, \sin\phi$ and their combinations.

Problem 3.2.
Consider the orbital angular momentum $\hat{L} = (\hat{L}_x, \hat{L}_y, \hat{L}_z)$ in Cartesian coordinates (x, y, z) and determine the commutators $[[[\hat{L}_x, \hat{L}_y], \hat{L}_x], \hat{L}_x], [[[\hat{L}_x, \hat{L}_y], \hat{L}_x], \hat{L}_y], [[[\hat{L}_x, \hat{L}_y],$ $\hat{L}_x], \hat{L}_z]$. Finally, determine the action of $\hat{L}^2 = \hat{L}_x^2 + \hat{L}_y^2 + \hat{L}_z^2$ on the combination $(\hat{x}[\hat{L}_y, \hat{z}] - \hat{y}[\hat{L}_x, \hat{z}] + \hat{z}[\hat{L}_x, \hat{y}])$.

Solution
To solve this problem, we need to consider the commutation rules of the angular momentum. In particular, we know that

$$\hat{L} \wedge \hat{L} = i\hbar\hat{L}$$

from which we get

$$[\hat{L}_x, \hat{L}_y] = \hat{L}_x\hat{L}_y - \hat{L}_y\hat{L}_x = i\hbar\hat{L}_z$$
$$[\hat{L}_z, \hat{L}_x] = \hat{L}_z\hat{L}_x - \hat{L}_x\hat{L}_z = i\hbar\hat{L}_y$$
$$[\hat{L}_y, \hat{L}_z] = \hat{L}_y\hat{L}_z - \hat{L}_z\hat{L}_y = i\hbar\hat{L}_x.$$

The above relations can be used to simplify the first commutator requested by the text

$$[[[\hat{L}_x, \hat{L}_y], \hat{L}_x], \hat{L}_x] = i\hbar[[\hat{L}_z, \hat{L}_x], \hat{L}_x] = -\hbar^2[\hat{L}_y, \hat{L}_x] = i\hbar^3\hat{L}_z.$$

When the last $\hat{L}_x$ is interchanged with $\hat{L}_y$, we get

$$[[[\hat{L}_x, \hat{L}_y], \hat{L}_x], \hat{L}_y] = i\hbar[[\hat{L}_z, \hat{L}_x], \hat{L}_y] = -\hbar^2[\hat{L}_y, \hat{L}_y] = 0.$$

Also, when the last $\hat{L}_x$ is interchanged with $\hat{L}_z$, we get

$$[[[\hat{L}_x, \hat{L}_y], \hat{L}_x], \hat{L}_z] = i\hbar[[\hat{L}_z, \hat{L}_x], \hat{L}_z] = -\hbar^2[\hat{L}_y, \hat{L}_z] = -i\hbar^3\hat{L}_x.$$

As for the second point, from the relation

$$[\hat{L}_x, \hat{y}] = [\hat{y}\hat{p}_z - \hat{z}\hat{p}_y, \hat{y}] = i\hbar\hat{z}$$

and its cyclic permutations, we find

$$\hat{x}[\hat{L}_y,\hat{z}] - \hat{y}[\hat{L}_x,\hat{z}] + \hat{z}[\hat{L}_x,\hat{y}] = i\hbar(\hat{x}^2 + \hat{y}^2 + \hat{z}^2).$$

Therefore, we see that the combination $\hat{x}[\hat{L}_y,\hat{z}] - \hat{y}[\hat{L}_x,\hat{z}] + \hat{z}[\hat{L}_x,\hat{y}]$ is directly proportional to the square of the distance from the origin of coordinates ($\hat{r}^2$) and is independent of the angles, i.e. when projected on the angular variables it is proportional to the spherical harmonic $Y_{0,0}(\theta,\phi)$. Since the spherical harmonics $Y_{l,m}$ are eigenfunctions of $\hat{L}^2$ with eigenvalues $\hbar^2 l(l+1)$, we find

$$\hat{L}^2(\hat{x}[\hat{L}_y,\hat{z}] - \hat{y}[\hat{L}_x,\hat{z}] + \hat{z}[\hat{L}_x,\hat{y}]) = 0.$$

Problem 3.3.

Consider the Hamiltonian of a plane rigid rotator

$$\hat{H}_1 = \frac{\hat{L}_z^2}{2I}$$

with $\hat{L}_z$ the z component of the orbital angular momentum and I the momentum of inertia. Then, consider the Hamiltonian of a free particle with anisotropic mass ($m_x \neq m_y$) moving on a two dimensional plane

$$\hat{H}_2 = \frac{\hat{p}_x^2}{2m_x} + \frac{\hat{p}_y^2}{2m_y}$$

with $\hat{p}_{x,y}$ the x,y component of the momentum operator:

- determine the lower bound of the product $\sigma_1^2\sigma_2^2$, with $\sigma_i^2 = \langle(\Delta\hat{H}_i)^2\rangle = \langle\hat{H}_i^2\rangle - \langle\hat{H}_i\rangle^2$ ($i=1,2$) and where the average $\langle...\rangle$ is meant on a generic state $|\psi\rangle$;
- determine the energy of the ground state of the Hamiltonian $\hat{H} = \hat{H}_1 + \hat{H}_2$;
- when $m_x = m_y$, determine the eigenvalues of $\hat{H} = \hat{H}_1 + \hat{H}_2$.

Solution
We use the results of Problem 2.32, where we have seen that the Heisenberg uncertainty relations can be written as

$$\langle(\Delta\hat{F})^2\rangle\langle(\Delta\hat{K})^2\rangle \geq \frac{1}{4}\langle\hat{M}\rangle^2$$

where $\hat{K}$ and $\hat{F}$ are two generic self-adjoint operators with commutation rule $[\hat{K},\hat{F}] = i\hat{M}$. We have to identify $\hat{K} = \hat{H}_1$ and $\hat{F} = \hat{H}_2$ and determine the commutator

$$[\hat{H}_1,\hat{H}_2] = \frac{\hat{L}_z}{2I}[\hat{L}_z,\hat{H}_2] + [\hat{L}_z,\hat{H}_2]\frac{\hat{L}_z}{2I} = \frac{i\hbar}{2I}(\hat{L}_z\hat{p}_x\hat{p}_y + \hat{p}_x\hat{p}_y\hat{L}_z)\left(\frac{1}{m_x} - \frac{1}{m_y}\right)$$

where we have used

$$[\hat{H}_2,\hat{L}_z] = \left[\frac{\hat{p}_x^2}{2m_x} + \frac{\hat{p}_y^2}{2m_y}, \hat{x}\hat{p}_y - \hat{y}\hat{p}_x\right] = -i\hbar\hat{p}_x\hat{p}_y\left(\frac{1}{m_x} - \frac{1}{m_y}\right)$$

$$[\hat{p}_s,\hat{p}_j] = \delta_{sj}1 \qquad [\hat{s},\hat{p}_j] = \delta_{sj}i\hbar1 \qquad (s,j = x,y)$$

and the Jacobi identity (see Problem 2.2). Therefore, the result is

$$\sigma_1^2 \sigma_2^2 = \langle (\Delta \hat{H}_1)^2 \rangle \langle (\Delta \hat{H}_2)^2 \rangle \geq \frac{1}{4} \left(\frac{\hbar}{2I} \right)^2 (\langle \psi | (\hat{L}_z \hat{p}_x \hat{p}_y + \hat{p}_x \hat{p}_y \hat{L}_z) | \psi \rangle)^2 \left(\frac{1}{m_x} - \frac{1}{m_y} \right)^2$$

and we see that we get 0 in the isotropic case. The same happens when we are in a state with zero $p_{x,y}$ or L_z. The ground state of the Hamiltonian $\hat{H} = \hat{H}_1 + \hat{H}_2$ is $\psi = \frac{1}{\sqrt{2\pi}}$, i.e. a state where both $\langle (\Delta \hat{H}_1)^2 \rangle$ and $\langle (\Delta \hat{H}_2)^2 \rangle$ are zero. Finally, in the isotropic case ($m_x = m_y = m$), $[\hat{H}_1, \hat{H}_2] = 0$ and the eigenvalues of $\hat{H}_1$, $\hat{H}_2$ are summed: $\hat{H}_1$ has a discrete spectrum with eigenvalues $\frac{\hbar^2 k^2}{2I}$, with k an integer number (see also Problem 5.1); $\hat{H}_2$ has a continuous spectrum with eigenvalues $\frac{p_x^2 + p_y^2}{2m}$. Therefore, the eigenvalues of $\hat{H}$ are

$$E_{k, p_x, p_y} = \frac{\hbar^2 k^2}{2I} + \frac{p_x^2 + p_y^2}{2m}.$$

Problem 3.4.
Let us consider a system whose Hamiltonian is

$$\hat{H} = \frac{\hat{L}_+ \hat{L}_-}{\hbar}$$

where $\hat{L}_\pm$ are the raising and lowering operators for the z component of the orbital angular momentum. At time $t = 0$, the system is described by the following wave function

$$\psi(\theta, \phi, 0) = A \sin \theta \sin \phi.$$

Expand the initial wave function on the spherical harmonics and determine the normalization constant. Finally, determine the time evolution of the wave function and find the time t at which the wave function is identical to the initial state, i.e. $\psi(\theta, \phi, t) = A \sin \theta \cos \phi$.

Solution
We first recall some useful formulae for the spherical harmonics $Y_{l,m}(\theta, \phi) = \langle \theta, \phi | l, m \rangle$

$$Y_{l,m}(\theta, \phi) = (-1)^m \sqrt{\frac{(2l+1)}{4\pi} \frac{(l-m)!}{(l+m)!}} e^{im\phi} P_l^m(\cos \theta) \qquad m \geq 0$$

$$Y_{l,m}(\theta, \phi) = Y_{l,-|m|}(\theta, \phi) = (-1)^{|m|} Y_{l,|m|}^*(\theta, \phi) \qquad m < 0$$

where $P_l^m(\cos \theta)$ is the associated Legendre polynomial of order l defined by

$$P_l^m(u) = (1 - u^2)^{m/2} \frac{d^m}{du^m} P_l(u) \qquad 0 \leq m \leq l$$

$$P_l(u) = \frac{1}{2^l l!} \frac{d^l}{du^l} \left[(u^2 - 1)^l \right].$$

For $l = m = 1$ we get

$$Y_{1,\pm 1}(\theta, \phi) = \mp \sqrt{\frac{3}{8\pi}} e^{\pm i\phi} \sin\theta.$$

Therefore, the normalized wave function can be written as

$$\psi(\theta, \phi, 0) = \frac{1}{\sqrt{2}}(Y_{1,1}(\theta, \phi) + Y_{1,-1}(\theta, \phi))$$

that implies $A = -i\sqrt{\frac{3}{4\pi}}$. To determine the time evolution, it is convenient to find the eigenvalues of the Hamiltonian. From the definition of the raising and lowering operators for the z component of the orbital angular momentum, we have

$$\hat{L}_+ = (\hat{L}_x + i\hat{L}_y) \qquad \hat{L}_- = (\hat{L}_x - i\hat{L}_y)$$

and, hence

$$\hat{L}_+\hat{L}_- = (\hat{L}_x^2 + \hat{L}_y^2 + \hbar\hat{L}_z) = (\hat{L}^2 - \hat{L}_z^2 + \hbar\hat{L}_z)$$

where $\hat{L}^2 = \hat{L}_x^2 + \hat{L}_y^2 + \hat{L}_z^2$ is the squared orbital angular momentum. The Hamiltonian is diagonal with respect to the basis given by the $Y_{l,m}$

$$\hat{H}|l,m\rangle = (\hbar l(l+1) - \hbar m^2 + \hbar m)|l,m\rangle$$

with eigenvalues

$$E_{l,m} = (\hbar l(l+1) - \hbar m^2 + \hbar m).$$

Therefore $E_{1,1} = 2\hbar$ for $Y_{1,1}$ and $E_{1,-1} = 0$ for $Y_{1,-1}$. The time evolution is obtained with the action of $e^{-\frac{i\hat{H}t}{\hbar}}$ on $\psi(\theta, \phi, 0)$

$$\psi(\theta, \phi, t) = e^{-\frac{i\hat{H}t}{\hbar}}\psi(\theta, \phi, 0) = \frac{i}{\sqrt{2}}(e^{-2it}Y_{1,1}(\theta, \phi) + Y_{1,-1}(\theta, \phi)).$$

The time t at which $\psi(\theta, \phi, t) = \psi(\theta, \phi, 0)$ is found by imposing that $e^{-2it} = 1$. This happens for the first time when $t = \pi$.

Problem 3.5.
In a constant magnetic field B_1, a spin $1/2$ particle evolves from time $t = 0$ to $t = T$ according to the Hamiltonian

$$\hat{H}(t) = -\mu\hat{\sigma}_z B_1$$

where $\hat{\sigma}_z$ is the usual Pauli matrix for the z component of the spin. At time $t = 0$ the probability that the spin is $+\hbar/2$ along the x axis is 1. Solve the Schrödinger equation for the spinor $|\psi(t)\rangle = \begin{pmatrix} a(t) \\ b(t) \end{pmatrix}$ and calculate $|\psi(T)\rangle$. After $t = T$, the time evolution proceeds in a discontinuous way with the Hamiltonian which depends on time according to the law

$$\begin{cases} \hat{H}(t) = -\mu\hat{\sigma}_z B_1 & t \leq T \\ \hat{H}(t) = +\mu\hat{\sigma}_z B_2 & t > T. \end{cases}$$

Calculate $|\psi(2T)\rangle$. Find B_2 such that a measurement of spin in the y direction at time $t = 2T$ must give $-\hbar/2$.

Solution
At time $t = 0$, since the probability to find the particle with spin projection $+\hbar/2$ along x is 1, the wave function must be an eigenstate of the Pauli matrix $\hat{\sigma}_x$

$$\hat{\sigma}_x|\psi(0)\rangle = |\psi(0)\rangle$$

with the Pauli matrices given by

$$\hat{\sigma}_x = \begin{pmatrix} 0 & 1 \\ 1 & 0 \end{pmatrix} \quad \hat{\sigma}_y = \begin{pmatrix} 0 & -i \\ i & 0 \end{pmatrix} \quad \hat{\sigma}_z = \begin{pmatrix} 1 & 0 \\ 0 & -1 \end{pmatrix}.$$

The normalized eigenstate is

$$|\psi(0)\rangle = \frac{1}{\sqrt{2}}\begin{pmatrix} 1 \\ 1 \end{pmatrix}.$$

In the time interval $0 \leq t \leq T$, time evolution is driven by $\hat{H}(t) = -\mu\hat{\sigma}_z B_1$, whose eigenvalues are $\mp\mu B_1$. The corresponding time evolution operator is diagonal

$$\hat{U}(0,t) = e^{-i\hat{H}t/\hbar} = \begin{pmatrix} e^{\frac{i\mu}{\hbar}B_1 t} & 0 \\ 0 & e^{-\frac{i\mu}{\hbar}B_1 t} \end{pmatrix}.$$

Therefore, we can apply $\hat{U}(0,t)$ to $|\psi(0)\rangle$

$$|\psi(t)\rangle = \hat{U}(0,t)|\psi(0)\rangle = \frac{e^{-\frac{i\mu}{\hbar}B_1 t}}{\sqrt{2}}\begin{pmatrix} e^{2\frac{i\mu}{\hbar}B_1 t} \\ 1 \end{pmatrix}.$$

When $t > T$, time evolution is driven by $\hat{H}(t) = \mu\hat{\sigma}_z B_2$, and the corresponding time evolution operator can be applied to $|\psi(T)\rangle = \hat{U}(0,T)|\psi(0)\rangle$

$$|\psi(t)\rangle = \hat{U}(T,t)|\psi(T)\rangle = \frac{e^{-\frac{i\mu}{\hbar}B_1 T + \frac{i\mu}{\hbar}B_2(t-T)}}{\sqrt{2}}\begin{pmatrix} e^{2\frac{i\mu}{\hbar}B_1 T - 2\frac{i\mu}{\hbar}B_2(t-T)} \\ 1 \end{pmatrix}.$$

A measurement of the spin projection along y is surely $-\hbar/2$ at time $2T$ when

$$|\psi(2T)\rangle = \frac{e^{i\alpha}}{\sqrt{2}}\begin{pmatrix} i \\ 1 \end{pmatrix}$$

meaning that, apart from an unimportant phase factor, we are exactly in the eigenstate with negative eigenvalue of the Pauli matrix given by $\hat{\sigma}_y$. The required condition is therefore

$$e^{\frac{2i\mu T}{\hbar}(B_1-B_2)} = i$$

and, hence

$$B_1 = B_2 + \frac{\pi\hbar}{4\mu T} + \frac{n\pi\hbar}{\mu T} \qquad n = 0, \pm 1, \pm 2, \ldots$$

Problem 3.6.
We consider a Stern-Gerlach apparatus allowing the orientation of the spin in a generic direction $\hat{n} = (\sin\theta\cos\phi, \sin\theta\sin\phi, \cos\theta)$, with arbitrary θ and ϕ. For the state with spin projection $+\frac{\hbar}{2}$ in the $\hat{n}$ direction, determine the values of θ and ϕ such that:

1) the probability that a measurement of the z component of the spin gives $+\frac{\hbar}{2}$ is $P(z, +) = \frac{1}{4}$;
2) the angle ϕ maximizes the probability $P(y, +)$ that a measurement of the y component of the spin gives $+\frac{\hbar}{2}$.

Solution
We start from the relation between the spin operator and the Pauli matrices

$$\hat{S} = \frac{\hbar}{2}\hat{\sigma}$$

where

$$\hat{\sigma}_x = \begin{pmatrix} 0 & 1 \\ 1 & 0 \end{pmatrix} \qquad \hat{\sigma}_y = \begin{pmatrix} 0 & -i \\ i & 0 \end{pmatrix} \qquad \hat{\sigma}_z = \begin{pmatrix} 1 & 0 \\ 0 & -1 \end{pmatrix}.$$

The spin matrix in the direction of $\hat{n} = (\sin\theta\cos\phi, \sin\theta\sin\phi, \cos\theta)$ is obtained from the scalar product

$$\hat{S}\cdot\hat{n} = \frac{1}{2}\hbar\begin{pmatrix} \cos\theta & \sin\theta\,e^{-i\phi} \\ \sin\theta\,e^{i\phi} & -\cos\theta \end{pmatrix}.$$

The eigenvectors of $\hat{S}\cdot\hat{n}$ are

$$|+,\hat{n}\rangle = \begin{pmatrix} \cos\left(\frac{\theta}{2}\right) \\ \sin\left(\frac{\theta}{2}\right)e^{i\phi} \end{pmatrix} \qquad |-,\hat{n}\rangle = \begin{pmatrix} -\sin\left(\frac{\theta}{2}\right)e^{-i\phi} \\ \cos\left(\frac{\theta}{2}\right) \end{pmatrix}.$$

The state $|+,\hat{n}\rangle$ represents the state with spin projection $+\frac{\hbar}{2}$ in the $\hat{n}$ direction. In this state, the probability that a measurement of the z component of the spin gives $+\frac{\hbar}{2}$, is the square modulus of the scalar product with

$$|+,z\rangle = \begin{pmatrix} 1 \\ 0 \end{pmatrix}$$

that is the eigenstate of $\hat{\sigma}_z$ with eigenvalue $+1$. The result is $\cos\left(\frac{\theta}{2}\right)$. Therefore, the condition $P(z,+) = \frac{1}{4}$ means that $\cos^2\left(\frac{\theta}{2}\right) = \frac{1}{4}$ and

$$\cos\left(\frac{\theta}{2}\right) = \pm\frac{1}{2}.$$

There are two possible solutions: $\frac{\theta}{2} = \frac{\pi}{3}$ and $\frac{\theta}{2} = \frac{2\pi}{3}$. Since $\theta \leq \pi$, the solution is $\theta = \frac{2\pi}{3}$. Using $\sin\left(\frac{\theta}{2}\right) = \frac{\sqrt{3}}{2}$, the state with spin projection $+\frac{\hbar}{2}$ in the $\hat{n}$ direction becomes

$$|\psi\rangle = \begin{pmatrix} \frac{1}{2} \\ \frac{\sqrt{3}}{2}e^{i\phi} \end{pmatrix}.$$

We now need to set the condition in 2) to determine ϕ. The probability that a measurement of the y component of the spin gives $+\frac{\hbar}{2}$, is the square modulus of the scalar product with

$$|+,y\rangle = \frac{1}{\sqrt{2}}\begin{pmatrix} 1 \\ i \end{pmatrix}$$

that is the eigenstate of $\hat{\sigma}_y$ with eigenvalue $+1$. Therefore, we find

$$P(y,+) = \frac{1}{8}(1 + \sqrt{3}\sin\phi)^2 + \frac{3}{8}\cos^2\phi.$$

The maximum is found from the equations

$$\begin{cases} \frac{dP}{d\phi} = \frac{\sqrt{3}}{4}\cos\phi(1 + \sqrt{3}\sin\phi) - \frac{3}{4}\cos\phi\sin\phi = \frac{\sqrt{3}}{4}\cos\phi = 0 \\ \frac{d^2P}{d\phi^2} = -\frac{\sqrt{3}}{4}\sin\phi < 0 \end{cases}$$

with solution $\phi = \frac{\pi}{2}$.

Problem 3.7.
At time $t = 0$, a quantum state $|\psi(0)\rangle$ is an eigenstate of $\hat{L}^2$, $\hat{L}_z$ with eigenvalues $2\hbar^2$ and 0 respectively. Determine the time evolution of $|\psi(0)\rangle$ according to the Hamiltonian $\hat{H} = \hat{L}_x$, with $\hat{L}_i$ the i-th component of the orbital angular momentum and $\hat{L}^2$ its square. When a measurement of the energy gives $-\hbar$, express the corresponding state as a linear combination of the eigenstates of $\hat{L}^2$ and $\hat{L}_z$.

Solution
The state $|\psi(0)\rangle$ is an eigenstate of the z component of the orbital angular momentum but not of the x component. It follows that the Hamiltonian is not diagonal with respect to the basis given by the spherical harmonics. We remark that the choice of diagonalizing $\hat{L}_z$ together with $\hat{L}^2$ is purely arbitrary: the commutation rules for the angular momentum are very general and, to perform calculations, one has to provide a matrix representation of the operators or to write them in terms of differential operators. Two standard examples are the Pauli matrices for the spin and the momentum operator written as $-i\hbar\nabla$. To solve our problem it is more convenient to

work with matrices. In this case, $\hat{L}_z$ is a diagonal matrix with eigenvalues $\hbar$, 0 and $-\hbar$

$$\hat{L}_z = \hbar \begin{pmatrix} 1 & 0 & 0 \\ 0 & 0 & 0 \\ 0 & 0 & -1 \end{pmatrix}$$

and its eigenvectors are

$$|1,1\rangle_z = \begin{pmatrix} 1 \\ 0 \\ 0 \end{pmatrix} \qquad |1,0\rangle_z = \begin{pmatrix} 0 \\ 1 \\ 0 \end{pmatrix} \qquad |1,-1\rangle_z = \begin{pmatrix} 0 \\ 0 \\ 1 \end{pmatrix}$$

and $|\psi(0)\rangle = |1,0\rangle_z$. The matrix representation of $\hat{L}_x$ is given by

$$\hat{L}_x = \frac{\hbar}{2} \begin{pmatrix} 0 & \sqrt{2} & 0 \\ \sqrt{2} & 0 & \sqrt{2} \\ 0 & \sqrt{2} & 0 \end{pmatrix}.$$

The eigenvalues of this matrix are $\hbar$, 0, $-\hbar$ and correspond to the eigenvectors

$$|1,1\rangle_x = \frac{1}{2} \begin{pmatrix} 1 \\ \sqrt{2} \\ 1 \end{pmatrix} \qquad |1,0\rangle_x = \frac{1}{\sqrt{2}} \begin{pmatrix} 1 \\ 0 \\ -1 \end{pmatrix} \qquad |1,-1\rangle_x = \frac{1}{2} \begin{pmatrix} -1 \\ \sqrt{2} \\ -1 \end{pmatrix}.$$

The time evolution operator is $e^{-i\hat{H}t/\hbar}$, and its action on a generic eigenstate with energy E_n is particularly simple, because it produces a phase $e^{-iE_n t/\hbar}$. Therefore, we need to expand our state $|\psi(0)\rangle$ as a linear combination of the eigenvectors $|1,m\rangle_x$ with $m = 0, \pm 1$

$$|\psi(0)\rangle = a|1,1\rangle_x + b|1,0\rangle_x + c|1,-1\rangle_x.$$

The coefficients are found to be $a = c = 1/\sqrt{2}$, $b = 0$. When acting with $e^{-i\hat{H}t/\hbar}$ on this state, the terms in the right hand side acquire a phase factor depending on their eigenvalues

$$|\psi(t)\rangle = \frac{1}{\sqrt{2}} (e^{-it}|1,1\rangle_x + e^{it}|1,-1\rangle_x).$$

From the text of the problem, we know that a measurement of $\hat{H}$ gives $-\hbar$. Soon after this measurement, the system is described by the state $|1,-1\rangle_x$ which can be decomposed as

$$|1,-1\rangle_x = a|1,1\rangle_z + b|1,0\rangle_z + c|1,-1\rangle_z$$

with the following coefficients: $a = c = -1/2$, $b = 1/\sqrt{2}$.

Problem 3.8.

An atom with the orbital angular momentum $l = 1$ is subject to a constant magnetic field $B = B(\sin\theta\cos\phi, \sin\theta\sin\phi, \cos\theta)$, where B is a constant parameter and θ, ϕ

give the direction of B. The atom is described by the following Hamiltonian

$$\hat{H} = \mu \hat{L} \cdot B$$

where μ is a constant magnetic moment. Characterize the energy spectrum.

Solution
We note that the eigenvalues of the scalar product $\hat{L} \cdot B$ are invariant under rotations. Therefore, we can choose a reference system such that the magnetic field is oriented along the z direction, i.e. $B = B(0,0,1)$. Consequently, the Hamiltonian becomes proportional to the third component of the orbital angular momentum, $\hat{L}_z$, whose eigenvalues are $\hbar, 0, -\hbar$. The eigenvalues of the Hamiltonian $\hat{H} = \mu B \hat{L}_z$ follow

$$E_{+1} = \mu B \hbar \qquad E_0 = 0 \qquad E_{-1} = -\mu B \hbar.$$

It is instructive to repeat the calculation without choosing the direction of B along z, and show that we get the same result. For $l = 1$ the angular momentum has the following matrix representation

$$\hat{L}_x = \frac{\hbar}{\sqrt{2}} \begin{pmatrix} 0 & 1 & 0 \\ 1 & 0 & 1 \\ 0 & 1 & 0 \end{pmatrix} \quad \hat{L}_y = \frac{\hbar}{\sqrt{2}} \begin{pmatrix} 0 & -i & 0 \\ i & 0 & -i \\ 0 & i & 0 \end{pmatrix} \quad \hat{L}_z = \hbar \begin{pmatrix} 1 & 0 & 0 \\ 0 & 0 & 0 \\ 0 & 0 & -1 \end{pmatrix}.$$

Therefore, writing down explicitly $\hat{H} = \mu \hat{L} \cdot B$, we get

$$\hat{H} = \mu (B_x \hat{L}_x + B_y \hat{L}_y + B_z \hat{L}_z) = \mu \hbar B \begin{pmatrix} \cos\theta & \frac{\sin\theta e^{-i\phi}}{\sqrt{2}} & 0 \\ \frac{\sin\theta e^{i\phi}}{\sqrt{2}} & 0 & \frac{\sin\theta e^{-i\phi}}{\sqrt{2}} \\ 0 & \frac{\sin\theta e^{i\phi}}{\sqrt{2}} & -\cos\theta \end{pmatrix}.$$

We can now calculate the characteristic polynomial $\det(\hat{H} - \lambda \, \mathbb{1})$. Setting $\tilde{\lambda} = \mu \hbar B \lambda$, we need to calculate

$$\det(\hat{H} - \lambda \, \mathbb{1}) = \mu \hbar B \det \begin{pmatrix} \cos\theta - \tilde{\lambda} & \frac{\sin\theta e^{-i\phi}}{\sqrt{2}} & 0 \\ \frac{\sin\theta e^{i\phi}}{\sqrt{2}} & -\tilde{\lambda} & \frac{\sin\theta e^{-i\phi}}{\sqrt{2}} \\ 0 & \frac{\sin\theta e^{i\phi}}{\sqrt{2}} & -\cos\theta - \tilde{\lambda} \end{pmatrix}.$$

The result is

$$\frac{\det(\hat{H} - \lambda \, \mathbb{1})}{\mu \hbar B} = -\tilde{\lambda}(\tilde{\lambda}^2 - \cos^2\theta) - \frac{(\cos\theta - \tilde{\lambda})\sin^2\theta}{2} - \frac{(-\cos\theta - \tilde{\lambda})\sin^2\theta}{2} =$$
$$= -\tilde{\lambda}(\tilde{\lambda}^2 - \sin^2\theta - \cos^2\theta) = -\tilde{\lambda}(\tilde{\lambda}^2 - 1).$$

The eigenvalues are the roots of the equation

$$\tilde{\lambda}(\tilde{\lambda}^2 - 1) = 0$$

with solutions $\tilde{\lambda} = 0, \pm 1$, that is the same result obtained previously.

Problem 3.9.

We consider a system with two particles, each one with spin $1/2$. Let the z components of the spins, $\hat{S}_{1z}$ and $\hat{S}_{2z}$, be diagonal, i.e. we use the vector basis $\left|\frac{1}{2}, S_{1z}\right\rangle_z \otimes \left|\frac{1}{2}, S_{2z}\right\rangle_z$ with $S_{1z}, S_{2z} = \pm\frac{1}{2}$ (to simplify matters, set $\hbar = 1$). For a general linear combination of the elements of this vector basis, determine the probability that a simultaneous measurement of $\hat{S}_{1z}$ and $\hat{S}_{2z}$ gives as result:

- $S_{1z} = +1/2$, $S_{2z} = +1/2$;
- $S_{1z} = +1/2$, $S_{2z} = -1/2$;
- $S_{1z} = -1/2$, $S_{2z} = +1/2$;
- $S_{1z} = -1/2$, $S_{2z} = -1/2$.

Then, determine the probability that a simultaneous measurement of $\hat{S}_{1y}$ (the y component of the spin for the first particle) and $\hat{S}_{2z}$ gives $S_{1y} = +1/2$ and $S_{2z} = +1/2$. Finally, determine the probability that a measurement of $\hat{S}_{2z}$ alone gives $-1/2$ as result.

Solution

We take the general linear combination of the four states $\left|\frac{1}{2}, S_{1z}\right\rangle_z \otimes \left|\frac{1}{2}, S_{2z}\right\rangle_z$ with $S_{1z}, S_{2z} = \pm\frac{1}{2}$

$$
|\psi\rangle = \alpha \left|\frac{1}{2}, \frac{1}{2}\right\rangle_z \otimes \left|\frac{1}{2}, \frac{1}{2}\right\rangle_z + \beta \left|\frac{1}{2}, \frac{1}{2}\right\rangle_z \otimes \left|\frac{1}{2}, -\frac{1}{2}\right\rangle_z +
$$
$$
\gamma \left|\frac{1}{2}, -\frac{1}{2}\right\rangle_z \otimes \left|\frac{1}{2}, \frac{1}{2}\right\rangle_z + \delta \left|\frac{1}{2}, -\frac{1}{2}\right\rangle_z \otimes \left|\frac{1}{2}, -\frac{1}{2}\right\rangle_z.
$$

The probability that a simultaneous measurement of $\hat{S}_{1z}$ and $\hat{S}_{2z}$ gives one of the four results reported in the text, is the square modulus of the coefficient multiplying the state with the desired values of the projections S_{1z}, S_{2z}

$$
P\left(S_{1z} = \frac{1}{2}, S_{2z} = \frac{1}{2}\right) = |\alpha|^2 \qquad P\left(S_{1z} = \frac{1}{2}, S_{2z} = -\frac{1}{2}\right) = |\beta|^2
$$

$$
P\left(S_{1z} = -\frac{1}{2}, S_{2z} = \frac{1}{2}\right) = |\gamma|^2 \qquad P\left(S_{1z} = -\frac{1}{2}, S_{2z} = -\frac{1}{2}\right) = |\delta|^2.
$$

To determine the probability associated with a measurement of $\hat{S}_{1y}$, we need to use the basis where the y component of the spin $\hat{S}_y$ is diagonal. The matrix representation of such observable is proportional to a Pauli matrix

$$
\hat{S}_y = \frac{1}{2} \begin{pmatrix} 0 & -i \\ i & 0 \end{pmatrix}
$$

with eigenvalues $\pm 1/2$ corresponding to the eigenvectors

$$
\left|\frac{1}{2}, \frac{1}{2}\right\rangle_y = \frac{1}{\sqrt{2}} \begin{pmatrix} -i \\ 1 \end{pmatrix} \qquad \left|\frac{1}{2}, -\frac{1}{2}\right\rangle_y = \frac{1}{\sqrt{2}} \begin{pmatrix} i \\ 1 \end{pmatrix}.
$$

Also, the eigenvectors corresponding to projections $\pm 1/2$ of the z component of the spin are

$$\left|\frac{1}{2},\frac{1}{2}\right\rangle_z = \begin{pmatrix} 1 \\ 0 \end{pmatrix} \qquad \left|\frac{1}{2},-\frac{1}{2}\right\rangle_z = \begin{pmatrix} 0 \\ 1 \end{pmatrix}.$$

Then, we need to write $\left|\frac{1}{2},\frac{1}{2}\right\rangle_z, \left|\frac{1}{2},-\frac{1}{2}\right\rangle_z$ in terms of $\left|\frac{1}{2},\frac{1}{2}\right\rangle_y, \left|\frac{1}{2},-\frac{1}{2}\right\rangle_y$

$$\left|\frac{1}{2},\frac{1}{2}\right\rangle_z = \frac{i}{\sqrt{2}}\left(\left|\frac{1}{2},\frac{1}{2}\right\rangle_y - \left|\frac{1}{2},-\frac{1}{2}\right\rangle_y\right)$$

$$\left|\frac{1}{2},-\frac{1}{2}\right\rangle_z = \frac{1}{\sqrt{2}}\left(\left|\frac{1}{2},\frac{1}{2}\right\rangle_y + \left|\frac{1}{2},-\frac{1}{2}\right\rangle_y\right).$$

To determine the probability to find $S_{1y} = 1/2$ and $S_{2z} = 1/2$, we consider in $|\psi\rangle$ only those states with $S_{2z} = 1/2$

$$|\psi\rangle_{S_{2z}=1/2} = \alpha\left|\frac{1}{2},\frac{1}{2}\right\rangle_z \otimes \left|\frac{1}{2},\frac{1}{2}\right\rangle_z + \gamma\left|\frac{1}{2},-\frac{1}{2}\right\rangle_z \otimes \left|\frac{1}{2},\frac{1}{2}\right\rangle_z$$

and expand $\left|\frac{1}{2},\frac{1}{2}\right\rangle_z, \left|\frac{1}{2},-\frac{1}{2}\right\rangle_z$ for the first particle in terms of $\left|\frac{1}{2},\frac{1}{2}\right\rangle_y, \left|\frac{1}{2},-\frac{1}{2}\right\rangle_y$

$$|\psi\rangle_{S_{2z}=1/2} = \frac{i}{\sqrt{2}}\alpha\left(\left|\frac{1}{2},\frac{1}{2}\right\rangle_y - \left|\frac{1}{2},-\frac{1}{2}\right\rangle_y\right) \otimes \left|\frac{1}{2},\frac{1}{2}\right\rangle_z +$$

$$\frac{1}{\sqrt{2}}\gamma\left(\left|\frac{1}{2},\frac{1}{2}\right\rangle_y + \left|\frac{1}{2},-\frac{1}{2}\right\rangle_y\right) \otimes \left|\frac{1}{2},\frac{1}{2}\right\rangle_z =$$

$$\frac{i\alpha+\gamma}{\sqrt{2}}\left|\frac{1}{2},\frac{1}{2}\right\rangle_y \otimes \left|\frac{1}{2},\frac{1}{2}\right\rangle_z + \frac{\gamma-i\alpha}{\sqrt{2}}\left|\frac{1}{2},-\frac{1}{2}\right\rangle_y \otimes \left|\frac{1}{2},\frac{1}{2}\right\rangle_z.$$

The answer to the second point is

$$P\left(S_{1y} = \frac{1}{2}, S_{2z} = \frac{1}{2}\right) = \frac{|\alpha|^2 + |\gamma|^2}{2}.$$

Finally, the probability that a measurement of $\hat{S}_{2z}$ gives $-1/2$, is

$$P\left(S_{2z} = -\frac{1}{2}\right) = |\beta|^2 + |\delta|^2$$

that is obtained taking the square of the scalar product between $|\psi\rangle$ and

$$|\psi\rangle_{S_{2z}=-1/2} = \beta\left|\frac{1}{2},\frac{1}{2}\right\rangle_z \otimes \left|\frac{1}{2},-\frac{1}{2}\right\rangle_z + \delta\left|\frac{1}{2},-\frac{1}{2}\right\rangle_z \otimes \left|\frac{1}{2},-\frac{1}{2}\right\rangle_z.$$

Problem 3.10.
Let us consider two particles with spins $S_1 = 3/2$ and $S_2 = 1$:

- at time $t = 0$, determine the wave function of the generic bound state with total spin $S = S_1 + S_2 = 1/2$ and projection along z equal to $S_z = S_{1z} + S_{2z} = 1/2$;
- determine the time evolution of the wave function $\psi(t)$ according to the Hamiltonian $\hat{H} = \hat{S}_{1z}\hat{S}_{2z}/\hbar$, and the time at which $\psi(t) = \psi(0)$;
- determine the time evolution of the average $\langle \hat{S}_{2+}\hat{S}_{2-} \rangle$.

In the above expressions, $\hat{S}_{iz}$ is the z component of the spin for the i-th particle, and $\hat{S}_{2\pm}$ are the raising and lowering operators for the z component of $\hat{S}_2$.

Solution
Let $|S_1, S_{1z}\rangle$ and $|S_2, S_{2z}\rangle$ be the eigenfunctions of the operators $\hat{S}_{1z}$ and $\hat{S}_{2z}$ respectively. We need to move to the basis $|S, S_z, S_1, S_2\rangle$ where the total spin $\hat{S} = \hat{S}_1 + \hat{S}_2$ and its projection along z are diagonal. The relation between the two bases is provided by

$$|S, S_z, S_1, S_2\rangle = \sum_{S_{1z}, S_{2z}} C^{S,S_z}_{S_1, S_{1z}, S_2, S_{2z}} |S_1, S_{1z}\rangle \otimes |S_2, S_{2z}\rangle$$

where we need to determine the coefficients $C^{S,S_z}_{S_1, S_{1z}, S_2, S_{2z}}$ (known as Clebsh-Gordan coefficients) in the case $S_1 = 3/2$, $S_2 = 1$, $S = 1/2$, $S_z = 1/2$. The degeneracy of the bound states for two particles with spins $S_1 = 3/2$ and $S_2 = 1$ is

$$d_{12} = (2S_1 + 1)(2S_2 + 1) = \left(2 \times \frac{3}{2} + 1\right)(2 \times 1 + 1) = 12$$

and the sum $S = S_1 + S_2$ can take the values $S = S_1 + S_2, S_1 + S_2 - 1, S_1 + S_2 - 2, ..., |S_2 - S_1|$. In our case $S = 5/2, 3/2, 1/2$. The state with the z component of the total spin $S_z = S_{1z} + S_{2z} = 1/2$ must be a linear combination of the states $|S_1, S_{1z}\rangle$, $|S_2, S_{2z}\rangle$ such that $S_z = S_{1z} + S_{2z} = 1/2$

$$\left|\frac{1}{2}, \frac{1}{2}, \frac{3}{2}, 1\right\rangle = \alpha \left|\frac{3}{2}, \frac{3}{2}\right\rangle \otimes |1, -1\rangle + \beta \left|\frac{3}{2}, \frac{1}{2}\right\rangle \otimes |1, 0\rangle + \gamma \left|\frac{3}{2}, -\frac{1}{2}\right\rangle \otimes |1, 1\rangle.$$

To find explicitly the coefficients α, β and γ, we act with $\hat{S}_+ = \hat{S}_{1+} \otimes \mathbb{1} + \mathbb{1} \otimes \hat{S}_{2+}$ on both members of the above equation: when acting on the left hand side, we consider the raising operator for the z component of the total spin, i.e. $\hat{S}_+$; when acting on the right hand side, we consider that $\hat{S}_+$ is the sum of $\hat{S}_{1+}$ and $\hat{S}_{2+}$ and that both of them act on the spin of a specific particle, leaving the other unchanged. The action of $\hat{S}_+$ on a generic state is

$$\hat{S}_+ |S, S_z\rangle = \hbar\sqrt{(S - S_z)(S + S_z + 1)} |S, S_z + 1\rangle$$

and gives zero when we act on the state with $S_z = S$. Therefore, we find

$$\hat{S}_+ \left| \frac{1}{2}, \frac{1}{2}, \frac{3}{2}, 1 \right\rangle = 0 = \sqrt{2}\alpha \left| \frac{3}{2}, \frac{3}{2} \right\rangle \otimes |1,0\rangle + \sqrt{3}\beta \left| \frac{3}{2}, \frac{3}{2} \right\rangle \otimes |1,0\rangle +$$
$$\sqrt{2}\beta \left| \frac{3}{2}, \frac{1}{2} \right\rangle \otimes |1,1\rangle + 2\gamma \left| \frac{3}{2}, \frac{1}{2} \right\rangle \otimes |1,1\rangle =$$
$$(\sqrt{2}\alpha + \sqrt{3}\beta) \left| \frac{3}{2}, \frac{3}{2} \right\rangle \otimes |1,0\rangle + (\sqrt{2}\beta + 2\gamma) \left| \frac{3}{2}, \frac{1}{2} \right\rangle \otimes |1,1\rangle.$$

Multiplying this equation by $|3/2, 3/2\rangle \otimes |1,0\rangle$ and $|3/2, 1/2\rangle \otimes |1,1\rangle$, and using the orthogonality and the normalization of the states, we end up with

$$\begin{cases} \sqrt{2}\alpha + \sqrt{3}\beta = 0 \\ \sqrt{2}\beta + 2\gamma = 0 \end{cases}$$

with solution $\alpha = -\sqrt{3/2}\beta, \gamma = -\beta/\sqrt{2}$. The condition that the state is normalized leads to $\beta = 1/\sqrt{3}$. Therefore, the answer to the first question is

$$\left| \frac{1}{2}, \frac{1}{2}, \frac{3}{2}, 1 \right\rangle = -\frac{1}{\sqrt{2}} \left| \frac{3}{2}, \frac{3}{2} \right\rangle \otimes |1,-1\rangle + \frac{1}{\sqrt{3}} \left| \frac{3}{2}, \frac{1}{2} \right\rangle \otimes |1,0\rangle - \frac{1}{\sqrt{6}} \left| \frac{3}{2}, -\frac{1}{2} \right\rangle \otimes |1,1\rangle.$$

The time evolution is obtained with the action of the time evolution operator $e^{-i\hat{H}t/\hbar}$, where $\hat{H} = \hat{S}_{1z}\hat{S}_{2z}/\hbar$. The result is

$$\left| \frac{1}{2}, \frac{1}{2}, \frac{3}{2}, 1, t \right\rangle = -\frac{e^{\frac{3it}{2}}}{\sqrt{2}} \left| \frac{3}{2}, \frac{3}{2} \right\rangle \otimes |1,-1\rangle +$$
$$\frac{1}{\sqrt{3}} \left| \frac{3}{2}, \frac{1}{2} \right\rangle \otimes |1,0\rangle - \frac{e^{\frac{it}{2}}}{\sqrt{6}} \left| \frac{3}{2}, -\frac{1}{2} \right\rangle \otimes |1,1\rangle.$$

The condition that $\psi(t) = \psi(0)$ leads to $t = 4\pi k$, with k an integer number. As for the time evolution of $\langle \hat{S}_{2+}\hat{S}_{2-} \rangle_t$, it is given by the formula

$$\frac{d\langle \hat{S}_{2+}\hat{S}_{2-} \rangle_t}{dt} = \frac{i}{\hbar} \langle [\hat{H}, \hat{S}_{2+}\hat{S}_{2-}] \rangle.$$

To compute the commutator in the right hand side, we recall that $[\hat{S}_i, \hat{S}_j] = i\hbar\varepsilon_{ijk}\hat{S}_k$, where ε_{ijk} is the Levi-Civita tensor and $i, j, k = x, y, z$. Such tensor is totally anti-symmetric and conventionally chosen in such a way that $\varepsilon_{xyz} = 1$. Recalling the definitions $\hat{S}_\pm = (\hat{S}_x \pm i\hat{S}_y)$, we get $[\hat{S}_\pm, \hat{S}_z] = \mp\hbar\hat{S}_\pm$. Using the Jacobi identity (see also Problem 2.2)

$$[\hat{A}, \hat{B}\hat{C}] = \hat{B}[\hat{A}, \hat{C}] + [\hat{A}, \hat{B}]\hat{C}$$

and the result $[\hat{S}_{1i}, \hat{S}_{2j}] = 0, \forall i, j$, we find

$$[\hbar\hat{H}, \hat{S}_{2+}\hat{S}_{2-}] = [\hat{S}_{1z}\hat{S}_{2z}, \hat{S}_{2+}\hat{S}_{2-}] = \hat{S}_{1z}[\hat{S}_{2z}, \hat{S}_{2+}\hat{S}_{2-}] =$$
$$\hat{S}_{1z}([\hat{S}_{2z}, \hat{S}_{2+}]\hat{S}_{2-} + \hat{S}_{2+}[\hat{S}_{2z}, \hat{S}_{2-}]) = \hbar\hat{S}_{1z}(\hat{S}_{2+}\hat{S}_{2-} - \hat{S}_{2+}\hat{S}_{2-}) = 0.$$

This implies that $\langle \hat{S}_{2+}\hat{S}_{2-} \rangle_t$ does not change in time, and its value can be found by averaging on the state at time $t = 0$

$$\langle \hat{S}_{2+}\hat{S}_{2-} \rangle_t = \langle \hat{S}_{2+}\hat{S}_{2-} \rangle_{t=0} = \left\langle \frac{1}{2},\frac{1}{2},\frac{3}{2},1,t=0 \middle| \hat{S}_{2+}\hat{S}_{2-} \middle| \frac{1}{2},\frac{1}{2},\frac{3}{2},1,t=0 \right\rangle.$$

Another possible way to determine $\langle \hat{S}_{2+}\hat{S}_{2-} \rangle_t$, is to average the operator $\hat{S}_{2+}\hat{S}_{2-}$ on the state at time t: also in this case we conclude that $\langle \hat{S}_{2+}\hat{S}_{2-} \rangle$ does not change in time. We now use the relation $\hat{S}_{2+}\hat{S}_{2-} = \hat{S}_2^2 - \hat{S}_{2z}^2 + \hbar\hat{S}_{2z}$, leading to

$$\hat{S}_{2+}\hat{S}_{2-}\left(-\frac{1}{\sqrt{2}}\left|\frac{3}{2},\frac{3}{2}\right\rangle \otimes |1,-1\rangle + \frac{1}{\sqrt{3}}\left|\frac{3}{2},\frac{1}{2}\right\rangle \otimes |1,0\rangle - \frac{1}{\sqrt{6}}\left|\frac{3}{2},-\frac{1}{2}\right\rangle \otimes |1,1\rangle \right) =$$

$$\frac{2\hbar^2}{\sqrt{3}}\left|\frac{3}{2},\frac{1}{2}\right\rangle \otimes |1,0\rangle - \frac{2\hbar^2}{\sqrt{6}}\left|\frac{3}{2},-\frac{1}{2}\right\rangle \otimes |1,1\rangle.$$

Finally, since the states are orthogonal, the value of $\langle \hat{S}_{2+}\hat{S}_{2-} \rangle_t$ at time $t = 0$ is given by

$$\langle \hat{S}_{2+}\hat{S}_{2-} \rangle_{t=0} = \left\langle \frac{1}{2},\frac{1}{2},\frac{3}{2},\frac{1}{2},t=0 \middle| \hat{S}_{2+}\hat{S}_{2-} \middle| \frac{1}{2},\frac{1}{2},\frac{3}{2},\frac{1}{2},t=0 \right\rangle = \hbar^2.$$

Problem 3.11.

Consider a one dimensional problem where pairs of particles (each one with spin 1/2) form bound states. The Hamiltonian for such a system is

$$\hat{H} = \frac{\hat{p}^2}{2m} + U(x)\hat{S}_1 \cdot \hat{S}_2$$

with m the effective mass and $\hat{S}_1$, $\hat{S}_2$ the spin operators for the two particles. The potential energy is $U(x) = 0$ for $x \leq 0$ and $U(x) = U_0 > 0$ for $x > 0$. For a given energy $E > \frac{U_0}{4}$, determine the reflection coefficient when the bound state is a triplet.

Solution
We need to rewrite the product $\hat{S}_1 \cdot \hat{S}_2$ in terms of the operators $\hat{S}_1^2$, $\hat{S}_2^2$ and $\hat{S}^2$. To this end, we write

$$\hat{S}^2 = \hat{S}_1^2 + \hat{S}_2^2 + 2\hat{S}_1 \cdot \hat{S}_2$$

from which we get

$$\hat{S}_1 \cdot \hat{S}_2 = \frac{1}{2}(\hat{S}^2 - \hat{S}_1^2 - \hat{S}_2^2).$$

For a fixed value of the total spin S, the solution of the Schrödinger equation (see Problem 2.23) has the form

$$\begin{cases} \psi(x) = e^{ik_1 x} + C_R e^{-ik_1 x} & x \leq 0 \\ \psi(x) = C_T e^{ik_2 x} & x > 0 \end{cases}$$

with

$$k_1^2 = \frac{2mE}{\hbar^2} \qquad k_2^2 = \frac{2m(E - \frac{1}{2}U_0(S(S+1) - S_1(S_1+1) - S_2(S_2+1)))}{\hbar^2}$$

where the constants C_R, C_T are determined from the condition of continuity of the wave function and its first derivative in $x = 0$. As for the spin part, there are $d_{12} = (2S_1 + 1)(2S_2 + 2) = 4$ bound states for the two particles with spin $S_{1,2} = 1/2$. The total spin for these bound states can take the values $S = S_1 + S_2, S_1 + S_2 - 1, S_1 + S_2 - 2, ..., |S_2 - S_1|$, that implies $S = 1, 0$ in our case. These states are (see also Problem 4.9)

$$\left|1,1,\frac{1}{2},\frac{1}{2}\right\rangle = \left|\frac{1}{2},\frac{1}{2}\right\rangle \otimes \left|\frac{1}{2},\frac{1}{2}\right\rangle$$

$$\left|1,0,\frac{1}{2},\frac{1}{2}\right\rangle = \frac{1}{\sqrt{2}}\left(\left|\frac{1}{2},\frac{1}{2}\right\rangle \otimes \left|\frac{1}{2},-\frac{1}{2}\right\rangle + \left|\frac{1}{2},-\frac{1}{2}\right\rangle \otimes \left|\frac{1}{2},\frac{1}{2}\right\rangle\right)$$

$$\left|1,-1,\frac{1}{2},\frac{1}{2}\right\rangle = \left|\frac{1}{2},-\frac{1}{2}\right\rangle \otimes \left|\frac{1}{2},-\frac{1}{2}\right\rangle$$

defining the so-called triplet (i.e. a state with spin 1 that is triple degenerate).

$$\left|0,0,\frac{1}{2},\frac{1}{2}\right\rangle = \frac{1}{\sqrt{2}}\left(\left|\frac{1}{2},\frac{1}{2}\right\rangle \otimes \left|\frac{1}{2},-\frac{1}{2}\right\rangle - \left|\frac{1}{2},-\frac{1}{2}\right\rangle \otimes \left|\frac{1}{2},\frac{1}{2}\right\rangle\right)$$

defines a singlet state (i.e. a state with spin 0 that is not degenerate). The reflection coefficient, in line with what we have found in previous problems (see Problem 2.23), is given by

$$R = |C_R|^2 = \left(\frac{k_1 - k_2}{k_1 + k_2}\right)^2$$

where $k_2^2 = \frac{2m\left(E - \frac{U_0}{4}\right)}{\hbar^2}$ for the triplet. Therefore, we find

$$R = \left(\frac{\sqrt{2mE} - \sqrt{2m(E - \frac{U_0}{4})}}{\sqrt{2mE} + \sqrt{2m(E - \frac{U_0}{4})}}\right)^2 = 4\frac{\left(4E - \frac{U_0}{2} - 4\sqrt{E(E - \frac{U_0}{4})}\right)^2}{U_0^2}.$$

Problem 3.12.

A hydrogen atom with nuclear spin $S_n = 1/2$ (vector basis given by $\left|\frac{1}{2},\frac{1}{2}\right\rangle_N$ and $\left|\frac{1}{2},-\frac{1}{2}\right\rangle_N$ and $\hbar = 1$ for simplicity) is found in a state with the orbital angular momentum $l = 2$. The total angular momentum operator is $\hat{F} = \hat{S}_n + \hat{S}_e + \hat{L}$, where $\hat{S}_e$ refers to the electron (spin $S_e = 1/2$) part. How many different quantum states are present? What are the possible values of the total angular momentum F? How many states belong to the different F? Denoting with M_F the eigenvalues of $\hat{F}_z$ (the projection of $\hat{F}$ along the z axis), write down the normalized states $|F, M_F\rangle = |3,3\rangle$ and $|F, M_F\rangle = |3,2\rangle$. Finally, determine $\langle 3,2|\hat{S}_e \cdot \hat{S}_N|3,2\rangle$.

Solution

For a given value of the angular momentum J, we have $2J + 1$ states and, hence, $l = 2$ means 5 states while the spin $1/2$ means 2 states. Therefore, we find $5 \times 2 \times 2 = 20$ states. The same number of states can also be seen as coming from the direct product of the spaces associated with the different angular momenta. When considering two angular momenta, J_1 and J_2, the sum of the two (J_{tot}) can take the values $J_{tot} = J_1 + J_2, J_1 + J_2 - 1, J_1 + J_2 - 2, ..., |J_2 - J_1|$. Therefore, when we sum $l = 2$ and $S_e = 1/2$, we have $5/2$ and $3/2$; if we further add S_n, we find $F = 3, 2, 1$. The case $F = 2$ is double degenerate. Counting these states, we have 7 states with $F = 3$, 10 states with $F = 2$, and 3 states with $F = 1$, for a total of $7 + 10 + 3 = 20$. The state with maximum F and M_F is unique

$$|3,3\rangle = \left|\frac{1}{2},\frac{1}{2}\right\rangle_N \otimes \left|\frac{1}{2},\frac{1}{2}\right\rangle_e \otimes |2,2\rangle_L.$$

Starting from $|3,3\rangle$ and using the lowering operator for the z component of the total angular momentum

$$\hat{F}_- = \hat{S}_{n-} + \hat{S}_{e-} + \hat{L}_-$$

we can obtain states with $M_F = 2$. For the lowering operator, we recall the property

$$\hat{J}_-|J,M_J\rangle = \sqrt{(J+M_J)(J-M_J+1)}|J,M_j-1\rangle.$$

Therefore, when acting with $\hat{F}_-$ on $|3,3\rangle$ and with $\hat{S}_{n,-} + \hat{S}_{e,-} + \hat{L}_-$ on $\left|\frac{1}{2},\frac{1}{2}\right\rangle_N \otimes \left|\frac{1}{2},\frac{1}{2}\right\rangle_e \otimes |2,2\rangle_L$, we find the following result

$$|3,2\rangle = \frac{1}{\sqrt{6}}\left(\left|\frac{1}{2},-\frac{1}{2}\right\rangle_N \otimes \left|\frac{1}{2},\frac{1}{2}\right\rangle_e \otimes |2,2\rangle_L +\right.$$
$$\left.\left|\frac{1}{2},\frac{1}{2}\right\rangle_N \otimes \left|\frac{1}{2},-\frac{1}{2}\right\rangle_e \otimes |2,2\rangle_L + 2\left|\frac{1}{2},\frac{1}{2}\right\rangle_N \otimes \left|\frac{1}{2},\frac{1}{2}\right\rangle_e \otimes |2,1\rangle_L\right).$$

As for the last point, we recall that

$$\hat{A} \cdot \hat{B} = \hat{A}_x\hat{B}_x + \hat{A}_y\hat{B}_y + \hat{A}_z\hat{B}_z = \frac{1}{2}\left(\hat{A}_+\hat{B}_- + \hat{A}_-\hat{B}_+\right) + \hat{A}_z\hat{B}_z$$

with $\hat{A}_\pm$, $\hat{B}_\pm$ the raising and lowering operators for the z component of $\hat{A}$, $\hat{B}$. Setting $\hat{A} = \hat{S}_N$ and $\hat{B} = \hat{S}_e$, we find that $\langle 3,2|\hat{S}_{ez} \cdot \hat{S}_{Nz}|3,2\rangle = 0$ and

$$\langle 3,2|\hat{S}_N \cdot \hat{S}_e|3,2\rangle = \frac{1}{2}\langle 3,2|\hat{S}_{N+}\hat{S}_{e-} + \hat{S}_{N-}\hat{S}_{e+}|3,2\rangle = \frac{1}{6}.$$

Problem 3.13.
Determine the operator $\hat{R}_{\alpha,z}$, allowing for a rotation of an angle α around the z axis. Determine the eigenstates of $\hat{R}_{\alpha,z}$ and rotate of an angle $\alpha = \frac{\pi}{2}$ the state $|\psi\rangle = \frac{|1,1\rangle + |1,-1\rangle}{\sqrt{2}}$, with $|1,m\rangle$ ($m = \pm 1, 0$) the eigenstates of the z component of the orbital angular momentum $\hat{L}_z$.

Solution
Symmetry transformations (including rotations) are described by unitary operators. In particular, we identify the orbital angular momentum as the operator generating

rotations. Therefore, for a generic rotation of an angle α around the n axis, we need to use the projection of the orbital angular momentum on such axis and define

$$\hat{R}_{\alpha,\hat{n}} = e^{-i\alpha\hat{n}\cdot\hat{L}/\hbar}$$

as the operator generating the rotation. In our case, the rotation is around the z axis and $\hat{R}_{\alpha,z} = e^{-i\alpha\hat{L}_z/\hbar}$. The matrix representation of the z component of the orbital angular momentum is

$$\hat{L}_z = \hbar \begin{pmatrix} 1 & 0 & 0 \\ 0 & 0 & 0 \\ 0 & 0 & -1 \end{pmatrix}$$

and it is not difficult to calculate $e^{-i\alpha\hat{L}_z/\hbar}$ because $\hat{L}_z$ is diagonal

$$\hat{R}_{\alpha,z} = \begin{pmatrix} e^{-i\alpha} & 0 & 0 \\ 0 & 0 & 0 \\ 0 & 0 & e^{i\alpha} \end{pmatrix}.$$

Therefore, the states $|1,m\rangle$ $(m = \pm 1, 0)$ are eigenstates of $\hat{R}_{\alpha,z}$ with eigenvalues $e^{\mp i\alpha}$ and

$$\hat{R}_{\alpha,z}|\psi\rangle = \frac{e^{-i\alpha}|1,1\rangle + e^{i\alpha}|1,-1\rangle}{\sqrt{2}}$$

from which

$$\hat{R}_{\frac{\pi}{2},z}|\psi\rangle = -i\frac{|1,1\rangle - |1,-1\rangle}{\sqrt{2}}.$$

Problem 3.14.
A quantum system with spin $S = 1/2$ has the following Hamiltonian

$$\hat{H} = \hat{L}_z + \frac{\hat{S}_x}{2}$$

with $\hat{L}_z$ the z component of the orbital angular momentum, and $\hat{S}_x$ the projection of the spin on the x axis. At time $t = 0$, a simultaneous measurement of $\hat{L}_y$ and $\hat{S}_y$ (the y components of the orbital angular momentum and the spin) gives $\hbar$ and $\hbar/2$:

- write down the state at $t = 0$ in the vector basis where $\hat{L}_z$, $\hat{S}_z$ are diagonal;
- determine the time evolution of the state at the generic time t;
- determine the first time at which a measurement of $\hat{L}_y$ and $\hat{S}_y$ gives the same values found at time $t = 0$.

Solution
The operator $\hat{L}_y$ has the following matrix representation

$$\hat{L}_y = \frac{i\hbar}{2} \begin{pmatrix} 0 & -\sqrt{2} & 0 \\ \sqrt{2} & 0 & -\sqrt{2} \\ 0 & \sqrt{2} & 0 \end{pmatrix}$$

and its eigenvectors are

$$|1,1\rangle_y = \frac{1}{2}\begin{pmatrix} 1 \\ i\sqrt{2} \\ -1 \end{pmatrix} \qquad |1,0\rangle_y = \frac{1}{\sqrt{2}}\begin{pmatrix} 1 \\ 0 \\ 1 \end{pmatrix} \qquad |1,-1\rangle_y = \frac{1}{2}\begin{pmatrix} 1 \\ -i\sqrt{2} \\ -1 \end{pmatrix}$$

corresponding to the eigenvalues $\hbar$, 0, $-\hbar$. The eigenstate of $\hat{S}_y$ with eigenvalue $\hbar/2$ is given by

$$\left|\frac{1}{2},\frac{1}{2}\right\rangle_y = \frac{1}{\sqrt{2}}\begin{pmatrix} -i \\ 1 \end{pmatrix}$$

and, therefore, the state at time $t = 0$ is

$$|\psi(0)\rangle = |1,1\rangle_y \otimes \left|\frac{1}{2},\frac{1}{2}\right\rangle_y = \frac{1}{2\sqrt{2}}\begin{pmatrix} 1 \\ i\sqrt{2} \\ -1 \end{pmatrix} \otimes \begin{pmatrix} -i \\ 1 \end{pmatrix} =$$

$$\frac{1}{2\sqrt{2}}\left[\begin{pmatrix} 1 \\ 0 \\ 0 \end{pmatrix} + i\sqrt{2}\begin{pmatrix} 0 \\ 1 \\ 0 \end{pmatrix} - \begin{pmatrix} 0 \\ 0 \\ 1 \end{pmatrix}\right] \otimes \left[-i\begin{pmatrix} 1 \\ 0 \end{pmatrix} + \begin{pmatrix} 0 \\ 1 \end{pmatrix}\right]$$

where we have expanded in the vector basis where $\hat{L}_z$, $\hat{S}_z$ are diagonal. To determine the time evolution, we need to expand the spin part in the vector basis where $\hat{S}_x$ is diagonal. The eigenstates of $\hat{S}_x$ are

$$\left|\frac{1}{2},\frac{1}{2}\right\rangle_x = \frac{1}{\sqrt{2}}\begin{pmatrix} 1 \\ 1 \end{pmatrix} \qquad\qquad \left|\frac{1}{2},-\frac{1}{2}\right\rangle_x = \frac{1}{\sqrt{2}}\begin{pmatrix} 1 \\ -1 \end{pmatrix}$$

and we can express $\left|\frac{1}{2},\frac{1}{2}\right\rangle_y$ in terms of $\left|\frac{1}{2},\pm\frac{1}{2}\right\rangle_x$ as

$$\left|\frac{1}{2},\frac{1}{2}\right\rangle_y = \frac{1}{\sqrt{2}}\begin{pmatrix} -i \\ 1 \end{pmatrix} = \frac{(1-i)}{2\sqrt{2}}\begin{pmatrix} 1 \\ 1 \end{pmatrix} - \frac{(1+i)}{2\sqrt{2}}\begin{pmatrix} 1 \\ -1 \end{pmatrix} =$$

$$\left(\frac{1-i}{2}\right)\left|\frac{1}{2},\frac{1}{2}\right\rangle_x - \left(\frac{1+i}{2}\right)\left|\frac{1}{2},-\frac{1}{2}\right\rangle_x.$$

The time evolution is

$$|\psi(t)\rangle = e^{-i\frac{\hat{L}_z t}{\hbar}} \otimes e^{-i\frac{\hat{S}_x t}{2\hbar}}\left(|1,1\rangle_y \otimes \left|\frac{1}{2},\frac{1}{2}\right\rangle_y\right) =$$

$$\left[\frac{e^{-it}}{2}\begin{pmatrix} 1 \\ 0 \\ 0 \end{pmatrix} + \frac{i\sqrt{2}}{2}\begin{pmatrix} 0 \\ 1 \\ 0 \end{pmatrix} - \frac{e^{it}}{2}\begin{pmatrix} 0 \\ 0 \\ 1 \end{pmatrix}\right] \otimes \left[\frac{(1-i)e^{-\frac{it}{4}}}{2\sqrt{2}}\begin{pmatrix} 1 \\ 1 \end{pmatrix} - \frac{(1+i)e^{\frac{it}{4}}}{2\sqrt{2}}\begin{pmatrix} 1 \\ -1 \end{pmatrix}\right]$$

where the operator $e^{-i\hat{L}_z t/\hbar}$ acts on the orbital part ($|1,1\rangle_y$) while $e^{-i\hat{S}_x t/2\hbar}$ acts on the spin part ($|\frac{1}{2},\frac{1}{2}\rangle_y$). The first time at which the state is equal to the initial one is found from $e^{-\frac{it}{4}} = 1$, leading to $t = 8\pi$.

Problem 3.15.

Consider a hydrogen atom with potential energy $V(r) = -\frac{1}{r}$ (the electron charge is such that $e = 1$) and the trial function $\psi(r) = Ne^{-(\frac{r}{a})^2}$, where r is the radial coordinate and N, a constants:

- determine the normalization constant N;
- determine the average value of the energy on the state $\psi(r)$;
- determine the optimal value of a using the variational method.

Solution
We start from the normalization condition for the trial function $\psi(r)$

$$\int_0^{+\infty} |\psi(r)|^2 r^2 dr \int_0^\pi \sin\theta d\theta \int_0^{2\pi} d\phi = 4\pi N^2 \int_0^{+\infty} r^2 e^{-\frac{2r^2}{a^2}} dr = 4\pi N^2 \frac{a^3}{8}\sqrt{\frac{\pi}{2}} = 1$$

where we have used the integral $\int_0^{+\infty} r^2 e^{-\frac{2r^2}{a^2}} dr = \frac{a^3}{8}\sqrt{\frac{\pi}{2}}$. The resulting value of N is $N = \sqrt{\frac{1}{a^3}}\left(\frac{2}{\pi}\right)^{\frac{3}{4}}$. We then proceed with the computation of the average energy. As for the kinetic energy, we rewrite the Laplacian operator as

$$\nabla^2 = \frac{1}{r^2}\frac{\partial}{\partial r}\left(r^2 \frac{\partial}{\partial r}\right) - \frac{\hat{L}^2}{\hbar^2 r^2}$$

where $\hat{L}^2$ is the squared orbital angular momentum. The trial function $\psi(r)$ has no angular part, meaning that it is proportional to the spherical harmonic $Y_{0,0}(\theta,\phi) = \frac{1}{\sqrt{4\pi}}$ that is an eigenfunction of $\hat{L}^2$ with zero eigenvalue. Therefore, the average kinetic energy is

$$T(a) = \langle\psi|\hat{T}|\psi\rangle = -\frac{\hbar^2}{2m}4\pi N^2 \int_0^{+\infty} r^2 e^{-\frac{r^2}{a^2}}\left(\frac{1}{r^2}\frac{d}{dr}\left(r^2 \frac{d}{dr}e^{-\frac{r^2}{a^2}}\right)\right) dr =$$

$$-\frac{\hbar^2}{2m}4\pi N^2 \int_0^{+\infty} r^2 \left(\frac{4r^2 - 6a^2}{a^4}\right)e^{-\frac{2r^2}{a^2}} dr = \frac{3}{2}\frac{\hbar^2}{ma^2}$$

where we have used $\int_0^{+\infty} r^4 e^{-\frac{2r^2}{a^2}} dr = \frac{3a^5}{32}\sqrt{\frac{\pi}{2}}$. As for the average value of the potential energy, we get

$$V(a) = \langle\psi|\hat{V}|\psi\rangle = -4\pi N^2 \int_0^{+\infty} re^{-\frac{2r^2}{a^2}} dr = -\sqrt{\frac{2}{\pi}}\frac{2}{a}$$

where we have used $\int_0^{+\infty} r e^{-\frac{2r^2}{a^2}} dr = \frac{a^2}{4}$. The average energy is

$$E(a) = T(a) + V(a) = \frac{3}{2} \frac{\hbar^2}{ma^2} - \sqrt{\frac{2}{\pi} \frac{2}{a}}.$$

Imposing the variational condition $\frac{dE(a)}{da} = 0$ (see also Problems 2.27 and 2.29), we find that

$$-3\frac{\hbar^2}{ma^3} + \sqrt{\frac{2}{\pi} \frac{2}{a^2}} = 0$$

and, hence, $a = \sqrt{\frac{\pi}{2} \frac{3}{2} \frac{\hbar^2}{m}}$.

Problem 3.16.

Consider a quantum system with the total (orbital plus spin) angular momentum $J = 3/2$:

- write down the matrix representation of the raising and lowering operators $\hat{J}_\pm$ for the z component of $\hat{J}$. To this end, make use of the vector basis where $\hat{J}_z$ is diagonal;
- write down the matrix representation of $\hat{J}_x$ and $\hat{J}_y$ (the x and y components of $\hat{J}$);
- determine the sum of the diagonal elements of the matrix representation of the angular momentum in the direction $\hat{n} = \frac{1}{\sqrt{3}}(1,1,1)$.

Solution

The matrix representation of $\hat{J}_z$ is a diagonal matrix with eigenvalues $M_J = -\frac{3\hbar}{2}$, $-\frac{\hbar}{2}, +\frac{\hbar}{2}, +\frac{3\hbar}{2}$, that is

$$\hat{J}_z = \frac{\hbar}{2} \begin{pmatrix} 3 & 0 & 0 & 0 \\ 0 & 1 & 0 & 0 \\ 0 & 0 & -1 & 0 \\ 0 & 0 & 0 & -3 \end{pmatrix}.$$

The action of the raising and lowering operators $\hat{J}_\pm$ on the generic eigenstate $|J, M_J\rangle$ is

$$\hat{J}_+|J, M_J\rangle = \hbar\sqrt{(J - M_J)(J + M_J + 1)}|J, M_J + 1\rangle$$
$$\hat{J}_-|J, M_J\rangle = \hbar\sqrt{(J + M_J)(J - M_J + 1)}|J, M_J - 1\rangle$$

and, hence

$$\hat{J}_+ = (\hat{J}_-)^\dagger = \hbar \begin{pmatrix} 0 & \sqrt{3} & 0 & 0 \\ 0 & 0 & 2 & 0 \\ 0 & 0 & 0 & \sqrt{3} \\ 0 & 0 & 0 & 0 \end{pmatrix}.$$

Once we know $\hat{J}_\pm$, $\hat{J}_x$ and $\hat{J}_y$ follow

$$\hat{J}_x = \frac{\hat{J}_+ + \hat{J}_-}{2} = \frac{\hbar}{2} \begin{pmatrix} 0 & \sqrt{3} & 0 & 0 \\ \sqrt{3} & 0 & 2 & 0 \\ 0 & 2 & 0 & \sqrt{3} \\ 0 & 0 & \sqrt{3} & 0 \end{pmatrix}$$

$$\hat{J}_y = \frac{\hat{J}_+ - \hat{J}_-}{2i} = \frac{\hbar}{2} \begin{pmatrix} 0 & -i\sqrt{3} & 0 & 0 \\ i\sqrt{3} & 0 & -2i & 0 \\ 0 & 2i & 0 & -i\sqrt{3} \\ 0 & 0 & \sqrt{3} & 0 \end{pmatrix}.$$

The sum of the diagonal elements of $\hat{J} \cdot \hat{n}$ is zero, since the trace of a matrix is invariant under rotations.

Problem 3.17.
Let us consider the three dimensional rotator with Hamiltonian

$$\hat{H} = \frac{-3\hbar \hat{L}_z + \hat{L}^2}{2I}$$

where $\hat{L}^2$ is the squared orbital angular momentum, $\hat{L}_z$ its component along the z direction, and I the momentum of inertia:

- for a given value l of the orbital angular momentum, determine the smallest energy available $E_{l,l}$ and its eigenstate;
- determine the ground state for the system, showing that it corresponds to the case $l = 1$;
- determine the eigenstate and eigenvalue for the first excited state.

Solution
The eigenstates of $\hat{H}$ are the spherical harmonics $Y_{l,m} = \langle \theta, \phi | l, m \rangle$, with the property

$$\hat{L}^2 |l, m\rangle = \hbar^2 l(l+1) |l, m\rangle \qquad \hat{L}_z |l, m\rangle = \hbar m |l, m\rangle .$$

Therefore, the eigenvalues of $\hat{H}$ are

$$E_{l,m} = \frac{\hbar^2}{2I} \left(l(l+1) - 3m \right).$$

For a given l, we find a minimum for $m = l$

$$E_{l,l} = \frac{\hbar^2}{2I} \left(l^2 - 2l \right).$$

The minimum $E_{l,l}$ is found when $l = 1$, $E_{1,1} = -\frac{\hbar^2}{2I}$, with eigenstate

$$Y_{1,1}(\theta, \phi) = -\sqrt{\frac{3}{8\pi}} e^{+i\phi} \sin \theta.$$

Finally, the first excited state is double degenerate and corresponds to $l = 2$, $m = 2$ or $l = 0$, $m = 0$ with energy $E_{0,0} = E_{2,2} = 0$. The two eigenstates are

$$Y_{0,0}(\theta, \phi) = \frac{1}{\sqrt{4\pi}} \qquad Y_{2,2}(\theta, \phi) = \frac{1}{4}\sqrt{\frac{15}{2\pi}} \sin^2 \theta e^{2i\phi}.$$

Problem 3.18.

Two particles of spin $S_1 = 1$ and $S_2 = 1/2$ form a bound state with total spin $S = 1/2$ and projection along the z axis $S_z = 1/2$. Find the probability to measure the different values of the projection along z and x for the spin of the second particle ($\hat{S}_{2z}$ and $\hat{S}_{2x}$). If a measurement of $\hat{S}_{2x}$ gives $1/2$, determine the possible outcomes of a simultaneous measurement of $\hat{S}_{1x}$ and the associated probabilities. For simplicity, assume $\hbar = 1$.

Solution

Let us represent the states with $|S_1, S_{1z}\rangle_z$ and $|S_2, S_{2z}\rangle_z$, where S_i is the spin of the i-th particle and S_{iz} its projection along the z axis. For the bound states, we use the total spin $S = S_1 + S_2$, its projection along the z axis, $S_z = S_{1z} + S_{2z}$, and the values of the spin S_1 and S_2: $|S, S_z, S_1, S_2\rangle_z$. The total number of states for two particles with spins $S_1 = 1$ and $S_2 = 1/2$ is

$$d_{12} = (2S_1 + 1)(2S_2 + 1) = (2 \times 1 + 1)\left(2 \times \frac{1}{2} + 1\right) = 6.$$

As already discussed in Problem 3.12, the same number of states can also be seen as coming from the direct product of the spaces associated with the different spins. When considering two spins, S_1 and S_2, the sum of the two can take the values $S = S_1 + S_2, S_1 + S_2 - 1, S_1 + S_2 - 2, ..., |S_2 - S_1|$, that implies $S = 3/2, S = 1/2$ in our case. Consequently, the degeneracy of the bound states is $d_S = d_{3/2} + d_{1/2} = 4 + 2 = 6$. The state with maximum S and S_z is unique

$$\left|\frac{3}{2}, \frac{3}{2}, 1, \frac{1}{2}\right\rangle_z = |1, 1\rangle_z \otimes \left|\frac{1}{2}, \frac{1}{2}\right\rangle_z.$$

If we act on the left hand side with the lowering operator $\hat{S}_- = \hat{S}_{1-} \otimes \mathbb{1} + \mathbb{1} \otimes \hat{S}_{2-}$, we get

$$\hat{S}_-\left|\frac{3}{2}, \frac{3}{2}, 1, \frac{1}{2}\right\rangle_z = \sqrt{3}\left|\frac{3}{2}, \frac{1}{2}, 1, \frac{1}{2}\right\rangle_z$$

from which (acting with $\hat{S}_{1-} \otimes \mathbb{1} + \mathbb{1} \otimes \hat{S}_{2-}$ on $|1, 1\rangle_z \otimes |\frac{1}{2}, \frac{1}{2}\rangle_z$), we get

$$\left|\frac{3}{2}, \frac{1}{2}, 1, \frac{1}{2}\right\rangle_z = \sqrt{\frac{2}{3}}|1, 0\rangle_z \otimes \left|\frac{1}{2}, \frac{1}{2}\right\rangle_z + \frac{1}{\sqrt{3}}|1, 1\rangle_z \otimes \left|\frac{1}{2}, -\frac{1}{2}\right\rangle_z.$$

At fixed $S_z = 1/2$, the state with $S = 1/2$ must be of the form

$$\left|\frac{1}{2}, \frac{1}{2}, 1, \frac{1}{2}\right\rangle_z = \alpha|1, 0\rangle_z \otimes \left|\frac{1}{2}, \frac{1}{2}\right\rangle_z + \beta|1, 1\rangle_z \otimes \left|\frac{1}{2}, -\frac{1}{2}\right\rangle_z,$$

with α and β real constants to be determined. Since the states $|\frac{3}{2}, \frac{1}{2}, 1, \frac{1}{2}\rangle_z$ and $|\frac{1}{2}, \frac{1}{2}, 1, \frac{1}{2}\rangle_z$ correspond to different values of the total spin S, they are orthogonal, that implies

$$0 = \left(\sqrt{\frac{2}{3}} \langle 1,0|_z \otimes \left\langle \frac{1}{2},\frac{1}{2} \right|_z + \frac{1}{\sqrt{3}} \langle 1,1|_z \otimes \left\langle \frac{1}{2},-\frac{1}{2} \right|_z \right) \times$$

$$\left(\alpha |1,0\rangle_z \otimes \left| \frac{1}{2},\frac{1}{2} \right\rangle_z + \beta |1,1\rangle_z \otimes \left| \frac{1}{2},-\frac{1}{2} \right\rangle_z \right) = \sqrt{\frac{2}{3}}\alpha + \frac{1}{\sqrt{3}}\beta.$$

This condition, together with the normalization condition $1 = \alpha^2 + \beta^2$, completely determines the state

$$\left| \frac{1}{2},\frac{1}{2},1,\frac{1}{2} \right\rangle = \frac{1}{\sqrt{3}} |1,0\rangle_z \otimes \left| \frac{1}{2},\frac{1}{2} \right\rangle_z - \sqrt{\frac{2}{3}} |1,1\rangle_z \otimes \left| \frac{1}{2},-\frac{1}{2} \right\rangle_z.$$

A measurement of $\hat{S}_{2z}$ on this state produces $1/2$ and $-1/2$ as result, with probabilities $P(1/2) = 1/3$ and $P(-1/2) = 2/3$. To get insight on the possible outcomes for the measurement of the x component of $\hat{S}_2$, we need to change basis, and express $\left| \frac{1}{2},\pm\frac{1}{2} \right\rangle_z$ in the basis where $\hat{S}_{2x}$ is diagonal. Therefore (see also Problem 3.14)

$$\left| \frac{1}{2},\frac{1}{2} \right\rangle_z = \frac{1}{\sqrt{2}} \left(\left| \frac{1}{2},\frac{1}{2} \right\rangle_x + \left| \frac{1}{2},-\frac{1}{2} \right\rangle_x \right) \quad \left| \frac{1}{2},-\frac{1}{2} \right\rangle_z = \frac{1}{\sqrt{2}} \left(\left| \frac{1}{2},\frac{1}{2} \right\rangle_x - \left| \frac{1}{2},-\frac{1}{2} \right\rangle_x \right)$$

where the subscript x means that we are dealing with the eigenstates of $\hat{S}_{2x}$. With this change of basis, the state becomes

$$\left| \frac{1}{2},\frac{1}{2},1,\frac{1}{2} \right\rangle_z = \left(\frac{1}{\sqrt{6}} |1,0\rangle_z - \sqrt{\frac{1}{3}} |1,1\rangle_z \right) \otimes \left| \frac{1}{2},\frac{1}{2} \right\rangle_x +$$

$$\left(\frac{1}{\sqrt{6}} |1,0\rangle_z + \sqrt{\frac{1}{3}} |1,1\rangle_z \right) \otimes \left| \frac{1}{2},-\frac{1}{2} \right\rangle_x.$$

We see that a measurement of $\hat{S}_{2x}$ gives $\pm 1/2$ with probability $1/2$. Also, if a measurement of $\hat{S}_{2x}$ gives $+1/2$, the outcome of a simultaneous measurement of $\hat{S}_{1x}$ is obtained by projecting $|1,0\rangle_z$ and $|1,1\rangle_z$ on the vector basis where $\hat{S}_{1x}$ is diagonal. The eigenvectors of $\hat{S}_{1x}$ are

$$|1,1\rangle_x = \frac{1}{2} \begin{pmatrix} 1 \\ \sqrt{2} \\ 1 \end{pmatrix} \quad |1,0\rangle_x = \frac{1}{\sqrt{2}} \begin{pmatrix} 1 \\ 0 \\ -1 \end{pmatrix} \quad |1,-1\rangle_x = \frac{1}{2} \begin{pmatrix} 1 \\ -\sqrt{2} \\ 1 \end{pmatrix}$$

and, hence

$$|1,1\rangle_z = \frac{1}{2} |1,1\rangle_x + \frac{1}{\sqrt{2}} |1,0\rangle_x + \frac{1}{2} |1,-1\rangle_x$$

$$|1,0\rangle_z = \frac{1}{\sqrt{2}} (|1,1\rangle_x - |1,-1\rangle_x).$$

The part of $\left|\frac{1}{2},\frac{1}{2},1,\frac{1}{2}\right\rangle_z$ with $S_{2x} = 1/2$ becomes

$$\left|\frac{1}{2},\frac{1}{2},1,\frac{1}{2}\right\rangle_{z,S_{2x}=1/2} = \left(-\frac{1}{\sqrt{6}}|1,0\rangle_x - \frac{1}{\sqrt{3}}|1,-1\rangle_x\right) \otimes \left|\frac{1}{2},\frac{1}{2}\right\rangle_x.$$

Therefore, given $S_{2x} = 1/2$, a simultaneous measurement of $\hat{S}_{1x}$ gives as result $S_{1x} = 1,0,-1$ with probabilities

$$P(S_{1x} = 1, S_{2x} = 1/2) = 0 \quad P(S_{1x} = 0, S_{2x} = 1/2) = \frac{1}{6} \quad P(S_{1x} = -1, S_{2x} = 1/2) = \frac{1}{3}.$$

Problem 3.19.

Two particles with spins $S_1 = 1$ and $S_2 = 1/2$ form a bound state. Construct the eigenstates for the total spin $\hat{S} = \hat{S}_1 + \hat{S}_2$ and its projection $\hat{S}_z$ along the z axis. If at time $t = 0$ the system is in the state with total spin $S = 1/2$ and projection $S_z = 1/2$, and if the Hamiltonian of the system is $\hat{H} = \hat{S}_{2z}$ (i.e. the projection of the second spin on the z axis), determine the probability that at a time $t > 0$ a measurement of the squared total spin and $\hat{S}_z$ gives $15/4$ and $1/2$ respectively. For simplicity, use $\hbar = 1$.

Solution

The total spin for these bound states can take the values $S = S_1 + S_2, S_1 + S_2 - 1, S_1 + S_2 - 2, ..., |S_2 - S_1|$, which means $S = \frac{3}{2}$ and $S = \frac{1}{2}$. Following the same procedures of previous problems (see Problems 3.10 and 3.18), we write down the unique state with maximum S and S_z

$$\left|\frac{3}{2},\frac{3}{2},1,\frac{1}{2}\right\rangle = |1,1\rangle \otimes \left|\frac{1}{2},\frac{1}{2}\right\rangle$$

and we act with the lowering operator, $\hat{S}_- = \hat{S}_{1-} \otimes \mathbb{1} + \mathbb{1} \otimes \hat{S}_{2-}$, to produce states with smaller S_z (see also Problems 3.12 and 3.18). The result is

$$\left|\frac{3}{2},\frac{1}{2},1,\frac{1}{2}\right\rangle = \sqrt{\frac{2}{3}}|1,0\rangle \otimes \left|\frac{1}{2},\frac{1}{2}\right\rangle + \frac{1}{\sqrt{3}}|1,1\rangle \otimes \left|\frac{1}{2},-\frac{1}{2}\right\rangle$$

$$\left|\frac{3}{2},-\frac{1}{2},1,\frac{1}{2}\right\rangle = \sqrt{\frac{2}{3}}|1,0\rangle \otimes \left|\frac{1}{2},-\frac{1}{2}\right\rangle + \frac{1}{\sqrt{3}}|1,-1\rangle \otimes \left|\frac{1}{2},\frac{1}{2}\right\rangle$$

$$\left|\frac{3}{2},-\frac{3}{2},1,\frac{1}{2}\right\rangle = |1,-1\rangle \otimes \left|\frac{1}{2},-\frac{1}{2}\right\rangle.$$

The state with $S = 1/2$ and $S_z = 1/2$ is orthogonal to $\left|\frac{3}{2},\frac{1}{2},1,\frac{1}{2}\right\rangle$ and is given by

$$\left|\frac{1}{2},\frac{1}{2},1,\frac{1}{2}\right\rangle = \frac{1}{\sqrt{3}}|1,0\rangle \otimes \left|\frac{1}{2},\frac{1}{2}\right\rangle - \sqrt{\frac{2}{3}}|1,1\rangle \otimes \left|\frac{1}{2},-\frac{1}{2}\right\rangle.$$

Again, using the lowering operator, we obtain

$$\left|\frac{1}{2},-\frac{1}{2},1,\frac{1}{2}\right\rangle = \frac{1}{\sqrt{3}}|1,0\rangle \otimes \left|\frac{1}{2},-\frac{1}{2}\right\rangle - \sqrt{\frac{2}{3}}|1,-1\rangle \otimes \left|\frac{1}{2},\frac{1}{2}\right\rangle.$$

To obtain the time evolution of the state with $S = 1/2$ and $S_z = 1/2$, we act with the time evolution operator $e^{-i\hat{H}t}$ on the wave function at time $t = 0$

$$\left|\frac{1}{2},\frac{1}{2},1,\frac{1}{2},t\right\rangle = e^{-i\hat{H}t}\left|\frac{1}{2},\frac{1}{2},1,\frac{1}{2}\right\rangle = \frac{e^{\frac{-it}{2}}}{\sqrt{3}}|1,0\rangle\otimes\left|\frac{1}{2},\frac{1}{2}\right\rangle - \sqrt{\frac{2}{3}}e^{\frac{it}{2}}|1,1\rangle\otimes\left|\frac{1}{2},-\frac{1}{2}\right\rangle.$$

The state corresponding to $S(S+1) = 15/4$ and $S_z = 1/2$ is $\left|\frac{3}{2},\frac{1}{2},1,\frac{1}{2}\right\rangle$. The desired probability is the square modulus of the projection of $\left|\frac{1}{2},\frac{1}{2},1,\frac{1}{2},t\right\rangle$ on $\left|\frac{3}{2},\frac{1}{2},1,\frac{1}{2}\right\rangle$

$$P\left(S(S+1) = \frac{15}{4}, S_z = \frac{1}{2}\right) = \left|\left\langle\frac{3}{2},\frac{1}{2},1,\frac{1}{2}\middle|\frac{1}{2},\frac{1}{2},1,\frac{1}{2},t\right\rangle\right|^2.$$

The result is

$$P\left(S(S+1) = \frac{15}{4}, S_z = \frac{1}{2}\right) = \frac{8}{9}\sin^2\left(\frac{t}{2}\right).$$

At time $t = 0$ this probability is zero because the states are orthogonal. It is because of the time evolution that we find a non zero projection on the state $\left|\frac{3}{2},\frac{1}{2},1,\frac{1}{2}\right\rangle$.

Problem 3.20.
A neutron with mass m has zero charge, spin $1/2$, and magnetic moment $g\mu_N$, where $g \approx -1.913$ and μ_N is the nuclear magneton $\mu_N = \frac{e\hbar}{2mc}$. At a time $t = 0$, the neutron is found in a state such that:

- its momentum is well defined and its value is $p = \hbar k$ ($k = $ const.);
- a measurement of $\hat{\sigma}_y$ ($\hat{S}_y = \frac{\hbar}{2}\hat{\sigma}_y$ is the y component of the spin and $\hat{\sigma}_y$ the Pauli matrix) gives 1 with probability $\frac{9}{10}$;
- $\langle\hat{\sigma}_z\rangle = 3/5$.

The neutron is plunged in a constant magnetic field B along the x axis. Write the wave function of the particle at time $t = 0$ and the Hamiltonian. Determine the time t (if it exists) at which there is a zero probability to find the system in the eigenstate of $\hat{\sigma}_y$ with eigenvalue -1.

Solution
From the first condition we infer that the wave function is a plane wave

$$\langle x|\psi(t = 0)\rangle = \frac{e^{ik\cdot x}}{(2\pi\hbar)^{\frac{3}{2}}}|\chi(t = 0)\rangle$$

where $|\chi(t = 0)\rangle$ is a spinor independent of the spatial coordinates. In the basis where $\hat{\sigma}_z$ is diagonal, the two possible states for the projection of the spin along the y direction are

$$\left|\frac{1}{2},\frac{1}{2}\right\rangle_y = \frac{1}{\sqrt{2}}\begin{pmatrix}1\\i\end{pmatrix} \qquad \left|\frac{1}{2},-\frac{1}{2}\right\rangle_y = \frac{1}{\sqrt{2}}\begin{pmatrix}1\\-i\end{pmatrix}.$$

At $t = 0$, the spinor $|\chi(t = 0)\rangle$ is a superposition of $|\frac{1}{2}, \frac{1}{2}\rangle_y$ and $|\frac{1}{2}, -\frac{1}{2}\rangle_y$

$$|\chi(t = 0)\rangle = c_+ \left|\frac{1}{2}, \frac{1}{2}\right\rangle_y + c_- \left|\frac{1}{2}, -\frac{1}{2}\right\rangle_y$$

where $c_\pm$ are projection constants with the properties $|c_+|^2 = \frac{9}{10}$, $|c_+|^2 + |c_-|^2 = 1$. Therefore, apart from an unimportant overall phase factor, we get

$$|\chi(t = 0)\rangle = \frac{3}{\sqrt{10}} \left|\frac{1}{2}, \frac{1}{2}\right\rangle_y + \frac{e^{i\phi}}{\sqrt{10}} \left|\frac{1}{2}, -\frac{1}{2}\right\rangle_y.$$

The relative phase ϕ is obtained from the last condition in the text

$$\frac{3}{5} = \langle\chi(t = 0)|\hat{\sigma}_z|\chi(t = 0)\rangle.$$

Indeed, the average $\langle\chi(t = 0)|\hat{\sigma}_z|\chi(t = 0)\rangle$ is equal to

$$\left(\frac{3}{2\sqrt{5}}(1 \quad -i) + \frac{e^{-i\phi}}{2\sqrt{5}}(1 \quad i)\right) \cdot \left(\frac{3}{2\sqrt{5}}\begin{pmatrix}1\\-i\end{pmatrix} + \frac{e^{i\phi}}{2\sqrt{5}}\begin{pmatrix}1\\i\end{pmatrix}\right) = \frac{3}{5}\cos\phi$$

from which we see that $\cos\phi = 1$, that implies $\phi = 2n\pi$ with n an integer number, and

$$|\psi(t = 0)\rangle = \frac{e^{ik\cdot x}}{(2\pi\hbar)^{\frac{3}{2}}} \frac{1}{\sqrt{5}}\begin{pmatrix}2\\i\end{pmatrix}.$$

There are two terms in the Hamiltonian: the first is the kinetic energy ($\frac{\hat{p}^2}{2m}$), and the second is the spin-field interaction ($g\mu_N B \frac{\hbar}{2}\hat{\sigma}_x$). Thus

$$\hat{H} = \frac{\hat{p}^2}{2m} + g\mu_N B\hat{S}_x = \frac{\hat{p}^2}{2m} + g\mu_N B\frac{\hbar}{2}\hat{\sigma}_x.$$

In order to answer the third question, we must evolve $|\chi(0)\rangle$. To this end, we expand $|\chi(0)\rangle$ in $\hat{H}$ eigenstates. The question concerns the spin. Since the field is directed along x, we expand in the eigenstates

$$\left|\frac{1}{2}, \frac{1}{2}\right\rangle_x = \frac{1}{\sqrt{2}}\begin{pmatrix}1\\1\end{pmatrix} \qquad \left|\frac{1}{2}, -\frac{1}{2}\right\rangle_x = \frac{1}{\sqrt{2}}\begin{pmatrix}1\\-1\end{pmatrix}.$$

To determine the coefficients of the expansion, we project $|\chi(0)\rangle$ on $|\frac{1}{2}, \frac{1}{2}\rangle_x$ and $|\frac{1}{2}, -\frac{1}{2}\rangle_x$

$$\left\langle\frac{1}{2}, \frac{1}{2}\right|_x \chi(0)\right\rangle = \left(\frac{1}{\sqrt{2}}(1 \quad 1)\right) \cdot \left(\frac{1}{\sqrt{5}}\begin{pmatrix}2\\i\end{pmatrix}\right) = \frac{(2+i)}{\sqrt{10}}$$

$$\left\langle \frac{1}{2},-\frac{1}{2}\middle| \chi(0) \right\rangle_x = \left(\frac{1}{\sqrt{2}}(1 \ -1) \right) \cdot \left(\frac{1}{\sqrt{5}} \binom{2}{i} \right) = \frac{(2-i)}{\sqrt{10}}$$

and, hence

$$|\chi(0)\rangle = \frac{1}{\sqrt{10}} \left\{ (2+i)\left|\frac{1}{2},\frac{1}{2}\right\rangle_x + (2-i)\left|\frac{1}{2},-\frac{1}{2}\right\rangle_x \right\}.$$

The time evolution is obtained with the action of the time evolution operator $e^{-i\hat{H}t/\hbar}$ on the wave function at time $t = 0$. Such operator is diagonal with respect to the eigenstates of $\hat{H}$ and, therefore, the eigenstates $\left|\frac{1}{2},\frac{1}{2}\right\rangle_x$ and $\left|\frac{1}{2},-\frac{1}{2}\right\rangle_x$ acquire the phase factor $e^{-iEt/\hbar}$, with $E = \pm\hbar\xi$ ($\xi = \frac{1}{2}g\mu_N B$) being the eigenvalues. The result is

$$|\chi(t)\rangle = \frac{1}{\sqrt{10}} \left\{ (2+i)e^{-i\xi t}\left|\frac{1}{2},\frac{1}{2}\right\rangle_x + (2-i)e^{i\xi t}\left|\frac{1}{2},-\frac{1}{2}\right\rangle_x \right\}.$$

The last question requires the projection of this wave function on the spinor $\left|\frac{1}{2},-\frac{1}{2}\right\rangle_y$. This turns out to be

$$\left\langle \frac{1}{2},-\frac{1}{2}\middle| \chi(t) \right\rangle_y = \frac{(2+i)e^{-i\xi t}}{\sqrt{10}}\left\langle \frac{1}{2},-\frac{1}{2}\middle|\frac{1}{2},\frac{1}{2}\right\rangle_{y,x} + \frac{(2-i)e^{i\xi t}}{\sqrt{10}}\left\langle \frac{1}{2},-\frac{1}{2}\middle|\frac{1}{2},-\frac{1}{2}\right\rangle_{y,x} =$$

$$\frac{(2+i)}{\sqrt{10}}e^{-i\xi t}\frac{(1+i)}{2} + \frac{(2-i)}{\sqrt{10}}e^{i\xi t}\frac{(1-i)}{2} =$$

$$\left\{ e^{-i\xi t}\frac{(1+3i)}{2\sqrt{10}} + e^{i\xi t}\frac{(1-3i)}{2\sqrt{10}} \right\}.$$

In order to have a vanishing projection, we must ensure

$$\cos(\xi t) + 3\sin(\xi t) = 0.$$

We take as auxiliary unknown $y = \cos(\xi t)$ and write

$$y + 3\sqrt{1-y^2} = 0$$

leading to $y = \pm\frac{3}{\sqrt{10}}$ and $t = \frac{\cos^{-1}\left(\frac{3}{\sqrt{10}}\right)}{\xi}$.

Problem 3.21.
A particle with spin $\frac{1}{2}$ and magnetic moment $\hat{\mu} = g\hat{S}$ ($\hat{S}$ is the spin operator) is subject to a constant magnetic field $B = (0, B_y, 0)$. Its Hamiltonian is given by

$$\hat{H} = -\hat{\mu} \cdot B = -g \cdot \hat{S} \cdot B = -g\hat{S}_y B_y.$$

At a time $t = 0$, the particle is in a state such that:

- the probability that a measurement of the z component of the spin gives $\frac{\hbar}{2}$ is $2/3$;
- the averages of the spin operators in the x and y directions ($\hat{S}_x$ and $\hat{S}_y$) are such that $\langle \hat{S}_x \rangle = \langle \hat{S}_y \rangle \geq 0$.

Determine the eigenvalues of the Hamiltonian. Knowing that the averages of the x and y components of the spin are equal, i.e. $\langle \hat{S}_x \rangle = \langle \hat{S}_y \rangle$, determine the state at $t = 0$, and evolve the wave function to a later time t.

Solution
We start by writing down a matrix representation for the Hamiltonian

$$\hat{H} = -g\hat{S} \cdot B = -gB_y \frac{\hbar}{2} \hat{\sigma}_y = -gB_y \frac{\hbar}{2} \begin{pmatrix} 0 & -i \\ i & 0 \end{pmatrix}$$

where $\hat{\sigma}_y$ is one of the Pauli matrices. The eigenvalues of the Hamiltonian are rapidly found

$$\det \begin{pmatrix} -E & \frac{igB_y\hbar}{2} \\ -\frac{igB_y\hbar}{2} & -E \end{pmatrix} = E^2 - \frac{\hbar^2 g^2 B_y^2}{4} = 0$$

from which we get $E_\pm = \mp gB_y\hbar/2$. It is then convenient to use the eigenstates of $\hat{S}_z = \frac{\hbar}{2}\hat{\sigma}_z$, even if the Hamiltonian $\hat{H}$ is not diagonal with respect to such a vector basis. The eigenstates of $\hat{\sigma}_z$ are

$$\left| \frac{1}{2}, \frac{1}{2} \right\rangle_z = \begin{pmatrix} 1 \\ 0 \end{pmatrix} \qquad \left| \frac{1}{2}, -\frac{1}{2} \right\rangle_z = \begin{pmatrix} 0 \\ 1 \end{pmatrix}$$

and we can always think of expressing $|\psi(0)\rangle$ in terms of $\left| \frac{1}{2}, \frac{1}{2} \right\rangle_z, \left| \frac{1}{2}, -\frac{1}{2} \right\rangle_z$

$$|\psi(0)\rangle = \alpha \left| \frac{1}{2}, \frac{1}{2} \right\rangle_z + \beta \left| \frac{1}{2}, -\frac{1}{2} \right\rangle_z$$

where α can be thought of as a positive real parameter ($\alpha > 0$). From the first condition given in the text, it follows that

$$\alpha = \left\langle \frac{1}{2}, \frac{1}{2} \middle| \psi(0) \right\rangle_z = \sqrt{\frac{2}{3}}$$

because $\left| \left\langle \frac{1}{2}, \frac{1}{2} \middle| \psi(0) \right\rangle_z \right|^2 = \frac{2}{3}$. Since $|\alpha|^2 + |\beta|^2 = 1$, we can write

$$\beta = \left\langle \frac{1}{2}, -\frac{1}{2} \middle| \psi(0) \right\rangle_z = \frac{e^{i\varphi}}{\sqrt{3}}$$

in terms of a yet unknown phase φ. One can find φ from the condition $\langle \hat{S}_x \rangle = \langle \hat{S}_y \rangle$. Let us start by computing $\langle \hat{S}_x \rangle$. The eigenstates of $\hat{S}_x$ are

$$\left| \frac{1}{2}, \frac{1}{2} \right\rangle_x = \frac{1}{\sqrt{2}} \begin{pmatrix} 1 \\ 1 \end{pmatrix} \qquad \left| \frac{1}{2}, -\frac{1}{2} \right\rangle_x = \frac{1}{\sqrt{2}} \begin{pmatrix} 1 \\ -1 \end{pmatrix}.$$

The states $\left|\frac{1}{2}, \pm\frac{1}{2}\right\rangle_z$ can be expanded in terms of $\left|\frac{1}{2}, \pm\frac{1}{2}\right\rangle_x$

$$\left|\frac{1}{2}, \frac{1}{2}\right\rangle_z = \frac{1}{\sqrt{2}}\left(\left|\frac{1}{2}, \frac{1}{2}\right\rangle_x + \left|\frac{1}{2}, -\frac{1}{2}\right\rangle_x\right)$$

$$\left|\frac{1}{2}, -\frac{1}{2}\right\rangle_z = \frac{1}{\sqrt{2}}\left(\left|\frac{1}{2}, \frac{1}{2}\right\rangle_x - \left|\frac{1}{2}, -\frac{1}{2}\right\rangle_x\right)$$

so that

$$\hat{S}_x|\psi(0)\rangle = \hat{S}_x \frac{1}{\sqrt{3}}\left\{\left(\left|\frac{1}{2}, \frac{1}{2}\right\rangle_x + \left|\frac{1}{2}, -\frac{1}{2}\right\rangle_x\right) + \frac{e^{i\varphi}}{\sqrt{2}}\left(\left|\frac{1}{2}, \frac{1}{2}\right\rangle_x - \left|\frac{1}{2}, -\frac{1}{2}\right\rangle_x\right)\right\} =$$

$$\frac{\hbar}{2\sqrt{3}}\left\{\left(\left|\frac{1}{2}, \frac{1}{2}\right\rangle_x - \left|\frac{1}{2}, -\frac{1}{2}\right\rangle_x\right) + \frac{e^{i\varphi}}{\sqrt{2}}\left(\left|\frac{1}{2}, \frac{1}{2}\right\rangle_x + \left|\frac{1}{2}, -\frac{1}{2}\right\rangle_x\right)\right\}$$

from which, using the orthogonality of states $\left|\frac{1}{2}, \pm\frac{1}{2}\right\rangle_x$, we get

$$\langle\hat{S}_x\rangle = \frac{\hbar\sqrt{2}}{3}\cos\varphi.$$

We proceed in a similar way for $\langle\hat{S}_y\rangle$. The eigenstates of $\hat{S}_y$ are

$$\left|\frac{1}{2}, \frac{1}{2}\right\rangle_y = \frac{1}{\sqrt{2}}\begin{pmatrix}1\\i\end{pmatrix} \qquad \left|\frac{1}{2}, -\frac{1}{2}\right\rangle_y = \frac{1}{\sqrt{2}}\begin{pmatrix}1\\-i\end{pmatrix}$$

and

$$\left|\frac{1}{2}, \frac{1}{2}\right\rangle_z = \frac{1}{\sqrt{2}}\left(\left|\frac{1}{2}, \frac{1}{2}\right\rangle_y + \left|\frac{1}{2}, -\frac{1}{2}\right\rangle_y\right)$$

$$\left|\frac{1}{2}, -\frac{1}{2}\right\rangle_z = -\frac{i}{\sqrt{2}}\left(\left|\frac{1}{2}, \frac{1}{2}\right\rangle_y - \left|\frac{1}{2}, -\frac{1}{2}\right\rangle_y\right)$$

so that

$$\hat{S}_y|\psi(0)\rangle = \hat{S}_y \frac{1}{\sqrt{3}}\left\{\left(\left|\frac{1}{2}, \frac{1}{2}\right\rangle_y + \left|\frac{1}{2}, -\frac{1}{2}\right\rangle_y\right) - \frac{ie^{i\varphi}}{\sqrt{2}}\left(\left|\frac{1}{2}, \frac{1}{2}\right\rangle_y - \left|\frac{1}{2}, -\frac{1}{2}\right\rangle_y\right)\right\} =$$

$$\frac{\hbar}{2\sqrt{3}}\left\{\left(\left|\frac{1}{2}, \frac{1}{2}\right\rangle_y - \left|\frac{1}{2}, -\frac{1}{2}\right\rangle_y\right) - \frac{ie^{i\varphi}}{\sqrt{2}}\left(\left|\frac{1}{2}, \frac{1}{2}\right\rangle_y + \left|\frac{1}{2}, -\frac{1}{2}\right\rangle_y\right)\right\}$$

and

$$\langle\hat{S}_y\rangle = \frac{\hbar\sqrt{2}}{3}\sin\varphi.$$

Imposing the condition $\langle\hat{S}_y\rangle = \langle\hat{S}_x\rangle \geq 0$, we obtain

$$\sin\varphi = \cos\varphi$$

and

$$\varphi = \frac{\pi}{4} + 2n\pi \quad n = 0, 1, 2, 3, \ldots$$

In conclusion, the initial state (if $n = 0$) is

$$|\psi(0)\rangle = \sqrt{\frac{2}{3}} \left|\frac{1}{2}, \frac{1}{2}\right\rangle_z + \frac{e^{i\frac{\pi}{4}}}{\sqrt{3}} \left|\frac{1}{2}, -\frac{1}{2}\right\rangle_z$$

and the average components of the spin in the x and y directions are given by

$$\langle \hat{S}_x \rangle = \langle \hat{S}_y \rangle = \frac{\hbar}{3}.$$

The time evolution of the wave function at any subsequent time t is

$$|\psi(t)\rangle = \left|\frac{1}{2}, \frac{1}{2}\right\rangle_y e^{-iE_+ t/\hbar} A + \left|\frac{1}{2}, -\frac{1}{2}\right\rangle_y e^{-iE_- t/\hbar} B$$

where

$$A = \left\langle \frac{1}{2}, \frac{1}{2}\Big|_y \psi(0)\right\rangle = \frac{3-i}{2\sqrt{3}}$$

and

$$B = \left\langle \frac{1}{2}, -\frac{1}{2}\Big|_y \psi(0)\right\rangle = \frac{1+i}{2\sqrt{3}}.$$

4

Central Force Field

Problem 4.1.
Let us consider a particle subject to the following three dimensional harmonic potential

$$U(r) = \frac{M\omega^2 r^2}{2}$$

with r the radial distance in spherical polar coordinates. Discuss the properties of the associated Schrödinger equation for $r \approx 0$ and $r \to +\infty$.

Solution
Our potential energy is dependent only on the absolute value of the distance from the origin. For this reason, we seek the solution by separating radial and angular variables

$$\psi_{n,l,m}(r,\theta,\phi) = \frac{R_{n,l}(r)}{r} Y_{l,m}(\theta,\phi).$$

The associated Schrödinger equation becomes

$$\left[-\frac{\hbar^2}{2M} \frac{d^2}{dr^2} + \frac{1}{2}M\omega^2 r^2 + \frac{\hbar^2 l(l+1)}{2Mr^2} - E_{n,l} \right] R_{n,l}(r) = 0.$$

If we define the new length scale $a = \sqrt{\hbar/(M\omega)}$, we can rewrite the equation in terms of the variables $x = r/a$, $\varepsilon = E_{n,l}/(\hbar\omega)$. Removing all quantum numbers subscripts for simplicity, the Schrödinger equation takes now the form

$$\left[\frac{d^2}{dx^2} - x^2 - \frac{l(l+1)}{x^2} + 2\varepsilon \right] R(x) = 0$$

which, when $x \approx 0$, yields

$$\left[\frac{d^2}{dx^2} - \frac{l(l+1)}{x^2} \right] R(x) = 0$$

Cini M., Fucito F., Sbragaglia M.: Solved Problems in Quantum and Statistical Mechanics.
DOI 10.1007/978-88-470-2315-4_4, © Springer-Verlag Italia 2012

where we discarded all the non divergent terms. All singularities are regular. The regular singularities are found in the coefficients of the differential equation (on which no derivatives act) written in the form $\frac{p_j}{x^{k-j}}$. k is the degree of the highest order derivative (in our case, we have a second order differential equation, so that $k = 2$) and $j = 1, ..., k$ is an index indicating the coefficient (see section 1.7 for a detailed discussion). If we make the ansatz $R(x) = x^\alpha$ and plug it back in the equation, we get

$$\alpha(\alpha - 1) - l(l+1) = 0$$

which has solutions $\alpha = -l, l+1$. The physical solution is the one without divergences in the origin, i.e. $R(x) = x^{l+1}$.

Let us now discuss the behaviour of the solution close to infinity $(r = +\infty)$. First of all, we need to change variable by setting $x = 1/t$. The derivatives transform like

$$\frac{d}{dx} = -t^2 \frac{d}{dt} \qquad \frac{d^2}{dx^2} = t^4 \frac{d^2}{dt^2} + 2t^3 \frac{d}{dt}$$

and the Schrödinger equation in the variable t becomes

$$\left(\frac{d^2}{dt^2} + \frac{2}{t}\frac{d}{dt} - \frac{1}{t^6} - \frac{l(l+1)}{t^2} + \frac{2\varepsilon}{t^4} \right) R(t) = 0$$

where we see that the singularities are not regular. A way to study the properties of the solution for $x \gg 1$, is to assume the functional form $R(x) = e^{S(x)}$, plug it back in the original equation, and neglect the term with the second derivative

$$\left(\frac{dS}{dx} \right)^2 - (x^2 - 2\varepsilon) = 0$$

so that $S(x) = S(x_0) \pm \int_{x_0}^x \sqrt{s^2 - 2\varepsilon}\, ds \approx \pm\frac{x^2}{2}$, because at infinity the term s^2 dominates over ε. We then choose the minus sign in the exponent $R(x) \approx e^{-x^2/2}$ to ensure a zero probability density function for large x. This is in line with the familiar solution of the one dimensional oscillator, since the three dimensional one is separable.

Problem 4.2.
Determine the energy spectrum for a particle with mass M subject to the following three dimensional central potential

$$\begin{cases} U(r) = 0 & r \le a \\ U(r) = +\infty & r > a \end{cases}$$

where a is a constant and r is the radial distance in spherical polar coordinates.

Solution
The solution to our problem is given by the wave function $\psi_{n,l,m}(r, \theta, \phi)$ that for $r \le a$ describes the motion of a free particle and such that $\psi_{n,l,m}(a, \theta, \phi) = 0$. We

seek a solution in the form

$$\psi_{n,l,m}(r,\theta,\phi) = \frac{R_{n,l}(r)}{r}Y_{l,m}(\theta,\phi)$$

where the term $1/r$ is important to remove the dependence on the first order derivative $\frac{2}{r}\frac{\partial}{\partial r}$ in the Laplacian operator. The corresponding Schrödinger equation for $R_{n,l}(r)$ is

$$-\frac{\hbar^2}{2M}\frac{d^2 R_{n,l}(r)}{dr^2} + \frac{\hbar^2 l(l+1)}{2Mr^2}R_{n,l}(r) = E_{n,l}R_{n,l}(r)$$

given that the spherical harmonics $Y_{l,m}(\theta,\phi)$ are eigenstates of the angular part of the Laplacian operator. Setting $k^2 = 2ME_{n,l}/\hbar^2$, and removing all quantum numbers subscripts for simplicity, we get

$$R'' - \left(\frac{l(l+1)}{r^2} - k^2\right)R = 0.$$

In the point $r = 0$ we find a regular singularity. The indicial exponent around this point is found by plugging $R = r^\lambda$ in the original equation, with the result $\lambda = l+1$ and $\lambda = -l$. The physical solution is the first one, because in this case we can guarantee the finiteness of the wave function when $r \to 0$ (remember that the radial wave function is R/r, and the behaviour for $r \approx 0$ is therefore given by r^l). We now remove such asymptotic behaviour by setting $R(r) = r^{l+1}y(r)$. The resulting Schrödinger equation for $y(r)$ becomes

$$y'' + \frac{2(l+1)}{r}y' + k^2 y = 0.$$

We can also study the behaviour of y close to infinity. To do that, we make use of the change of variable $r = 1/t$ and find the equation for $y(t)$

$$y'' - \frac{2l}{t}y' + \frac{k^2}{t^4}y = 0$$

where we see that the singularity in $t = 0$ $(r = +\infty)$ is irregular. We then seek the solution in the form $y(t) = e^{ik/t}F(t)$ to remove the singularity in t^4 and find the indicial equation. The equation for $F(t)$ is

$$F'' + \left(-\frac{2ik}{t^2} - \frac{2l}{t}\right)F' + \left(\frac{2ik}{t^3} + \frac{2ikl}{t^3}\right)F = 0.$$

Setting $F(t) = t^\beta$, the resulting indicial equation has solution $\beta = 1+l$. Plugging all these results back in the original equation for the function $R(r) = r^{l+1}e^{ikr}F(r)$, we find

$$F'' + \left(2ik + \frac{2(l+1)}{r}\right)F' + \frac{2ik(l+1)}{r}F = 0.$$

The change of variable $x = -2ikr$ allows us to obtain the confluent hypergeometric equation

$$xF'' + (2(l+1) - x)F' - (l+1)F = 0$$

defining the function $F(l+1, 2(l+1)|x)$. We therefore write the general solution as

$$\psi_{n,l,m}(r,\theta,\phi) = \frac{1}{r}R_{n,l}(r)Y_{l,m}(\theta,\phi) = r^l e^{ikr} F(l+1, 2(l+1)| - 2ikr)Y_{l,m}(\theta,\phi) =$$

$$\frac{1}{\sqrt{r}}\left(\frac{2}{k}\right)^{l+\frac{1}{2}} \Gamma\left(l+\frac{3}{2}\right) J_{l+\frac{1}{2}}(kr)Y_{l,m}(\theta,\phi).$$

The $J_{l+1/2}$ are the Bessel functions of half-integral order, whose relation with the hypergeometric function is

$$J_{l+\frac{1}{2}}(kr) = \frac{1}{\Gamma(l+\frac{3}{2})}\left(\frac{1}{2}kr\right)^{l+\frac{1}{2}} e^{ikr} F(l+1, 2(l+1)| - 2ikr).$$

The quantization rules are set by the boundary condition in $r = a$

$$J_{l+\frac{1}{2}}(ka) = 0$$

from which $k = X_{n,l}/a$ that means $E_{n,l} = \hbar^2 X_{n,l}^2/(2Ma^2)$. $X_{n,l}$ are the zeros of the Bessel functions which can be determined once we know the value of l. For example, when $l = 0$, the function $J_{1/2}(ka)$ is

$$J_{1/2}(ka) = \sqrt{\frac{2ka}{\pi}}\frac{\sin(ka)}{ka}$$

and the zeros correspond to $X_{n,0} = n\pi = ka$, with n an integer number. Therefore, one gets $E_{n,0} = \hbar^2(n\pi)^2/(2Ma^2)$. The other quantization rules (for $l > 0$) can be obtained from

$$j_l(x) = \sqrt{\frac{\pi}{2x}}J_{l+1/2}(x)$$

with $j_l(x)$ the spherical Bessel functions with the property

$$j_l(x) = (-x)^l \left(\frac{1}{x}\frac{d}{dx}\right)^l \frac{\sin x}{x}.$$

Problem 4.3.
Determine the minimum value of $U_0 > 0$ such that a particle with mass m subject to the following central potential

$$\begin{cases} U(r) = -U_0 & r \le a & \text{(region } I\text{)} \\ U(r) = 0 & r > a & \text{(region } II\text{)} \end{cases}$$

has s-type bound states. In the above expressions, a is a constant and r is the radial distance in spherical polar coordinates.

Solution
When the particle is in the bound state of s-type, the orbital angular momentum is $l = 0$. We then seek a solution with a negative energy E: we set $\varepsilon = -E > 0$ and write the wave function as

$$\psi(r,\theta,\phi) = \frac{R(r)}{r}Y_{0,0}(\theta,\phi).$$

The corresponding Schrödinger equation for the radial part $R(r)$ becomes

$$\begin{cases} \frac{d^2 R_I}{dr^2} + k_1^2 R_I = 0 & r \leq a \\ \frac{d^2 R_{II}}{dr^2} - k_2^2 R_{II} = 0 & r > a \end{cases}$$

where $k_1^2 = 2m(U_0 - \varepsilon)/\hbar^2$ and $k_2^2 = 2m\varepsilon/\hbar^2$. We note that the first order derivative $\frac{2}{r}\frac{\partial}{\partial r}$ of the Laplacian cancels out, due to the choice of $R(r)/r$ as the radial part of the wave function. The solutions of the above equations, satisfying finiteness of the wave function in $r = 0, +\infty$, are

$$\begin{cases} R_I(r) = A\sin(k_1 r) & r \leq a \\ R_{II}(r) = Be^{-k_2 r} & r > a. \end{cases}$$

Then, we need to impose the continuity conditions for the wave function and its derivative in $r = a$. These conditions are equivalent to

$$\frac{R_I'(a)}{R_I(a)} = \frac{R_{II}'(a)}{R_{II}(a)}$$

yielding $k_1 \cot(k_1 a) = -k_2$. We note that such a kind of condition is the same we find when we solve the one dimensional problem of a particle in a rectangular potential well with an infinite energy barrier on one side (see Problem 2.12). If we set $x = k_1 a$ and $y = k_2 a$, the solution is given by the intersection of the two curves

$$\begin{cases} x^2 + y^2 = \frac{2ma^2 U_0}{\hbar^2} = \tilde{R}^2 \\ y = -x\cot x. \end{cases}$$

The smallest x such that $-x\cot x = 0$ is $x = \pi/2$. The radius $\tilde{R}$ has to take this value to find an intersection corresponding to a bound state with zero energy. That corresponds to

$$U_0 = \frac{\pi^2 \hbar^2}{8ma^2}.$$

For the values

$$\tilde{R}^2 = \frac{2ma^2 U_0}{\hbar^2} > \pi^2/4$$

we have other bound states with negative energies.

Problem 4.4.

A hydrogen atom with Hamiltonian $\hat{H} = \hat{T} + \hat{V}$, where $\hat{T}$ is the kinetic energy operator and $V(r) = -\frac{e^2}{r}$ the potential energy, is in the ground state:

- determine the average value of the modulus of the force $F = \frac{e^2}{r^2}$ between the electron and the nucleus, and the one of the associated potential energy $V(r)$;
- starting from the average energy $\langle \hat{H} \rangle$, determine the average value of the kinetic energy $\langle \hat{T} \rangle$ and $\langle 2\hat{T} + \hat{V} \rangle$.

Solution

We start from the wave function of the ground state

$$\psi(r, \theta, \phi) = R_{1,0}(r) Y_{0,0}(\theta, \phi) = \frac{1}{\sqrt{\pi a^3}} e^{-\frac{r}{a}}$$

with a the Bohr radius. We then make use of the following integrals

$$\int_0^{+\infty} e^{-x} dx = 1 \qquad \int_0^{+\infty} x e^{-x} dx = 1$$

to obtain the average values of $\hat{F}$ and $\hat{V}$

$$\langle \hat{F} \rangle = \left\langle \frac{e^2}{\hat{r}^2} \right\rangle = \int \psi^2(r, \theta, \phi) \frac{e^2}{r^2} d^3r = \frac{4\pi e^2}{\pi a^3} \int_0^{+\infty} e^{-\frac{2r}{a}} dr = \frac{2e^2}{a^2}$$

$$\langle \hat{V} \rangle = -\left\langle \frac{e^2}{\hat{r}} \right\rangle = -\int \psi^2(r, \theta, \phi) \frac{e^2}{r} d^3r = -\frac{4\pi e^2}{\pi a^3} \int_0^{+\infty} r e^{-\frac{2r}{a}} dr = -\frac{e^2}{a}.$$

Since we know the energy of the ground state for the hydrogen atom, $\langle \hat{H} \rangle = -\frac{e^2}{2a}$, we find that

$$\langle \hat{T} \rangle = \langle \hat{H} \rangle - \langle \hat{V} \rangle = \frac{e^2}{2a}$$

and

$$\langle 2\hat{T} + \hat{V} \rangle = 0.$$

This result is a particular case of the so-called *virial theorem*, i.e. a general equation relating the average of the total kinetic energy with that of the total potential energy for physical systems consisting of particles bound by potential forces. In the most general case, such relation has the form

$$2\langle \psi | \hat{T} | \psi \rangle = \langle \psi | \hat{r} \cdot \nabla | \psi \rangle$$

which can be simplified when treating potential forces with spherical symmetry and proportional to r^n (our case corresponds to $n = -1$). In such cases, we find

$$r \cdot \nabla = r \frac{d}{dr} V(r) = nV(r)$$

so that

$$2\langle\psi|\hat{T}|\psi\rangle = n\langle\psi|\hat{V}|\psi\rangle.$$

If we know that the state ψ has an average energy equal to E

$$\langle\psi|\hat{T}|\psi\rangle + \langle\psi|\hat{V}|\psi\rangle = E$$

we can find the average of the kinetic and potential energy as

$$\langle\psi|\hat{T}|\psi\rangle = \frac{n}{n+2}E \quad \langle\psi|\hat{V}|\psi\rangle = \frac{2}{n+2}E$$

which, for $n = -1$, is consistent with the previous result.

Problem 4.5.
Determine the energy spectrum for a particle with mass M subject to the following three dimensional central potential

$$U(r) = \frac{1}{2}M\omega^2 r^2$$

with ω a constant and r the radial distance in spherical polar coordinates. Discuss the properties of the solution in spherical and Cartesian coordinates.

Solution
As usual with centrally symmetric potentials, we seek the solution in the form

$$\psi_{n,l,m}(r,\theta,\phi) = \frac{R_{n,l}(r)}{r}Y_{l,m}(\theta,\phi).$$

The resulting three dimensional Schrödinger equation for the radial part $R_{n,l}(r)$ becomes

$$\left(-\frac{\hbar^2}{2M}\frac{d^2}{dr^2} + \frac{\hbar^2}{2M}\frac{l(l+1)}{r^2} + \frac{1}{2}M\omega^2 r^2\right)R_{n,l}(r) = E_{n,l}R_{n,l}(r)$$

or, equivalently

$$\left(\frac{d^2}{dz^2} - z^2 - \frac{l(l+1)}{z^2} + 2\varepsilon\right)R(z) = 0$$

where $z = \sqrt{M\omega/\hbar}\,r$ and $\varepsilon = E_{n,l}/(\hbar\omega)$, and where we have removed all quantum numbers subscripts for simplicity. The equation has a regular singularity in $z = 0$ and an irregular one in $z = +\infty$, suggesting that the solution for the differential equation is in the class of the confluent hypergeometric functions. Close to the origin ($z = 0$), we set $R(z) \approx z^\lambda$ and find the solutions $\lambda = l+1, -l$. The first solution is the physical one because it does not diverge in $z = 0$. Moreover, for $z \gg 1$, the solution is of the form $R(z) \approx e^{-z^2/2}$. This can be seen by neglecting the terms z^{-2} and ε with respect to z^2 (see also Problem 4.1). Therefore, the general solution can be sought in the form

$$R(z) = z^{l+1}e^{-z^2/2}F(z).$$

The original equation becomes

$$F'' + 2\left(\frac{l+1}{z} - z\right)F' + (2\varepsilon - 2l - 3)F = 0.$$

To find the usual confluent hypergeometric form, we need to define the new variable $x = z^2$ and obtain

$$xF'' + \left(l + \frac{3}{2} - x\right)F' + \left(\frac{1}{2}\varepsilon - \frac{1}{2}l - \frac{3}{4}\right)F = 0.$$

The solution is $F = F(-\varepsilon/2 + l/2 + 3/4, l + 3/2 | z^2)$. We recall that the general confluent hypergeometric function $F(A, C | z)$ is defined by the series

$$F(A, C | z) = 1 + \frac{A}{C}\frac{z}{1!} + \frac{A(A+1)}{C(C+1)}\frac{z^2}{2!} + \dots$$

and that the series reduces to a polynomial of degree $|A|$ when $A = -n$, with n a non negative integer (see also Problem 2.25). Therefore, to preserve the convergent behaviour of the solution at infinity, we have to set

$$-n = -\frac{\varepsilon}{2} + \frac{l}{2} + \frac{3}{4}.$$

The corresponding eigenvalues for the Hamiltonian are

$$E_{n,l} = \hbar\omega\left(2n + l + \frac{3}{2}\right).$$

We note that we can write the potential also in Cartesian coordinates (x, y, z not to be confused with the above variables)

$$U(x, y, z) = \frac{1}{2}M\omega^2 r^2 = \frac{1}{2}M\omega^2(x^2 + y^2 + z^2).$$

The associated Schrödinger equation allows the separation of variables. The resulting picture is that of three independent one dimensional harmonic oscillators whose Hamiltonian eigenvalues are

$$E_{n_x, n_y, n_z} = \hbar\omega\left(n_x + n_y + n_z + \frac{3}{2}\right)$$

with n_x, n_y, n_z non negative integers. The eigenstates are therefore

$$\psi_{n_x, n_y, n_z}(x, y, z) = C_{n_x}C_{n_y}C_{n_z}e^{-a^2 r^2/2}H_{n_x}(ax)H_{n_y}(ay)H_{n_z}(az)$$

with

$$C_n = \frac{1}{2^{n/2}\sqrt{n!}}\left(\frac{M\omega}{\hbar\pi}\right)^{1/4} \qquad a = \sqrt{\frac{M\omega}{\hbar}}$$

and where

$$H_n(\xi) = (-1)^n e^{\xi^2} \frac{d^n}{d\xi^n} e^{-\xi^2}$$

represents the n-th order Hermite polynomial.

Problem 4.6.

The three dimensional quantum harmonic oscillator with Hamiltonian

$$\hat{H} = \frac{\hat{p}^2}{2m} + \frac{m}{2} \left[\omega^2(\hat{x}^2 + \hat{z}^2) + \Omega^2 \hat{y}^2 \right]$$

is slightly anisotropic, with $\lambda = \frac{\Omega - \omega}{\omega} \ll 1$. The system is prepared in the ground state:

- using Cartesian coordinates, write down the wave function $\psi(x,y,z)$ of the ground state;
- using spherical polar coordinates (r, θ, ϕ), expand $\psi(r, \theta, \phi)$ to first order in λ;
- analyzing the wave function obtained in the previous point, write down the probability to find $l = 0$ as a function of the radial distance r.

Solution

The Hamiltonian may be seen as that of two harmonic oscillators with frequency ω (along x and z) and one harmonic oscillator with frequency Ω (along y). Variables can be separated and the wave function can be sought as the product of functions

$$\psi_{n_x,n_y,n_z}(x,y,z) = \psi_{n_x}(x)\psi_{n_y}(y)\psi_{n_z}(z)$$

with n_x, n_y, n_z non negative integers. The ground state is obtained when $n_x = n_y = n_z = 0$

$$\psi_{0,0,0}(x,y,z) = \left(\frac{m\omega}{\pi\hbar}\right)^{\frac{1}{2}} \left(\frac{m\Omega}{\pi\hbar}\right)^{\frac{1}{4}} e^{-F(x,y,z)}$$

with the function $F(x,y,z)$ defined by

$$F(x,y,z) = \frac{x^2 + z^2}{2x_0^2} + \frac{y^2}{2y_0^2}$$

and $x_0^2 = \frac{\hbar}{m\omega}$, $y_0^2 = \frac{\hbar}{m\Omega}$. The associated energy is

$$E_{0,0,0} = \hbar\left(\omega + \frac{\Omega}{2}\right).$$

As for the second point, we need to start from the function $F(x,y,z)$ written in spherical polar coordinates

$$F(x,y,z) = \frac{x^2 + y^2 + z^2}{2x_0^2} + y^2 \left(\frac{1}{2y_0^2} - \frac{1}{2x_0^2}\right) =$$
$$\frac{r^2}{2x_0^2}\left(1 + \frac{y^2}{r^2}x_0^2\left(\frac{1}{y_0^2} - \frac{1}{x_0^2}\right)\right) = \frac{r^2}{2x_0^2}\left[1 + \lambda \frac{y^2}{r^2}\right].$$

Setting $y = r \sin \theta \sin \phi$, we find the following results

$$F(r, \theta, \phi) = \frac{r^2}{2x_0^2} \left(1 + \lambda \sin^2 \theta \sin^2 \phi\right)$$

$$\psi_{0,0,0}(r, \theta, \phi) = \left(\frac{m\omega}{\pi\hbar}\right)^{\frac{1}{2}} \left(\frac{m\Omega}{\pi\hbar}\right)^{\frac{1}{4}} e^{-\frac{r^2}{2x_0^2}} e^{-\frac{r^2}{2x_0^2} \lambda \sin^2 \theta \sin^2 \phi}.$$

The isotropic limit is given by $\lambda \to 0$

$$\psi_{0,0,0}(r, \theta, \phi) = \psi_0^{(iso)}(r) = \left(\frac{m\omega}{\pi\hbar}\right)^{\frac{3}{4}} e^{-\frac{r^2}{2x_0}}.$$

When λ is small, we can expand the wave function in Taylor series

$$\psi_{0,0,0}(r, \theta, \phi) = \psi_0^{(iso)}(r) \left(1 + \frac{\lambda}{4}\right) \left(1 - \lambda \frac{r^2}{2x_0^2} \sin^2 \theta \sin^2 \phi\right) \approx$$

$$\psi_0^{(iso)}(r) \left(1 + \lambda \left(\frac{1}{4} - \frac{r^2}{2x_0^2} \sin^2 \theta \sin^2 \phi\right)\right)$$

where the factor $\left(1 + \frac{\lambda}{4}\right)$ comes from

$$\Omega^{\frac{1}{4}} = (\omega(1 + \lambda))^{\frac{1}{4}} \approx \omega^{\frac{1}{4}} \left(1 + \frac{\lambda}{4}\right).$$

As for the third point, the probability to measure $l = 0$ is obtained by taking the square of the projection of $\psi_{0,0,0}$ on $Y_{0,0}$

$$P(r, l = 0) = \left| \int_0^\pi \sin \theta d\theta \int_0^{2\pi} Y_{0,0}(\theta, \phi) \psi_{0,0,0}(r, \theta, \phi) d\phi \right|^2 =$$

$$\left| \frac{\psi_0^{(iso)}(r)}{\sqrt{4\pi}} \left[4\pi + \pi\lambda - \frac{2\lambda r^2 \pi}{3x_0^2}\right] \right|^2;$$

where we have used the integral $\int_0^\pi \sin^3 \theta d\theta \int_0^{2\pi} \sin^2 \phi d\phi = \frac{4}{3}\pi$.

Problem 4.7.

Let us study a hydrogen atom and neglect its nuclear spin. We denote by n the principal quantum number, by l, m the angular quantum numbers, and by $|\frac{1}{2}, \pm\frac{1}{2}\rangle$ the eigenstates of the z component of the electron spin $\hat{S}_z$. A complete set of bound states is given by $|n, l, m\rangle \otimes |\frac{1}{2}, \pm\frac{1}{2}\rangle$. The atom is prepared in the state

$$|\psi\rangle = \frac{|1, 0, 0\rangle \otimes |\frac{1}{2}, \frac{1}{2}\rangle + |2, 1, 1\rangle \otimes |\frac{1}{2}, -\frac{1}{2}\rangle + |2, 1, 0\rangle \otimes |\frac{1}{2}, \frac{1}{2}\rangle}{\sqrt{3}};$$

- determine the average of the z component of the spin $\langle \psi | \hat{S}_z | \psi \rangle$ and of the energy $\langle \psi | \hat{H} | \psi \rangle$;
- determine the standard deviation σ_E of the energy defined by

$$\sigma_E = \sqrt{\langle \psi | \hat{H}^2 | \psi \rangle - (\langle \psi | \hat{H} | \psi \rangle)^2}.$$

Solution

The average of the z component of the spin is easily found, once we recall that

$$\hat{S}_z |n,l,m\rangle \otimes \left| \frac{1}{2}, \pm\frac{1}{2} \right\rangle = \pm\frac{\hbar}{2} |n,l,m\rangle \otimes \left| \frac{1}{2}, \pm\frac{1}{2} \right\rangle$$

so that, using the orthogonality of the states, we find

$$\langle \psi | \hat{S}_z | \psi \rangle = \frac{1}{3} \left(\langle 1,0,0| \otimes \left\langle \frac{1}{2}, \frac{1}{2} \right| + \langle 2,1,1| \otimes \left\langle \frac{1}{2}, -\frac{1}{2} \right| + \langle 2,1,0| \otimes \left\langle \frac{1}{2}, \frac{1}{2} \right| \right) \times$$

$$\hat{S}_z \left(|1,0,0\rangle \otimes \left| \frac{1}{2}, \frac{1}{2} \right\rangle + |2,1,1\rangle \otimes \left| \frac{1}{2}, -\frac{1}{2} \right\rangle + |2,1,0\rangle \otimes \left| \frac{1}{2}, \frac{1}{2} \right\rangle \right) =$$

$$\frac{1}{3}\hbar \left(\frac{1}{2} - \frac{1}{2} + \frac{1}{2} \right) = \frac{\hbar}{6}.$$

The energy spectrum for the hydrogen atom is given by

$$E_n = -\frac{e^2}{2n^2 a} \qquad n = 1,2,3,\dots$$

with a the Bohr radius. Therefore, we obtain (we express the result in terms of the Rydberg $R = \frac{e^2}{2a}$)

$$\hat{H} | \psi \rangle = -\frac{R}{\sqrt{3}} \left(|1,0,0\rangle \otimes \left| \frac{1}{2}, \frac{1}{2} \right\rangle + \frac{1}{4} |2,1,1\rangle \otimes \left| \frac{1}{2}, -\frac{1}{2} \right\rangle + \frac{1}{4} |2,1,0\rangle \otimes \left| \frac{1}{2}, \frac{1}{2} \right\rangle \right)$$

from which

$$\langle \psi | \hat{H} | \psi \rangle = -\frac{R}{3} \left(1 + \frac{2}{4} \right) = -\frac{R}{2}.$$

For the square of the Hamiltonian, we find

$$\hat{H}^2 | \psi \rangle = \frac{R^2}{\sqrt{3}} \left(|1,0,0\rangle \otimes \left| \frac{1}{2}, \frac{1}{2} \right\rangle + \frac{1}{16} |2,1,1\rangle \otimes \left| \frac{1}{2}, -\frac{1}{2} \right\rangle + \frac{1}{16} |2,1,0\rangle \otimes \left| \frac{1}{2}, \frac{1}{2} \right\rangle \right)$$

from which

$$\langle \psi | \hat{H}^2 | \psi \rangle = \frac{R^2}{3} \left(1 + \frac{2}{16} \right) = \frac{3R^2}{8}.$$

The associated standard deviation is

$$\sigma_E = \sqrt{\langle \psi | \hat{H}^2 | \psi \rangle - (\langle \psi | \hat{H} | \psi \rangle)^2} = \frac{R}{\sqrt{8}}.$$

Problem 4.8.
A particle with unitary mass is subject to a centrally symmetric force with associated potential

$$V(r) = \frac{\alpha}{r^\beta}$$

where α is a real number, β an integer number, and r the radial distance in spherical polar coordinates. Determine the largest value of β and the corresponding values of α such that the system has positive energies in the discrete spectrum. In this case, analyze the wave function in a neighborhood of the origin. To simplify matters, make use of atomic units (see Problem 5.15).

Solution
As we know from the theory, the discrete spectrum is characterized by all those energy levels which represent bound states. In such a case, the region of motion is bounded, the resulting wave function can be normalized, and the probability to find the particle at infinity tends to zero. To study the problem, we need to replace the potential V with the effective potential V_{eff} given by the sum of V and the centrifugal term

$$V_{eff}(r) = \frac{\alpha}{r^\beta} + \frac{l(l+1)}{2r^2}$$

which is different for states with different l. We see that the motion of a particle in a spherically symmetric field, is equivalent to a one dimensional motion under the effect of the potential energy V_{eff}. To have positive values of the energy in the discrete spectrum, V_{eff} cannot be zero at infinity but it must be a positive constant (or infinity). Also, for $r = 0$, $V_{eff}(0) = \pm\infty$ depending on the values of α and β. Therefore, our requirements are:

- the effective potential is positive at infinity, $V(+\infty) > 0$;
- the effective potential has a local minimum for a finite r.

To verify the existence of a local maximum or minimum, we take the first derivative of V_{eff}, and set it to zero in the point r_0

$$-\frac{\beta\alpha}{(l(l+1))} = r_0^{\beta-2}:$$

- the case $\beta > 2$ and $\alpha < 0$ gives $V(0) = -\infty$ and $V(+\infty) = 0$. We immediately see that r_0 is a local maximum. Consequently, this case cannot describe bound states;
- the case $\alpha > 0$ (still with $\beta > 2$) is not producing bound states because $V(0) = +\infty$, $V(+\infty) = 0$ and the first derivative of the potential is zero for an imaginary r_0;

- when $\beta = 2$ (with arbitrary α) the effective potential is monotonic without a local minimum. No bound states are possible;
- when $\beta = 1$, we find $V(+\infty) = 0$ and we have to rule out this case because the discrete spectrum, if any, has no positive energies;
- the case $\beta = 0$ produces $V(0) = +\infty, V(+\infty) = 0$ and the effective potential has not a minimum for a finite r_0;
- with $\beta = -1$, $\alpha < 0$ we find $V(0) = +\infty, V(+\infty) = -\infty$. Again, we do not find a minimum in V_{eff} and, consequently, the Hamiltonian has not a discrete spectrum;
- finally, the case $\beta = -1$ and $\alpha > 0$ satisfies our requirements.

The stationary Schrödinger equation can be analyzed by setting $\psi(r, \theta, \phi) = \frac{R(r)}{r} Y_{l,m}(\theta, \phi)$ and eliminating the spherical harmonics $Y_{l,m}(\theta, \phi)$ using their orthogonality properties. When working with atomic units (see Problem 5.15), we get

$$\left[\frac{d^2}{dr^2} + 2E - 2\alpha r - \frac{l(l+1)}{r^2} \right] R(r) = 0.$$

We have regular singularities in $r = 0$, and we can seek the solution in the form $R(r) \approx r^\lambda$. Plugging this back in the original equation, we have

$$\lambda(\lambda - 1) r^{\lambda-2} + 2E r^\lambda - 2\alpha r^{\lambda+1} - l(l+1) r^{\lambda-2} = 0.$$

The second and third terms are subleading with respect to $r^{\lambda-2}$ when $r \to 0$, so that $\lambda = -l, l+1$. The physical solution corresponds to the non divergent solution in the origin, i.e. $\lambda = l+1$.

Problem 4.9.
Characterize the ground state for the Helium atom. In particular, determine the value of the total spin and determine its degeneracy when the interaction between the two electrons is vanishingly small. Repeat the exercise for the first excited energy level. Finally, provide qualitative arguments to characterize the energy of the first excited state when the interaction between the electrons is taken into account. Make use of the wave functions of the first two atomic levels

$$\phi_{1s}(r) = \frac{1}{\sqrt{\pi}} \left(\frac{Z}{a} \right)^{\frac{3}{2}} e^{-\frac{Zr}{a}} \qquad \phi_{2s}(r) = \frac{1}{\sqrt{4\pi}} \left(\frac{Z}{2a} \right)^{\frac{3}{2}} \left(2 - \frac{Zr}{a} \right) e^{-\frac{Zr}{2a}}$$

with $Z = 2$ the atomic number and a the characteristic atomic length scale (the Bohr radius).

Solution
We first neglect the interaction between the two electrons. When acting on a function of the coordinates, the Hamiltonian takes the form

$$\hat{H} = -\frac{\hbar^2}{2m} (\nabla_1^2 + \nabla_2^2) - Ze^2 \left(\frac{1}{\hat{r}_1} + \frac{1}{\hat{r}_2} \right).$$

In this approximation, the problem is that of two independent electrons, each one subject to the Coulomb central field $U(r) = -Ze^2/r$. The solution is known from the theory of the hydrogen-like atoms, leading to the energy spectrum

$$\varepsilon_n = -\frac{Z^2 e^2}{2an^2} \qquad n = 1, 2, 3, \ldots$$

Therefore, the energy of the ground state of the Helium atom is

$$E_1 = 2\varepsilon_1 = -\frac{Z^2 e^2}{a}.$$

The wave function is characterized by the symmetric (s) and antisymmetric (a) linear combinations (i.e. it takes a $\pm$ sign under the interchange of the electrons) of $\phi_{1s}(r_1)$ and $\phi_{1s}(r_2)$. For the ground state, the antisymmetric combination vanishes, and we are left with the symmetric orbital wave function

$$\psi_s(r_1, r_2) = \phi_{1s}(r_1)\phi_{1s}(r_2) = \frac{1}{\pi}\left(\frac{Z}{a}\right)^3 e^{-\frac{Z(r_1+r_2)}{a}}.$$

The total wave function is $\psi(r_1, r_2) = \psi_s(r_1, r_2)\chi$, where χ refers to the spin part. Since the total wave function has to be antisymmetric, the spin part has to be antisymmetric

$$\chi = \frac{1}{\sqrt{2}}\left(\left|\frac{1}{2}, \frac{1}{2}\right\rangle_1 \otimes \left|\frac{1}{2}, -\frac{1}{2}\right\rangle_2 - \left|\frac{1}{2}, -\frac{1}{2}\right\rangle_1 \otimes \left|\frac{1}{2}, \frac{1}{2}\right\rangle_2\right)$$

that is a singlet state (i.e. a state with spin 0 that is not degenerate).

The first excited energy state is a bit more complicated. One electron is in the state $1s$ ($n = 1, l = 0$), while the other in $2s$ ($n = 2, l = 0$). The possible orbital wave functions are

$$\Phi_s(r_1, r_2) = \frac{1}{\sqrt{2}}(\phi_{1s}(r_1)\phi_{2s}(r_2) + \phi_{1s}(r_2)\phi_{2s}(r_1))$$

$$\psi(r_1, r_2) = \Phi_s(r_1, r_2)\chi$$

where $\Phi_s(r_1, r_2)$ is symmetric and defines the *parahelium* states, while $\Phi_a(r_1, r_2)$ is antisymmetric and defines the *orthohelium* states. As for the total wave function (orbital motion plus spin), $\Phi_s(r_1, r_2)$ is multiplied by the singlet state previously described, while $\Phi_a(r_1, r_2)$ is multiplied by one of the following spin wave functions

$$\chi_1 = \left|\frac{1}{2}, \frac{1}{2}\right\rangle_1 \otimes \left|\frac{1}{2}, \frac{1}{2}\right\rangle_2$$

$$\chi_2 = \frac{1}{\sqrt{2}}\left(\left|\frac{1}{2}, \frac{1}{2}\right\rangle_1 \otimes \left|\frac{1}{2}, -\frac{1}{2}\right\rangle_2 + \left|\frac{1}{2}, -\frac{1}{2}\right\rangle_1 \otimes \left|\frac{1}{2}, \frac{1}{2}\right\rangle_2\right)$$

$$\chi_3 = \left|\frac{1}{2}, -\frac{1}{2}\right\rangle_1 \otimes \left|\frac{1}{2}, -\frac{1}{2}\right\rangle_2,$$

defining the so-called triplet (i.e. a state with spin 1 that is triple degenerate). The energy of the parahelium and orthohelium states is the same, unless we introduce the interaction potential between the electrons, $U_{int}(r_1, r_2) = e^2/(|r_1 - r_2|)$. The wave function $\Phi_a(r_1, r_2)$ is zero when $r_1 = r_2 = 0$, while $\Phi_s(r_1, r_2)$ has a maximum in this point. This means that the electrons preferentially stay either away from the nuclear region (orthohelium) or close to it (parahelium). The contribution of the interaction potential comes from

$$\langle \hat{U}_{int} \rangle = \int \int \Phi^*(r_1, r_2) \frac{e^2}{|r_1 - r_2|} \Phi(r_1, r_2) d^3 r_1 d^3 r_2$$

which can be calculated for the orthohelium ($\Phi = \Phi_a$) and the parahelium states ($\Phi = \Phi_s$). The result is that $\langle \hat{U}_{int} \rangle$ is positive in sign and smaller for the orthohelium. This happens because in the orthohelium case the spatial wave function is antisymmetric, the electrons tend to stay away from each other, and this reduces the repulsive effect of U_{int}. We conclude that the total energy is lower for the orthohelium state.

Problem 4.10.

An atom has two electrons with modulus of the charge e, spin $S = 1/2$, and with the orbital angular momentum $l = 0$. The nuclear charge is e. Characterize the energy spectrum when the two electrons occupy the lowest energy levels available, and in the limit where the interaction potential between the electrons is negligible. Determine the wave function of the system and verify its normalization. Finally, determine the probability P_ε to find both electrons in a spherical region of radius ε centered in the nucleus.

Solution

From the point of view of Quantum Mechanics, the two electrons have to be considered as indistinguishable particles. Particles with half-odd-integer spin obey the Fermi-Dirac statistics and their resulting wave function has to be antisymmetric when the two particles are interchanged. Such property is a direct consequence of the Pauli principle which states that, in a system composed of identical fermions, there cannot be two particles with the same quantum numbers. The antisymmetric combination of identical objects is in fact zero.

In the case of our problem, the wave functions are those of the hydrogen atom $(\psi(r) = R_{n,l}(r) Y_{l,m}(\theta, \phi) \otimes |\frac{1}{2}, S_z/\hbar\rangle)$ with $n = 1, l = m = 0, S_z = \pm \hbar/2$

$$\psi_1(r_1) = R_{1,0}(r_1) Y_{0,0}(\theta_1, \phi_1) \otimes \left|\frac{1}{2}, \frac{1}{2}\right\rangle = 2a^{-3/2} e^{-r_1/a} \frac{1}{\sqrt{4\pi}} \otimes \left|\frac{1}{2}, \frac{1}{2}\right\rangle$$

$$\psi_2(r_2) = R_{1,0}(r_2) Y_{0,0}(\theta_2, \phi_2) \otimes \left|\frac{1}{2}, -\frac{1}{2}\right\rangle = 2a^{-3/2} e^{-r_2/a} \frac{1}{\sqrt{4\pi}} \otimes \left|\frac{1}{2}, -\frac{1}{2}\right\rangle$$

where $a = \hbar^2/m_e e^2$ is the Bohr radius (m_e is the mass of the electron) and where $|\frac{1}{2}, \pm\frac{1}{2}\rangle$ refer to the eigenstates of $\hat{S}_z$, i.e. the operator describing the z component of the spin with eigenvalues $\pm\hbar/2$. The antisymmetric combination of $\psi_1(r_1)$ and $\psi_2(r_2)$ is

$$\psi(r_1, r_2) = \frac{1}{\sqrt{2}}(\psi_1(r_1)\psi_2(r_2) - \psi_1(r_2)\psi_2(r_1)).$$

We then verify the normalization $N^2 = \int\int |\psi(r_1, r_2)|^2 d^3 r_1 d^3 r_2$

$$N^2 = \frac{16a^{-6}}{(4\pi)^2} \left(\int_0^\pi \sin\theta d\theta \int_0^{2\pi} d\phi \right)^2 \left(\int_0^{+\infty} r^2 e^{-2r/a} dr \right)^2 = 1$$

where we have used that $\int_0^{+\infty} x^2 e^{-2x} dx = \frac{1}{4}$. Finally, the probability to find the two electrons in a spherical region of radius ε is given by the integral of $|\psi(r_1, r_2)|^2$ in the region $0 \le r_{1,2} \le \varepsilon$

$$P_\varepsilon = 16a^{-6} \int_0^\varepsilon r_1^2 e^{-2r_1/a} dr_1 \int_0^\varepsilon r_2^2 e^{-2r_2/a} dr_2 = 16 \left(\int_0^{\frac{\varepsilon}{a}} x^2 e^{-2x} dx \right)^2.$$

A direct calculation shows that

$$P_\varepsilon = 16 \left(\lim_{\beta \to 2} \frac{d^2}{d\beta^2} \int_0^{\frac{\varepsilon}{a}} e^{-x\beta} dx \right)^2 = 16 \left(e^{-2\varepsilon/a} \left(-\frac{\varepsilon^2}{2a^2} - \frac{\varepsilon}{2a} - \frac{1}{4} \right) + \frac{1}{4} \right)^2.$$

Problem 4.11.
Let us consider the following spherical wave function

$$\psi(r) = \frac{e^{\pm ik\cdot r}}{r}$$

where r is the radial distance in spherical coordinates, $k = kr/r$ the wave vector and k its modulus. Determine:

- the density flux for the probability density function;
- the number of particles in the unit time passing through the spherical surface of radius R for a given $k = kr/r$.

Solution
The wave vector $k = kr/r$ is in the same direction of r. Consequently, the wave function can be written as a function of the radial distance r and the modulus k

$$\psi(r) = \frac{e^{\pm ikr}}{r}.$$

The density flux for the probability density function is obtained from the usual density flux

$$J = \frac{i\hbar}{2m} (\psi\nabla\psi^* - \psi^*\nabla\psi)$$

by considering its radial component

$$J_r = \frac{\hbar}{2mi}\left[\psi^*\frac{d\psi}{dr} - \psi\frac{d\psi^*}{dr}\right].$$

An explicit calculation shows that

$$J_r = \frac{\hbar}{2mi}\left[\frac{e^{\mp ikr}}{r}\left(\pm ik\frac{e^{\pm ikr}}{r} - \frac{e^{\pm ikr}}{r^2}\right) - \frac{e^{\pm ikr}}{r}\left(\mp ik\frac{e^{\mp ikr}}{r} - \frac{e^{\mp ikr}}{r^2}\right)\right] = \pm\frac{\hbar k}{mr^2}.$$

The number of particles in the unit time passing (exiting or entering) through the spherical surface of radius R is just the integral of J_r over such surface

$$N = \int_0^{2\pi} d\phi \int_0^\pi j_R R^2 \sin\theta d\theta = \pm\frac{4\pi\hbar k}{m}.$$

Problem 4.12.

Consider a hydrogen atom and determine the probability distribution function for the momentum in the ground state (1s). With such result, determine the average kinetic energy, and compare the result with the one obtained in the position space (see Problem 5.15). For simplicity, make use of atomic units.

Solution
The wave function of the ground (1s) state has principal quantum number $n = 1$ and the orbital angular momentum $l = 0$

$$\psi_{1,0,0}(r,\theta,\phi) = R_{1,0}(r)Y_{0,0}(\theta,\phi) = 2e^{-r}\frac{1}{\sqrt{4\pi}}$$

where we have used atomic units. In this way, the variable r is dimensionless. We recover physical units (see also Problem 5.15) by introducing the proper length scale (a, the Bohr radius) and energy scale (e^2/a, the atomic energy). To find the distribution of the momentum, we have to determine the generic mode (whose absolute value is denoted with p_r) in Fourier space starting from the wave function

$$\psi_{1,0,0}(p_r) = \frac{1}{\sqrt{\pi}}\int e^{-(i\mathbf{p}\cdot\mathbf{r}+r)}d^3r = 2\sqrt{\pi}\int_0^{+\infty}\int_{-1}^{+1} e^{-(ip_r r\cos\theta + r)}r^2\,dr\,d\cos\theta =$$

$$\frac{2\sqrt{\pi}}{ip_r}\left(\int_0^{+\infty} e^{-(-ip_r r + r)}r\,dr - \int_0^{+\infty} e^{-(ip_r r + r)}r\,dr\right) =$$

$$\frac{2\sqrt{\pi}}{ip_r}\left(\frac{1}{(-ip_r+1)^2} - \frac{1}{(ip_r+1)^2}\right) = \frac{8\sqrt{\pi}}{(p_r^2+1)^2}$$

and take its square modulus

$$|\psi_{1,0,0}(p_r)|^2 = \frac{64\pi}{(1+p_r^2)^4}.$$

The integral of such quantity gives the normalization of $\psi_{1,0,0}(p_r)$

$$\int_0^{+\infty} |\psi_{1,0,0}(p_r)|^2 d^3 p_r = 256\pi^2 \int_0^{+\infty} \frac{p_r^2}{(1+p_r^2)^4} dp_r = 8\pi^3$$

where we have used the indefinite integral

$$\int \frac{x^2}{(1+x^2)^4} dx = -\frac{x}{6(1+x^2)^3} + \frac{x}{24(1+x^2)^2} + \frac{x}{16(1+x^2)} + \frac{\arctan x}{16} + const.$$

The normalized probability distribution function between $p_r = 0$ and $p_r = +\infty$ is

$$P(p_r) = \frac{|\psi_{1,0,0}(p_r)|^2 4\pi p_r^2}{8\pi^3} = \frac{32 p_r^2}{\pi(1+p_r^2)^4}.$$

The kinetic term in the three dimensional Schrödinger equation (with mass $m = 1$ and $\hbar = 1$) may be written as (radial part plus angular part) $\hat{T} = \frac{\hat{p}_r^2}{2} + \frac{\hat{L}^2}{2r^2}$, with $\hat{L}^2$ the squared orbital angular momentum. Moreover, the angular term for the ground state is zero, due to the fact that it is proportional to the spherical harmonic $Y_{0,0}(\theta, \phi)$ that is an eigenstate of $\hat{L}^2$ with eigenvalue 0. Therefore, using the indefinite integral

$$\int \frac{x^4}{(1+x^2)^4} dx = \frac{x}{6(1+x^2)^3} - \frac{7x}{24(1+x^2)^2} + \frac{x}{16(1+x^2)} + \frac{\arctan x}{16} + const.$$

we find

$$\langle \hat{T} \rangle = \frac{1}{2} \langle \hat{p}_r^2 \rangle = \frac{1}{2} \int_0^{+\infty} P(p_r) p_r^2 dp_r = \frac{16}{\pi} \int_0^{+\infty} \frac{p_r^4}{(1+p_r^2)^4} dp_r = \frac{1}{2}.$$

Such value is in agreement with the kinetic energy determined in Problem 5.15, where we will use the representation of the wave functions in the position space.

5

Perturbation Theory and WKB Method

Problem 5.1.

A plane rigid rotator has the following Hamiltonian

$$\hat{H}_0 = \frac{\hat{L}_z^2}{2I}$$

where I is the momentum of inertia and $\hat{L}_z = -i\hbar\frac{d}{d\phi}$ is the component of the angular momentum in the z direction (ϕ is the azimuthal angle). A small perturbation

$$\hat{H}_1 = \lambda\cos(2\hat{\phi}) \qquad \lambda \ll 1$$

is applied to the rotator. Determine the average value of $\hat{H}_1$ on each unperturbed eigenstate. Then, determine the off-diagonal elements of the perturbation matrix between the degenerate states. Finally, find the first order correction to the energy of the ground state and the splitting induced on the energy of the first excited state.

Solution

We first determine the eigenstates of $\hat{H}_0$. To do that, we have to solve the differential equation

$$-\frac{\hbar^2}{2I}\frac{d\psi_k(\phi)}{d\phi} = \varepsilon_k\psi_k(\phi) \qquad 0 \le \phi \le 2\pi$$

leading to the following normalized eigenstates

$$\psi_k(\phi) = \frac{e^{ik\phi}}{\sqrt{2\pi}}$$

with k an integer number. The values k and $-k$ correspond to the same eigenvalue of the energy

$$\varepsilon_k = \hbar^2\frac{k^2}{2I}$$

Cini M., Fucito F., Sbragaglia M.: Solved Problems in Quantum and Statistical Mechanics.
DOI 10.1007/978-88-470-2315-4_5, © Springer-Verlag Italia 2012

meaning that all the states show a degeneracy, with the only exception of $k = 0$. The average value of the perturbation $\hat{H}_1$ on a generic eigenstate is given by

$$\langle k|\hat{H}_1|k\rangle = \frac{\lambda}{2\pi} \int_0^{2\pi} e^{-ik\phi} \cos(2\phi) e^{ik\phi} d\phi = \frac{\lambda}{2\pi} \int_0^{2\pi} \cos(2\phi) d\phi = 0$$

and, in particular, for the ground state ($k = 0$) we find $\langle 0|\hat{H}_1|0\rangle = 0$. The off-diagonal matrix elements between degenerate states are

$$\langle -k|\hat{H}_1|k\rangle = \frac{\lambda}{2\pi} \int_0^{2\pi} e^{2ik\phi} \cos(2\phi) d\phi = \frac{1}{2}\lambda \delta_{k1} \quad k > 0$$

as we can see by writing $\cos(2\phi) = (e^{2i\phi} + e^{-2i\phi})/2$ and using the fact that the integral of the exponential function $e^{\pm 2i\phi + 2ik\phi}$ over a period is zero, unless $k = 1$. As a consequence, the perturbation matrix is zero with the only exception of the first excited state. In the latter case, we can represent such matrix in the following way

$$\hat{H}_1^{(1)} = \frac{\lambda}{2} \begin{pmatrix} 0 & 1 \\ 1 & 0 \end{pmatrix}.$$

We conclude that the ground state and the states with $|k| > 1$ are unperturbed. The first excited state with energy ε_1 splits in two states with energies $\hbar^2 \frac{1}{2I} \pm \frac{1}{2}\lambda$.

Problem 5.2.
A quantum system is characterized by a discrete spectrum whose eigenstates are $|\psi_L^{(0)}\rangle$ and $|\psi_R^{(0)}\rangle$, corresponding to the energies $E_L = 0$ and $E_R = M$ respectively. In such a vector basis, the matrix representation of the unperturbed Hamiltonian is

$$\hat{H}_0 = \begin{pmatrix} 0 & 0 \\ 0 & M \end{pmatrix}.$$

A perturbation $\hat{V}$ is switched on and the new Hamiltonian is

$$\hat{H} = \hat{H}_0 + \hat{V} = \begin{pmatrix} 0 & m \\ m & M \end{pmatrix}.$$

Using perturbation theory, determine the first and second order corrections to the eigenvalues of $\hat{H}_0$. To do that, you can define a parameter $\lambda = \frac{m}{M}$ and assume that $\lambda \ll 1$.

Solution
The problem is non degenerate because $|\psi_L^{(0)}\rangle$ and $|\psi_R^{(0)}\rangle$ correspond to different energies. Also, the perturbation can be written as $\lambda \hat{V}'$ where

$$\hat{V}' = \begin{pmatrix} 0 & M \\ M & 0 \end{pmatrix}.$$

The unperturbed eigenstates of $\hat{H}_0$ are

$$|\psi_L^{(0)}\rangle = \begin{pmatrix} 1 \\ 0 \end{pmatrix} \quad |\psi_R^{(0)}\rangle = \begin{pmatrix} 0 \\ 1 \end{pmatrix}$$

corresponding to the unperturbed energies $E_L^{(0)} = 0$ and $E_R^{(0)} = M$. The first order corrections to the eigenvalues correspond to the diagonal elements of the perturbation matrix $V_{nk}' = \langle \psi_n^{(0)} | \hat{V}' | \psi_k^{(0)} \rangle$, with $n, k = L, R$

$$\langle \psi_L^{(0)} | \hat{V}' | \psi_L^{(0)} \rangle = \langle \psi_R^{(0)} | \hat{V}' | \psi_R^{(0)} \rangle = 0.$$

For the second order corrections in λ, we need to use

$$\Delta E_n^{(2)} = \sum_{k \neq n} \frac{|V_{nk}'|^2}{E_n^{(0)} - E_k^{(0)}}.$$

The matrix elements of interest are

$$V_{LR}' = V_{RL}' = M$$

from which we find

$$\Delta E_L^{(2)} = -\frac{M^2}{M} = -M \quad \Delta E_R^{(2)} = \frac{M^2}{M} = M$$

and

$$E_L \cong E_L^{(0)} + \lambda \Delta E_L^{(1)} + \lambda^2 \Delta E_L^{(2)} = -\frac{m^2}{M}$$

$$E_R \cong E_R^{(0)} + \lambda \Delta E_R^{(1)} + \lambda^2 \Delta E_R^{(2)} = M + \frac{m^2}{M}.$$

Problem 5.3.

Let us consider a quantum system with Hamiltonian $\hat{H}$ such that

$$\begin{cases} \hat{H}|0\rangle = 0 \\ \hat{H}|n\rangle = n!\, Q|n\rangle \quad n = 1, 2, 3, \dots \end{cases}$$

where $|n\rangle$ are the eigenstates ($\langle n|k\rangle = \delta_{nk}$) and $Q > 0$ a positive constant. Using perturbation theory, find the first order correction $\Delta E_0^{(1)}$ to the energy of the ground state induced by the perturbation

$$\hat{V} = V \sum_{n=0}^{+\infty} \alpha^{n/2} \left(|n\rangle\langle 0| + |0\rangle\langle n| \right)$$

where $V > 0$. Also, determine the second order correction $\Delta E_0^{(2)}$ to the energy of the ground state induced by the same perturbation. What is the value of α such that the total correction $\Delta E_0^{(1)} + \Delta E_0^{(2)}$ is zero?

Solution
The ground state is $|0\rangle$ and has zero energy. The first order correction is

$$\Delta E_0^{(1)} = \langle 0|\hat{V}|0\rangle = 2V.$$

As for the second order correction, it can be found with the Taylor series of the exponential function

$$\Delta E_0^{(2)} = -\sum_{n=1}^{\infty} \frac{|\langle 0|\hat{V}|n\rangle|^2}{Qn!} = -\frac{V^2}{Q}\sum_{n=1}^{\infty} \frac{\alpha^n}{n!} = -\frac{V^2}{Q}(e^{\alpha} - 1).$$

The value of α such that the total correction is $\Delta E_0^{(1)} + \Delta E_0^{(2)} = 0$ is given by the equation

$$2V = \frac{V^2}{Q}(e^{\alpha} - 1)$$

so that

$$\alpha = \ln\left(\frac{2Q+V}{V}\right).$$

Problem 5.4.
A one dimensional quantum harmonic oscillator with Hamiltonian

$$\hat{H}_0 = \left(\hat{a}^{\dagger}\hat{a} + \frac{1}{2}1\right)\hbar\omega = \left(\hat{n} + \frac{1}{2}1\right)\hbar\omega$$

is subject to a small perturbation $\hat{V} = \lambda\hat{a}^{\dagger}\hat{a}^{\dagger}\hat{a}\hat{a}$. Determine the first order perturbative corrections to the energy spectrum. Then, compute the exact solution and compare it with the result of perturbation theory.

Solution
We need to write the perturbation in a more convenient way. We know the commutation relations of the creation and annihilation operators $\hat{a}^{\dagger}$ and $\hat{a}$

$$[\hat{a}, \hat{a}^{\dagger}] = \hat{a}\hat{a}^{\dagger} - \hat{a}^{\dagger}\hat{a} = 1$$

and we can rewrite $\hat{V}$ in the following way

$$\hat{V} = \lambda\hat{a}^{\dagger}\hat{a}^{\dagger}\hat{a}\hat{a} = \lambda\hat{a}^{\dagger}(\hat{a}\hat{a}^{\dagger} - 1)\hat{a} = \lambda(\hat{n}^2 - \hat{n})$$

with $\hat{n} = \hat{a}^{\dagger}\hat{a}$ the number operator. We note that the perturbation is diagonal with respect to the eigenstates ($|n\rangle$) of the harmonic oscillator

$$\hat{V}|n\rangle = \lambda(\hat{n}^2 - \hat{n})|n\rangle = \lambda(n^2 - n)|n\rangle.$$

The first order correction to the energy spectrum is therefore the exact solution to our problem.

Problem 5.5.
A quantum particle with mass M moves in a two dimensional infinite potential well

$$\begin{cases} V(x,y) = 0 & |x| \leq \frac{L}{2}, |y| \leq \frac{L}{2} \\ V(x,y) = +\infty & |x| \geq \frac{L}{2}, |y| \geq \frac{L}{2}. \end{cases}$$

Determine the eigenstates and eigenvalues for the Hamiltonian. Imagine to construct states with four electrons (with spin $1/2$). What is the energy of the ground state? At some point, a perturbation

$$\hat{H}'(\hat{x}, \hat{y}) = \lambda \hat{X}(\hat{x}) \hat{Y}(\hat{y})$$

is switched on. In the above expression, $\hat{X}(\hat{x})$ and $\hat{Y}(\hat{y})$ are generic operators, $\hat{x}$ and $\hat{y}$ the position operators in the two coordinates, and λ a small parameter. Construct the perturbation matrix for the case of a single electron occupying the first excited state. In the special case

$$\hat{H}'(\hat{x}, \hat{y}) = \lambda \frac{\hat{x}\hat{y}}{L^2}$$

determine the first order correction to such a state.

Solution
We need to solve the Schrödinger equation in the two dimensional potential well. This is similar to the three dimensional case solved in Problem 2.10. When variables are separated in the Schrödinger equation, we find the eigenfunctions

$$\psi_{n,m}(x,y) = u_n(x)u_m(y)$$

with n, m positive integers and $u_n(x)$ correctly normalized

$$u_n(x) = \sqrt{\frac{2}{L}} \sin\left(\frac{\pi n}{L}\left(x + \frac{L}{2}\right)\right).$$

The eigenvalues are

$$E_{n,m} = \frac{\hbar^2 \pi^2}{2ML^2}(n^2 + m^2).$$

We note that the ground state (denoted with $E_{1,1}$ and corresponding to the case $n = m = 1$) is non degenerate. The first excited state (denoted with $E_{1,2}$ and corresponding to the case $n + m = 3$) is double degenerate. When we have four electrons with spin $1/2$, two of them go in the ground state, while the remaining two in the first excited state, with a total energy

$$E_4 = 2(E_{1,1} + E_{1,2}).$$

When we have a single electron in the first excited state, a vector basis is given by

$$\begin{cases} \langle x,y|1\rangle = u_2(x)u_1(y) \\ \langle x,y|2\rangle = u_1(x)u_2(y) \end{cases}$$

with the property $\langle 1|2 \rangle = 0$, $\langle 1|1 \rangle = \langle 2|2 \rangle = 1$. The representation of the perturbation matrix in such a vector basis is

$$\hat{H}' = \begin{pmatrix} \lambda \langle u_2|\hat{X}|u_2 \rangle \langle u_1|\hat{Y}|u_1 \rangle & \lambda \langle u_2|\hat{X}|u_1 \rangle \langle u_1|\hat{Y}|u_2 \rangle \\ \lambda \langle u_1|\hat{X}|u_2 \rangle \langle u_2|\hat{Y}|u_1 \rangle & \lambda \langle u_1|\hat{X}|u_1 \rangle \langle u_2|\hat{Y}|u_2 \rangle \end{pmatrix}.$$

Therefore, when we have $\hat{H}'(\hat{x},\hat{y}) = \lambda \frac{\hat{x}\hat{y}}{L^2}$, we need to consider the following integrals

$$\frac{2}{L}\int_{-\frac{L}{2}}^{+\frac{L}{2}} \sin\left(\frac{\pi}{L}\left(x+\frac{L}{2}\right)\right) \sin\left(\frac{2\pi}{L}\left(x+\frac{L}{2}\right)\right) x\, dx = -\frac{16}{9\pi^2}L$$

$$\frac{2}{L}\int_{-\frac{L}{2}}^{+\frac{L}{2}} \sin\left(\frac{2\pi}{L}\left(x+\frac{L}{2}\right)\right) \sin\left(\frac{2\pi}{L}\left(x+\frac{L}{2}\right)\right) x\, dx = 0.$$

The second integral is zero because the integrand is the product of an even function $(\sin^2\left(\frac{2\pi}{L}\left(x+\frac{L}{2}\right)\right))$ and an odd one (x). The first integral is non zero and can be done using the following results

$$\int_0^\pi \sin\xi \sin(2\xi)\xi\, d\xi = 2\int_0^\pi \sin^2\xi \cos\xi\, \xi\, d\xi = -\frac{2}{3}\int_0^\pi \sin^3\xi\, d\xi$$

and

$$\int_0^\pi \sin^3\xi\, d\xi = \int_0^\pi \sin\xi\, d\xi - \int_0^\pi \sin\xi \cos^2\xi\, d\xi = \left(-\cos\xi + \frac{1}{3}\cos\xi\right)\Big|_0^\pi = \frac{4}{3}.$$

Therefore, the matrix representation of $\hat{H}'$ is

$$\hat{H}' = \lambda\left(\frac{16}{9\pi^2}\right)^2 \begin{pmatrix} 0 & 1 \\ 1 & 0 \end{pmatrix}$$

with eigenvalues $\Delta E_\pm^{(1)} = \pm\lambda(\frac{16}{9\pi^2})^2$. We see that the perturbation removes the degeneracy completely, and the energy of the first excited state splits in two levels with energies $E_{1,2} + \Delta E_\pm^{(1)}$.

Problem 5.6.
A particle is confined in a one dimensional potential well in the segment $0 \leq x \leq L$ and is under the effect of the perturbation $V(x) = V_0\cos(\frac{\pi x}{L})$. Determine the first order correction to the energy of the ground state and the first excited state.

Solution
As we know from the solution of the Schrödinger equation in the one dimensional potential well (see Problem 2.10), the normalized eigenstates are given by

$$\langle x|n \rangle = \psi_n(x) = \sqrt{\frac{2}{L}}\sin\left(\frac{n\pi x}{L}\right) \qquad n = 1,2,3,\ldots$$

and they are not degenerate. When applying time independent perturbation theory to determine the energy correction, we have to compute the general integral

$$\langle n|\hat{V}|n\rangle = \Delta E_n^{(1)} = \frac{2V_0}{L} \int_0^{\frac{L}{2}} \sin^2\left(\frac{n\pi x}{L}\right) \cos\left(\frac{\pi x}{L}\right) dx = \frac{2V_0}{\pi} \int_0^{\frac{\pi}{2}} \sin^2(nt)\cos t\, dt.$$

For $n = 1$ (the ground state), the integral is simple

$$\Delta E_1^{(1)} = \frac{2V_0}{\pi} \int_0^{\frac{\pi}{2}} \sin^2 t \cos t\, dt = \frac{2V_0}{\pi} \int_0^1 \sin^2 t\, d(\sin t) = \frac{2V_0}{\pi} \int_0^1 x^2\, dx = \frac{2V_0}{3\pi}.$$

For $n = 2$, we find

$$\Delta E_2^{(1)} = \frac{2V_0}{\pi} \int_0^{\frac{\pi}{2}} \sin^2(2t)\cos t\, dt = \frac{8V_0}{\pi} \int_0^{\frac{\pi}{2}} \sin^2 t \cos^3 t\, dt =$$
$$\frac{8V_0}{\pi} \int_0^1 \sin^2 t \cos^2 t\, d(\sin t) = \frac{8V_0}{\pi} \int_0^1 x^2(1-x^2)\, dx = \frac{16V_0}{15\pi}.$$

Problem 5.7.
A quantum system is described by the following Hamiltonian

$$\hat{H} = \hat{H}_0 + \hat{H}'$$

where

$$\hat{H}_0 = \begin{pmatrix} 1 & 0 \\ 0 & -1 \end{pmatrix}$$

and $\hat{H}'$ has to be considered as a small perturbation with the following representation

$$\hat{H}' = \eta \begin{pmatrix} 0 & 1 \\ 1 & 0 \end{pmatrix}$$

with η a small parameter. Determine the correction to the energy of the ground state up to second order in perturbation theory. Then, find the exact solution and compare it with the result of perturbation theory.

Solution
First of all, we need to compute the eigenstates and eigenvectors of the unperturbed Hamiltonian $\hat{H}_0$. We determine the characteristic polynomial $p(\lambda)$ associated with $\hat{H}_0$

$$p(\lambda) = \det(\hat{H}_0 - \lambda\mathbb{1}) = \det\begin{pmatrix} 1-\lambda & 0 \\ 0 & -1-\lambda \end{pmatrix} = -(1-\lambda^2).$$

Solving the equation $p(\lambda) = 0$ we find the eigenvalues $\lambda = \pm 1$ and the corresponding eigenstates

$$|+1\rangle = \begin{pmatrix} 1 \\ 0 \end{pmatrix} \qquad |-1\rangle = \begin{pmatrix} 0 \\ 1 \end{pmatrix}$$

which satisfy

$$\hat{H}_0|-1\rangle = -|-1\rangle \quad \hat{H}_0|1\rangle = +|1\rangle.$$

The ground state is $|-1\rangle$ with energy $E_{-1} = -1$; the other is the excited state with energy $E_1 = +1$. To determine the first order correction $\Delta E_{-1}^{(1)}$ to the ground state, we need to compute the matrix element $\langle -1|\hat{H}'|-1\rangle$. The result is

$$\Delta E_{-1}^{(1)} = \langle -1|\hat{H}'|-1\rangle = \eta \begin{pmatrix} 0 & 1 \end{pmatrix} \begin{pmatrix} 0 & 1 \\ 1 & 0 \end{pmatrix} \begin{pmatrix} 0 \\ 1 \end{pmatrix} = 0.$$

The first non zero correction for the ground state energy comes at the second order in perturbation theory. To compute it, we make use of the expression

$$\Delta E_{-1}^{(2)} = \sum_{n \neq -1} \frac{|\langle n|\hat{H}'|-1\rangle|^2}{E_{-1} - E_n}$$

where with $|n\rangle$ we mean a generic state different from the ground state. The previous expression is particularly simple in our case, because we have only two states

$$\Delta E_{-1}^{(2)} = \frac{|\langle 1|\hat{H}'|-1\rangle|^2}{E_{-1} - E_1} = -\frac{\eta^2}{2}$$

where we have used that

$$\langle 1|\hat{H}'|-1\rangle = \langle -1|\hat{H}'|1\rangle = \eta \begin{pmatrix} 0 & 1 \end{pmatrix} \begin{pmatrix} 0 & 1 \\ 1 & 0 \end{pmatrix} \begin{pmatrix} 1 \\ 0 \end{pmatrix} = \eta.$$

To answer the last question, we note that we can rewrite the full Hamiltonian as

$$\hat{H} = \hat{H}_0 + \hat{H}' = \begin{pmatrix} 1 & \eta \\ \eta & -1 \end{pmatrix}$$

and determine the characteristic polynomial

$$p^{(\eta)}(\lambda) = \det(\hat{H} - \lambda \mathbb{1}) = \det \begin{pmatrix} 1 - \lambda & \eta \\ \eta & -1 - \lambda \end{pmatrix} = -(1 - \lambda^2) - \eta^2$$

from which we extract the eigenvalues (solving $p^{(\eta)}(\lambda) = 0$) as

$$E_{-1}^{(\eta)} = -\sqrt{1 + \eta^2} \quad E_1^{(\eta)} = +\sqrt{1 + \eta^2}.$$

The energy of the ground state can be expanded for small η, and we find

$$E_{-1}^{(\eta)} = -\sqrt{1 + \eta^2} \approx -1 - \frac{\eta^2}{2} + \mathcal{O}(\eta^4)$$

in agreement with the result of perturbation theory previously found.

Problem 5.8.
Let us consider the two dimensional harmonic oscillator with Hamiltonian

$$\hat{H}_0 = \frac{\hat{p}_x^2 + \hat{p}_y^2}{2M} + \frac{1}{2}M\omega^2(\hat{x}^2 + \hat{y}^2).$$

Write down the eigenstates of $\hat{H}_0$. Then, using Cartesian coordinates, write down the eigenfunctions of the lowest three energy levels in terms of the eigenfunctions of the one dimensional harmonic oscillator. At some point, the perturbation

$$\hat{H}' = \lambda \hat{x}^2 \hat{y}^2$$

is switched on. Determine the first order correction induced by such perturbation on the energy of the second excited state. You can make use of the following identities

$$(\hat{x}^2)_{0,0} = \frac{1}{2}x_0^2 \quad (\hat{x}^2)_{1,1} = \frac{3}{2}x_0^2 \quad (\hat{x}^2)_{2,2} = \frac{5}{2}x_0^2 \quad (\hat{x}^2)_{0,2} = \frac{1}{\sqrt{2}}x_0^2$$

where $x_0 = \sqrt{\frac{\hbar}{M\omega}}$ and where we have defined the generic matrix element of the squared position operator $(\hat{x}^2)_{n,m} = \langle n|\hat{x}^2|m\rangle$ (see also Problem 2.4), with $|n\rangle$, $|m\rangle$ the eigenstates of the one dimensional case. Also, you can consider that $\hat{x}^2$ has zero matrix elements when evaluated between two states with opposite parity.

Solution
The Hamiltonian $\hat{H}_0$ is the sum of two Hamiltonians, each one representing a one dimensional harmonic oscillator. The corresponding energy is the sum of the two energies

$$E_{m,n}^{(0)} = \hbar\omega(m+n+1) = \hbar\omega\left(m+\frac{1}{2}\right) + \hbar\omega\left(n+\frac{1}{2}\right)$$

with m and n non negative integers. Let us take the eigenfunctions ψ_n of the one dimensional harmonic oscillator with coordinate x

$$\langle x|n\rangle = \psi_n(\xi) = C_n H_n(\xi)e^{-\xi^2/2}$$

where

$$\xi = \sqrt{\frac{M\omega}{\hbar}}x \quad C_n = \frac{1}{2^{n/2}\sqrt{n!}}\left(\frac{M\omega}{\hbar\pi}\right)^{1/4}$$

and where

$$H_n(\xi) = (-1)^n e^{\xi^2}\frac{d^n}{d\xi^n}e^{-\xi^2}$$

represents the n-th order Hermite polynomial. These states have a well defined parity, i.e. they are even (odd) for n even (odd).

We now construct the eigenstates of $\hat{H}_0$ as the product of the eigenstates of the one dimensional case. The ground state is

$$\Psi_{0,0}(x,y) = \psi_0(x)\psi_0(y).$$

The first excited state is double degenerate, with a vector basis given by

$$\Psi_{1,0}(x,y) = \psi_1(x)\psi_0(y) \qquad \Psi_{0,1}(x,y) = \psi_0(x)\psi_1(y).$$

The second excited state is triple degenerate, with a vector basis given by

$$\Psi_{1,1}(x,y) = \psi_1(x)\psi_1(y)$$

$$\Psi_{2,0}(x,y) = \psi_2(x)\psi_0(y)$$

$$\Psi_{0,2}(x,y) = \psi_0(x)\psi_2(y).$$

Therefore, the representation of the perturbation matrix in such a vector basis is

$$\lambda \begin{pmatrix} (\hat{x}^2)^2_{1,1} & 0 & 0 \\ 0 & (\hat{x}^2)_{2,2}(\hat{x}^2)_{0,0} & (\hat{x}^2)^2_{2,0}(\hat{x}^2)^2_{0,2} \\ 0 & (\hat{x}^2)^2_{2,0}(\hat{x}^2)^2_{0,2} & (\hat{x}^2)_{2,2}(\hat{x}^2)_{0,0} \end{pmatrix} = \lambda x_0^4 \begin{pmatrix} \frac{9}{4} & 0 & 0 \\ 0 & \frac{5}{4} & \frac{1}{2} \\ 0 & \frac{1}{2} & \frac{5}{4} \end{pmatrix}.$$

The perturbation removes the degeneracy completely. The energy eigenvalue corresponding to the state $\Psi_{1,1}(x,y)$ is $3\hbar\omega + \frac{9}{4}\lambda x_0^4$, with $\frac{9}{4}\lambda x_0^4$ the energy correction induced by the perturbation. The other two states ($\Psi_{2,0}(x,y)$ and $\Psi_{0,2}(x,y)$) mix together, with resulting energies $3\hbar\omega + \frac{7}{4}\lambda x_0^4$ and $3\hbar\omega + \frac{3}{4}\lambda x_0^4$.

Problem 5.9.
A three dimensional rigid rotator has Hamiltonian

$$\hat{H}_0 = \frac{\hat{L}^2}{2I}$$

where $\hat{L}^2$ is the square of the orbital angular momentum and I the momentum of inertia. The rotator is subject to the following perturbation

$$\hat{H}' = \lambda\sqrt{\frac{3}{4\pi}}\cos\hat{\theta}$$

where θ is the azimuthal angle and λ is a small parameter. Discuss a possible physical interpretation for the perturbation when the rigid rotator possesses an electric dipole. Determine the effect of $\hat{H}'$ on the ground state of the rotator.

Solution
A possible physical interpretation for the perturbation is that of a coupling energy between the electric dipoles (say d) and a given uniform electric field (say E) directed along the z axis. The coupling between the dipole and the electric field produces the scalar product $d \cdot E$ and this explains the $\cos\theta$ in the perturbation. A similar situation will be analyzed in the context of equilibrium statistical mechanics in Problem 7.31.

As for the unperturbed Hamiltonian, its eigenstates are given by the spherical harmonics $Y_{l,m}(\theta,\phi) = \langle \theta,\phi|l,m\rangle$ which are in fact eigenstates of $\hat{L}^2$

$$\hat{H}|l,m\rangle = E_{l,m}|l,m\rangle$$

with the energies $E_{l,m} = \frac{\hbar^2 l(l+1)}{2I}$. The unperturbed ground state is given by the constant spherical harmonic

$$\langle \theta,\phi|0,0\rangle = Y_{0,0}(\theta,\phi) = \frac{1}{\sqrt{4\pi}}$$

and the first order correction by

$$\Delta E_0^{(1)} = \langle 0,0|\hat{H}'|0,0\rangle = \frac{\lambda}{4\pi}\sqrt{\frac{3}{4\pi}}\int_0^{2\pi}d\phi\int_0^{\pi}\sin\theta\cos\theta d\theta = 0.$$

Therefore, we need to consider the second order effect to find a non zero correction. This can be easily understood because the perturbation is proportional to the spherical harmonic

$$Y_{1,0}(\theta,\phi) = \sqrt{\frac{3}{4\pi}}\cos\theta$$

and the spherical harmonics are orthogonal. Therefore, when applying the formula for the second order correction in perturbation theory

$$\Delta E_0^{(2)} = \sum_{|l,m\rangle\neq|0,0\rangle}\frac{|\langle 0,0|\delta\hat{H}|l,m\rangle|^2}{E_{0,0}-E_{l,m}}$$

the only term producing a non zero effect is that with $|l,m\rangle = |1,0\rangle$

$$\Delta E_0^{(2)} = -\frac{|\langle 0,0|\hat{H}'|1,0\rangle|^2}{\frac{2\hbar^2}{2I}} = -\frac{I\lambda^2}{\hbar^2}\left|\int\frac{1}{\sqrt{4\pi}}Y_{1,0}^2(\theta,\phi)d\Omega\right|^2 = -\frac{I\lambda^2}{4\pi\hbar^2}.$$

Problem 5.10.
A measurement of the energy

$$\hat{H} = \hat{S}_x$$

for a particle with spin $1/2$ gives surely $H = 1/2$. In the above expression, $\hat{S}_x$ is the x component of the spin. Assuming $\hbar = 1$, determine the first and second order energy corrections on this state induced by the perturbation

$$\delta\hat{H} = \varepsilon\hat{S}_+\hat{S}_-$$

where $\hat{S}_\pm$ are the raising and lowering operators for the z component of the spin.

Solution
We need to determine the perturbative corrections to the eigenvalue $1/2$. In the vector basis where the z component of the spin is diagonal, the matrix representation

of the Hamiltonian can written as

$$\hat{H} = \hat{S}_x = \frac{1}{2}\hat{\sigma}_x = \frac{1}{2}\begin{pmatrix} 0 & 1 \\ 1 & 0 \end{pmatrix}$$

where $\hat{\sigma}_x$ is one of the Pauli matrices. The eigenstates corresponding to the eigenvalues ± 1 of $\hat{\sigma}_x$ are

$$\left| \frac{1}{2}, \frac{1}{2} \right\rangle = \frac{1}{\sqrt{2}}\begin{pmatrix} 1 \\ 1 \end{pmatrix} \qquad \left| \frac{1}{2}, -\frac{1}{2} \right\rangle = \frac{1}{\sqrt{2}}\begin{pmatrix} 1 \\ -1 \end{pmatrix}.$$

From the properties of the raising and lowering operators for the z component of the spin

$$\hat{S}_\pm = \hat{S}_x \pm i\hat{S}_y = \frac{(\hat{\sigma}_x \pm i\hat{\sigma}_y)}{2}$$

$$\hat{\sigma}_y = \begin{pmatrix} 0 & -i \\ i & 0 \end{pmatrix} \quad \hat{S}_+ = \begin{pmatrix} 0 & 1 \\ 0 & 0 \end{pmatrix} \quad \hat{S}_- = \begin{pmatrix} 0 & 0 \\ 1 & 0 \end{pmatrix}$$

we can write the perturbation as

$$\delta\hat{H} = \varepsilon\hat{S}_+\hat{S}_- = \varepsilon\begin{pmatrix} 1 & 0 \\ 0 & 0 \end{pmatrix}.$$

The first order correction to the energy of $\left| \frac{1}{2}, \frac{1}{2} \right\rangle$ is

$$\Delta E_{\frac{1}{2}}^{(1)} = \left\langle \frac{1}{2}, \frac{1}{2} \right| \varepsilon\hat{S}_+\hat{S}_- \left| \frac{1}{2}, \frac{1}{2} \right\rangle = \frac{\varepsilon}{2}\begin{pmatrix} 1 & 1 \end{pmatrix}\begin{pmatrix} 1 & 0 \\ 0 & 0 \end{pmatrix}\begin{pmatrix} 1 \\ 1 \end{pmatrix} = \frac{1}{2}\varepsilon.$$

The second order correction is

$$\Delta E_{\frac{1}{2}}^{(2)} = \sum_{n \neq \frac{1}{2}} \frac{|\langle \frac{1}{2}, n| \delta\hat{H} |\frac{1}{2}, \frac{1}{2}\rangle|^2}{E_{\frac{1}{2}} - E_n}$$

and the summation is particularly simple in our case, because we have only two states

$$\Delta E_{\frac{1}{2}}^{(2)} = \frac{\left| \frac{\varepsilon}{2}\begin{pmatrix} 1 & 1 \end{pmatrix}\begin{pmatrix} 1 & 0 \\ 0 & 0 \end{pmatrix}\begin{pmatrix} 1 \\ -1 \end{pmatrix} \right|^2}{\frac{1}{2} - (-\frac{1}{2})} = \frac{1}{4}\varepsilon^2.$$

Problem 5.11.

A particle without spin is found in a quantum eigenstate of the operators $\hat{L}^2$ and $\hat{L}_z$, with $\hat{L}$ the orbital angular momentum and $\hat{L}_i$ its i-th component ($i = x, y, z$).

Determine the average values of $\hat{L}_x$, $\hat{L}_y$, $\hat{L}_x^2$, $\hat{L}_y^2$. Then, given the Hamiltonian

$$\hat{H} = \hat{L}_x^2 + \hat{L}_y^2$$

determine the time evolution of the generic eigenstate. At some point, the perturbation

$$\delta\hat{H} = \varepsilon\hat{L}_z \qquad \varepsilon \ll 1$$

is switched on. Using first order perturbation theory, determine the correction to the energy of the ground state for the case $l = 1$.

Solution
The state of the particle $|l,m\rangle$ is entirely specified by the quantum numbers l and m such that

$$\hat{L}^2 |l,m\rangle = \hbar^2 l(l+1) |l,m\rangle$$

$$\hat{L}_z |l,m\rangle = \hbar m |l,m\rangle .$$

Before computing the average values of the operators given in the text, let us recall some useful properties of the angular momentum. The commutation rules are

$$[\hat{L}_i, \hat{L}_j] = i\varepsilon_{ijk}\hbar\hat{L}_k$$

where ε_{ijk} is the Levi-Civita tensor and $i, j, k = x, y, z$. Such tensor is totally antisymmetric and conventionally chosen in such a way that $\varepsilon_{xyz} = 1$. The raising ($\hat{L}_+$) and lowering ($\hat{L}_-$) operators for the z component of the orbital angular momentum are defined by

$$\hat{L}_{\pm} = (\hat{L}_x \pm i\hat{L}_y) \qquad \hat{L}_x = \frac{(\hat{L}_+ + \hat{L}_-)}{2} \qquad \hat{L}_y = -i\frac{(\hat{L}_+ - \hat{L}_-)}{2}.$$

The associated commutation rules are given by

$$[\hat{L}_{\pm}, \hat{L}_z] = \mp\hbar\hat{L}_{\pm} \qquad [\hat{L}_+, \hat{L}_-] = 2\hbar\hat{L}_z.$$

Finally, the action of $\hat{L}_{\pm}$ on the states $|l,m\rangle$ is

$$\hat{L}_{\pm} |l,m\rangle = \hbar\sqrt{(l\mp m)(l\pm m+1)} |l,m\pm 1\rangle .$$

Let us now compute the expectation values for $\hat{L}_x$, $\hat{L}_y$, $\hat{L}_x^2$, $\hat{L}_y^2$. For $\hat{L}_x$ and $\hat{L}_y$, we get

$$\langle l,m|\hat{L}_x|l,m\rangle = \frac{1}{2}\langle l,m|(\hat{L}_+ + \hat{L}_-)|l,m\rangle =$$
$$\frac{\hbar}{2}\sqrt{(l-m)(l+m+1)}\langle l,m|l,m+1\rangle +$$
$$\frac{\hbar}{2}\sqrt{(l+m)(l-m+1)}\langle l,m|l,m-1\rangle = 0$$

and similarly for $\hat{L}_y$. The expectation values of $\hat{L}_x^2$ and $\hat{L}_y^2$ are non zero. To see that, it is useful to rewrite $\hat{L}_x^2$ and $\hat{L}_y^2$ as

$$\hat{L}_x^2 = \frac{(\hat{L}_+ + \hat{L}_-)^2}{4} = \frac{(\hat{L}_+^2 + \hat{L}_-^2 + \hat{L}_+\hat{L}_- + \hat{L}_-\hat{L}_+)}{4} = \frac{(\hat{L}_+^2 + \hat{L}_-^2 + 2\hat{L}_-\hat{L}_+ + 2\hbar\hat{L}_z)}{4}$$

$$\hat{L}_y^2 = -\frac{(\hat{L}_+ - \hat{L}_-)^2}{4} = -\frac{(\hat{L}_+^2 + \hat{L}_-^2 - \hat{L}_+\hat{L}_- - \hat{L}_-\hat{L}_+)}{4} = \frac{(-\hat{L}_+^2 - \hat{L}_-^2 + 2\hat{L}_-\hat{L}_+ + 2\hbar\hat{L}_z)}{4}.$$

When averaged on the state $|l,m\rangle$, $\hat{L}_\pm^2$ give zero. Therefore, we have

$$\langle l,m|\hat{L}_x^2|l,m\rangle = \langle l,m|\hat{L}_y^2|l,m\rangle = \frac{1}{2}\langle l,m|(\hat{L}_-\hat{L}_+ + \hbar\hat{L}_z)|l,m\rangle =$$
$$\frac{\hbar^2}{2}(m + (l-m)(l+m+1)) = \frac{\hbar^2}{2}(l(l+1) - m^2).$$

This is the correct result because $\hat{L}^2 = \hat{L}_x^2 + \hat{L}_y^2 + \hat{L}_z^2$, and the average value of $\hat{L}_x^2 + \hat{L}_y^2$ must coincide with the one of $\hat{L}^2 - \hat{L}_z^2$

$$\langle l,m|\hat{L}^2 - \hat{L}_z^2|l,m\rangle = \hbar^2(l(l+1) - m^2).$$

To determine the state at time t (say $|l,m,t\rangle$), we have to apply the time propagator $e^{-i\hat{H}t/\hbar}$ to $|l,m\rangle$

$$|l,m,t\rangle = e^{-i\hat{H}t/\hbar}|l,m\rangle = e^{-i\hbar(l(l+1)-m^2)t}|l,m\rangle.$$

As for the last point, we first have to determine the unperturbed energy spectrum

$$E_{l,m} = \hbar^2(l(l+1) - m^2).$$

For $l = 1$ we find $E_{1,1} = E_{1,-1} = \hbar^2, E_{1,0} = 2\hbar^2$. Therefore, the ground state for $l = 1$ has energy $\hbar^2$ and is double degenerate. To find the correction to the energy in the degenerate case, we have to determine the eigenvalues of the matrix whose elements are $\langle 1,\pm 1|\delta\hat{H}|1,\pm 1\rangle$. Our case is particularly simple because the perturbation is diagonal. The matrix representation of $\hat{L}_z$ and the eigenvectors of interest are

$$\hat{L}_z = \hbar \begin{pmatrix} 1 & 0 & 0 \\ 0 & 0 & 0 \\ 0 & 0 & -1 \end{pmatrix} \qquad |1,1\rangle = \begin{pmatrix} 1 \\ 0 \\ 0 \end{pmatrix} \qquad |1,-1\rangle = \begin{pmatrix} 0 \\ 0 \\ 1 \end{pmatrix}$$

and the perturbation matrix is

$$\begin{pmatrix} \langle 1,1|\delta\hat{H}|1,1\rangle & \langle 1,1|\delta\hat{H}|1,-1\rangle \\ \langle 1,-1|\delta\hat{H}|1,+1\rangle & \langle 1,-1|\delta\hat{H}|1,-1\rangle \end{pmatrix} = \varepsilon\hbar \begin{pmatrix} 1 & 0 \\ 0 & -1 \end{pmatrix}$$

that is a diagonal matrix with eigenvalues $\pm \varepsilon \hbar$. Therefore, the perturbation removes the degeneracy with a resulting energy splitting

$$E_{1,1}^{(\varepsilon)} = E_{1,1} + \Delta E_{1,1} = \hbar^2 + \varepsilon \hbar$$

$$E_{1,-1}^{(\varepsilon)} = E_{1,-1} + \Delta E_{1,-1} = \hbar^2 - \varepsilon \hbar.$$

Problem 5.12.
A two dimensional quantum system in the (x,y) plane is described by the following Hamiltonian

$$\hat{H} = \frac{\hat{p}_x^2 + \hat{p}_y^2}{2m} + \frac{1}{2}m(\omega_x^2 \hat{x}^2 + \omega_y^2 \hat{y}^2) + V(e^{\lambda \hat{\rho}} - 1)$$

representing a harmonic oscillator with mass m and frequencies ω_x, ω_y, plus a perturbation

$$\delta \hat{H} = V(e^{\lambda \hat{\rho}} - 1)$$

where V is a constant, $\lambda \ll 1$ a perturbation parameter, and $\hat{\rho} = \hat{a}_x \hat{a}_y + \hat{a}_y^\dagger \hat{a}_x^\dagger$. In our notation, $\hat{a}_s^\dagger$ and $\hat{a}_s$ $(s = x, y)$ are the creation and annihilation operators for the one dimensional oscillator

$$\hat{H}_s = \frac{\hat{p}_s^2}{2m} + \frac{1}{2}m\omega_s^2 \hat{s}^2 = \hbar \omega \left(\hat{a}_s^\dagger \hat{a}_s + \frac{1}{2}\mathbb{1} \right).$$

Using perturbation theory in λ and ignoring effects higher than $\mathcal{O}(\lambda^2)$, determine the correction to the energy of the ground state of the harmonic oscillator. Then, determine the value of V for which such effect is zero.

Solution
If we expand the perturbation in Taylor series up to $\mathcal{O}(\lambda^2)$, we get

$$\delta \hat{H} = V(e^{\lambda \hat{\rho}} - 1) = V \left(\lambda \hat{\rho} + \frac{\lambda^2 \hat{\rho}^2}{2} + \dots \right)$$

When applying perturbation theory, we have to consider the effects of the first term $(\lambda \hat{\rho})$ up to the second order in perturbation theory. Consistently, the effects of the second term $(\frac{\lambda^2 \hat{\rho}^2}{2})$ are considered up to the first order in perturbation theory. The unperturbed Hamiltonian represents two independent quantum harmonic oscillators, that implies we have stationary states of type $|n_x, n_y\rangle = |n_x\rangle|n_y\rangle$ $(n_{x,y} = 0, 1, 2, \dots)$ with energy

$$E_{n_x, n_y} = \hbar \omega_x \left(n_x + \frac{1}{2} \right) + \hbar \omega_y \left(n_y + \frac{1}{2} \right).$$

It follows that the ground state corresponds to $n_x = n_y = 0$. Let us then evaluate the effect of the perturbation term $\lambda \hat{\rho}$ on this state. The first order correction is zero because

$$\Delta E_{\lambda \hat{\rho}}^{(1)} = \langle 0,0|\lambda V \hat{\rho}|0,0\rangle = \lambda V \langle 0,0|\hat{a}_x \hat{a}_y + \hat{a}_y^\dagger \hat{a}_x^\dagger|0,0\rangle = \lambda V \langle 0,0|1,1\rangle = 0$$

where we have used the orthogonality of states ($\langle 0|1 \rangle = 0$) and the properties of the creation and annihilation operators

$$\hat{a}_s^\dagger |0\rangle = |1\rangle \qquad \hat{a}_s |0\rangle = 0 \qquad s = x, y.$$

A non zero effect is obtained at the second order in perturbation theory

$$\Delta E_{\lambda\hat{\rho}}^{(2)} = \frac{|\langle 0,0|\lambda V \hat{\rho}|1,1\rangle|^2}{E_{0,0} - E_{1,1}} = -\frac{\lambda^2 V^2}{\hbar(\omega_x + \omega_y)}.$$

Concerning the term $\frac{\lambda^2 \hat{\rho}^2}{2}$, as explained before, we only need to consider its effect up to first order in perturbation theory. When evaluating $\hat{\rho}^2$, we get

$$\hat{\rho}^2 = \hat{a}_x \hat{a}_y \hat{a}_x \hat{a}_y + \hat{a}_x \hat{a}_y \hat{a}_y^\dagger \hat{a}_x^\dagger + \hat{a}_y^\dagger \hat{a}_x^\dagger \hat{a}_x \hat{a}_y + \hat{a}_y^\dagger \hat{a}_x^\dagger \hat{a}_y^\dagger \hat{a}_x^\dagger$$

and we see that, when averaged on the ground state, the only term producing a non zero result is $\hat{a}_x \hat{a}_y \hat{a}_y^\dagger \hat{a}_x^\dagger$. Therefore, the first order perturbative correction is

$$\Delta E_{\frac{\lambda^2 \hat{\rho}^2}{2}}^{(1)} = \frac{\lambda^2 V}{2} \langle 0,0|\hat{a}_x \hat{a}_y \hat{a}_y^\dagger \hat{a}_x^\dagger|0,0\rangle = \frac{\lambda^2 V}{2}.$$

Summing all the contributions, we get

$$\Delta E_{tot} = \Delta E_{\lambda\hat{\rho}}^{(2)} + \Delta E_{\frac{\lambda^2 \hat{\rho}^2}{2}}^{(1)} = -\frac{\lambda^2 V^2}{\hbar(\omega_x + \omega_y)} + \frac{\lambda^2 V}{2}.$$

We see that the effect is zero when $\Delta E_{tot} = 0$, that implies $V = \frac{\hbar(\omega_x + \omega_y)}{2}$.

Problem 5.13.
A one dimensional quantum harmonic oscillator with charge q (consider $\hbar = m = \omega = 1$) is in the ground state $\psi_0(x)$. At some point, a uniform electric field directed along the positive x direction is switched on. Determine the probability that the oscillator is found in the excited states of the new Hamiltonian.

Solution
When there is no electric field, the wave function for the ground state of the harmonic oscillator is

$$\psi_0(x) = \frac{1}{\pi^{\frac{1}{4}}} e^{-\frac{1}{2}x^2}.$$

As discussed in Problem 2.33, the eigenstates of the one dimensional harmonic oscillator with charge q under the effect of the uniform electric field E_x, are written as

$$\psi_n(y) = \frac{1}{\sqrt{2^n n! \sqrt{\pi}}} e^{-\frac{1}{2}y^2} H_n(y) = \frac{(-1)^n}{\sqrt{2^n n! \sqrt{\pi}}} e^{\frac{1}{2}y^2} \frac{d^n e^{-y^2}}{dy^n}$$

where $y = x - qE_x = x - 2A$ is a shifted coordinate ($A = qE_x/2$). When prepared in the state $\psi_0(x)$, the probability that the harmonic oscillator is in one of the excited states $\psi_n(y)$, is given by the square of the projection coefficient of $\psi_n(y)$ on $\psi_0(x)$. For the projection coefficient, we can integrate by parts n times to get

$$c_n = \int_{-\infty}^{+\infty} \psi_0(x)\psi_n(y)dx = \frac{(-1)^n}{\sqrt{2^n n!}\sqrt{\pi}} e^{-2A^2} \int_{-\infty}^{+\infty} e^{-2Ax} \frac{d^n e^{-x^2+4Ax}}{dx^n} dx =$$

$$\frac{(-1)^n e^{-2A^2}}{\sqrt{2^n n!}\sqrt{\pi}} \left\{ \left[\left(\frac{d^n e^{-x^2+4Ax}}{dx^n} \right) e^{-2xA} - \left(\frac{d^{n-1} e^{-x^2+4Ax}}{dx^{n-1}} \right) \frac{de^{-2xA}}{dx} + \cdots \right] \Big|_{-\infty}^{+\infty} + \right.$$

$$(-1)^n \int_{-\infty}^{+\infty} \left(\frac{d^n e^{-2xA}}{d^n x} \right) e^{-x^2+4Ax} dx \bigg\} = \frac{(-1)^n 2^n A^n e^{-2A^2}}{\sqrt{2^n n!}\sqrt{\pi}} \int_{-\infty}^{+\infty} e^{-x^2+2Ax} dx.$$

Using the integral $\int_{-\infty}^{+\infty} e^{-x^2+2Ax} dx = \sqrt{\pi} e^{A^2}$ we get

$$c_n = e^{-A^2} \frac{(-1)^n}{\sqrt{n!}} (\sqrt{2}A)^n.$$

The desired probability is

$$P_{0,n} = |c_n|^2 = \frac{1}{n!} (2A^2)^n e^{-2A^2} = \frac{1}{n!} \lambda^n e^{-\lambda}$$

that is a Poisson distribution with mean value $\lambda = 2A^2$.

Problem 5.14.

A one dimensional harmonic oscillator with charge q, mass M, and frequency ω is subject to a uniform electric field

$$E(t) = \frac{A}{\sqrt{\pi}\tau} e^{-\left(\frac{t}{\tau}\right)^2}$$

with A, τ constants. From the point of view of Classical Mechanics, determine the momentum transferred from the perturbation to the oscillator from time $t = -\infty$ to time $t = +\infty$.

Then, suppose to treat the system as a quantum mechanical one. If at time $t = -\infty$ the oscillator is in the fundamental state, determine the transition probability that it will be in the first excited state at $t = +\infty$.

Solution
The Hamiltonian of the system is

$$\hat{H} = \frac{\hat{p}^2}{2M} + \frac{M\omega^2 \hat{x}^2}{2} - \frac{Aq\hat{x}}{\sqrt{\pi}\tau} e^{-\left(\frac{t}{\tau}\right)^2}.$$

Using the derivative of the potential $V(x,t) = -qE(t)x$ with respect to x and changing its sign, we can determine the force acting on the oscillator. The integral of such a force gives the classical momentum transferred from the electric field to the harmonic oscillator

$$P = \int_{-\infty}^{+\infty} F(t)\,dt = \int_{-\infty}^{+\infty} qE(t)\,dt = \frac{qA}{\sqrt{\pi}\tau} \int_{-\infty}^{+\infty} e^{-(\frac{t}{\tau})^2}\,dt = qA.$$

As for the transition probability between a generic initial state $|n\rangle$ and a final state $|m\rangle$, it is

$$w_{m,n} = \frac{1}{\hbar^2} \left| \int_{-\infty}^{+\infty} V_{mn}(t) e^{i\frac{(E_m-E_n)}{\hbar}t}\,dt \right|^2$$

where E_n, E_m represent the eigenvalues of the states $|n\rangle, |m\rangle$ and $V_{mn}(t)$ is the matrix element of the potential $V(t)$ evaluated on the unperturbed states

$$V_{mn}(t) = -\int_{-\infty}^{+\infty} \psi_m^*(x) qE(t) x \psi_n(x)\,dx = -qE(t)\langle m|\hat{x}|n\rangle.$$

We specialize these formulae to our case, using $(E_1 - E_0)/\hbar = \omega$, and determining the matrix element making use of the creation and annihilation operators

$$\langle 1|\hat{x}|0\rangle = \sqrt{\frac{\hbar}{2M\omega}} \langle 1|(\hat{a}+\hat{a}^\dagger)|0\rangle = \sqrt{\frac{\hbar}{2M\omega}}.$$

The desired transition probability is

$$w_{1,0} = \frac{q^2 A^2}{2\pi M \omega \hbar} \left| \int_{-\infty}^{+\infty} e^{-(\frac{t}{\tau})^2 + i\omega t} \frac{dt}{\tau} \right|^2 = \frac{P^2}{2M\omega\hbar} e^{-\frac{\omega^2\tau^2}{2}}.$$

Depending on the value of τ, representing the characteristic time of the perturbation, we have different physical scenarios. If $\tau \gg 1/\omega$, the characteristic time of the perturbation is much larger than the oscillation time, and the oscillator perceives a slowly varying perturbation. This is the adiabatic case: the transition probability is small and, in the limit of an infinitely slow perturbation, it tends to zero (the system is then found in a stationary state). If $\tau \ll 1/\omega$ we are in the opposite case: the perturbation is fast, its derivative with respect to time is very large, and the result is

$$w_{1,0} = \frac{P^2}{2M\omega\hbar}$$

that is the ratio between the classical energy and the quantum mechanical one. For the perturbation theory to be valid, this probability must be much smaller than 1, the latter being the probability to remain in the original ground state.

Problem 5.15.

Let us consider a hydrogen-like atom in the ground state with nuclear charge equal to Ze (e is the absolute value of the electron charge). Determine the average values of the kinetic and potential energies. Using the first order corrections coming from perturbation theory, determine the energy variation when the nuclear charge changes from Ze to $(Z+1)e$. Finally, compare the result of perturbation theory with the exact result. To simplify matters, you can make use of atomic units.

Solution
The ground state for an atom with nuclear charge Ze has quantum numbers $n = 1$ (principal quantum number) and $l = 0$, $m = 0$ (angular momentum quantum numbers). If we use atomic units we get

$$\psi_{1,0,0}(r,\theta,\phi) = R_{1,0}(r)Y_{0,0}(\theta,\phi) = 2Z^{\frac{3}{2}}e^{-Zr}\frac{1}{\sqrt{4\pi}}.$$

To obtain dimensional quantities, we have to multiply by the Bohr radius, $a = \hbar^2/(m_e e^2) = 0,529 \times 10^{-8}$ cm, where m_e is the mass of the electron. The atomic unit of energy is $E_{atomic} = e^2/a = \frac{m_e e^4}{\hbar^2} = 4,46 \times 10^{-11}$erg $= 27,21$ electronvolt (eV). To go to atomic units we set $e = m_e = \hbar = 1$. Let us then verify the normalization of the wave function

$$\int |\psi_{1,0,0}|^2 d^3r = 4Z^3 \int_0^{+\infty} e^{-2Zr}r^2 dr = \frac{1}{2}\lim_{\beta \to 1}\frac{d^2}{d\beta^2}\int_0^{+\infty} e^{-r\beta}dr = 1.$$

The radial probability distribution function between $r = 0$ and $r = +\infty$ is given by $P(r) = 4Z^3 r^2 e^{-2Zr}$. Due to the normalization of the wave function, $P(r)$ is correctly normalized to unity when we integrate between $r=0$ and $r=+\infty$, i.e. $\int_0^{+\infty}P(r)dr = 1$. The potential energy is $U(r) = -Z/r$ and its expectation value on the ground state is

$$\langle \hat{U} \rangle = 4Z^3 \int_0^{+\infty} e^{-2Zr}\left(-\frac{Z}{r}\right)r^2 dr = -4Z^4 \int_0^{+\infty} e^{-2Zr}r\, dr =$$
$$Z^2 \lim_{\beta \to 1}\frac{d}{d\beta}\int_0^{+\infty} e^{-r\beta}dr = Z^2 \lim_{\beta \to 1}\left(-\frac{1}{\beta^2}\right)\int_0^{+\infty} e^{-r}dr = -Z^2.$$

The kinetic energy operator is

$$\hat{T} = -\frac{1}{2}\nabla^2 = -\frac{1}{2}\left(\frac{1}{r^2}\frac{\partial}{\partial r}\left(r^2\frac{\partial}{\partial r}\right) - \frac{\hat{L}^2}{\hbar^2 r^2}\right)$$

where the last term includes the square of the orbital angular momentum operator, whose eigenfunctions are the spherical harmonics $Y_{l,m}(\theta,\phi)$ with eigenvalues $\hbar^2 l(l+1)$. In our case $l = 0$, and the term $\frac{\hat{L}^2}{\hbar^2 r^2}$ is not present. The average value of the kinetic energy is therefore

$$\langle \hat{T} \rangle = 4Z^3 \int_0^{+\infty} e^{-Zr}r^2\left(-\frac{1}{2}\frac{1}{r^2}\frac{d}{dr}\left(r^2\frac{d}{dr}\right)\right)e^{-Zr}dr =$$
$$2Z^3 \int_0^{+\infty} e^{-2Zr}(2Zr - Z^2 r^2)dr = Z^2 \int_0^{+\infty} re^{-r}dr - \frac{Z^2}{4}\int_0^{+\infty} e^{-r}r^2 dr =$$
$$Z^2 - \frac{Z^2}{4}\lim_{\beta \to 1}\frac{d^2}{d\beta^2}\int_0^{+\infty} e^{-\beta r}dr = \frac{Z^2}{2}$$

that is a positive value, as it should be. To verify these calculations, we can set $Z = 1$, and see if we recover the result for the ground state ($n = 1, l = 0, m = 0$) of the hydrogen atom whose energy is $E_1(Z = 1) = -1/2$. Summing the kinetic and potential contributions, we get

$$\langle \hat{H} \rangle = \langle \hat{T} \rangle + \langle \hat{U} \rangle = E_1 = -1/2$$

that is the correct result.

When the nuclear charge changes from Z to $Z + 1$, the associated potential may be seen as the original potential with charge Z plus a perturbation U', defined by

$$U' = U_{final} - U_{initial} = -\frac{(Z+1)}{r} + \frac{Z}{r} = -\frac{1}{r}.$$

Using first order perturbation theory, we know that the correction to the energy of the ground state is

$$\Delta E^{(1)}(Z) = \langle \hat{U}' \rangle = -Z$$

that is the same integral done before with $-1/r$ instead of $-Z/r$. At the same time, we also know the exact result for the ground state

$$\Delta E(Z) = E_{final} - E_{initial} = -\frac{(Z+1)^2}{2} + \frac{Z^2}{2} = -Z - \frac{1}{2}.$$

We observe that for large Z the perturbative calculation agrees very well with the exact result.

Problem 5.16.

A Hydrogen atom is placed in an external electric field given by the following potential

$$V(r, \theta, t) = -\frac{1}{\pi} \frac{B \tau r \cos \theta}{t^2 + \tau^2}$$

where B, τ are constants, and θ, r are the inclination angle and the radial distance in spherical polar coordinates respectively. If at time $t = -\infty$ the atom is in the ground state, determine the probability that it will be in the state $2p$ at $t = +\infty$. To simplify matters, you can make use of atomic units.

Solution
We directly apply the formula for the transition probability

$$P_{m,0} = \left| \int_{-\infty}^{+\infty} \langle 2, 1, m | \hat{V} | 1, 0, 0 \rangle e^{i\omega t} dt \right|^2 =$$

$$\left| \langle 2, 1, m | \hat{r} \cos \hat{\theta} | 0, 0, 0 \rangle \int_{-\infty}^{+\infty} \frac{B\tau}{\pi} \frac{e^{i\omega t}}{t^2 + \tau^2} dt \right|^2.$$

In fact, the quantum numbers of the state $2p$ are $n = 2, l = 1, m = 0, \pm 1$. The difference between the two energies (in atomic units the energy is $E_n = -1/(2n^2)$) is: $\omega = E_2 - E_1 = -1/8 - (-1/2) = 3/8$. We now calculate $\langle 2, 1, m | \hat{r} \cos \hat{\theta} | 1, 0, 0 \rangle$.

First of all, we note that the state with $l = 1$ is triple degenerate

$$\psi_{2,1,1}(r,\theta,\phi) = R_{2,1}(r)Y_{1,1}(\theta,\phi) = \frac{1}{8\sqrt{\pi}}re^{-\frac{r}{2}}e^{i\phi}\sin\theta$$

$$\psi_{2,1,0}(r,\theta,\phi) = R_{2,1}(r)Y_{1,0}(\theta,\phi) = \frac{1}{4\sqrt{2\pi}}re^{-\frac{r}{2}}\cos\theta$$

$$\psi_{2,1,-1}(r,\theta,\phi) = R_{2,1}(r)Y_{1,-1}(\theta,\phi) = \frac{1}{8\sqrt{\pi}}re^{-\frac{r}{2}}e^{-i\phi}\sin\theta.$$

The wave function for the ground state has not an angular dependence, due to the fact that the spherical harmonic for $l = 0$ is constant

$$\psi_{1,0,0}(r) = \frac{1}{\sqrt{\pi}}e^{-r}.$$

The matrix element $\langle 2,1,m|\hat{r}\cos\hat{\theta}|1,0,0\rangle$ is zero if we use $\psi_{2,1,\pm1}$, i.e. the states with the z component of the angular momentum equal to ±1, because V and $\psi_{1,0,0}$ do not depend on ϕ and the integral of $e^{\pm i\phi}$ over a period is zero. The matrix element with the state whose z component of the angular momentum is zero gives

$$\langle 2,1,0|\hat{r}\cos\hat{\theta}|1,0,0\rangle = \frac{1}{2\sqrt{2}}\int_0^{+\infty}r^4e^{-\frac{3r}{2}}dr\int_{-1}^{+1}\cos^2\theta\,d(\cos\theta) =$$

$$\frac{1}{3\sqrt{2}}\int_0^{+\infty}r^4e^{-\frac{3r}{2}}dr = \left(\frac{2}{3}\right)^5\frac{1}{3\sqrt{2}}\lim_{\beta\to1}\frac{d^4}{d\beta^4}\int_0^{+\infty}e^{-r\beta}dr = \frac{2^7\sqrt{2}}{3^5}.$$

Substituting this result back in the formula for the transition probability, we obtain

$$P_{1,0} = \frac{B^2\tau^2}{\pi^2}\frac{2^{15}}{3^{10}}\left|\int_{-\infty}^{+\infty}\frac{e^{i\omega t}}{(t+i\tau)(t-i\tau)}dt\right|^2 = \frac{B^2\tau^2}{\pi^2}\frac{2^{15}}{3^{10}}\left|2i\pi\lim_{t\to i\tau}\frac{e^{i\omega t}}{(t+i\tau)}\right|^2 = \frac{B^2 2^{15}}{3^{10}}e^{-\frac{3}{4}\tau}$$

where we have used the method of residuals, choosing a contour surrounding the region Ω of the complex plane where $\Im(\Omega) > 0$.

Problem 5.17.
For a given energy ε, the WKB method leads to the following expression for the transmission coefficient through a potential barrier $V(x)$

$$T \approx e^{-\frac{2}{\hbar}\int_{x_1}^{x_2}\sqrt{2m(V(x)-\varepsilon)}dx}$$

where x_1, x_2 satisfy the condition $V(x_1) = V(x_2) = \varepsilon$. Using the WKB method, give an estimate of the transmission coefficient for an electron with charge e and mass m through the potential barrier of Fig. 5.1: the one dimensional potential is $V(x) = 0$ for $x < 0$ and $V(x) = V_0 - eE_x x$ for $x \geq 0$, where E_x is a constant field and V_0 the potential energy barrier in $x = 0$. You can assume that the energy is $\varepsilon < V_0$ (see Fig. 5.1).

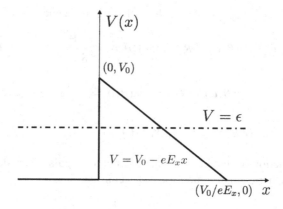

Fig. 5.1 An electron with constant energy ε and charge e has a non zero probability to pass through the one dimensional potential barrier that is $V(x) = 0$ for $x < 0$ and $V(x) = V_0 - eE_x x$ for $x \geq 0$, where E_x is a constant field and V_0 the potential energy barrier in $x = 0$. In Problem 5.17 we determine the transmission coefficient through this potential energy barrier using the WKB method

Solution

The transmission coefficient can be computed using the same ideas of Problems 2.22, 2.23, 2.31, plus the WKB approximation when determining the wave function. The result for the transmission coefficient for the penetration through the potential barrier is the one given in the text

$$T \approx e^{-\frac{2}{\hbar} \int_{x_1}^{x_2} \sqrt{2m(V(x)-\varepsilon)}dx}.$$

The points x_1, x_2 are found from the condition $V(x_1) = V(x_2) = \varepsilon$, yielding $x_1 = 0, x_2 = \frac{V_0-\varepsilon}{eE_x}$. Therefore, we can compute the integral in the exponential function of the previous formula

$$\int_0^{\frac{V_0-\varepsilon}{eE_x}} \sqrt{2m(V_0 - \varepsilon - eE_x x)}dx = \sqrt{2meE_x} \int_0^{\frac{V_0-\varepsilon}{eE_x}} \sqrt{\left(\frac{V_0-\varepsilon}{eE_x} - x\right)} dx =$$

$$\frac{2}{3}\sqrt{2meE_x}\left(\frac{V_0-\varepsilon}{eE_x}\right)^{\frac{3}{2}}.$$

The resulting transmission coefficient is

$$T \approx e^{-\frac{4}{3}\frac{\sqrt{2m}}{eE_x\hbar}(V_0-\varepsilon)^{\frac{3}{2}}}.$$

Problem 5.18.

Using the Bohr-Sommerfeld quantization rule, determine the energy levels of a one dimensional harmonic oscillator with unitary frequency and mass m. Comment on the final result.

Solution
The Bohr-Sommerfeld quantization rule is given by

$$\oint \sqrt{2m(E_n - V)}dx = 2\int_{x_1}^{x_2} \sqrt{2m\left(E_n - \frac{1}{2}mx^2\right)}dx = \left(n + \frac{1}{2}\right)2\pi\hbar$$

where x_1, x_2 are the turning points of the classical motion and where, in the second integral, we have used the harmonic potential $V(x) = \frac{1}{2}mx^2$. To solve the integral, we use

$$\int \sqrt{a + cx^2}dx = \frac{x}{2}\sqrt{a + cx^2} + \frac{a}{2}\int \frac{1}{\sqrt{a + cx^2}}dx$$

and get

$$\int_{x_1}^{x_2} \sqrt{2m\left(E_n - \frac{1}{2}mx^2\right)}dx = \frac{1}{2}E_n\left(2m\int_{x_1}^{x_2} \frac{1}{\sqrt{2m\left(E_n - \frac{1}{2}mx^2\right)}}dx\right) = \frac{1}{2}E_nT = E_n\pi.$$

This is true because the first term in the integration formula is zero, since at the turning points we have $E_n = V(x_1) = V(x_2)$. Moreover, $T = 2\pi/\omega = 2\pi$ is the relation between the period of motion and the (unitary) frequency, with T defined as the time needed to go from x_1 to x_2 and come back

$$2m\int_{x_1}^{x_2} \frac{1}{\sqrt{2m\left(E_n - \frac{1}{2}mx^2\right)}}dx = 2m\int_{x_1}^{x_2} \frac{1}{p}dx = 2\int_{x_1}^{x_2} \frac{1}{v}dx = 2\int_{x_1}^{x_2} dt = T$$

where $p = mv = \sqrt{2m(E_n - V)}$ with p, v the momentum and velocity of the particle. The final result is therefore

$$E_n = \left(n + \frac{1}{2}\right)\hbar.$$

We note that such semi-classical result is the exact result for the quantum harmonic oscillator. All higher other corrections in the WKB approximation are indeed zero in this case.

Problem 5.19.
Using the Bohr-Sommerfeld quantization rule, determine the energy spectrum for a free particle with mass m in a one dimensional infinite potential well of width a. Compare with the exact result (see Problem 2.10).

Solution
We start by considering the potential well localized in the region $0 \leq x \leq a$. The momentum is conserved and its absolute value is equal to p_n. Due to the reflection from the wall (say the wall in $x = a$), momentum undergoes a change from p_n to $-p_n$. The Bohr-Sommerfeld quantization rule yields

$$\oint p_n dx = p_n \int_0^a dx - p_n \int_a^0 dx = 2p_n \int_0^a dx = \left(n + \frac{1}{2}\right)h$$

so that

$$p_n = \frac{h}{2a}\left(n + \frac{1}{2}\right) \qquad n = 0, 1, 2, \dots$$

The associated energy is therefore

$$E_n = \frac{p_n^2}{2m} = \frac{h^2}{8ma^2}\left(n + \frac{1}{2}\right)^2.$$

The energy spectrum for a particle in a one dimensional infinite potential well can be calculated exactly (see Problem 2.10)

$$E_n = \frac{p_n^2}{2m} = \frac{h^2}{8ma^2}n^2.$$

As we see, opposite to the result in Problem 5.18, the Bohr-Sommerfeld quantization rule does not provide an exact result but just an approximate one.

Problem 5.20.
The potential $U(x)$ is characterized by two symmetrical wells separated by a barrier (see Fig. 5.2). With an impenetrable barrier, the energy levels correspond to the case of a single particle in one well (region I or region II). The ground state of such case has energy E_0, and a passage through the potential barrier results in a splitting of the ground state into two energy levels. Using the WKB method, give an estimate of such effect.

Solution
The subject of this problem is the quantum tunnelling, i.e. the quantum mechanical phenomenon for which a particle passes through a potential energy barrier that

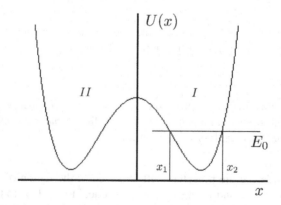

Fig. 5.2 A potential barrier with two symmetric wells. Due to the tunneling effect through the potential barrier, the ground state of the single particle motion in one well receives a correction. Problems 5.20 and 5.21 characterize this quantum mechanical effect

classically could not overcome, because its total kinetic energy is lower than the potential energy of the barrier itself (see Fig. 5.2, where the energy E_0 is lower than $U(x)$ in the origin). In Quantum Mechanics it exists a finite probability to pass through such barrier which goes to zero as the barrier gets higher. Here we want to characterize such probability using the WKB method. Neglecting the tunneling through the barrier, we can solve the Schrödinger equation describing the motion of the particle in one well (say in the region I of Fig. 5.2)

$$\hat{H}\psi_0(x) = E_0\psi_0(x).$$

The wave function $\psi_0(x)$ is such that its square modulus is normalized to 1 when integrated between 0 and $+\infty$, i.e. $\psi_0(x)$ solves the one dimensional problem with an infinite potential barrier in $x = 0$. When the probability of tunneling is considered, the particle can go through the barrier and enter into the region II, where its wave function is described by $\psi_0(-x)$. The correct zeroth order approximation for the wave function, taking into account the tunneling effect, is given by the symmetric and antisymmetric combinations of $\psi_0(x)$ and $\psi_0(-x)$

$$\begin{cases} \psi_1(x) = \frac{1}{\sqrt{2}}[\psi_0(x) + \psi_0(-x)] \\ \psi_2(x) = \frac{1}{\sqrt{2}}[\psi_0(x) - \psi_0(-x)]. \end{cases}$$

Being the potential $U(x)$ symmetric, the wave functions have a well defined parity when $x \rightarrow -x$. Since the probability of the tunneling is small, the variation in the energy levels is small as well. Moreover, in the region I we have $\psi_0(x) \gg \psi_0(-x)$, while in the region II we have the opposite, $\psi_0(-x) \gg \psi_0(x)$. From these considerations, it follows that the product $\psi_0(x)\psi_0(-x)$ is a vanishingly small quantity in both regions and the symmetric/antisymmetric combinations shown above are correctly normalized

$$\int_{-\infty}^{+\infty} |\psi_1(x)|^2 dx \approx \frac{1}{2}\left[\int_0^{+\infty} |\psi_0(x)|^2 dx + \int_{-\infty}^0 |\psi_0(-x)|^2 dx\right] = 1$$

$$\int_{-\infty}^{+\infty} |\psi_2(x)|^2 dx \approx \frac{1}{2}\left[\int_0^{+\infty} |\psi_0(x)|^2 dx + \int_{-\infty}^0 |\psi_0(-x)|^2 dx\right] = 1.$$

Let us now consider the region I and write the corresponding Schrödinger equation for ψ_0 and ψ_1

$$\begin{cases} \frac{d^2\psi_0}{dx^2} + \frac{2m}{\hbar^2}(E_0 - U)\psi_0 = 0 \\ \frac{d^2\psi_1}{dx^2} + \frac{2m}{\hbar^2}(E_1 - U)\psi_1 = 0. \end{cases}$$

If we are interested in deriving a formula for the difference $E_1 - E_0$, we can multiply the first equation by ψ_1, the second by ψ_0, and subtract the two expressions

$$-\psi_1 \frac{d^2 \psi_0}{dx^2} + \psi_0 \frac{d^2 \psi_1}{dx^2} + \frac{2m}{\hbar^2}(E_1 - E_0)\psi_1 \psi_0 = 0.$$

Integrating between 0 and $+\infty$, we find

$$(E_1 - E_0)\frac{2m}{\hbar^2} \int_0^{+\infty} \psi_1(x)\psi_0(x)dx = \int_0^{+\infty} \left(\psi_1 \frac{d^2 \psi_0}{dx^2} - \psi_0 \frac{d^2 \psi_1}{dx^2}\right)dx =$$
$$\left(\psi_1 \frac{d\psi_0}{dx} - \psi_0 \frac{d\psi_1}{dx}\right)\Bigg|_0^{+\infty}.$$

In the region I, $\psi_0(-x)$ is vanishingly small, and we can further simplify

$$\int_0^{+\infty} \psi_1(x)\psi_0(x)dx = \int_0^{+\infty} \frac{1}{\sqrt{2}}[\psi_0(x) + \psi_0(-x)]\psi_0(x)dx \approx$$
$$\frac{1}{\sqrt{2}} \int_0^{+\infty} |\psi_0(x)|^2 dx = \frac{1}{\sqrt{2}}.$$

Moreover, $\psi_0(\pm\infty) = 0$ because ψ_0 represents a bound state. From the definition of ψ_1 it follows that

$$\psi_1(0) = \sqrt{2}\psi_0(0)$$
$$\frac{d\psi_1(0)}{dx} = \lim_{x \to 0} \frac{1}{\sqrt{2}}\left[\frac{d\psi_0(x)}{dx} - \frac{d\psi_0(-x)}{d(-x)}\right] = 0.$$

Therefore, we obtain

$$E_0 - E_1 = \frac{\hbar^2}{m}\psi_0(0)\frac{d\psi_0(x)}{dx}\Bigg|_{x=0}.$$

For $E_2 - E_0$ we find a similar expression with the sign changed

$$E_2 - E_0 = \frac{\hbar^2}{m}\psi_0(0)\frac{d\psi_0(x)}{dx}\Bigg|_{x=0}.$$

The explicit form of $\psi_0(0)$ and its derivative is obtained with the WKB method. The point $x = 0$ is in the forbidden classical region, because $U(0) > E_0$. For a generic point x in this region, the wave function is

$$\psi_0(x) = \sqrt{\frac{\omega}{2\pi v(x)}}e^{-\frac{1}{\hbar}\int_x^{x_1} |p(y)|dy}$$

with

$$p(y) = \sqrt{2m(E_0 - U(y))}.$$

Its derivative is

$$\frac{d\psi_0(x)}{dx} = \left\{-\frac{1}{2}\sqrt{\frac{\omega}{2\pi v^3(x)}}\frac{dv(x)}{dx} + \sqrt{\frac{\omega}{2\pi v(x)}}\left[\frac{d}{dx}\left(-\frac{1}{\hbar}\int_x^{x_1}|p(y)|dy\right)\right]\right\}F(x,x_1) =$$

$$\left(-\frac{1}{2}\sqrt{\frac{\omega}{2\pi v^3(x)}}\frac{dv(x)}{dx} + \sqrt{\frac{\omega}{2\pi v(x)}}\frac{|p(x)|}{\hbar}\right)F(x,x_1)$$

where $F(x,x_1) = e^{-\frac{1}{\hbar}\int_x^{x_1}|p(y)|dy}$. In these formulae, $v(x) = |p(x)|/m$ and ω is the angular frequency of the classical period of motion, satisfying

$$T = \frac{2\pi}{\omega} = 2m\int_{x_1}^{x_2}\frac{dx}{\sqrt{2m(E_0 - U(x))}}$$

where x_1, x_2 are the turning points of the classical motion deduced from $U(x_1) = U(x_2) = E_0$. If we specialize to the case where $x = 0$ is a maximum of the potential, we find

$$\frac{dv(x)}{dx}\bigg|_{x=0} = 0$$

and, plugging all this in the formula for the energy, we get

$$E_2 - E_1 = \frac{\omega\hbar}{\pi}e^{-\frac{2}{\hbar}\int_0^{x_1}|p(y)|dy} = \frac{\omega\hbar}{\pi}e^{-\frac{1}{\hbar}\int_{-x_1}^{+x_1}|p(y)|dy}$$

giving the energy difference of the splitting in terms of the integral of the modulus of the momentum $|p(y)|$ in the classically forbidden region. In Problem 5.21 we will apply this formula to a biquadratic potential.

Problem 5.21.
Specialize the formulae obtained in Problem 5.20 to the ground state of the potential

$$U(x) = g^2\frac{(x^2 - a^2)^2}{8}.$$

In the above expression, g and a are constants. To determine the turning points (i.e. the points x_1 and x_2 in Fig. 5.2) and the angular frequency for the classical motion, you can approximate the potential with a quadratic form around its minimum. Moreover, you can treat $\hbar$ as a small parameter.

Solution
The potential $U(x)$ is very similar to the one shown in Fig. 5.2 with two minima located at $(\pm a, 0)$ and a maximum in $(0, g^2a^4/8)$. When we increase the parameter g, the height of potential barrier increases and the two wells become separated when $g \to +\infty$. It is easy to see that, around the two minima, the potential $U(x)$ is well approximated by a quadratic harmonic potential. To show that, let us take the minimum located at $x = a$ and set $y = x - a$. In the limit $y \to 0$, we can rewrite the potential as

$$U(x) = \frac{g^2}{8}(x^2 - a^2)^2 = \frac{g^2}{8}(x+a)^2(x-a)^2 = \frac{g^2}{8}(y+2a)^2y^2 \approx \frac{1}{2}g^2a^2y^2 = \frac{1}{2}m\omega^2y^2$$

from which we see that the angular frequency is such that $ga = \sqrt{m}\omega$. The energy of the ground state is $E_0 = \hbar\omega/2$ and the turning points of the classical motion are given by the condition $E_0 = U(x)$. For the region I (see Fig. 5.2), we get

$$\frac{1}{2}m\omega^2(x-a)^2 = E_0 = \frac{1}{2}\hbar\omega$$

yielding $x^I_{1,2} = a \pm \sqrt{\hbar/(m\omega)}$. The turning points in the region II are determined in a similar way, with the result $x^{II}_{1,2} = -a \pm \sqrt{\hbar/(m\omega)}$. Therefore, the tunneling phenomenon takes place between the points $-a + \sqrt{\hbar/(m\omega)}$ and $a - \sqrt{\hbar/(m\omega)}$. As discussed in the last formula of Problem 5.20, the energy splitting is

$$E_2 - E_1 = \frac{\omega\hbar}{\pi}e^{-\frac{2}{\hbar}I(x_1)}$$

where

$$I(x_1) = \int_0^{+x_1} |p(y)|dy = \int_0^{a-\sqrt{\frac{\hbar}{m\omega}}} \sqrt{2m(U(y)-E_0)}\,dy.$$

Treating $\hbar$ as a small parameter, the integral $I(x_1)$ can be computed as follows

$$I(x_1) = \int_0^{a-\sqrt{\frac{\hbar}{m\omega}}} \sqrt{2m\left(\frac{g^2}{8}(y^2-a^2)^2 - \frac{1}{2}\hbar\omega\right)}\,dy =$$

$$\int_0^{a-\sqrt{\frac{\hbar}{m\omega}}} \frac{m\omega(a^2-y^2)}{2a} \sqrt{1 - \frac{4a^2\hbar}{m\omega(a^2-y^2)^2}}\,dy \approx$$

$$\int_0^{a-\sqrt{\frac{\hbar}{m\omega}}} \frac{m\omega(a^2-y^2)}{2a}\,dy - \int_0^{a-\sqrt{\frac{\hbar}{m\omega}}} \frac{a\hbar}{(a^2-y^2)}\,dy =$$

$$\int_0^{a-\sqrt{\frac{\hbar}{m\omega}}} \left(\frac{m\omega a}{2} - \frac{m\omega y^2}{2a} - \frac{\hbar}{2(a-y)} - \frac{\hbar}{2(a+y)}\right)dy =$$

$$\frac{m\omega a^2}{3} - \frac{1}{2}\hbar + \frac{1}{2}\hbar\ln\sqrt{\frac{\hbar}{m\omega a^2}} - \frac{1}{2}\hbar\ln\left(2 - \sqrt{\frac{\hbar}{m\omega a^2}}\right) + \mathcal{O}(\hbar^{\frac{3}{2}}) =$$

$$\frac{m\omega a^2}{3} - \frac{1}{2}\hbar + \frac{\hbar}{2}\ln\sqrt{\frac{\hbar}{4m\omega a^2}} + \mathcal{O}(\hbar^{\frac{3}{2}}).$$

Plugging this result back in the formula for the energy splitting, we get

$$E_2 - E_1 = \frac{\omega\hbar}{\pi}e^{-\frac{2}{\hbar}I(x_1)} \approx \frac{ga\hbar}{\sqrt{m}\pi}\sqrt{\frac{4m\omega a^2}{\hbar}}e^{-\frac{2m\omega a^2}{3\hbar}} = 2\sqrt{\frac{g^3 a^5\hbar}{m}}e^{-\frac{2m\omega a^2}{3\hbar}}.$$

Part III

Statistical Mechanics – Problems

6

Thermodynamics and Microcanonical Ensemble

Problem 6.1.

We know that the free energy $F(T,V,N)$ of a thermodynamic system is extensive. Show that

$$N\left(\frac{\partial F}{\partial N}\right)_{T,V} + V\left(\frac{\partial F}{\partial V}\right)_{T,N} = Nf = F$$

with f the free energy density expressed in suitable variables. Given this result, from the differential properties of $F(T,V,N)$, show that

$$\Phi = N\mu$$

with Φ the Gibbs potential defined as $\Phi = F + PV$. In the above expression, μ is the chemical potential properly defined in terms of $F(T,V,N)$.

Solution

The fact that the free energy is an extensive thermodynamic potential means that

$$F(T,V,N) = Nf(T,v)$$

where $v = V/N$ is the specific volume and $f(T,v)$ the free energy density, which is a function of the specific volume v and the temperature T. From the derivatives of F we know that

$$N\left(\frac{\partial F}{\partial N}\right)_{T,V} = N\left(f(T,v) + N\left(\frac{\partial f}{\partial v}\right)_T \left(\frac{\partial v}{\partial N}\right)_V\right) = N\left(f(T,v) - N\left(\frac{\partial f}{\partial v}\right)_T \frac{V}{N^2}\right)$$

where we have considered that $f(T,v)$ depends on N because $v = V/N$. By the same token, we can write

$$V\left(\frac{\partial F}{\partial V}\right)_{T,N} = VN\left(\frac{\partial f}{\partial v}\right)_T \left(\frac{\partial v}{\partial V}\right)_N = V\left(\frac{\partial f}{\partial v}\right)_T .$$

Cini M., Fucito F., Sbragaglia M.: Solved Problems in Quantum and Statistical Mechanics.
DOI 10.1007/978-88-470-2315-4_6, © Springer-Verlag Italia 2012

Summing up the previous two equations, we obtain

$$N\left(\frac{\partial F}{\partial N}\right)_{T,V} + V\left(\frac{\partial F}{\partial V}\right)_{T,N} = Nf = F$$

that is the desired result. From the differential properties of F, we also know that

$$dF = -SdT - PdV + \mu dN$$

that implies

$$P = -\left(\frac{\partial F}{\partial V}\right)_{T,N} \qquad \mu = \left(\frac{\partial F}{\partial N}\right)_{T,V}$$

which can be used in the previous identity to find

$$N\left(\frac{\partial F}{\partial N}\right)_{T,V} = N\mu = F + PV.$$

This answers the second question because $\Phi = F + PV$.

Problem 6.2.

A thermodynamic system has an internal energy E, a pressure P, a volume V, a chemical potential μ, and a number of particles N. Assume that the entropy of such system is extensive and prove the relation

$$S = N\left(\frac{\partial S}{\partial N}\right)_{E,V} + V\left(\frac{\partial S}{\partial V}\right)_{E,N} + E\left(\frac{\partial S}{\partial E}\right)_{V,N}.$$

Using such result and the first law of thermodynamics, prove the Gibbs-Duhem equation

$$Nd\mu = -SdT + VdP.$$

Solution

Due to the extensive nature of the entropy function, we can rewrite it as

$$S(E,V,N) = Ns(e,v)$$

where $v = \frac{V}{N}$ and $e = \frac{E}{N}$ are the specific volume and specific energy respectively. Therefore, the function s represents the entropy density. Given the above functional relation, we can evaluate some derivatives of interest

$$V\left(\frac{\partial S}{\partial V}\right)_{E,N} = V\left(\frac{\partial s}{\partial v}\right)_e$$

$$E\left(\frac{\partial S}{\partial E}\right)_{V,N} = E\left(\frac{\partial s}{\partial e}\right)_v$$

$$N\left(\frac{\partial S}{\partial N}\right)_{E,V} = Ns(e,v) - V\left(\frac{\partial s}{\partial v}\right)_e - E\left(\frac{\partial s}{\partial e}\right)_v.$$

If we sum up the previous three relations, we obtain

$$S = N\left(\frac{\partial S}{\partial N}\right)_{E,V} + V\left(\frac{\partial S}{\partial V}\right)_{E,N} + E\left(\frac{\partial S}{\partial E}\right)_{V,N}$$

that is the desired result. We can now eliminate the derivatives using the first law of thermodynamics

$$\left(\frac{\partial S}{\partial V}\right)_{E,N} = \frac{P}{T} \qquad \left(\frac{\partial S}{\partial E}\right)_{V,N} = \frac{1}{T} \qquad \left(\frac{\partial S}{\partial N}\right)_{E,V} = -\frac{\mu}{T}$$

and obtain

$$S = -\frac{\mu N}{T} + \frac{PV}{T} + \frac{E}{T}.$$

If we multiply by the temperature T and differentiate both sides, we get

$$d(TS) = TdS + SdT = -d(\mu N) + d(PV) + dE.$$

Again, we can use the first law of thermodynamics in its differential form, $TdS = dE + PdV - \mu dN$, to further simplify as

$$Nd\mu = -SdT + VdP$$

that is the final result.

Problem 6.3.
Prove the following Maxwell relations

$$\left(\frac{\partial S}{\partial V}\right)_T = \left(\frac{\partial P}{\partial T}\right)_V \qquad \left(\frac{\partial T}{\partial P}\right)_S = \left(\frac{\partial V}{\partial S}\right)_P \qquad \left(\frac{\partial S}{\partial P}\right)_T = -\left(\frac{\partial V}{\partial T}\right)_P$$

valid for a thermodynamic system with a constant number of particles. To solve this problem, use the first law of thermodynamics $dE = TdS - PdV$.

Solution
Following the hint given by the text, we can obtain

$$\left(\frac{\partial E}{\partial S}\right)_V = T \qquad \left(\frac{\partial E}{\partial V}\right)_S = -P$$

that implies

$$\left(\frac{\partial}{\partial V}\left(\frac{\partial E}{\partial S}\right)_V\right)_S = \left(\frac{\partial T}{\partial V}\right)_S \qquad \left(\frac{\partial}{\partial S}\left(\frac{\partial E}{\partial V}\right)_S\right)_V = -\left(\frac{\partial P}{\partial S}\right)_V$$

from which we find (using Schwartz lemma for mixed partial derivatives) the following identity

$$\left(\frac{\partial T}{\partial V}\right)_S = -\left(\frac{\partial P}{\partial S}\right)_V.$$

We can then use the definition of the Jacobian for the variables (x,y) and (r,s)

$$J(x,y) = \frac{\partial(x,y)}{\partial(r,s)} = \left(\frac{\partial x}{\partial r}\right)_s \left(\frac{\partial y}{\partial s}\right)_r - \left(\frac{\partial x}{\partial s}\right)_r \left(\frac{\partial y}{\partial r}\right)_s$$

to rewrite the previous relation as

$$\frac{\partial(T,S)}{\partial(V,S)} = \frac{\partial(P,V)}{\partial(V,S)}.$$

Obviously, the relation between the variables (T,S) and (P,V) can be expressed also in terms of other variables (r,s), not necessarily equal to (V,S). To do that, it is sufficient to multiply such relation by $\frac{\partial(V,S)}{\partial(r,s)}$

$$\frac{\partial(T,S)}{\partial(r,s)} = \frac{\partial(P,V)}{\partial(r,s)}$$

where r and s have to be chosen properly. By choosing these variables as $(V,T),(P,S)$ and (P,T) respectively, we can prove the desired relations

$$\left(\frac{\partial S}{\partial V}\right)_T = \frac{\partial(S,T)}{\partial(V,T)} = -\frac{\partial(T,S)}{\partial(V,T)} = -\frac{\partial(P,V)}{\partial(V,T)} = \frac{\partial(P,V)}{\partial(T,V)} = \left(\frac{\partial P}{\partial T}\right)_V$$

$$\left(\frac{\partial T}{\partial P}\right)_S = \frac{\partial(T,S)}{\partial(P,S)} = \frac{\partial(P,V)}{\partial(P,S)} = \left(\frac{\partial V}{\partial S}\right)_P$$

$$\left(\frac{\partial S}{\partial P}\right)_T = \frac{\partial(S,T)}{\partial(P,T)} = -\frac{\partial(T,S)}{\partial(P,T)} = -\frac{\partial(P,V)}{\partial(P,T)} = -\left(\frac{\partial V}{\partial T}\right)_P.$$

Problem 6.4.

Consider a statistical system with a constant number of particles. Using the various thermodynamic potentials, express the specific heat at constant volume C_V, and the one at constant pressure C_P, in terms of the thermal expansion coefficient α, the isothermal compressibility κ_T, and the adiabatic compressibility κ_S, given by

$$\alpha = \frac{1}{V}\left(\frac{\partial V}{\partial T}\right)_P \qquad \kappa_T = -\frac{1}{V}\left(\frac{\partial V}{\partial P}\right)_T \qquad \kappa_S = -\frac{1}{V}\left(\frac{\partial V}{\partial P}\right)_S.$$

Solution
From the definition of C_V we can write

$$C_V = T\left(\frac{\partial S}{\partial T}\right)_V = T\frac{\partial(S,V)}{\partial(T,V)} = T\frac{\partial(S,V)}{\partial(T,P)}\frac{\partial(T,P)}{\partial(T,V)} =$$

$$T\frac{\left(\frac{\partial S}{\partial T}\right)_P \left(\frac{\partial V}{\partial P}\right)_T - \left(\frac{\partial S}{\partial P}\right)_T \left(\frac{\partial V}{\partial T}\right)_P}{\left(\frac{\partial V}{\partial P}\right)_T}$$

where we can use the definition of the specific heat at constant pressure $C_P = T \left(\frac{\partial S}{\partial T} \right)_P$ and get

$$C_V = C_P - T \frac{\left(\frac{\partial S}{\partial P} \right)_T \left(\frac{\partial V}{\partial T} \right)_P}{\left(\frac{\partial V}{\partial P} \right)_T}.$$

The infinitesimal variation of the Gibbs potential Φ gives

$$d\Phi = -SdT + VdP$$

that implies

$$\left(\frac{\partial \Phi}{\partial T} \right)_P = -S \qquad \left(\frac{\partial \Phi}{\partial P} \right)_T = V$$

and also

$$\left(\frac{\partial}{\partial P} \left(\frac{\partial \Phi}{\partial T} \right)_P \right)_T = -\left(\frac{\partial S}{\partial P} \right)_T \qquad \left(\frac{\partial}{\partial T} \left(\frac{\partial \Phi}{\partial P} \right)_T \right)_P = \left(\frac{\partial V}{\partial T} \right)_P.$$

The use of Schwartz lemma for mixed partial derivatives leads to

$$\left(\frac{\partial S}{\partial P} \right)_T = -\left(\frac{\partial V}{\partial T} \right)_P$$

which is a Maxwell relation that we can plug in the equation obtained previously

$$C_V = C_P + T \frac{\left(\frac{\partial V}{\partial T} \right)_P^2}{\left(\frac{\partial V}{\partial P} \right)_T} = C_P - TV \frac{\alpha^2}{\kappa_T}.$$

We now need another equation relating C_V and C_P. To this end, we make use of the adiabatic compressibility κ_S, and write it in terms of the Jacobians

$$-\kappa_S V = \left(\frac{\partial V}{\partial P} \right)_S = \frac{\partial(V,S)}{\partial(V,T)} \frac{\partial(P,T)}{\partial(P,S)} \frac{\partial(V,T)}{\partial(P,T)} = \frac{\left(\frac{\partial S}{\partial T} \right)_V}{\left(\frac{\partial S}{\partial T} \right)_P} \left(\frac{\partial V}{\partial P} \right)_T$$

that implies

$$\frac{\kappa_S}{\kappa_T} = \frac{C_V}{C_P}.$$

This is the second equation we were looking for. We can now solve the coupled equations

$$\begin{cases} C_V - C_P = -TV \frac{\alpha^2}{\kappa_T} \\ \frac{C_V}{C_P} = \frac{\kappa_S}{\kappa_T} \end{cases}$$

with the final result

$$C_P = \frac{TV\alpha^2}{\kappa_T - \kappa_S} \qquad C_V = \frac{\kappa_S}{\kappa_T} \frac{TV\alpha^2}{\kappa_T - \kappa_S}.$$

Problem 6.5.

Using the various thermodynamic potentials for a gas with a fixed number of particles, prove the following identities

$$\left(\frac{\partial T}{\partial P}\right)_H = \frac{T^2}{C_P}\left[\frac{\partial(V/T)}{\partial T}\right]_P$$

$$\left(\frac{\partial E}{\partial V}\right)_T = T^2\left[\frac{\partial(P/T)}{\partial T}\right]_V$$

$$\left(\frac{\partial E}{\partial S}\right)_T = -P^2\left[\frac{\partial(T/P)}{\partial P}\right]_V.$$

Moreover, prove that the first of these identities is exactly zero for a classical ideal gas. For the first identity, make use of a relation between H, T and P of the type $f(H,T,P) = 0$, with f unknown. For the second and third identity, use the first law of thermodynamics. Also, make use of Maxwell relations.

Solution

Following the hint given by the text, we start from a general relation between H, T and P in the form

$$f(H,T,P) = 0$$

with f unknown. Differentiating both sides we find

$$df = 0 = \left(\frac{\partial f}{\partial H}\right)_{T,P} dH + \left(\frac{\partial f}{\partial T}\right)_{H,P} dT + \left(\frac{\partial f}{\partial P}\right)_{H,T} dP$$

that implies

$$\left(\frac{\partial H}{\partial T}\right)_P = -\left(\frac{\partial f}{\partial T}\right)_{H,P} \bigg/ \left(\frac{\partial f}{\partial H}\right)_{T,P}$$

$$\left(\frac{\partial H}{\partial P}\right)_T = -\left(\frac{\partial f}{\partial P}\right)_{H,T} \bigg/ \left(\frac{\partial f}{\partial H}\right)_{T,P}$$

$$\left(\frac{\partial T}{\partial P}\right)_H = -\left(\frac{\partial f}{\partial P}\right)_{H,T} \bigg/ \left(\frac{\partial f}{\partial T}\right)_{H,P}$$

which may be combined as

$$\left(\frac{\partial T}{\partial P}\right)_H = -\left(\frac{\partial T}{\partial H}\right)_P \left(\frac{\partial H}{\partial P}\right)_T.$$

With the identity $\left(\frac{\partial H}{\partial T}\right)_P = C_P$, we have

$$\left(\frac{\partial T}{\partial P}\right)_H = -\frac{1}{C_P}\left(\frac{\partial H}{\partial P}\right)_T.$$

We now start from the differential expression of the enthalpy

$$dH = TdS + VdP$$

and expand dS in terms of the independent variables T and P

$$dH = T\left[\left(\frac{\partial S}{\partial T}\right)_P dT + \left(\frac{\partial S}{\partial P}\right)_T dP\right] + VdP$$

from which we immediately find

$$\left(\frac{\partial H}{\partial P}\right)_T = V + T\left(\frac{\partial S}{\partial P}\right)_T.$$

Also, one of Maxwell relations tells us that $\left(\frac{\partial S}{\partial P}\right)_T = -\left(\frac{\partial V}{\partial T}\right)_P$, so that

$$\left(\frac{\partial H}{\partial P}\right)_T = -T^2\left[\frac{\partial(V/T)}{\partial T}\right]_P$$

with the final result

$$\left(\frac{\partial T}{\partial P}\right)_H = -\frac{1}{C_P}\left(\frac{\partial H}{\partial P}\right)_T = \frac{T^2}{C_P}\left[\frac{\partial(V/T)}{\partial T}\right]_P.$$

We also note that for a classical ideal gas, the equation of state gives $PV = NkT$, so that $\frac{V}{T} = \frac{Nk}{P}$ and

$$\left[\frac{\partial(V/T)}{\partial T}\right]_P = 0.$$

As for the second identity, we start from

$$dE = TdS - PdV$$

and differentiate with respect to V at constant temperature

$$\left(\frac{\partial E}{\partial V}\right)_T = T\left(\frac{\partial S}{\partial V}\right)_T - P.$$

Again, one Maxwell relation gives

$$\left(\frac{\partial S}{\partial V}\right)_T = \left(\frac{\partial P}{\partial T}\right)_V$$

from which we can obtain immediately the final result

$$\left(\frac{\partial E}{\partial V}\right)_T = T^2\left[\frac{\partial(P/T)}{\partial T}\right]_V.$$

Finally, for the third identity, we can use a similar procedure. Starting from

$$dE = TdS - PdV$$

we can differentiate with respect to S at constant temperature

$$\left(\frac{\partial E}{\partial S}\right)_T = T - P\left(\frac{\partial V}{\partial S}\right)_T$$

that can be further simplified using the Maxwell relations

$$\left(\frac{\partial V}{\partial S}\right)_T = \left(\frac{\partial T}{\partial P}\right)_V$$

with the final result

$$\left(\frac{\partial E}{\partial S}\right)_T = -P^2\left[\frac{\partial(T/P)}{\partial P}\right]_V.$$

Problem 6.6.
Unlike an ideal gas, which cools down during an adiabatic expansion, a one dimensional rubber band (with spring constant K and rest position $x_0 = 0$) is increasing its temperature when elongated in an adiabatic way. Write down the first law of thermodynamics for this case, looking at possible similarities with the case of the ideal gas. If the rubber band is elongated isothermally, what happens to the entropy? For the first part make sure that the signs are appropriate, according to experimental observations. In the second part, use Maxwell-type relations derived from the appropriate thermodynamic potential.

Solution
We know from the first law of thermodynamics that $TdS = \delta Q = dE - dW$. When studying an ideal gas, it is appropriate to define the work on the system by $dW = -PdV$, with P the pressure and V the volume. This is in agreement with the fact that the work done on the system (positive sign) reduces the volume occupied by the gas. This means that for an adiabatic transformation ($\delta Q = 0$) we have $dW = -PdV = dE$, i.e. the gas heats up ($dE > 0$) during a compression ($dV < 0$). For the rubber band, given the elongation x, we have a restoring force $F = -Kx$. The associated work on the system is $dW = Kxdx$ so that, for an adiabatic transformation, we have

$$dE = Kxdx$$

meaning that the energy increases upon elongation. Therefore, the first law of thermodynamics is

$$TdS = \delta Q = dE - Kxdx.$$

Let us now face the second point. The appropriate thermodynamic potential is the free energy F whose variation is such that

$$dF = dW - SdT = Kxdx - SdT$$

from which

$$\left(\frac{\partial F}{\partial T}\right)_x = -S \qquad \left(\frac{\partial F}{\partial x}\right)_T = Kx.$$

Let us derive the first expression with respect to x, the second with respect to T and set them equal (using Schwartz lemma for mixed partial derivatives). The result is

$$\left(\frac{\partial S}{\partial x}\right)_T = -K\left(\frac{\partial x}{\partial T}\right)_x = 0$$

which is nothing but a Maxwell-type relation for the rubber band. Such relation implies that during an isothermal elongation the entropy stays constant.

Problem 6.7.
Let E and M be the internal energy and the magnetization of some material immersed in a magnetic field H. Prove that, for the specific heat at constant H, the following relation

$$C_H = \left(\frac{\partial E}{\partial T}\right)_H - H\left(\frac{\partial M}{\partial T}\right)_H$$

holds.

Solution
Let us start by writing down the first law of thermodynamics for the magnetic system. The work done to increase the magnetization by dM is $dW = HdM$. As a consequence, if we choose T and M as variables to describe the system, the first law gives

$$TdS = dE(T,M) - H(T,M)dM = \left(\frac{\partial E}{\partial T}\right)_M dT + \left(\frac{\partial E}{\partial M}\right)_T dM - H(T,M)dM.$$

Let us stress that the situation is the analogue of an homogeneous fluid described by two variables among P, V, T. In our case, the variable H (that is the analogue of P for the fluid) is intensive whereas the extensive one is given by the magnetization M (the analogue of the volume V). We therefore find

$$C_M = T\left(\frac{\partial S}{\partial T}\right)_M = \left(\frac{\partial E}{\partial T}\right)_M$$

that is the quantity of heat that we have to supply in order to increase of a degree the temperature of the system at constant magnetization, i.e. the analogue of the specific heat at constant volume. If, instead of the variables T and M, we choose T and H, we get

$$TdS = dE(T,H) - HdM(T,H) =$$
$$\left(\frac{\partial E}{\partial T}\right)_H dT + \left(\frac{\partial E}{\partial H}\right)_T dH - H\left(\frac{\partial M}{\partial T}\right)_H dT - H\left(\frac{\partial M}{\partial H}\right)_T dH$$

from which, at constant H, we immediately find

$$C_H = T \left(\frac{\partial S}{\partial T} \right)_H = \left(\frac{\partial E}{\partial T} \right)_H - H \left(\frac{\partial M}{\partial T} \right)_H.$$

Problem 6.8.

Let us consider the first law of thermodynamics for a system with volume V, number of particles N, energy E, and entropy $S = S(E,V,N)$. Using the definition of the chemical potential as

$$\mu = \left(\frac{\partial F}{\partial N} \right)_{T,V}$$

with the free energy given by $F = E - TS$, and the relation

$$\left(\frac{\partial S}{\partial E} \right)_{N,V} = \frac{1}{T}$$

prove that

$$\frac{\mu}{T} = - \left(\frac{\partial S}{\partial N} \right)_{E,V}.$$

To solve the problem, use the differential expression of $S = S(E,V,N)$ with the volume V and the temperature T kept constant.

Solution

Following the hint, we start from the differential expression of $S = S(E,V,N)$

$$dS(E,V,N) = \left(\frac{\partial S}{\partial E} \right)_{V,N} dE + \left(\frac{\partial S}{\partial V} \right)_{E,N} dV + \left(\frac{\partial S}{\partial N} \right)_{E,V} dN.$$

Now, we can follow the variation of S with both T and V kept constant

$$(dS)_{T,V} = \left(\frac{\partial S}{\partial E} \right)_{N,V} (dE)_{T,V} + \left(\frac{\partial S}{\partial N} \right)_{E,V} (dN)_{T,V}$$

from which

$$\left(\frac{\partial S}{\partial N} \right)_{T,V} = \left(\frac{\partial S}{\partial E} \right)_{N,V} \left(\frac{\partial E}{\partial N} \right)_{T,V} + \left(\frac{\partial S}{\partial N} \right)_{E,V}.$$

As given by the text, we also can use

$$\left(\frac{\partial S}{\partial E} \right)_{N,V} = \frac{1}{T}$$

and, hence

$$T \left(\frac{\partial S}{\partial N} \right)_{T,V} = \left(\frac{\partial E}{\partial N} \right)_{T,V} + T \left(\frac{\partial S}{\partial N} \right)_{E,V}.$$

which finally gives

$$-T\left(\frac{\partial S}{\partial N}\right)_{E,V} = \left(\frac{\partial E}{\partial N}\right)_{T,V} - T\left(\frac{\partial S}{\partial N}\right)_{T,V} = \left(\frac{\partial}{\partial N}(E-TS)\right)_{T,V} = \left(\frac{\partial F}{\partial N}\right)_{T,V} = \mu$$

that is

$$\frac{\mu}{T} = -\left(\frac{\partial S}{\partial N}\right)_{E,V}$$

which is the desired result.

Problem 6.9.
A statistical system is composed of N independent distinguishable particles. Each one of these particles has only two energy levels, E_1 and E_2, such that $E_2 - E_1 = \varepsilon > 0$. Choose a suitable ground state for the energy and write down the total energy as a function of the temperature T. Finally, discuss the limits $T \to 0$ and $T \to +\infty$.

Solution
We set the ground state to have zero energy, $E_1 = 0$. As a consequence, we find that $E_2 = \varepsilon$. A general state is completely specified once we assign the set $\{n_j\}, j = 1,\ldots,N$, where $n_j = 0$ or 1 indicates if the j-th particle is in the ground state or in the excited one, respectively. Using this convention, the expression for the total energy is

$$E = \sum_{j=1}^{N} n_j \varepsilon = m\varepsilon$$

where m is the occupation number for the second energy level, i.e. the number of particles having energy ε. In order to compute the energy as a function of the temperature, we need to use the relation

$$\left(\frac{\partial S}{\partial E}\right)_N = \frac{1}{T}$$

where the entropy $S = k \ln \Omega$ requires the knowledge of the number of microstates Ω accessible to the system. This can be computed by simply considering all the possible ways to choose m objects out of N

$$\Omega(m,N) = \frac{N!}{(N-m)!m!}.$$

Using the Stirling approximation, we find

$$S(m,N) = k \ln \Omega(m,N) \approx k[N \ln N - (N-m)\ln(N-m) - m\ln m]$$

$$\frac{1}{T} = \left(\frac{\partial S}{\partial E}\right)_N = \frac{1}{\varepsilon}\left(\frac{\partial S}{\partial m}\right)_N = \frac{k}{\varepsilon}\ln\left(\frac{N-m}{m}\right)$$

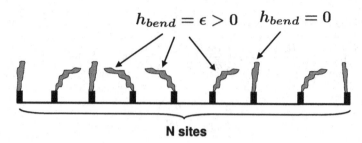

$h_{bend} = \epsilon > 0$ $h_{bend} = 0$

N sites

Fig. 6.1 A one dimensional chain composed of $N \gg 1$ localized sites. Each site is occupied by a polymer with two energy states: it can be straight (with energy $h_{bend} = \epsilon = 0$) or it can bend (on the right or on the left) with energy $h_{bend} = \epsilon > 0$. The thermodynamic properties of this system are discussed in Problem 6.10

from which we get

$$E = m\epsilon = \frac{N\epsilon}{1 + e^{\epsilon/kT}}.$$

In the limit $T \to 0$, we find $E \to 0$ meaning that only the ground state is occupied by the particles. In the limit $T \to +\infty$, we obtain $E \to N\epsilon/2$ indicating that both energy levels are equally populated.

Problem 6.10.

We consider a one dimensional chain composed of $N \gg 1$ localized sites. Each site is occupied by a polymer with two energy states: it can be straight (with energy $h_{bend} = 0$) or it can bend (on the right or on the left) with energy $h_{bend} = \epsilon > 0$, independently of the bending direction (see Fig. 6.1). Compute the entropy of the system, $S(E,N)$, for a fixed total bending energy $E = m\epsilon$ (m is an integer number such that $m \gg 1$). Also, determine the internal energy as a function of the temperature and the resulting heat capacity, $C_N = \left(\frac{\partial E}{\partial T}\right)_N$, under the assumption that $(N - m) \gg 1$. Finally, determine the behaviour of the internal energy in the limit of low and high temperatures.

Solution

We know that the total bending energy is fixed and equal to $m\epsilon$. This means that we have m bended polymers. Therefore, we have to consider all the possible ways to extract m distinguishable (the sites are localized) objects out of N, i.e. $\frac{N!}{m!(N-m)!}$. We also have to consider the degeneracy (that is 2 in this case) associated with the positive bending energy, since the polymer can bend on the right and on the left. Therefore, the total number of microstates $\Omega(E,N)$ associated with the total energy $E = m\epsilon$ is

$$\Omega(E,N) = \frac{N!}{m!(N-m)!} 2^m.$$

We then apply the Boltzmann entropy formula and use the Stirling approximation for the factorials to get

$$S(E,N) = k \ln \Omega(E,N) \approx$$
$$kN \left[\frac{m}{N} \ln 2 - \frac{m}{N} \ln \left(\frac{m}{N} \right) - \left(1 - \frac{m}{N} \right) \ln \left(1 - \frac{m}{N} \right) \right] =$$
$$kN \left[\frac{E}{N\varepsilon} \ln 2 - \frac{E}{N\varepsilon} \ln \left(\frac{E}{N\varepsilon} \right) - \left(1 - \frac{E}{N\varepsilon} \right) \ln \left(1 - \frac{E}{N\varepsilon} \right) \right].$$

The temperature is given by the derivative of the entropy

$$\frac{1}{T} = \left(\frac{\partial S}{\partial E} \right)_N = \frac{k}{\varepsilon} \left[\ln 2 - \ln \left(\frac{E}{N\varepsilon} \right) + \ln \left(1 - \frac{E}{N\varepsilon} \right) \right]$$

from which

$$\ln \left(\frac{N\varepsilon}{E} - 1 \right) = \beta\varepsilon - \ln 2$$

so that

$$E = \frac{2N\varepsilon}{e^{\beta\varepsilon} + 2}.$$

The heat capacity becomes

$$C_N = \left(\frac{\partial E}{\partial T} \right)_N = 2Nk(\beta\varepsilon)^2 \frac{e^{\beta\varepsilon}}{(2 + e^{\beta\varepsilon})^2}.$$

In the limit of low temperatures we find

$$\lim_{\beta \to +\infty} E = 0$$

indicating that only the ground state ($\varepsilon = 0$) is populated. In the limit of high temperatures we find

$$\lim_{\beta \to 0} E = \frac{2}{3} N\varepsilon.$$

We note that if we change the degeneracy from 2 to 1, all the results coincide with those of Problem 6.9.

Problem 6.11.
We want to study the thermodynamic properties of a magnetic system with unitary volume. Such system is characterized by the following constitutive equations for the magnetization and internal energy as a function of the temperature T and magnetic field H

$$M(T,H) = Nm \left[\coth \left(\frac{mH}{kT} \right) - \frac{kT}{mH} \right]$$

$$E(T,H) = C_M T$$

with C_M, m and N constants. Working in the limit $mH \gg kT$, find the relation between the temperatures and magnetizations of two generic thermodynamic states (say 1 and 2) connected by an adiabatic transformation. Finally, in the same limit $mH \gg kT$, give an estimate of the entropy $S(E, M)$ once we know the value of the energy (E_0) and magnetization (M_0) of a given reference state.

Solution

The first part of the problem deals with an adiabatic transformation. For this reason, we start from the first law of thermodynamics for the magnetic system

$$\delta Q = dE - H dM$$

and require no heat exchange, i.e. $0 = \delta Q = dE - H dM$. Since $dE = C_M dT$, we get

$$H(M, T) dM = dE = C_M dT$$

and we need to extract the function $H(M, T)$ from our constitutive equation for the magnetization. In the limit $mH \gg kT$ we obtain

$$\coth\left(\frac{mH}{kT}\right) \approx 1$$

and

$$M(T, H) = Nm\left(1 - \frac{kT}{mH}\right)$$

from which we can find H as a function of M and T

$$H(M, T) = \frac{NkT}{Nm - M}.$$

Therefore, the corresponding adiabatic transformation is characterized by the following differential relation

$$\frac{C_M}{Nk}\frac{dT}{T} = \frac{dM}{Nm - M}$$

which can be integrated between (T_1, M_1) and (T_2, M_2) with the following result

$$\left(\frac{T_1}{T_2}\right)^{C_M/Nk} = \frac{Nm - M_2}{Nm - M_1}.$$

As for the second point, we start from the first law of thermodynamics in its differential form

$$dS = \frac{dE}{T} - \frac{H}{T} dM$$

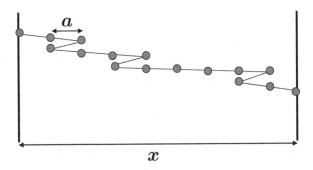

Fig. 6.2 A one dimensional rubber band is modelled with a chain of $N+1$ molecules. From one molecule we can establish a link with the successive one by moving a step forward or backward. The characteristic length of the link between two molecules is a, and the distance between the first and last molecule is x. In Problem 6.12 we show how to predict the thermodynamic properties of this system starting from the microcanonical ensemble

and integrate between (E_0, M_0) and a generic state with energy E and magnetization M

$$S(E,M) = C_M \int_{E_0}^{E} \frac{dE}{E} - Nk \int_{M_0}^{M} \frac{dM}{Nm - M} =$$
$$C_M \ln\left(\frac{E}{E_0}\right) + Nk \ln\left(\frac{Nm - M}{Nm - M_0}\right) + S(E_0, M_0).$$

Problem 6.12.
We want to describe the elasticity of a rubber band with a very simple one dimensional model characterized by a chain of $N+1$ molecules (see Fig. 6.2). From one molecule we can establish a link with the successive one by moving a step forward or backward, with no difference from the energetical point of view (i.e. the internal energy is only dependent on the total number N). The characteristic length of the link between two molecules is a, and the distance between the first and last molecule is x. Find the entropy for the system. Suppose that, for a small variation dx, we can write the work on the system as $dW = -g\,dx$, with g a tension needed to keep the distance x. At a fixed N, the number of all the possible pairs (N_+, N_-) consistent with a given macrostate at constant energy must be computed. Finally, find the relation between the temperature and the tension, and determine the sign of the latter.

Solution
The number of links realized with forward steps must be properly related to the number of backward steps in such a way that

$$x = a(N_+ - N_-)$$

plus the condition that

$$N = N_+ + N_-$$

because the molecules are $N + 1$, and the links are N. Therefore, we find that

$$\begin{cases} N_+ = \frac{1}{2}\left(N + \frac{x}{a}\right) \\ N_- = \frac{1}{2}\left(N - \frac{x}{a}\right). \end{cases}$$

The entropy is dependent on the number of total configurations, i.e. all the possible ways to extract N_+ forward steps and N_- backward steps out of N, with the constraint that $N = N_+ + N_-$. Therefore, we can write

$$S(x,N) = k\ln\left(\frac{N!}{(N-N_+)!N_+!}\right) = k\ln\left(\frac{N!}{N_-!N_+!}\right) = k\ln\left(\frac{N!}{(\frac{N}{2}-\frac{x}{2a})!(\frac{N}{2}+\frac{x}{2a})!}\right)$$

from which, using the Stirling approximation, we get

$$\frac{S(x,N)}{k} \approx N\ln N - \frac{1}{2}\left(N + \frac{x}{a}\right)\ln\left(\frac{N}{2} + \frac{x}{2a}\right) - \frac{1}{2}\left(N - \frac{x}{a}\right)\ln\left(\frac{N}{2} - \frac{x}{2a}\right).$$

From the first law of thermodynamics, we know that

$$T dS = dE - dW = dE + g dx$$

and, since the internal energy is only dependent on N, we can write

$$g = T\left(\frac{\partial S}{\partial x}\right)_N$$

leading to

$$g = -\frac{kT}{2a}\ln\left(\frac{N + \frac{x}{a}}{N - \frac{x}{a}}\right).$$

We immediately see that the tension g has negative sign, because $a > 0$, $x > 0$ and $\ln\left(\frac{N+\frac{x}{a}}{N-\frac{x}{a}}\right) > 0$.

Problem 6.13.
A statistical system is composed of N particles with spin $\frac{1}{2}$, immersed in a magnetic field H. The particles are fixed in their positions and possess a magnetic moment μ. The Hamiltonian of such system is

$$\mathscr{H} = -\mu H \sum_{i=1}^{N} \sigma_i$$

where $\sigma_i = \pm 1$. Determine the entropy, the energy, the specific heat, and the magnetization. Finally, defining the susceptibility as

$$\chi = \left(\frac{\partial M}{\partial H} \right)_{T,N}$$

prove Curie law, i.e. that χ is inversely proportional to the temperature when $H \to 0$.

Solution
Let us set $\varepsilon = \mu H$ and let $N_\pm$ be the number of particles with $\sigma_i = \pm 1$. The Hamiltonian can be written as

$$\mathcal{H} = E = -\mu H \sum_{i=1}^{N} \sigma_i = -\varepsilon N_+ + \varepsilon N_- = -\varepsilon N_+ + \varepsilon (N - N_+) = \varepsilon N - 2\varepsilon N_+$$

from which

$$N_+ = \frac{1}{2} \left(N - \frac{E}{\varepsilon} \right) \qquad N_- = \frac{1}{2} \left(N + \frac{E}{\varepsilon} \right).$$

The entropy is connected to the number of total configurations, i.e. all the possible ways to extract N_+ and N_- spins out of N, with the constraint that $N = N_+ + N_-$. Therefore, we can write

$$S(E,H,N) = k \ln \left(\frac{N!}{N_+!(N - N_+)!} \right) = k \ln \left(\frac{N!}{N_+! N_-!} \right) \approx$$

$$k \left(N \ln N - \left(\frac{N}{2} - \frac{E}{2\mu H} \right) \ln \left(\frac{N}{2} - \frac{E}{2\mu H} \right) - \left(\frac{N}{2} + \frac{E}{2\mu H} \right) \ln \left(\frac{N}{2} + \frac{E}{2\mu H} \right) \right)$$

where we have used the Stirling approximation for the factorials. The dependence on the temperature is found with

$$\frac{1}{T} = \left(\frac{\partial S}{\partial E} \right)_{H,N} = \frac{k}{2\mu H} \ln \frac{\left(N - \frac{E}{\mu H} \right)}{\left(N + \frac{E}{\mu H} \right)}$$

from which we extract the energy

$$E = -N\mu H \tanh \left(\frac{\mu H}{kT} \right)$$

and the specific heat

$$C = \left(\frac{\partial E}{\partial T} \right)_{H,N} = \frac{N\mu^2 H^2}{kT^2} \left(1 - \tanh^2 \left(\frac{\mu H}{kT} \right) \right).$$

Finally, the magnetization is given by

$$M = \mu (N_+ - N_-) = -\frac{E}{H} = N\mu \tanh \left(\frac{\mu H}{kT} \right)$$

and the susceptibility by

$$\chi = \left(\frac{\partial M}{\partial H}\right)_{T,N} = \frac{N\mu^2}{kT}\left(1 - \tanh^2\left(\frac{\mu H}{kT}\right)\right).$$

When the magnetic field is small, we find $\tanh^2\left(\frac{\mu H}{kT}\right) \approx 0$ and

$$\lim_{H\to 0}\chi = \frac{N\mu^2}{kT}$$

that is Curie law.

Problem 6.14.
Let us assume that the air rises in the atmosphere adiabatically like an ideal gas. Determine the way the temperature T changes as a function of the height z and express the final result in terms of the gravitational acceleration g, the ratio of the specific heats at constant pressure and volume, the mass m of each particle, and the Boltzmann constant k. To solve the problem, consider an infinitesimal cylinder of air and write down the equation determining the mechanical equilibrium, from which you can extract the variation of the pressure P in terms of the ratio P/T. Finally, combine this result with the variation of the temperature with respect to the pressure obtained from the equation of state and that for an adiabatic change ($PV^\gamma =$const., with γ the ratio of the specific heats) for a simple gas to get the desired result.

Solution
Let us consider the air inside a cylinder with height dz and base S. In order to obtain the condition of mechanical equilibrium, the pressure must be a function of the height, $P(z)$. The force acting on the inferior base is $P(z)S$, while that acting on the upper base is $-P(z+dz)S$. Also, gravity acts on the cylinder with a force equal to $-m\rho(z)Sdzg$, where $\rho(z)$ stands for the density in the number of particles and m is the mass of each particle. The mechanical equilibrium requires that

$$-P(z+dz)S + P(z)S - \rho(z)Sdzgm = 0$$

leading to a differential equation for $P(z)$

$$\frac{P(z+dz) - P(z)}{dz} = \frac{dP(z)}{dz} = -gm\rho(z).$$

The equation of state for an ideal gas can be used to find

$$\rho(z) = \frac{P(z)}{kT(z)}$$

which can be substituted in the previous expression to yield

$$\frac{dP(z)}{dz} = -\frac{gm}{k}\frac{P(z)}{T(z)}.$$

After differentiating the equation of state, and substituting Nk using this equation again, we find

$$VdP + PdV = NkdT = \frac{PV}{T}dT.$$

Dividing by PV, we get

$$\frac{dP}{P} + \frac{dV}{V} = \frac{dT}{T}.$$

Also, the condition that $PV^\gamma =$const. leads to

$$\frac{dP}{P} + \gamma\frac{dV}{V} = 0$$

that, combined with the previous one, gives

$$\frac{dT}{T} = \frac{dP}{P} + \frac{dV}{V} = \frac{dP}{P} - \frac{1}{\gamma}\frac{dP}{P}$$

or, equivalently

$$\left(\frac{dT}{dP}\right) = \left(\frac{\gamma-1}{\gamma}\right)\frac{T}{P}.$$

We now combine this result with the expression obtained previously for the derivative of the pressure with respect to height

$$\frac{dT}{dz} = \left(\frac{dT}{dP}\right)\left(\frac{dP}{dz}\right) = \left(\frac{1-\gamma}{\gamma}\right)\frac{mg}{k}$$

that is the desired result.

Problem 6.15.
Suppose we are able to measure the thermal expansion coefficient at constant pressure

$$\alpha = \frac{1}{V}\left(\frac{\partial V}{\partial T}\right)_P$$

for a thermodynamic fluid with a constant number of particles. Determine, under the same conditions, the derivative of the entropy with respect to the pressure, $\left(\frac{\partial S}{\partial P}\right)_T$.

Solution
We can solve the problem starting from the Gibbs potential defined as $\Phi = F + PV$, with F the free energy. From the differential form of Φ we get

$$d\Phi = -SdT + VdP$$

and we find immediately

$$\left(\frac{\partial \Phi}{\partial T}\right)_P = -S \qquad \left(\frac{\partial \Phi}{\partial P}\right)_T = V.$$

Then, using Schwartz lemma for mixed partial derivatives, we find the following identity

$$\left(\frac{\partial S}{\partial P}\right)_T = -\left(\frac{\partial}{\partial P}\left(\frac{\partial \Phi}{\partial T}\right)_P\right)_T = -\left(\frac{\partial}{\partial T}\left(\frac{\partial \Phi}{\partial P}\right)_T\right)_P = -\left(\frac{\partial V}{\partial T}\right)_P$$

and the final result

$$\left(\frac{\partial S}{\partial P}\right)_T = -\alpha V.$$

Problem 6.16.
A physical system is composed of N distinguishable particles, and each particle can be found in a state with energy 0 or $\varepsilon > 0$. The excited state has degeneracy $d = 4$ while the ground state is not degenerate. The total energy of the system is given by $E = n\varepsilon$, with n a positive integer ($n \leq N$). Write down the number of microstates corresponding to the macrostate with energy $E = n\varepsilon$. Then, identify the temperature T and compute the ratio of the occupation numbers for the two energy levels as a function of T and ε. Verify the limit of high temperatures in the final result.

Solution
The total energy is $E = n\varepsilon$ and the number of microstates related to this energy is given by all the possible ways to choose n distinguishable objects out of N

$$\Omega_n = \binom{N}{n} = \frac{N!}{n!(N-n)!}.$$

We also know that the excited energy state has a degeneracy equal to 4, i.e. whenever a particle occupies the energy level ε, this can happen in 4 different ways. Therefore, the number of microstates is given by

$$\Omega(E,N) = \Omega_n 4^n = \binom{N}{n} 4^n$$

where n depends on the energy E. For large N, we can use the Stirling approximation ($N! \approx N^N e^{-N}$) and write

$$S(E,N) = k \ln \Omega(E,N) \approx -Nk\left[\alpha \ln \alpha + (1-\alpha)\ln(1-\alpha) - \alpha \ln 4\right]$$

where we have defined $\alpha = \frac{n}{N} = \frac{E}{N\varepsilon}$. The temperature T is given by

$$\frac{1}{kT} = \frac{1}{k}\left(\frac{\partial S}{\partial E}\right)_N = \frac{1}{N\varepsilon k}\left(\frac{\partial S}{\partial \alpha}\right)_N = \frac{1}{\varepsilon}\ln\left(\frac{4(1-\alpha)}{\alpha}\right).$$

We can explicitly invert this expression for α, and find the occupation numbers

$$n = N\alpha = N\frac{4e^{-\beta\varepsilon}}{1+4e^{-\beta\varepsilon}} \qquad n_0 = N - n = N\frac{1}{1+4e^{-\beta\varepsilon}}$$

where $\beta = \frac{1}{kT}$. Their ratio is

$$\frac{n}{n_0} = \frac{4e^{-\beta\varepsilon}}{1 + 4e^{-\beta\varepsilon}}.$$

In the limit of high temperatures ($\beta \ll 1$), we get $\frac{n}{n_0} \approx 4/5$, meaning that the five levels available (one level with energy 0 and four levels with energy ε) are equally populated. Note that if we change the degeneracy from 4 to 1, all the results coincide with those of Problem 6.9.

Problem 6.17.
Two independent statistical systems (1 and 2) are both characterized by N energy levels and m_1, m_2 indistinguishable quanta distributed in these levels ($m_{1,2} \gg 1$). The energy of both systems is proportional to the number of associated quanta

$$E_1 = \alpha_1 m_1 \qquad E_2 = \alpha_2 m_2$$

with $\alpha_1, \alpha_2 > 0$. Write down the number of microstates and determine the entropies for both systems . Then, suppose we establish a contact between the two systems so that they reach some equilibrium condition without exchanging quanta. In this situation, determine the relation between m_1, m_2, α_1 and α_2.

Solution
The number of states available for each system is

$$\Omega(m_{1,2}, N) = \frac{(N - 1 + m_{1,2})!}{m_{1,2}!(N-1)!}$$

that is the way to distribute $m_{1,2}$ indistinguishable objects into N levels. If we interpret the energy levels as boxes, we have $N + 1$ partitioning lines delimiting these boxes. The first and last partitioning lines can be considered fixed (we are then left with $N - 1$ of them) and we need to find the total number of arrangements for $N - 1$ partitioning lines and $m_{1,2}$ quanta (see Fig. 6.3), for a total of $(N - 1 + m_{1,2})$ objects. The total number of these arrangements is $(N - 1 + m_{1,2})!$. Also, a permutation of two internal partitioning lines or quanta does not change the configuration (they are indistinguishable), and this is the reason for the presence of $(N - 1)!$ and $m_{1,2}!$ in the denominator. The entropy of both systems is given by the Boltzmann formula

$$\begin{aligned} S(m_{1,2}, N) &= k \ln \Omega(m_{1,2}, N) = k(\ln(N - 1 + m_{1,2})! - \ln(N-1)! - \ln m_{1,2}!) \approx \\ & k(\ln(N + m_{1,2})! - \ln N! - \ln m_{1,2}!) \approx \\ & k[(N + m_{1,2})\ln(N + m_{1,2}) - N\ln N - m_{1,2}\ln m_{1,2}] \end{aligned}$$

from which we can extract the temperatures

$$\frac{1}{T_{1,2}} = \left(\frac{\partial S}{\partial E_{1,2}}\right)_N = \frac{1}{\alpha_{1,2}}\left(\frac{\partial S}{\partial m_{1,2}}\right)_N = \frac{k}{\alpha_{1,2}}\ln\left(\frac{N + m_{1,2}}{m_{1,2}}\right).$$

N+1 Vertical walls

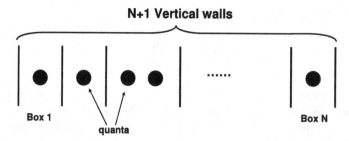

Box 1   Box N

quanta

Fig. 6.3 The arrangement of a given number of energy quanta into N levels can be thought of as a combinatorial problem where we distribute some indistinguishable objects (quanta) in N distinguishable boxes (energy levels). The boxes are characterized by indistinguishable partitioning lines (or 'vertical walls') separating the energy levels. For the technical details see Problem 6.17

The condition of thermal equilibrium is $T_1 = T_2$ so that

$$\frac{\alpha_1}{\alpha_2} = \frac{\ln\left(\frac{N+m_1}{m_1}\right)}{\ln\left(\frac{N+m_2}{m_2}\right)}$$

is the desired relation between m_1, m_2, α_1 and α_2.

Problem 6.18.
A one dimensional harmonic oscillator has the energy (in some suitable units) $\varepsilon = n + \frac{1}{2}$, where the positive integer n represents the number of energy quanta associated with the oscillator. Let us consider N one dimensional distinguishable harmonic oscillators with a fixed energy E and compute their total energy density E/N, entropy density S/N, and temperature T. Make the assumption that, in the limit of large N, the density E/N is finite, while $E - \frac{N}{2}$ and $E + \frac{N}{2}$ are both very large. Finally, analyze the high temperature limit of the energy, verifying the consistency with the equipartition theorem.

Solution
We have to determine all the microstates of the system with total energy

$$E = \sum_{i=1}^{N} \varepsilon_i = \sum_{i=1}^{N} \left(n_i + \frac{1}{2}\right)$$

where n_i are the energy quanta associated with the i-th oscillator. We can write the previous equation as

$$\sum_{i=1}^{N} n_i = E - \frac{N}{2} = Q$$

where Q is an integer number because it is a sum of integers. We can now think that the N oscillators are boxes where we have to distribute Q quanta. The N boxes imply the presence of $N+1$ partitioning lines: if we keep fixed the first and last partitioning lines, we have to determine the total number of arrangements for $N-1$ (indistinguishable) partitioning lines and Q (indistinguishable) quanta (see also Problem 6.17). The total number of these arrangements is

$$\Omega(E,N) = \frac{(Q+N-1)!}{Q!(N-1)!} = \frac{(E+\frac{N}{2}-1)!}{(E-\frac{N}{2})!(N-1)!}$$

and the entropy is given by the Boltzmann formula

$$S(E,N) = k\ln\Omega(E,N).$$

In the thermodynamic limit, for a finite density E/N, we can write (using the Stirling approximation for the factorials)

$$\frac{S(E,N)}{N} \approx k\left[\left(\frac{E}{N}+\frac{1}{2}\right)\ln\left(\frac{E}{N}+\frac{1}{2}\right) - \left(\frac{E}{N}-\frac{1}{2}\right)\ln\left(\frac{E}{N}-\frac{1}{2}\right)\right]$$

from which we compute the temperature

$$\frac{1}{T} = \left(\frac{\partial S}{\partial E}\right)_N = k\ln\left(\frac{E/N+\frac{1}{2}}{E/N-\frac{1}{2}}\right).$$

This allows us to express the energy density in terms of T

$$\frac{E}{N} = \frac{1}{2}\coth\left(\frac{1}{2kT}\right).$$

This result shows that the quantum mechanical oscillators do not obey the (classical) *equipartition theorem* for the energy. Only in the limit of high temperatures, when $T \gg 1$, we can use the Taylor expansion $\coth\frac{1}{x} \approx x$ $(x \gg 1)$ to find

$$\frac{E}{N} \approx kT.$$

This is in agreement with the equipartition theorem (see also Problem 7.20), which assigns to each degree of freedom an energy contribution equal to $kT/2$. In our case, we have N one dimensional oscillators, each one with a position and a momentum. The total energy would then be

$$E = \frac{kT}{2}\times N\times(1+1) = NkT$$

which is in agreement with the high temperature limit previously obtained.

Problem 6.19.

An ultrarelativistic gas with $N \gg 1$ particles is in a volume V. The total energy is

$$E = c \sum_{i=1}^{N} p_i$$

where p_i is the momentum of the i-th particle with $p_i = |p_i|$ its absolute value, and where c is the speed of light. The total energy is fixed and the particles are indistinguishable. Give an estimate for the entropy and write down the equation of state. Finally, determine the specific heat at constant pressure C_P. If D is the region of the phase space such that $\sum_{i=1}^{N} |p_i| \leq E/c$, the following integral

$$\mathscr{I}_N(E) = \int_D \prod_{i=1}^{N} p_i^2 dp_i = \frac{2^N}{(3N)!} \left(\frac{E}{c} \right)^{3N}$$

may be useful.

Solution

In the limit when $N \gg 1$ we can approximate (to determine the thermodynamic properties of the system) the integral over the surface at constant energy with the integral in the volume enclosed by that surface. In other words, the number of states is well approximated by

$$\Omega(E,V,N) \approx \Sigma(E,V,N) = \frac{1}{h^{3N}} \int_D \prod_{i=1}^{N} d^3 q_i \, d^3 p_i.$$

If we use the hint given in the text of the problem and the fact that the particles are indistinguishable, we immediately obtain that

$$\Omega(E,V,N) \approx \Sigma(E,V,N) = (4\pi V)^N \frac{\mathscr{I}_N(E)}{h^{3N} N!} = \frac{1}{N!(3N)!} \left(\frac{8\pi V E^3}{h^3 c^3} \right)^N.$$

Let us now use the Stirling approximation to evaluate the entropy $S(E,V,N)$

$$S(E,V,N) = k \ln \Omega(E,V,N) \approx k \ln \Sigma(E,V,N) \approx Nk \left[\ln \left(\tilde{b} \frac{V E^3}{N^4} \right) \right] + 4Nk$$

where $\tilde{b} = \frac{8\pi}{27 h^3 c^3}$. In order to obtain the temperature and the pressure, we have to compute the derivatives of the entropy

$$\frac{1}{T} = \left(\frac{\partial S}{\partial E} \right)_{V,N} = \frac{3Nk}{E} \qquad P = T \left(\frac{\partial S}{\partial V} \right)_{E,N} = \frac{NkT}{V}.$$

We find that the equation of state is exactly the same as that of an ideal gas: $PV = NkT$. The energy is $E = 3NkT$. The latter result is in agreement with the classical equipartition theorem predicting that (see Problem 7.20)

$$E = c \sum_{i=1}^{N} p_i = \sum_{i=1}^{N} \left\langle p_i \left(\frac{\partial H}{\partial p_i} \right) \right\rangle = 3NkT$$

where H is the Hamiltonian of the system, $\langle ... \rangle$ is meant as an ensemble average, and where the derivative with respect to p_i is performed by keeping fixed all the other variables. Using the equation of state and the internal energy, we can also write the entropy as a function of (T,P,N)

$$S(T,P,N) = Nk \ln \left(\frac{27 k^4 \tilde{b} T^4}{P} \right) + 4Nk.$$

This expression is useful to determine the specific heat at constant pressure

$$C_P = T \left(\frac{\partial S}{\partial T} \right)_{P,N} = 4Nk.$$

Problem 6.20.
Let us consider 2 systems A and B, each one composed of 2 distinguishable particles. Consider that the total energy for the system is $E_{TOT} = E_A + E_B = 5$ in some suitable units. A and B are in thermal equilibrium and are separated by a rigid wall not allowing for particles and energy exchange. Compute the entropy when $E_A = 3$ and $E_B = 2$. Repeat the same calculations when the energy exchange between the two systems is allowed. In the calculations, consider that each particle energy can only be a positive integer number.

Solution
Let us indicate with (q_1, q_2) a microstate of two particles where the first particle has energy q_1 and the second one has energy q_2. For the system A, all the possible microstates are given by $(3,0)$, $(0,3)$, $(2,1)$, $(1,2)$. Similarly, for the system B, we find the following set of microstates: $(2,0)$, $(0,2)$ and $(1,1)$. Therefore, for the system A the number of microstates is $\Omega_A(E_A = 3) = 4$, while for the system B we have $\Omega_B(E_B = 2) = 3$. When particles and energy exchange are not allowed, the total number of microstates is simply given by the product

$$\Omega(E_A + E_B = 5) = \Omega_A(E_A = 3) \times \Omega_B(E_B = 2) = 12$$

and the entropy is $S = k \ln 12$. In the second case we have an energy exchange. All the energies of A and B are those that satisfy $E_{TOT} = E_A + E_B = 5$. They are: $(E_A = 5, E_B = 0)$, $(E_A = 0, E_B = 5)$, $(E_A = 4, E_B = 1)$, $(E_A = 1, E_B = 4)$, $(E_A = 3, E_B = 2)$, $(E_A = 2, E_B = 3)$. For each one of these cases, the microstates are identified

$(E_A = 5, E_B = 0): A: (5,0), (0,5), (4,1), (1,4), (3,2), (2,3) \quad B: (0,0);$
$\Omega_A(E_A + E_B = 5) = \Omega_A(E_A = 5) \times \Omega_B(E_B = 0) = 6$

$(E_A = 0, E_B = 5): A : (0,0), B : (5,0), (0,5), (4,1), (1,4), (3,2), (2,3);$
$\Omega_A(E_A + E_B = 5) = \Omega_B(E_A = 0) \times \Omega_A(E_B = 5) = 6$

$(E_A = 4, E_B = 1): A : (4,0), (0,4), (3,1), (1,3), (2,2) \quad B : (1,0), (0,1);$
$\Omega_A(E_A + E_B = 5) = \Omega_A(E_A = 4) \times \Omega_B(E_B = 1) = 10$

$(E_A = 1, E_B = 4): A : (1,0), (0,1), B : (4,0), (0,4), (3,1), (1,3), (2,2);$
$\Omega_A(E_A + E_B = 5) = \Omega_A(E_A = 1) \times \Omega_B(E_B = 4) = 10$

$(E_A = 3, E_B = 2): A : (3,0), (0,3), (2,1), (1,2) \quad B : (2,0), (0,2), (1,1);$
$\Omega_A(E_A + E_B = 5) = \Omega_A(E_A = 3) \times \Omega_B(E_B = 2) = 12$

$(E_A = 2, E_B = 3): A : (2,0), (0,2), (1,1) \quad B : (3,0), (0,3), (2,1), (1,2);$
$\Omega_A(E_A + E_B = 5) = \Omega_A(E_A = 2) \times \Omega_B(E_B = 3) = 12.$

Therefore, the total number of states is given by

$$\Omega(E_{TOT} = 5) = \sum_{i=0}^{5} \Omega(E_A = i) \times \Omega(E_B = E_{TOT} - i) = 56$$

and the entropy by $S = k \ln 56$. We remark that in this case we can obtain the same number of states by considering all the possible ways to distribute 5 indistinguishable quanta into 4 distinguishable levels, that is

$$\Omega(E_{TOT} = 5) = \frac{(5+4-1)!}{3!5!} = \frac{8!}{3!5!} = 8 \times 7 = 56.$$

Problem 6.21.
Starting from the line at zero energy, and working in the two dimensional phase space of a classical plane rotator, draw the lines at constant energy producing cells with volume h in the phase space. Determine the energy of these states and compare them with the eigenvalues of the corresponding quantum mechanical problem.

Solution
The Hamiltonian of the classical plane rotator is $H = \frac{p^2}{2I}$, where p is the momentum associated with the rotation angle θ and I is the momentum of inertia. The variable θ is periodic and the phase space is the strip of the plane (θ, p) between $-\pi$ and π. In Fig. 6.4 we draw the lines at constant energy (that implies p=const.) corresponding to the momenta p_1, p_2, p_3. We obtain cells with volume h if the relations $2\pi p_1 = h$, $2\pi p_2 = 2h$, etc. hold. To summarize, we have found $p_n = n\hbar$ (n is an integer number) and the corresponding energies

$$E_n = \frac{n^2 \hbar^2}{2I}.$$

For the plane rotator in the quantum mechanical case we have

$$\hat{H} = \frac{\hat{L}_z^2}{2I} = -\frac{\hbar^2}{2I} \frac{\partial^2}{\partial \theta^2}$$

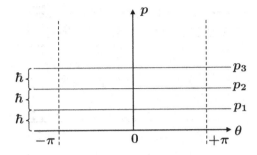

Fig. 6.4 In the phase space of a plane rotator, we draw the lines at constant energy producing cells with volume h. In this way, we find different momenta $p_n = n\hbar$ (n is an integer number) and the corresponding energies $E_n = \frac{n^2\hbar^2}{2I}$, with I the momentum of inertia of the plane rotator. For further details see Problem 6.21

where $\hat{L}_z$ is the orbital angular momentum in the direction of the rotation axis. An explicit calculation (see also Problem 5.1) shows that the eigenstates for the energy are

$$F_n(\theta) = \frac{1}{\sqrt{2\pi}} e^{in\theta}$$

with n an integer number. The eigenvalues are

$$E_n = \frac{n^2\hbar^2}{2I}$$

that is the very same result obtained with the above construction in the classical phase space.

Problem 6.22.
A statistical system is composed of R indistinguishable quanta distributed in N energy levels in such a way that we do not find empty levels ($R > N \gg 1$). Indicating with R_i the number of quanta in the i-th level, the total energy of the system is $R = R_1 + R_2 + \ldots + R_N$. Show that the total number of microstates is

$$\binom{R-1}{N-1}.$$

When $N = 3$ and $R = 6$, write down explicitly all the configurations. Finally, in the general case with $R > N \gg 1$, compute the entropy of the system and determine the equilibrium temperature. Try to establish an analogy between this physical system and the combinatorial problem of arranging R indistinguishable objects in N boxes.

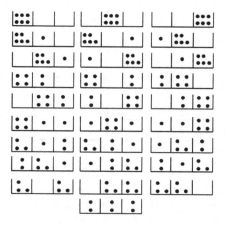

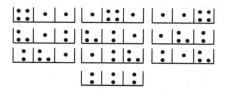

Fig. 6.5 The distribution of 6 indistinguishable quanta in 3 distinguishable energy levels. In the top panel we show all the possible 28 microstates. The requirement that there are no empty energy levels reduces the number of microstates to 10 (bottom panel). All the details are reported in Problem 6.22

Solution

We can think that the N levels are boxes delimited by $N+1$ partitioning lines. If we keep fixed the first and last partitioning lines and create all the possible arrangements of $R+N-1$ elements (quanta or partitioning lines), the number of microstates is evaluated as

$$\binom{N+R-1}{N-1} = \binom{N+R-1}{R} = \frac{(N+R-1)!}{R!(N-1)!}$$

where the denominator accounts for all the permutations realized by shuffling the R quanta and the $N-1$ partitioning lines (see also Problems 6.17 and 6.18). Unfortunately, this is not enough for our purposes, because we know from the text that no empty levels are present, i.e. all the partitioning lines have to be placed in between the quanta. This means that during an arrangement of R quanta, we only have

$R-1$ interspacings to place the partitioning lines. Therefore, the correct number of microstates is given by all the possible ways to select $N-1$ interspacings out of $R-1$

$$\binom{R-1}{N-1} = \frac{(R-1)!}{(R-N)!(N-1)!}$$

that is the expression reported in the text. Another alternative procedure is the following: first of all, we can think to occupy each of the N boxes with a single quantum; the remaining $R-N$ quanta can be freely arranged in the N boxes, for a total of

$$\binom{N+R-N-1}{N-1} = \binom{R-1}{N-1} = \frac{(R-1)!}{(R-N)!(N-1)!}$$

microstates. The case with $N=3$ and $R=6$ is treated explicitly in Fig. 6.5. For the sake of completeness, we first report the most general case (top panel of the figure) where the number of microstates is

$$\binom{N+R-1}{R} = 28$$

and where we also find empty levels. The condition that there is no empty level reduces (see lower panel of Fig. 6.5) the number of microstates to

$$\binom{R-1}{N-1} = 10.$$

In the case with $R > N \gg 1$, the general form of the entropy is

$$S(N,R) = k \ln \left(\frac{(R-1)!}{(R-N)!(N-1)!} \right) \approx k[-(R-N)\ln(R-N) + R\ln R - N\ln N]$$

where we have used the Stirling approximation for the factorials. We then find the temperature as

$$\frac{1}{T} = \left(\frac{\partial S}{\partial E} \right)_N = \left(\frac{\partial S}{\partial R} \right)_N = k \ln \frac{R}{(R-N)}.$$

Problem 6.23.
Let us consider N one dimensional classical harmonic oscillators with the same mass m and frequency ω in the microcanonical ensemble. Determine the internal energy in the limit $N \gg 1$.

Solution
Let us start from the Hamiltonian of the system

$$H_N = \sum_{i=1}^{N} \left(\frac{p_i^2}{2m} + \frac{1}{2} m\omega^2 q_i^2 \right)$$

where p_i and q_i are the momentum and position for the i-th oscillator. In order to compute the number of states at fixed energy E, we should perform the integral

$$\Omega(E,N) = \int_{H_N=E} \frac{d^N q \, d^N p}{h^N}$$

but, when determining the thermodynamic properties in the limit $N \gg 1$, we can replace the surface integral with the volume integral

$$\Sigma(E,N) = \int_{H_N \leq E} \frac{d^N q \, d^N p}{h^N}.$$

We next introduce the new variables Y_i ($i = 1, 2, ..., 2N$) with the property

$$p_j = \sqrt{2m} Y_j \qquad q_j = \sqrt{\frac{2}{m\omega^2}} Y_{N+j} \qquad j = 1, 2, ..., N$$

in such a way that the previous volume integral becomes

$$\Sigma(E,N) = (\sqrt{2m})^N \left(\sqrt{\frac{2}{m\omega^2}} \right)^N \frac{1}{h^N} \int_D \left(\prod_{i=1}^{2N} dY_i \right)$$

where D is the region such that

$$\sum_{i=1}^{N} \left(\frac{p_i^2}{2m} + \frac{1}{2} m\omega^2 q_i^2 \right) = \sum_{i=1}^{2N} Y_i^2 \leq E.$$

In other words, in the variables Y_i, the region D is just a $2N$ dimensional sphere with radius $\sqrt{E}$. We recall that the volume of the $2N$ dimensional sphere with radius R is

$$V_{2N} = \frac{\pi^N}{N\Gamma(N)} R^{2N}$$

and we find the following result for $\Sigma(E,N)$

$$\Sigma(E,N) = \left(\frac{2}{h\omega} \right)^N \int_D \left(\prod_{i=1}^{2N} dx_i \right) = \left(\frac{2}{h\omega} \right)^N \frac{\pi^N}{N\Gamma(N)} E^N.$$

Therefore, the entropy is given by

$$S(E,N) \approx Nk \ln E + C$$

where C is a constant independent of E. The internal energy is found through the relation

$$\frac{1}{T} = \left(\frac{\partial S}{\partial E} \right)_N = \frac{Nk}{E}$$

leading to

$$E = NkT$$

which is in agreement with the equipartition theorem assigning an energy contribution equal to $kT/2$ to each degree of freedom (see Problems 7.20, 7.21 and 7.18). The quantum mechanical analogue of this situation is discussed in Problem 6.18.

Problem 6.24.

N atoms are arranged regularly in N localized lattice sites so as to form a perfect crystal. If one moves n atoms (with the condition $1 \ll n \ll N$) from lattice sites to localized interstices of the lattice, this becomes an imperfect crystal with n defects. The number M of interstitial sites into which an atom can enter is of the same order of magnitude as N. Let ε be the energy necessary to move an atom from a lattice site to an interstitial one. Show that, in the equilibrium state at temperature T, the following relation

$$n \approx \sqrt{NM} e^{-\frac{\varepsilon}{2kT}}$$

is valid.

Solution
The number of states is

$$\Omega(n,M,N) = \frac{N!}{n!(N-n)!} \frac{M!}{n!(M-n)!}$$

where the first term accounts for all the possible ways to choose n atoms out of N available sites, and the second one for all the possible ways in which these n atoms can be arranged in the M interstices. We remark that the atoms are treated as distinguishable, because they occupy localized sites or interstices. Using the Stirling approximation and the Boltzmann formula $S = k \ln \Omega$, we find

$$S(n,M,N) \approx k[N\ln N + M\ln M - 2n\ln n - (N-n)\ln(N-n) - (M-n)\ln(M-n)]$$

and for the inverse temperature

$$\frac{1}{T} = \left(\frac{\partial S}{\partial E}\right)_{N,M} = \frac{1}{\varepsilon}\left(\frac{\partial S}{\partial n}\right)_{N,M} = \frac{k}{\varepsilon}\ln\left(\frac{(N-n)(M-n)}{n^2}\right)$$

leading to

$$\frac{(N-n)(M-n)}{n^2} = e^{\frac{\varepsilon}{kT}}$$

which is an equation for n. Instead of solving exactly this equation, we consider the condition $N,M \gg n$, so that we can approximate $(N-n)(M-n) \approx NM$ and finally obtain

$$n = \sqrt{NM} e^{-\frac{\varepsilon}{2kT}}.$$

Problem 6.25.

A statistical system is composed of two ultrarelativistic particles moving in a segment of length L. Write down the Hamiltonian for the system and compute the volume of the phase space enclosed by the surface at constant energy E.

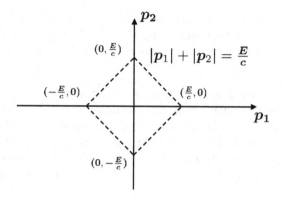

Fig. 6.6 The phase space region at constant energy for 2 ultrarelativistic particles moving in a segment of length L. Further technical details are reported in Problem 6.25

Solution

The energy for an ultrarelativistic particle is $E = |p|c$, with p the momentum and c the speed of light. In the case of two particles, the total energy is

$$H(p_1, p_2) = c(|p_1| + |p_2|).$$

The volume of the phase space enclosed by the region at constant E is

$$\Sigma(E, L) = \int_0^L dq_1 \int_0^L dq_2 \int_{H(p_1,p_2) \leq E} dp_1 dp_2 = L^2 \int_{H(p_1,p_2) \leq E} dp_1 dp_2.$$

If we set $x = cp/E$ in the Hamiltonian, we obtain

$$\Sigma(E, L) = \frac{E^2 L^2}{c^2} \int_{|x_1| + |x_2| \leq 1} dx_1 dx_2.$$

In the (x_1, x_2) plane we can identify the 4 regions (see also Fig. 6.6)

$$\begin{cases} x_1 = -x_2 + 1 & \text{1-st region} \\ x_1 = x_2 - 1 & \text{2-nd region} \\ x_1 = -x_2 - 1 & \text{3-rd region} \\ x_1 = x_2 + 1 & \text{4-th region} \end{cases}$$

that constitute a rhombic region whose total area can be easily computed as

$$\int_{|x_1| + |x_2| \leq 1} dx_1 dx_2 = 2$$

so that $\Sigma(E, L) = \frac{2E^2 L^2}{c^2}$.

Problem 6.26.

Consider a free gas with N particles and internal energy E inside a container of volume V. Starting from the Sackur-Tetrode formula for the entropy

$$S(E,V,N) = Nk \left\{ \frac{5}{2} - \ln \left[\left(\frac{3\pi\hbar^2}{m} \right)^{3/2} \frac{N^{\frac{5}{2}}}{VE^{\frac{3}{2}}} \right] \right\}$$

find the free energy F, the enthalpy H, and the Gibbs potential Φ.

Solution

To find the solution we must express the potentials in the right variables

$$F = F(T,V,N) \quad H = H(S,P,N) \quad \Phi = \Phi(P,T,N).$$

Let us start with the free energy whose variables are T,V,N

$$F(T,V,N) = E(S(T,V,N),V,N) - TS(T,V,N).$$

With this notation we have emphasized that F is a natural function of T,V,N while the entropy is a natural function of E, V, N. We must then write E, S in terms of T, V, N to proceed further. From the Sackur-Tetrode formula we get

$$E(S,V,N) = \left(\frac{3\pi\hbar^2}{m} \right) \frac{N^{\frac{5}{3}}}{V^{\frac{2}{3}}} e^{\frac{2S}{3Nk} - \frac{5}{3}}.$$

The first law of thermodynamics is $dE = TdS - PdV + \mu dN$, from which we find

$$T = \left(\frac{\partial E}{\partial S} \right)_{V,N} = \left(\frac{3\pi\hbar^2}{m} \right) \frac{N^{\frac{5}{3}}}{V^{\frac{2}{3}}} e^{\frac{2S}{3Nk} - \frac{5}{3}} \frac{2}{3Nk} = \frac{2E}{3Nk}.$$

This is the equipartition theorem which states that each degree of freedom contributes $kT/2$ to the energy. In turn, plugging this expression for E in the Sackur-Tetrode entropy will give

$$S(T,V,N) = Nk \left\{ \frac{5}{2} + \ln \left[\left(\frac{mkT}{2\pi\hbar^2} \right)^{\frac{3}{2}} \frac{V}{N} \right] \right\}.$$

Putting everything together

$$F(T,V,N) = E - TS = \frac{3NkT}{2} - \frac{5NkT}{2} - \ln \left[\left(\frac{mkT}{2\pi\hbar^2} \right)^{\frac{3}{2}} \frac{V}{N} \right] =$$

$$NkT \left\{ \ln \left[\left(\frac{2\pi\hbar^2}{mkT} \right)^{\frac{3}{2}} \frac{N}{V} \right] - 1 \right\}.$$

Let us now consider the enthalpy. Using once again the first law of thermodynamics, we find

$$-P = \left(\frac{\partial E}{\partial V}\right)_{S,N} = -\frac{2E}{3V}$$

from which $E = \frac{3}{2}PV$ and

$$H(S,P,N) = E + PV = \frac{5}{2}PV.$$

This is not yet enough, since H is a function of S, P, N. We must express V in terms of S, P, N. In the formula for the energy in terms of the entropy we previously got starting from the Sackur-Tetrode entropy, we substitute $E = 3/2PV$ and we get

$$V = \left(\frac{2\pi\hbar^2}{m}\right)^{\frac{3}{5}} \frac{N}{P^{\frac{3}{5}}} e^{\frac{2S}{5Nk}-1}.$$

Putting all of this in the formula for the enthalpy, we find

$$H(S,P,N) = \frac{5}{2}\left(\frac{2\pi\hbar^2}{m}\right)^{\frac{3}{5}} NP^{\frac{2}{5}} e^{\frac{2S}{5Nk}-1}$$

which is what we were looking for. At last let us consider the Gibbs potential, $\Phi(P,T,N)$. We use the equation of state of an ideal gas

$$-P = \left(\frac{\partial F}{\partial V}\right)_{T,N} = -\frac{NkT}{V}.$$

Then, taking the free energy and substituting $V = NkT/P$ we get

$$\Phi(P,T,N) = F(T,V(P,T,N),N) + PV(P,T,N) = NkT\left\{\ln\left[\left(\frac{2\pi\hbar^2}{mkT}\right)^{\frac{3}{2}}\frac{P}{kT}\right]\right\}.$$

7

Canonical Ensemble

Problem 7.1.

A classical gas in a volume V is composed of N independent and indistinguishable particles. The single particle Hamiltonian is $H = \frac{p^2}{2m}$, with m the mass of the particle and p the absolute value of the momentum. Moreover, for each particle, we find 2 internal energy levels: a ground state with energy 0 and degeneracy g_1, and an excited state with energy $E > 0$ and degeneracy g_2. Determine the canonical partition function and the specific heat C_V as a function of the temperature T. Analyze the limit of low temperatures and comment on the final result.

Solution

When computing the canonical partition function, we have to consider the continuous part of the Hamiltonian ($\frac{p^2}{2m}$), plus the contribution coming from the internal degrees of freedom. The partition function for the N particles is then the product of N single particle partition functions

$$Q_N(T,V,N) = \frac{V^N}{h^{3N} N!} \left(\int_{-\infty}^{+\infty} e^{-\beta \frac{p^2}{2m}} dp \right)^{3N} \left(g_1 + g_2 e^{-\beta E} \right)^N =$$

$$\frac{V^N}{h^{3N} N!} \left(\frac{2\pi m}{\beta} \right)^{\frac{3N}{2}} \left(g_1 + g_2 e^{-\beta E} \right)^N$$

due to the well known formula $\int_{-\infty}^{+\infty} e^{-ax^2} dx = \sqrt{\frac{\pi}{a}}$. The factorial $N!$ accounts for the classical indistinguishability of the particles. From the partition function we compute the energy

$$U = -\left(\frac{\partial \ln Q_N}{\partial \beta} \right)_{V,N} = \frac{3}{2} NkT + N \frac{g_2 E e^{-E/kT}}{g_1 + g_2 e^{-E/kT}}$$

Cini M., Fucito F., Sbragaglia M.: Solved Problems in Quantum and Statistical Mechanics.
DOI 10.1007/978-88-470-2315-4_7, © Springer-Verlag Italia 2012

and the specific heat

$$C_V = \left(\frac{\partial U}{\partial T}\right)_{V,N} = \frac{3}{2}Nk + \frac{d}{dT}\left(\frac{Ng_2Ee^{-E/kT}}{g_1+g_2e^{-E/kT}}\right) = \frac{3}{2}Nk + \frac{Ng_1g_2E^2e^{E/kT}}{kT^2(g_2+g_1e^{E/kT})^2}.$$

We note that the temperature appears only in the contribution of the internal degrees of freedom. For low temperatures, we find

$$\frac{g_1g_2E^2e^{E/kT}}{kT^2(g_2+g_1e^{E/kT})^2} \approx \frac{g_1g_2E^2}{kT^2g_1^2e^{E/kT}} \to 0$$

that is the expected result because in such limit only the ground state (the one with 0 energy) is populated.

Problem 7.2.

The Hamiltonian matrix for a quantum system can be written as

$$\hat{H} = -g\frac{B}{\sqrt{2}}\begin{pmatrix} 0 & 1 & 0 \\ 1 & 0 & 1 \\ 0 & 1 & 0 \end{pmatrix} \qquad g > 0.$$

Discuss a possible physical meaning of the Hamiltonian and compute the canonical partition function and the average energy as a function of the temperature.

Solution

We note that the Hamiltonian can be written as $\hat{H} = -gB\hat{L}_x$ with $\hat{L}_x$ the x component of the orbital angular momentum in three dimensions

$$\hat{L}_x = \frac{\hbar}{\sqrt{2}}\begin{pmatrix} 0 & 1 & 0 \\ 1 & 0 & 1 \\ 0 & 1 & 0 \end{pmatrix}.$$

We therefore interpret the Hamiltonian as that of a particle with spin 1 placed in a magnetic field with strength B and directed along the x axis. The computation of the canonical partition function directly follows from the eigenvalues of the Hamiltonian

$$E_1 = -gB \qquad E_2 = 0 \qquad E_3 = gB.$$

The resulting canonical partition function is

$$Q(\beta) = \sum_{i=1}^{3} e^{-\beta E_i} = e^{\beta gB} + 1 + e^{-\beta gB}$$

from which we extract the average energy with a suitable derivative

$$U = -\frac{d\ln Q(\beta)}{d\beta} = -\frac{gB(e^{\beta gB} - e^{-\beta gB})}{e^{\beta gB} + 1 + e^{-\beta gB}}.$$

Problem 7.3.
Let us consider N indistinguishable non interacting particles placed in a segment of length a. Find the corresponding quantum mechanical partition function and determine the free energy and the internal energy in the limit of low temperatures. Finally, assuming that the energy levels are infinitesimally spaced, give an estimate for the free energy and the internal energy as functions of the temperature. For simplicity, when computing the degeneracy factor for a given configuration in the energy space, use the classical treatment for indistinguishable particles.

Solution
We can write the partition function for the N particles as

$$Q_N(T,a,N) = \sum_E e^{-\frac{E}{kT}}$$

where E are the eigenvalues of the Hamiltonian of the system. We can explicitly write down the following conditions for the total energy E and the number N

$$E = \sum_\varepsilon n_\varepsilon \varepsilon \qquad N = \sum_\varepsilon n_\varepsilon$$

where n_ε is the number of particles with eigenvalue ε. Therefore, we can write the partition function in the following way

$$Q_N(T,a,N) = \tilde{\sum}_{\{n_\varepsilon\}} g\{n_\varepsilon\} e^{-\frac{1}{kT}\sum_\varepsilon n_\varepsilon \varepsilon}$$

with $g\{n_\varepsilon\}$ the number of possible ways to select a set $\{n_\varepsilon\}$ out of N objects, and where $\tilde{\sum}_{\{n_\varepsilon\}}$ is meant to be the sum over all the possible configurations satisfying $N = \sum_\varepsilon n_\varepsilon$. The degeneracy factor $g\{n_\varepsilon\}$ depends on the statistics and, in the approximation requested by the text, is equal to

$$g\{n_\varepsilon\} = \frac{N!}{(n_1!n_2!n_3!\cdots)}.$$

We further have to divide by $N!$ in order to take into account the indistinguishability of the particles, i.e. we have to use the Gibbs factor. Therefore, the degeneracy coefficient is $g\{n_\varepsilon\} = \prod_\varepsilon \frac{1}{n_\varepsilon!}$. The partition function is

$$Q_N(T,a,N) = \tilde{\sum}_{\{n_\varepsilon\}} \prod_\varepsilon \frac{1}{n_\varepsilon!}\left(e^{-\frac{\varepsilon}{kT}}\right)^{n_\varepsilon} = \frac{1}{N!}\tilde{\sum}_{\{n_\varepsilon\}} N!\left(\prod_\varepsilon \frac{1}{n_\varepsilon!}\left(e^{-\frac{\varepsilon}{kT}}\right)^{n_\varepsilon}\right) =$$
$$\frac{1}{N!}\left(\sum_\varepsilon e^{-\frac{\varepsilon}{kT}}\right)^N = \frac{1}{N!}Q_1^N(T,a)$$

where we have used the generalization of the binomial formula, and where we have denoted with $Q_1(T,a)$ the partition function of the single particle. To clarify the above steps in the calculation, set $x_\varepsilon = e^{-\varepsilon/kT}$ and consider the simple case where

$\varepsilon = 1, 2$ (*N* particles with 2 energy levels). The partition function is

$$Q_N(T, a, N) = \sum_{k_1=0}^{N} \frac{1}{k_1!(N-k_1)!} x_1^{k_1} x_2^{N-k_1}$$

where we recognize, on the right hand side, the binomial formula

$$(x_1 + x_2)^N = \sum_{k_1=0}^{N} \binom{N}{k_1} x_1^{k_1} x_2^{N-k_1} = \sum_{k_1=0}^{N} \frac{N!}{k_1!(N-k_1)!} x_1^{k_1} x_2^{N-k_1} = \tilde{\sum}_{\{k_i\}} N! \left(\prod_i \frac{1}{k_i!} x_i^{k_i} \right)$$

with $\tilde{\sum}_{\{k_i\}}$ the sum over all configurations such that $k_1 + k_2 = N$. Therefore, we have

$$Q_N(T, a, N) = \frac{(x_1 + x_2)^N}{N!} = \frac{Q_1^N(T, a)}{N!}.$$

Going back to the most general case, let us concentrate on the energy levels of the single particle to compute the related partition function. The eigenvalues of the Hamiltonian for a free particle in a segment a directly come from the solution of the stationary Schrödinger equation (see also Problem 2.10 in the section of Quantum Mechanics) for the eigenstate $\psi_n(x)$

$$-\frac{\hbar^2}{2m} \frac{d^2 \psi_n(x)}{dx^2} = \varepsilon_n \psi_n(x).$$

The requirement that $\psi_n(x)$ is zero at $x = 0$ and $x = a$ leads to the following normalized eigenstates

$$\psi_n(x) = \sqrt{\frac{2}{a}} \sin\left(\frac{n\pi x}{a}\right) \quad n = 1, 2, 3, \ldots$$

and gives rise to the discrete energy spectrum

$$\varepsilon_n = \frac{\pi^2 \hbar^2 n^2}{2ma^2} = \alpha n^2 \quad \alpha = \frac{\pi^2 \hbar^2}{2ma^2} \quad n = 1, 2, 3, \ldots$$

The resulting partition function of the single particle for low temperatures may be written as

$$Q_1(T, a) = \sum_{n=1}^{+\infty} e^{-\frac{\alpha n^2}{kT}} = e^{-\frac{\alpha}{kT}} + (e^{-\frac{\alpha}{kT}})^4 + \cdots \approx e^{-\frac{\alpha}{kT}}$$

from which we compute the free energy

$$F = -kT \ln Q_N = -kT \ln\left(\frac{Q_1^N}{N!}\right) \approx kTN \ln N + \frac{N\pi^2 \hbar^2}{2ma^2}$$

and the internal energy

$$U = F + TS = F - T\left(\frac{\partial F}{\partial T}\right)_{a,N} = -\left(\frac{\partial \ln Q_N}{\partial \beta}\right)_{a,N} \approx \frac{N\pi^2 \hbar^2}{2ma^2}.$$

When the energy levels are infinitesimally spaced, we can approximate the sum with an integral as

$$Q_1(T,a) = \sum_{n=1}^{+\infty} e^{-\frac{an^2}{kT}} = \sum_{n=0}^{+\infty} e^{-\frac{an^2}{kT}} - 1 \approx \int_0^{+\infty} e^{-\frac{ax^2}{kT}} dx - 1 =$$

$$\sqrt{\frac{kT}{\alpha}} \int_0^{+\infty} e^{-y^2} dy - 1 = \sqrt{\frac{ma^2 kT}{2\pi\hbar^2}} - 1$$

from which we compute the free energy

$$F = kT \ln N! - NkT \ln \left(\sqrt{\frac{ma^2 kT}{2\pi\hbar^2}} - 1 \right)$$

and the internal energy

$$U = - \left(\frac{\partial \ln Q_N}{\partial \beta} \right)_{a,N} = \frac{1}{2} \frac{NkT}{1 - \sqrt{\frac{2\pi\hbar^2}{ma^2 kT}}}.$$

Problem 7.4.
A physical system is composed of N distinguishable spins assuming two possible values ± 1. These two values correspond to the energy levels $\pm\varepsilon$, respectively. Compute the total energy E using the Boltzmann formula and the microcanonical ensemble. Finally, compare the results with those in the canonical ensemble.

Solution
Let us call $N_\pm$ the number of particles with spin orientation ± 1. Then, we have to consider the equations determining the total number of particles and the total energy for the system

$$\begin{cases} N = N_+ + N_- \\ E = (N_+ - N_-)\varepsilon \end{cases}$$

from which we extract $N_\pm$ as

$$\begin{cases} N_+ = \frac{1}{2} \left(N + \frac{E}{\varepsilon} \right) \\ N_- = \frac{1}{2} \left(N - \frac{E}{\varepsilon} \right). \end{cases}$$

For a given E and N, the number of states is therefore given by all the possible ways to extract N_+ (or N_-) objects out of N, with the constraint that $N = N_+ + N_-$. This number is

$$\Omega(E,N) = \frac{N!}{N_+!(N - N_+)!} = \frac{N!}{N_+!N_-!} = \frac{N!}{\left[\frac{1}{2}\left(N + \frac{E}{\varepsilon}\right)\right]! \left[\frac{1}{2}\left(N - \frac{E}{\varepsilon}\right)\right]!}$$

from which, with $N \gg 1$ and the use of the Stirling approximation, we obtain

$$S(E,N) = k\ln\Omega(E,N) \approx k\left[N\ln N - \frac{1}{2}\left(N+\frac{E}{\varepsilon}\right)\ln\left(N+\frac{E}{\varepsilon}\right) - \frac{1}{2}\left(N-\frac{E}{\varepsilon}\right)\ln\left(N-\frac{E}{\varepsilon}\right) + N\ln 2\right]$$

and

$$\frac{1}{T} = \left(\frac{\partial S}{\partial E}\right)_N = -\frac{k}{2\varepsilon}\ln\left(N+\frac{E}{\varepsilon}\right) + \frac{k}{2\varepsilon}\ln\left(N-\frac{E}{\varepsilon}\right)$$

leading to the internal energy

$$E = -N\varepsilon\tanh(\beta\varepsilon).$$

Let us repeat the same calculation in the canonical ensemble. The partition function for the single spin is obtained by summing $e^{-\beta H}$ over the two possible energy states ($H = \pm\varepsilon$)

$$Q_1(T) = \sum_{m=\pm 1} e^{-m\beta\varepsilon} = e^{\beta\varepsilon} + e^{-\beta\varepsilon} = 2\cosh(\beta\varepsilon)$$

and the total (N particles) partition function is given by

$$Q_N(T,N) = Q_1^N(T)$$

from which we determine the free energy, the entropy and the average internal energy as

$$F = -\frac{1}{\beta}\ln Q_N = -NkT\ln(2\cosh(\beta\varepsilon))$$

$$S = -\left(\frac{\partial F}{\partial T}\right)_N = Nk(\ln(2\cosh(\beta\varepsilon)) - \beta\varepsilon\tanh(\beta\varepsilon))$$

$$U = F + TS = -\left(\frac{\partial \ln Q_N}{\partial \beta}\right)_N = -N\varepsilon\tanh(\beta\varepsilon).$$

The result is basically the same as the one obtained with the microcanonical ensemble, with the exception that the energy E is replaced by the average energy $U = \langle E \rangle$.

Problem 7.5.
A point with mass m moves along the x axis under the effect of a conservative force with the following potential

$$V(q) = \begin{cases} \frac{1}{2}m\omega^2(q+a)^2 & q \le -a \\ 0 & -a < q < a \\ \frac{1}{2}m\omega^2(q-a)^2 & q \ge a \end{cases}$$

with a a positive constant. When the system is in contact with a reservoir at temperature T, find the canonical partition function, the average energy and the specific heat. Finally, comment on the limit $a \to 0$.

Solution
We compute the canonical partition function

$$Q_1(\beta) = \frac{1}{h} \int_{-\infty}^{+\infty} e^{-\beta \frac{p^2}{2m}} dp \int_{-\infty}^{+\infty} e^{-\beta V(q)} dq$$

where the integral in q can be done by dividing in three different sub intervals $([-\infty, -a], [-a, +a]$ and $[a, +\infty])$

$$Q_1(\beta) = \frac{1}{h} \int_{-\infty}^{+\infty} e^{-\beta \frac{p^2}{2m}} dp \left[\int_{-\infty}^{-a} e^{-\beta \frac{m\omega^2}{2}(q+a)^2} dq + \int_{-a}^{+a} dq + \int_{a}^{+\infty} e^{-\beta \frac{m\omega^2}{2}(q-a)^2} dq \right] .$$

The integral over momenta is immediate. The one over the space coordinates can be further simplified by setting $y = q \pm a$

$$Q_1(\beta) = \frac{1}{h} \left(\frac{2\pi m}{\beta} \right)^{1/2} \left[2 \int_0^{+\infty} e^{-\beta \frac{m\omega^2}{2} y^2} dy + 2a \right] = \frac{2\pi}{h\beta\omega} + \frac{2a}{h} \left(\frac{2\pi m}{\beta} \right)^{1/2}$$

where we have used the well known Gaussian integral $\int_{-\infty}^{+\infty} e^{-aq^2} dq = \sqrt{\frac{\pi}{a}}$. The average energy is

$$U = -\frac{d \ln Q_1(\beta)}{d\beta} = \frac{1}{2} kT + \frac{\pi(kT)^2}{2\pi kT + 2a\omega(2\pi kT m)^{1/2}}$$

which can be rearranged as

$$U = \frac{kT}{2} + \frac{AT^2}{BT + DT^{1/2}}$$

with $A = \pi k^2$, $B = 2\pi k$ and $D = 2a\omega(2\pi km)^{1/2}$. The specific heat is

$$C = \frac{dU}{dT} = \frac{k}{2} + \frac{ABT^2 + \frac{3}{2}ADT^{3/2}}{(BT + DT^{1/2})^2}.$$

We note that, in the limit $a \to 0$, the potential $V(q)$ becomes exactly quadratic and the Hamiltonian corresponds to a one dimensional harmonic oscillator. This can be explicitly verified in the above results, because in that limit we find

$$D \to 0 \quad A/B = k/2$$

so that

$$U \to kT \quad C \to k$$

which are the internal energy and the specific heat for a classical one dimensional harmonic oscillator.

Problem 7.6.

Let us consider a statistical system with N states, with energies $\varepsilon_n = n\varepsilon$, $n = 0, \ldots, N-1$. The system is in contact with a reservoir at temperature T. Determine the probability that the system is in the state with energy ε_n and verify the final result.

Solution

The canonical partition function is written as

$$Q_N(T,N) = \sum_{n=0}^{N-1} e^{-n\beta\varepsilon} = \sum_{n=0}^{N-1} x^n = \sum_{n=0}^{+\infty} x^n - \sum_{n=N}^{+\infty} x^n =$$

$$\frac{1}{1-x} - \sum_{m=0}^{+\infty} x^{m+N} = \frac{1}{1-x} - \frac{x^N}{1-x} = \frac{1-x^N}{1-x} = \frac{1-e^{-N\beta\varepsilon}}{1-e^{-\beta\varepsilon}}$$

where we have set $x = e^{-\beta\varepsilon}$. The probability associated with an energy level is

$$P(\varepsilon_n) = \frac{e^{-\beta\varepsilon_n}}{Q_N(T,N)} = \frac{e^{-n\beta\varepsilon}(1-e^{-\beta\varepsilon})}{1-e^{-N\beta\varepsilon}}.$$

It is simple to verify that

$$\sum_{n=0}^{N-1} P(\varepsilon_n) = 1$$

which proves the validity of final result, i.e. the fact that the probability is properly normalized.

Problem 7.7.

N independent and distinguishable particles move in a one dimensional segment between $q = 0$ and $q = L$. Determine the equation of state of the system, given the following single particle Hamiltonian

$$H = \frac{p^2}{2m} - \alpha \ln\left(\frac{q}{L_0}\right) \qquad \alpha > 0.$$

In the above expression, α is a constant giving the strength of the potential $V(q) = -\alpha \ln\left(\frac{q}{L_0}\right)$ and L_0 is a characteristic length scale. Determine the pressure for very low temperatures and comment on the limit $\alpha \to 0$.

Solution

We determine the single particle partition function by integrating in the phase space

$$Q_1(T,L) = \frac{1}{h} \int_{-\infty}^{+\infty} e^{-\beta\frac{p^2}{2m}} dp \int_0^L e^{\alpha\beta\ln(\frac{q}{L_0})} dq = \frac{1}{h}\sqrt{\frac{2m\pi}{\beta}} \frac{L^{\alpha\beta+1}}{\alpha\beta+1} \frac{1}{L_0^{\alpha\beta}}.$$

Due to independence, the N particles partition function is the product of N single particle partition functions, i.e. $Q_N(T,L,N) = Q_1^N(T,L)$. The pressure can be computed from the derivative of the free energy with respect to the 'volume', i.e. L in our case. Since the free energy is $F = -kT \ln Q_N$, the pressure follows

$$P = -\left(\frac{\partial F}{\partial L}\right)_{T,N} = KT \left(\frac{\partial \ln Q}{\partial L}\right)_{T,N} = \frac{NKT}{L}\left(1 + \frac{\alpha}{KT}\right).$$

In the limit of low temperatures, we find a pressure contribution different from zero

$$P = \frac{N\alpha}{L}.$$

This is possible because, even without thermal fluctuations, we have a non zero energy contribution coming from the potential $V(q)$ in the Hamiltonian. The pressure becomes zero when $\alpha \to 0$ and such potential contribution disappears.

Problem 7.8.
A physical system is characterized by two energy levels: the first one has energy E_1 and degeneracy g_1, while the second one has energy E_2 and degeneracy g_2. Prove that the entropy S can be written as

$$S = -k\left[p_1 \ln\left(\frac{p_1}{g_1}\right) + p_2 \ln\left(\frac{p_2}{g_2}\right)\right]$$

where p_i with $i = 1,2$ is the probability that the system is found in the i-th energy level. Use the connection between the entropy S and the average energy U and free energy F. Remember that both probabilities have to satisfy $p_1 + p_2 = 1$.

Solution
From the definition of free energy we know that

$$\frac{S}{k} = \beta(U - F)$$

where U is the average internal energy and F the free energy. For the average internal energy we can use

$$U = p_1 E_1 + p_2 E_2$$

while, for the free energy, we know that $-\beta F = \ln Q$, where Q is the partition function of the system. Therefore, we have

$$\frac{S}{k} = \beta p_1 E_1 + \beta p_2 E_2 + \ln Q.$$

We also note that the probability associated with the i-th level is $p_i = \frac{g_i e^{-\beta E_i}}{Q}$, so that

$$\beta E_i + \ln Q = -\ln\left(\frac{p_i}{g_i}\right).$$

Plugging this result in the equation for S, and recalling that $p_1 + p_2 = 1$, we obtain

$$\frac{S}{k} = \beta p_1 E_1 + \beta p_2 E_2 + \ln Q =$$

$$- (p_1 + p_2) \ln Q - p_1 \ln \left(\frac{p_1}{g_1} \right) - p_2 \ln \left(\frac{p_2}{g_2} \right) + \ln Q =$$

$$- p_1 \ln \left(\frac{p_1}{g_1} \right) - p_2 \ln \left(\frac{p_2}{g_2} \right)$$

that is the desired result.

Problem 7.9.
The canonical partition function for a fluid, whose molecules possess two characteristic frequencies ω and Ω, is given by

$$Q(\beta) = \frac{1}{(1 - e^{-\beta \hbar \omega})(1 - e^{-\beta \hbar \Omega})} \quad .$$

Find the average internal energy U, the entropy S, and the specific heat C. Finally, determine the low temperature limit of the entropy.

Solution
The internal energy, the entropy and the specific heat can be written as suitable derivatives of the canonical partition function $Q(\beta)$. The internal energy is

$$U = -\frac{d \ln Q(\beta)}{d\beta} = \frac{\hbar \omega}{e^{\beta \hbar \omega} - 1} + \frac{\hbar \Omega}{e^{\beta \hbar \Omega} - 1}.$$

The entropy is given by

$$S = \frac{U}{T} + k \ln Q = \frac{\hbar}{T} \left(\frac{\omega}{e^{\beta \hbar \omega} - 1} + \frac{\Omega}{e^{\beta \hbar \Omega} - 1} \right)$$
$$- k \left[\ln(1 - e^{-\beta \hbar \omega}) + \ln(1 - e^{-\beta \hbar \Omega}) \right].$$

In the limit of low temperatures (large β) the exponential functions are $\ll 1$ and the entropy goes to zero exponentially. Moreover, using the identity $\frac{d}{dT} = -k\beta^2 \frac{d}{d\beta}$, we can determine the specific heat

$$C = \frac{dU}{dT} = k\beta^2 \left[\frac{\hbar^2 \omega^2 e^{\beta \hbar \omega}}{(e^{\beta \hbar \omega} - 1)^2} + \frac{\hbar^2 \Omega^2 e^{\beta \hbar \Omega}}{(e^{\beta \hbar \Omega} - 1)^2} \right].$$

Problem 7.10.
Consider a system composed of 2 filaments (a, b) intersecting one with each other so as to form a double helix. On these filaments there are N sites forming energy bonds in such a way that the i-th site of the filament a can link only with the i-th site of the filament b (see Fig. 7.1). For an open bond, the energy of the system increases by a

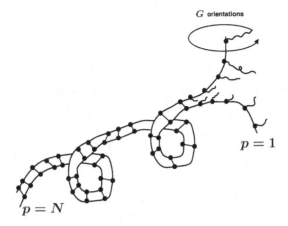

Fig. 7.1 We show a system composed of 2 filaments (a,b) intersecting one with each other so as to form a double helix. Each filament possesses N sites forming energy bonds with the other filament's sites. When a bond connecting the different sites is open, it releases an energy $\varepsilon > 0$. In Problem 7.10 we study the statistical mechanics of such system in the framework of the canonical ensemble

factor $\varepsilon > 0$. The system presents various configurations, including the configuration with all closed bonds, or configurations with open bonds from site 1 up to site p ($p < N$) and all the others closed. The last bond ($p = N$) cannot be opened. Moreover, for each site, there is a degeneracy $G > 1$ due to a rotational freedom around the site itself. Write down the canonical partition function and determine the average number of open bonds $\langle p \rangle$. Is there any critical temperature T_c above which the canonical ensemble is meaningless for large N?

Finally, after defining the variable $x = G^2 e^{-\beta\varepsilon}$ and setting $x = 1 + \eta$, find the behaviour of $\langle p \rangle$ for small η and compute the linear response function

$$\left(\frac{\partial \langle p \rangle}{\partial \eta}\right)_N$$

when $\eta \to 0$. For large N, determine the power law dependence on N of the linear response function.

Solution
Let us start by determining the partition function. Let us call p the number of open bonds. From the text we know that $0 \le p \le N - 1$. The partition function is

$$Q_N(T,N) = \sum_{p=0}^{N-1} G^{2p} e^{-\beta p\varepsilon} = \sum_{p=0}^{N-1} x^p = \frac{1 - x^N}{1 - x} \qquad x = G^2 e^{-\beta\varepsilon}$$

where we have assumed $x \neq 1$. For the convergence of the previous series at large N, it is important to have

$$G^2 e^{-\beta \varepsilon} < 1$$

from which

$$T < T_c = \frac{\varepsilon}{2k \ln G}$$

where T_c is a critical temperature above which the canonical ensemble does not give a finite partition function for large N. The average of p is evaluated as

$$\langle p \rangle = \frac{\sum_{p=0}^{N-1} p x^p}{\sum_{p=0}^{N-1} x^p} = x \frac{d}{dx} \ln \left(\sum_{p=0}^{N-1} x^p \right) = \frac{N x^N}{x^N - 1} - \frac{x}{x - 1}.$$

If we set $x = 1 + \eta$, we find

$$\frac{N x^N}{x^N - 1} - \frac{x}{x - 1} = \frac{N(1 + \eta)^N}{(1 + \eta)^N - 1} - \frac{1 + \eta}{\eta} \approx$$

$$\frac{N + N^2 \eta + \frac{N^2(N-1)}{2} \eta^2 + \frac{N^2(N-1)(N-2)}{6} \eta^3 \cdots}{N \eta + \frac{N(N-1)}{2} \eta^2 + \frac{N(N-1)(N-2)}{6} \eta^3 \cdots} - \frac{1 + \eta}{\eta}$$

that implies

$$\langle p \rangle \approx \frac{N - 1}{2} + \frac{N^2 - 1}{12} \eta + \mathcal{O}(\eta^2).$$

The requested linear response, when $\eta \to 0$, is

$$\left(\frac{\partial \langle p \rangle}{\partial \eta} \right)_N = \frac{N^2 - 1}{12}$$

which grows up with the second power of N for $N \gg 1$.

Problem 7.11.

Let us take the classical Hamiltonian

$$H = vp + \left(\frac{A}{p} \right)^2 + B^2 p^4 (q - vt)^2$$

where A, B, v are constants. Moving to a quantum mechanical description, determine the eigenvalues for the energy and the canonical partition function. To solve this problem, we suggest to simplify the form of the Hamiltonian with a canonical transformation whose generating function is given by

$$F(q, Q, t) = \lambda \frac{q - vt}{Q}$$

with the choice $\lambda = \frac{1}{B\sqrt{2M}}$, where M has the physical dimensions of a mass.

Solution
The canonical transformation generated by $F(q,Q,t)$ produces a change in the co-ordinates from (p,q,t) to (P,Q,t) such that

$$p = \left(\frac{\partial F}{\partial q} \right)_{Q,t} \qquad P = -\left(\frac{\partial F}{\partial Q} \right)_{q,t}.$$

The new Hamiltonian is given by

$$\tilde{H} = H + \left(\frac{\partial F}{\partial t} \right)_{q,Q}.$$

Since $p = \left(\frac{\partial F}{\partial q} \right)_{Q,t} = \frac{\lambda}{Q}$, $P = -\left(\frac{\partial F}{\partial Q} \right)_{q,t} = \lambda \frac{q-vt}{Q^2}$, $\left(\frac{\partial F}{\partial t} \right)_{q,Q} = -\frac{\lambda v}{Q}$, we find that

$$p = \frac{\lambda}{Q} \qquad q - vt = \frac{PQ^2}{\lambda}$$

and

$$\tilde{H} = H + \left(\frac{\partial F}{\partial t} \right)_{q,Q} = \frac{A^2 Q^2}{\lambda^2} + \lambda^2 B^2 P^2.$$

Finally, using the suggested value of λ, we find

$$\tilde{H} = \frac{P^2}{2M} + 2MB^2 A^2 Q^2 = \frac{P^2}{2M} + \frac{1}{2} M \omega^2 Q^2$$

that is not time dependent and corresponds to a one dimensional harmonic oscillator with angular frequency $\omega = 2AB$ and mass M. The eigenvalues of the Hamiltonian are

$$E_n = \left(n + \frac{1}{2} \right) \hbar \omega \qquad n = 0,1,2,\ldots$$

and the canonical partition function is

$$Q(T) = \sum_{n=0}^{+\infty} e^{-\beta \hbar \omega (n+\frac{1}{2})} = \frac{e^{-\beta \hbar \omega/2}}{1 - e^{-\beta \hbar \omega}} = \frac{1}{2 \sinh \left(\frac{\beta \hbar \omega}{2} \right)}.$$

Problem 7.12.
A cylinder with radius a (see Fig. 7.2) is positively charged, due to the fact that it re-leases negative charges in a larger cylinder with radius R. Thermal fluctuations tend to move away the negative charges, while the electrostatic attraction makes them move back. If the number of negative charges is N and if we neglect the repulsion between them, the Hamiltonian for the N negative charges (in suitable units) is

$$H = \sum_{i=1}^{N} \left[\frac{p_i^2}{2m} + 2e^2 n \ln \left(\frac{r_i}{L} \right) \right]$$

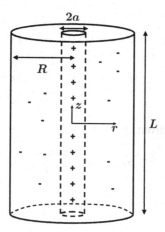

Fig. 7.2 A cylinder with radius a is positively charged, due to the fact that it releases negative charges in a larger cylinder with radius R. Thermal fluctuations tend to move away the negative charges, while the electrostatic attraction makes them move back. Thermodynamic equilibrium is described in the framework of the canonical ensemble in Problem 7.12

where r_i is the radial distance ($a \le r_i \le R$) of the i-th charge, e the absolute value of the charge, L the height of the cylinders and $n = \frac{N}{L}$.

Compute the canonical partition function. Also, determine the probability density function $p(r)$ associated with the radial position of a single charge. Finally, determine the average radial position and, more generally, the momentum of order ℓ, $\langle r^\ell \rangle$.

Solution
The canonical partition function is

$$
Q_N(T, L, N) = \frac{1}{h^{3N} N!} \int \left(\prod_{i=1}^{N} d^3 p_i d^3 q_i \right) \exp \left[-\beta \sum_{i=1}^{N} \left(\frac{p_i^2}{2m} + 2e^2 n \ln \left(\frac{r_i}{L} \right) \right) \right] \approx
$$

$$
\left(\frac{2\pi L}{N \lambda^3} \right)^N L^{2\beta e^2 n N} \left[\int_a^R r^{1 - 2e^2 n / kT} \, dr \right]^N =
$$

$$
\left(\frac{2\pi L}{N \lambda^3} \right)^N L^{2\beta e^2 n N} \left[\frac{R^{2(1 - e^2 n / kT)} - a^{2(1 - e^2 n / kT)}}{2(1 - e^2 n / kT)} \right]^N
$$

where

$$
\lambda = \frac{h}{\sqrt{2\pi m k T}}
$$

is the thermal length scale and where we have used the Stirling approximation for the factorials, i.e. $N! \approx N^N$. The probability density function associated with the radial position is

$$p(r) = \frac{re^{-(2e^2n/kT)\ln(r/L)}}{\int_a^R re^{-(2e^2n/kT)\ln(r/L)}dr} =$$

$$2\left(1 - \frac{e^2n}{kT}\right)\frac{r^{1-2e^2n/kT}}{R^{2(1-e^2n/kT)} - a^{2(1-e^2n/kT)}} = Ar^{1-2e^2n/kT}$$

where we have used the constant A defined as

$$A = 2\left(1 - \frac{e^2n}{kT}\right)\frac{1}{R^{2(1-e^2n/kT)} - a^{2(1-e^2n/kT)}}.$$

The average position becomes

$$\langle r \rangle = \int_a^R r\, p(r)\, dr = \left(\frac{AkT}{3kT - 2e^2n}\right)\left(R^{3-2e^2n/kT} - a^{3-2e^2n/kT}\right)$$

and the momentum of order ℓ

$$\langle r^\ell \rangle = \int_a^R r^\ell\, p(r)\, dr = \left(\frac{AkT}{(2+\ell)kT - 2e^2n}\right)\left(R^{2+\ell-2e^2n/kT} - a^{2+\ell-2e^2n/kT}\right).$$

Problem 7.13.

Let us consider a one dimensional chain composed of N rings. Each ring possesses r different configurations of energy E_i and length x_i ($i = 1, 2, ..., r$). Moreover, the chain is subject to a force $F > 0$ at its ends, in such a way that the energy for the single ring in the i-th configuration is

$$H_i = E_i - Fx_i.$$

The whole system is in thermal equilibrium at constant temperature T. Write down the partition function and the average length $\langle L \rangle$ of the chain. Specialize to the case with $r = 2$ and, assuming that $E_2 > E_1$ and $x_2 > x_1$, write down the average length for $F = 0$. When $F = 0$, analyze the limit of low and high temperatures and give an estimate of the characteristic temperature that separates the two regimes. Finally, compute the linear response

$$\chi = \left(\frac{\partial \langle L \rangle}{\partial F}\right)_{T,N}$$

showing that $\lim_{F \to 0} \chi > 0$. Comment on the physical meaning of these results.

Solution

From the energy of the single ring in the i-th configuration

$$H_i = E_i - Fx_i$$

we can write down the partition function for the system

$$Q_N(T,F,N) = \left[\sum_{i=1}^{r} e^{-\beta(E_i - Fx_i)} \right]^N$$

where we explicitly keep the dependence on F to be used later in the differentiation of the partition function. As for the average length, we can write it as a suitable derivative of the partition function

$$\langle L \rangle = \frac{1}{\beta} \left(\frac{\partial \ln Q_N}{\partial F} \right)_{T,N}.$$

For the case $r = 2$ we find

$$Q_N(T,F,N) = \left[e^{-\beta(E_1 - Fx_1)} + e^{-\beta(E_2 - Fx_2)} \right]^N$$

$$\langle L \rangle = \frac{1}{\beta} \left(\frac{\partial \ln Q_N}{\partial F} \right)_{T,N} = N \frac{x_1 e^{-\beta(E_1 - Fx_1)} + x_2 e^{-\beta(E_2 - Fx_2)}}{e^{-\beta(E_1 - Fx_1)} + e^{-\beta(E_2 - Fx_2)}}.$$

The limit $F \to 0$ is evaluated as

$$\langle L \rangle = N \frac{x_1 + x_2 e^{-\beta(E_2 - E_1)}}{1 + e^{-\beta(E_2 - E_1)}}.$$

If $\delta = E_2 - E_1 \ll kT$, we find

$$\langle L \rangle = N \frac{(x_1 + x_2)}{2}$$

while, for $\delta \gg kT$, we get

$$\langle L \rangle = N(x_1 + x_2 e^{-\beta \delta})$$

meaning that the transition between the high and low temperature regimes takes place when $\delta \approx kT$. In order to compute the linear response of the system, we need to expand $\langle L \rangle$ in series of F. Neglecting the second order contributions, we find that

$$\langle L \rangle = N \frac{x_1 e^{-\beta(E_1 - Fx_1)} + x_2 e^{-\beta(E_2 - Fx_2)}}{e^{-\beta(E_1 - Fx_1)} + e^{-\beta(E_2 - Fx_2)}} =$$

$$N \frac{x_1 e^{-\beta E_1}(1 + \beta Fx_1) + x_2 e^{-\beta E_2}(1 + \beta Fx_2)}{e^{-\beta E_1}(1 + \beta Fx_1) + e^{-\beta E_2}(1 + \beta Fx_2)} + \mathcal{O}(F^2) =$$

$$N \frac{x_1 (1 + \beta Fx_1) + x_2 e^{-\beta \delta}(1 + \beta Fx_2)}{(1 + \beta Fx_1) + e^{-\beta \delta}(1 + \beta Fx_2)} + \mathcal{O}(F^2) =$$

$$N \frac{(x_1 + x_2 e^{-\beta \delta}) + \beta F(x_1^2 + x_2^2 e^{-\beta \delta})}{(1 + e^{-\beta \delta})\left[1 + \frac{\beta F(x_1 + x_2 e^{-\beta \delta})}{(1 + e^{-\beta \delta})}\right]} + \mathcal{O}(F^2) =$$

$$N \frac{(x_1 + x_2 e^{-\beta \delta}) + \beta F(x_1^2 + x_2^2 e^{-\beta \delta})}{(1 + e^{-\beta \delta})}\left[1 - \frac{\beta F(x_1 + x_2 e^{-\beta \delta})}{(1 + e^{-\beta \delta})}\right] + \mathcal{O}(F^2) =$$

$$A + \beta FN \frac{(x_1 - x_2)^2 e^{-\beta \delta}}{(1 + e^{-\beta \delta})^2} + \mathcal{O}(F^2)$$

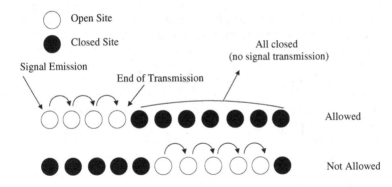

Fig. 7.3 We show a one dimensional array of lattice sites with two possible states: open and closed sites. A signal is produced by open sites and can propagate from left to right through consecutive open sites. In Problem 7.14 we characterize the canonical partition function of such system

where in the positive constant A we have collected all the terms independent of F. It follows that

$$\chi = \left(\frac{\partial \langle L \rangle}{\partial F}\right)_{T,N} = \beta N \frac{(x_1 - x_2)^2 e^{-\beta \delta}}{(1 + e^{-\beta \delta})^2} + \mathscr{O}(F)$$

$$\lim_{F \to 0} \chi = \beta N \frac{(x_1 - x_2)^2 e^{-\beta \delta}}{(1 + e^{-\beta \delta})^2} > 0$$

meaning that an increase of the force produces an elongation of the chain, that is the analogue of the thermodynamic relation $-\left(\frac{\partial V}{\partial P}\right)_{T,N} > 0$ valid for a gas.

Problem 7.14.

A one dimensional array of N lattice sites is in thermal equilibrium at temperature T. These sites can be closed (with energy 0) or open (with energy ε). A signal is produced from open sites and can travel from left to right. The signal can be transmitted from an open site, only if its nearest neighbor is open (see Fig. 7.3). The first site of the array always produces a signal that can travel up to a given point of the lattice. From that point on, there cannot be any other production of signals. Find the partition function of the system and the average number of open sites $\langle n \rangle$. At low temperatures, show that this quantity is independent of N. Finally, compute the fluctuation of the number of open sites $\langle (\Delta n)^2 \rangle = \langle n^2 \rangle - \langle n \rangle^2$. When counting all states, explicitly consider the configuration with all closed sites.

Solution
From the information given in the text, we know that there are only N available configurations with length $n = 0, 1, 2, 3, ..., N$. They are all the subarrays of open sites that allow for the transmission of the signal from the beginning of the array up

to the n-th site ($n = 1, 2, 3, ..., N$), plus the configuration with all closed sites ($n = 0$). If we set $x = e^{-\beta \varepsilon}$, the partition function may be written as

$$Q_N(T,N) = \sum_{n=0}^{N} e^{-\beta n \varepsilon} = \sum_{n=0}^{N} x^n = \sum_{n=0}^{+\infty} x^n - \sum_{n=N+1}^{+\infty} x^n =$$

$$\frac{1}{1-x} - \sum_{m=0}^{+\infty} x^{m+(N+1)} = \frac{1-x^{N+1}}{1-x} = \frac{1-e^{-\beta(N+1)\varepsilon}}{1-e^{-\beta \varepsilon}}.$$

The average number of open sites is

$$\langle n \rangle = \frac{\sum_{n=0}^{N} n e^{-\beta n \varepsilon}}{Q_N} = x \frac{d}{dx} \ln Q_N = x \frac{d}{dx} \ln \sum_{n=0}^{N} x^n = x \frac{d}{dx} \ln \frac{1-x^{N+1}}{1-x} =$$

$$x \frac{d}{dx} \ln(1 - x^{N+1}) - x \frac{d}{dx} \ln(1-x) = \frac{x}{1-x} - \frac{(N+1)x^{N+1}}{1-x^{N+1}} =$$

$$\frac{e^{-\beta \varepsilon}}{1-e^{-\beta \varepsilon}} - \frac{(N+1)e^{-\beta(N+1)\varepsilon}}{1-e^{-\beta(N+1)\varepsilon}}.$$

In the limit of low temperatures we have $x = e^{-\beta \varepsilon} \ll 1$. This means

$$\langle n \rangle \approx x = e^{-\beta \varepsilon}$$

which is independent of N. The reason for this independence is that, in the limit of low temperatures, only those configurations with low energy are available: these configurations have a small number of open sites.

As for the fluctuations of the number of open sites, we directly derive them from the partition function as

$$\langle (\Delta n)^2 \rangle = x \frac{d}{dx} \left(x \frac{d}{dx} \ln Q_N \right) = x \frac{d}{dx} \langle n \rangle = \frac{\sum_{n=0}^{N} n^2 x^n}{Q_N} - \left(\frac{\sum_{n=0}^{N} n x^n}{Q_N} \right)^2 =$$

$$\frac{x}{(1-x)^2} - \frac{(N+1)^2 x^{N+1}}{(1-x^{N+1})^2}.$$

In the limit of low temperatures, we find $x = e^{-\beta \varepsilon} \ll 1$, so that $\langle (\Delta n)^2 \rangle \approx x + 2x^2$ is independent of N. Again, this is due to the low temperature limit, where the system tends to occupy the lowest energy levels corresponding to a small number of open sites.

Problem 7.15.

Determine the energy fluctuations $\langle (\Delta E)^2 \rangle$ for a system of N independent one dimensional harmonic oscillators with frequency ω, mass m, and subject to a constant gravitational acceleration g along the direction of oscillation. Make use of the canonical ensemble in the classical limit.

Solution

If we call p the momentum and q the spatial coordinate, the Hamiltonian of the single oscillator is

$$H = \frac{p^2}{2m} + \frac{1}{2}m\omega^2 q^2 + mgq.$$

The canonical partition function is determined by performing first the integral over the momenta

$$Q_1(T) = \frac{1}{h} \int_{-\infty}^{+\infty} e^{-\beta \frac{p^2}{2m}} \, dp \int_{-\infty}^{+\infty} e^{-\frac{1}{2}\beta m\omega^2 q^2 - \beta mgq} \, dq =$$

$$\frac{1}{h}\sqrt{\frac{2\pi m}{\beta}} \int_{-\infty}^{+\infty} e^{-\frac{1}{2}\beta m\omega^2 q^2 - \beta mgq} \, dq.$$

For the integral in the coordinate q, we rearrange the exponential as

$$\int_{-\infty}^{+\infty} e^{-\frac{1}{2}\beta m\omega^2 q^2 - \beta mgq} \, dq = \int_{-\infty}^{+\infty} e^{-\frac{1}{2}\beta m\omega^2 \left(q + \frac{g}{\omega^2}\right)^2 + \frac{1}{2}\beta m \frac{g^2}{\omega^2}} \, dq =$$

$$e^{\frac{1}{2}\beta m \frac{g^2}{\omega^2}} \int_{-\infty}^{+\infty} e^{-\frac{1}{2}\beta m\omega^2 \left(q + \frac{g}{\omega^2}\right)^2} \, dq$$

where we can set $y = q + \frac{g}{\omega^2}$, so that

$$e^{\frac{1}{2}\beta m \frac{g^2}{\omega^2}} \int_{-\infty}^{+\infty} e^{-\frac{1}{2}\beta m\omega^2 y^2} \, dy = e^{\frac{1}{2}\beta m \frac{g^2}{\omega^2}} \sqrt{\frac{2\pi}{m\beta\omega^2}}.$$

The final result for the single oscillator partition function is

$$Q_1(T) = \frac{1}{h} \int_{-\infty}^{+\infty} e^{-\beta \frac{p^2}{2m}} \, dp \int_{-\infty}^{+\infty} e^{-\frac{1}{2}\beta m\omega^2 q^2 - \beta mgq} \, dq =$$

$$\frac{1}{h}\sqrt{\frac{2\pi m}{\beta}} \sqrt{\frac{2\pi}{m\beta\omega^2}} e^{\frac{\beta mg^2}{2\omega^2}} = \frac{2\pi}{h\beta\omega} e^{\frac{\beta mg^2}{2\omega^2}}$$

while, for the total partition function, we get $Q_N(T,N) = Q_1^N(T)$. The average energy is

$$U = -\left(\frac{\partial \ln Q_N}{\partial \beta}\right)_N = -N\frac{d}{d\beta}\left(\frac{m\beta g^2}{2\omega^2} + \ln\left(\frac{2\pi}{h\omega}\right) - \ln\beta\right) =$$

$$-\frac{Nmg^2}{2\omega^2} + \frac{N}{\beta} = -\frac{Nmg^2}{2\omega^2} + NkT.$$

The specific heat is then computed as $C = \left(\frac{\partial U}{\partial T}\right)_N = Nk$, from which we find the fluctuations

$$\langle (\Delta E)^2 \rangle = kT^2 C = Nk^2 T^2.$$

We note that we have obtained the same specific heat of a collection of N harmonic oscillators without the effect of gravity. This can be easily understood in terms of the Hamiltonian, that we can rewrite as

$$H = \frac{p^2}{2m} + \frac{1}{2}m\omega^2 q^2 + mgq = \frac{p^2}{2m} + \frac{1}{2}m\omega^2(q+x_0)^2 - \frac{mg^2}{2\omega^2}$$

with $x_0 = \frac{g}{\omega^2}$. This is nothing but the Hamiltonian of the harmonic oscillator whose equilibrium position is $q = -x_0$, plus the constant energy

$$E_0 = -\frac{mg^2}{2\omega^2}.$$

This means that the average energy is exactly the one of the harmonic oscillator plus the constant E_0, which is not affecting the specific heat.

Problem 7.16.
A one dimensional chain is hung on a ceiling. One of its extremes is fixed, while the other holds a mass M (see Fig. 7.4). Gravity is acting along the negative z direction. The chain is formed by two kinds of distinguishable rings: they are ellipses with the major axis oriented vertically or horizontally. The major and minor axes have lengths $l+a$ and $l-a$ respectively. The number of rings is fixed to N and the chain is in thermal equilibrium at temperature T. Find the average energy U and the average length $\langle L \rangle$ of the chain. Finally, determine the linear response function ($F = Mg$)

$$\chi = \left(\frac{\partial \langle L \rangle}{\partial F}\right)_{T,N}$$

in the limit of high temperatures.

Solution
If n is the number of rings with vertical major axis, the number of rings with vertical minor axis must be $N - n$. The total length is then

$$L = (l+a)n + (l-a)(N-n)$$

and the associated energy

$$E(n) = -MgL = -E_1 n - E_2(N-n)$$

where $E_1 = Mg(l+a)$ and $E_2 = Mg(l-a)$. Therefore, the canonical partition function is given by

$$Q_N(T,N) = \sum_{n=0}^{N} g_n e^{-\beta E(n)} = \sum_{n=0}^{N} \frac{N!}{n!(N-n)!} e^{n\beta E_1 + (N-n)\beta E_2}$$

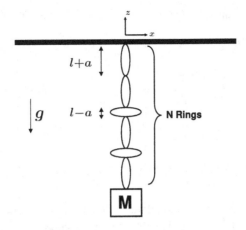

Fig. 7.4 A massless chain is hung on a ceiling and holds a mass M, with gravity acting along the negative z direction. The rings are ellipses with the major axis oriented vertically or horizontally. The total number of rings is fixed, while their arrangement (i.e. the length of the chain) changes as a function of the temperature. The canonical partition function is computed in Problems 7.16 and 7.17

where we have explicitly considered the appropriate degeneracy ($g_n = \frac{N!}{n!(N-n)!}$) for each configuration. We recognize in the partition function the binomial representation

$$(p+q)^N = \sum_{n=0}^{N} \frac{N!}{n!(N-n)!} p^n q^{N-n}$$

with $p = e^{\beta E_1}$ and $q = e^{\beta E_2}$. Therefore, we obtain the following result for the partition function

$$Q_N(T,N) = \left(e^{\beta E_1} + e^{\beta E_2} \right)^N.$$

Another possible approach to compute the partition function is discussed in Problem 7.17. The average energy is

$$U = -\left(\frac{\partial \ln Q_N}{\partial \beta} \right)_N = -N \frac{d}{d\beta} \ln \left(e^{\beta E_1} + e^{\beta E_2} \right) = -N \frac{E_1 e^{\beta E_1} + E_2 e^{\beta E_2}}{e^{\beta E_1} + e^{\beta E_2}} =$$
$$- NMg \left[l + a \tanh(\beta Mga) \right].$$

The average length is given by

$$\langle L \rangle = -\frac{1}{Mg} \left(\frac{\partial \ln Q_N}{\partial \beta} \right)_N = -\frac{U}{Mg}.$$

When $\beta \to +\infty$ we obtain the lowest energy state, that is $\langle L \rangle \approx N(l+a)$. In fact, the lowest energy state imposes the maximum length of the chain, given the minus sign in the relation between U and $\langle L \rangle$. On the other hand, in the limit $\beta \to 0$, we obtain

$$\langle L \rangle \approx Nl + N\beta Mga^2 = Nl + N\beta Fa^2.$$

In this limit, the linear response function is

$$\chi = \left(\frac{\partial \langle L \rangle}{\partial F} \right)_{T,N} = \frac{Na^2}{kT}$$

that is basically the same result we obtain (Curie law for magnetism) in the magnetic system of Problem 6.13 with the identification $a = \mu$.

Problem 7.17.
Consider the same physical situation of Problem 7.16. Show an alternative way to determine the partition function. Then, determine the average length $\langle L \rangle$ and give an estimate for the linear response function

$$\chi = \left(\frac{\partial \langle L \rangle}{\partial F} \right)_{T,N}$$

with $F = Mg$, showing that $\chi > 0$.

Solution
If we think that the origin of the z axis is located at the upper end of the chain (see Fig. 7.4) the total potential energy of the system is

$$E = -MgL$$

with L the total length of the chain, which is dependent on the number of rings with vertical major axis. If the chain is formed by n rings with vertical major axis (and $N - n$ rings with vertical minor axis), the length of the chain and the energy are

$$L = (l+a)n + (l-a)(N-n)$$

$$E = -Mg((l+a)n + (l-a)(N-n)) = -F((l+a)n + (l-a)(N-n)).$$

We can think of each ring contributing to the total gravitational potential energy with an energy $E_{\pm} = -Mg(l \pm a) = -F(l \pm a)$, depending on its vertical axis. This means that the whole system can be seen as composed of N rings with two energy states each. The partition function is then evaluated as

$$Q_N(T,F,N) = Q_1^N(T,F) = \left(e^{\beta F(l+a)} + e^{\beta F(l-a)} \right)^N = \left(2e^{Fl\beta} \cosh(aF\beta) \right)^N$$

where we have explicitly kept into account the dependence on F to be used later in the differentiation of the partition function. With respect to Problem 7.16, this is an alternative way to compute the partition function and gives the same result. The average length is given by

$$\langle L \rangle = \frac{1}{\beta} \left(\frac{\partial \ln Q_N}{\partial F} \right)_{T,N} = \frac{1}{\beta} \left(\frac{\partial \left(NFl\beta + N \ln(2\cosh(aF\beta)) \right)}{\partial F} \right)_{T,N} =$$

$$Nl + Na \tanh(aF\beta).$$

We remark that we would have obtained the same result starting from the expression of the total average energy

$$\langle L \rangle = -\frac{U}{F}$$

and $U = -\left(\frac{\partial \ln Q_N}{\partial \beta} \right)_{F,N}$. Finally, we compute the linear response function as

$$\chi = \left(\frac{\partial \langle L \rangle}{\partial F} \right)_{T,N} = \frac{Na^2\beta}{\cosh^2(aF\beta)}$$

from which we see that $\chi > 0$, i.e. the chain is elongating if the force increases.

Problem 7.18.
A two dimensional gas confined in the (x,y) plane is characterized by N non interacting particles in thermal equilibrium at temperature T. The Hamiltonian of the single particle is

$$H = \frac{1}{2m}(p_x^2 + p_y^2) + \frac{1}{2}m\omega^2 \left[a(x^2 + y^2) + 2bxy \right]$$

where p_x, p_y are the components of the momentum and m, ω, a and b are constants ($a > 0$ and $a^2 > b^2$). Compute the canonical partition function and the specific heat for the system.

Solution
The particles are non interacting so that the total partition function is

$$Q_N(T,N) = Q_1^N(T)$$

where Q_1 is the partition function of the single particle

$$Q_1(T) = \frac{1}{h^2} \int e^{-\beta H} \, dx\, dy\, dp_x\, dp_y = \frac{1}{h^2} I_p I_q$$

$$I_p = \int_{-\infty}^{+\infty} e^{-\beta \frac{p_x^2}{2m}} dp_x \int_{-\infty}^{+\infty} e^{-\beta \frac{p_y^2}{2m}} dp_y = \frac{2\pi m}{\beta}$$

$$I_q = \int_{-\infty}^{+\infty} e^{-\beta a m \omega^2 \frac{y^2}{2}} dy \int_{-\infty}^{+\infty} e^{-\beta m \omega^2 \frac{(ax^2 + 2bxy)}{2}} dx.$$

Using the integral $\int_{-\infty}^{+\infty} e^{-(Ax^2+2Bx)}dx = e^{B^2/A}\left(\frac{\pi}{A}\right)^{1/2}$, we easily obtain

$$I_q = \frac{2\pi}{\beta m \omega^2}\left(\frac{1}{a^2-b^2}\right)^{1/2}$$

leading to

$$Q_N(T,N) = \left(\frac{\beta h \omega}{2\pi}\right)^{-2N}(a^2-b^2)^{-N/2}.$$

In order to determine the specific heat C, we first need to compute the internal energy

$$U = -\left(\frac{\partial \ln Q_N}{\partial \beta}\right)_N = 2NkT$$

from which

$$C = \left(\frac{\partial U}{\partial T}\right)_N = 2Nk$$

which is in agreement with the classical equipartition of the energy (see Problem 7.20).

Problem 7.19.
Compute the average energy and the specific heat for a system of N distinguishable particles of mass m in d dimensions with the following Hamiltonian

$$H = \sum_{i=1}^{N}\frac{p_i^2}{2m} + \omega\sum_{i=1}^{N}q_i^b$$

where b is a positive integer and ω a positive constant independent of the temperature. In the above expression, p_i and q_i refer to the d dimensional modulus of the momentum and position for the i-th particle. Determine the internal energy and give the corresponding prediction for the equipartition theorem. When working with a generic d dimensional vector $x = (x_1, x_2, ..., x_d)$, make use of the following volume integral

$$\mathscr{I}(\ell,\phi,d) = \int e^{-\ell(x_1^2+...+x_d^2)^{\phi/2}}d^dx = G(\phi,d)\ell^{-d/\phi}$$

where $G(\phi,d)$ is a constant which is not useful to determine the average energy and the specific heat.

Solution
We start from the canonical partition function

$$Q_N(T,N) = Q_1^N(T)$$

where $Q_1(T)$ is the single particle partition function

$$Q_1(T) = \frac{1}{h^d}\int e^{-\beta\frac{(p_1^2+...+p_d^2)}{2m}}d^dp\int e^{-\beta\omega(q_1^2+...+q_d^2)^{b/2}}d^dq.$$

The integral over the momentum variables is the product of d Gaussian integrals (each one equal to $\sqrt{2\pi mkT}$) while the integral over the position coordinates corresponds to the case $\ell = \beta\omega$ and $\phi = b$ of the integral, $\mathscr{I}$, given above. Therefore, the total partition function is given by

$$Q_N(T,N) = \frac{[G(b,d)]^N}{h^{dN}} \left(\frac{\omega^{1/b}}{(2\pi m)^{1/2}} \right)^{-Nd} \beta^{-N(d/2+d/b)}.$$

We can extract the energy and the specific heat as follows

$$U = -\left(\frac{\partial \ln Q_N}{\partial \beta} \right)_N = Nd \left(\frac{1}{2} + \frac{1}{b} \right) kT$$

$$C = \left(\frac{\partial U}{\partial T} \right)_N = Nd \left(\frac{1}{2} + \frac{1}{b} \right) k.$$

We see that each degree of freedom in the momentum space contributes with a term $kT/2$ to the total energy, while those in the position space with a term kT/b. This is indeed consistent with the general result (we will discuss it in Problem 7.20)

$$\left\langle \sum_{i=1}^{N} \frac{p_i^2}{2m} \right\rangle = \frac{1}{2} \sum_{i=1}^{N} \sum_{k=1}^{d} \left\langle p_{ik} \left(\frac{\partial H}{\partial p_{ik}} \right) \right\rangle = \frac{NdkT}{2}$$

$$\left\langle \sum_{i=1}^{N} \omega q_i^b \right\rangle = \frac{1}{b} \sum_{i=1}^{N} \sum_{k=1}^{d} \left\langle q_{ik} \left(\frac{\partial H}{\partial q_{ik}} \right) \right\rangle = \frac{NdkT}{b}$$

where the derivative with respect to p_{ik}, q_{ik} is performed by keeping fixed all the other variables.

Problem 7.20.
Compute the canonical partition function for a gas with N particles under the effect of a one dimensional harmonic potential in the quantum mechanical case. Determine the specific heat and show that the classical limit is in agreement with the equipartition theorem for the energy.

Solution
The partition function is

$$Q_N(T,N) = Q_1^N(T) = \left(\sum_{n=0}^{+\infty} e^{-(n+\frac{1}{2})\frac{\hbar\omega}{kT}} \right)^N = e^{-\frac{N\hbar\omega}{2kT}} \left(\sum_{n=0}^{+\infty} \left(e^{-\frac{\hbar\omega}{kT}} \right)^n \right)^N =$$

$$e^{-\frac{N\hbar\omega}{2kT}} \left(\frac{1}{1-e^{-\frac{\hbar\omega}{kT}}} \right)^N = \left(\frac{1}{2\sinh\left(\frac{\hbar\omega}{2kT} \right)} \right)^N$$

so that we can compute the free energy

$$F = -kT \ln Q_N = NkT \ln \left(2 \sinh \left(\frac{\hbar \omega}{2kT} \right) \right)$$

and the entropy

$$S = - \left(\frac{\partial F}{\partial T} \right)_N = -Nk \ln \left(2 \sinh \left(\frac{\hbar \omega}{2kT} \right) \right) + \frac{N\hbar\omega}{2T} \frac{1}{\tanh \left(\frac{\hbar\omega}{2kT} \right)}.$$

The internal energy $U = F + TS$ and the specific heat $C = \left(\frac{\partial U}{\partial T} \right)_N$ are given by

$$U = \frac{N\hbar\omega}{2} \frac{1}{\tanh \left(\frac{\hbar\omega}{2kT} \right)} \qquad C = Nk \left(\frac{\hbar\omega}{2kT} \right)^2 \frac{1}{\sinh^2 \left(\frac{\hbar\omega}{2kT} \right)}.$$

In the limit of high temperatures $C \approx Nk$. This is exactly the result obtained using the equipartition theorem, i.e. a contribution to the average energy of $kT/2$ for each degree of freedom (see Problem 7.20). In our case, for each oscillator, we have one degree of freedom for the position and another for the momentum leading to

$$N \left(\frac{kT}{2} + \frac{kT}{2} \right) = NkT.$$

This can be made more rigorous if we consider a generic non interacting system composed of N particles in one dimension (for simplicity) with Hamiltonian

$$H = \sum_{i=1}^{N} H_i = \sum_{i=1}^{N} \left(T(p_i) + U(q_i) \right)$$

where T and U are the kinetic and the potential energy while $p = (p_1, p_2, ..., p_N)$ and $q = (q_1, q_2, ..., q_N)$ represent the momenta and positions of the whole system. If x_i and x_j are generic phase space variables (momentum or position), we find the following identity

$$\left\langle x_i \left(\frac{\partial H}{\partial x_j} \right) \right\rangle = \frac{\int x_i \left(\frac{\partial H}{\partial x_j} \right) e^{-\frac{H}{kT}} d^N p \, d^N q}{\int e^{-\frac{H}{kT}} d^N p \, d^N q} = \frac{-kT x_i e^{-\frac{H}{kT}} \Big|_{(x_j)_1}^{(x_j)_2} + kT \int \left(\frac{\partial x_i}{\partial x_j} \right) e^{-\frac{H}{kT}} d^N p \, d^N q}{\int e^{-\frac{H}{kT}} d^N p \, d^N q}$$

where the derivative with respect to x_i is done by keeping fixed all the other variables. In the above expression, we have used the integration by parts and denoted with $(x_j)_2$, $(x_j)_1$ the values of the variable x_j at the boundaries. If x_j is a momentum, we have a boundary value for the kinetic energy $T((x_j)_{1,2}) = +\infty$; if it is a position, we have $U((x_j)_{1,2}) = +\infty$ to ensure particles confinement. This is enough to safely assume that the boundary term is zero

$$x_i e^{-\frac{H}{kT}} (-kT) \Big|_{(x_j)_1}^{(x_j)_2} = 0$$

and obtain the following result

$$\left\langle x_i \left(\frac{\partial H}{\partial x_j} \right) \right\rangle = kT \frac{\int \left(\frac{\partial x_i}{\partial x_j} \right) e^{-\frac{H}{kT}} d^N p \, d^N q}{\int e^{-\frac{H}{kT}} d^N p \, d^N q} = \delta_{ij} kT.$$

If we use the Hamiltonian of a collection of N harmonic oscillators with mass m and frequency ω

$$H = \sum_{i=1}^{N} \left(\frac{p_i^2}{2m} + \frac{1}{2} m\omega^2 q_i^2 \right)$$

and the fact that $p_i \left(\frac{\partial H}{\partial p_i} \right) = \frac{p_i^2}{m}$, $q_i \left(\frac{\partial H}{\partial q_i} \right) = m\omega^2 q_i^2$, we can write

$$\left\langle \sum_{i=1}^{N} \left(\frac{p_i^2}{m} + m\omega^2 q_i^2 \right) \right\rangle = 2NkT.$$

This implies the following relation for the average energy

$$\langle H \rangle = \left\langle \sum_{i=1}^{N} \left(\frac{p_i^2}{2m} + \frac{1}{2} m\omega^2 q_i^2 \right) \right\rangle = NkT.$$

Problem 7.21.
Consider the one dimensional quantum harmonic oscillator with mass m and frequency ω in the canonical ensemble. Starting from the density matrix of the system, find the correct probability density function associated with the position x and discuss the following limits: $kT \gg \hbar\omega$ and $\hbar\omega \gg kT$. Finally, analyze the average value of the energy and compare the quantum mechanical and classical cases.

Solution
We start from the density matrix written as

$$\hat{\rho}(T) = \sum_{n=0}^{+\infty} e^{-\frac{\varepsilon_n}{kT}} |n\rangle\langle n|$$

with $|n\rangle$ the eigenstates and $\varepsilon_n = \hbar\omega(n + 1/2)$. The partition function may be written as

$$Q(T) = Tr(\hat{\rho}(T)) = \int \langle x|\hat{\rho}|x\rangle dx = \int \sum_{n=0}^{+\infty} e^{-\frac{\varepsilon_n}{kT}} \psi_n^2(x) dx$$

where we have used the completeness of the eigenstates of the position operator, i.e. $\int |x\rangle\langle x|dx = \mathbb{1}$. The requested probability density function is

$$P(x) = A \sum_{n=0}^{+\infty} e^{-\frac{\varepsilon_n}{kT}} \psi_n^2(x)$$

where the constant A is needed for the normalization and where we have used $\psi_n^2(x)$ instead of $|\psi_n(x)|^2$ because the wave functions are real. The probability density function may be found if we are able to sum the series. We will not do that directly, but we will instead calculate

$$\frac{dP(x)}{dx} = 2A \sum_{n=0}^{+\infty} e^{-\frac{\varepsilon_n}{kT}} \psi_n \frac{d\psi_n}{dx}.$$

Obviously $d\psi_n/dx = \frac{i}{\hbar}\hat{p}\psi_n$, where $\hat{p}$ is the momentum operator and where, from now on, we will make use of the creation and annihilation operators $\hat{a}^\dagger$, $\hat{a}$. Let us then recall the relation between $\hat{x}$, $\hat{p}$ and $\hat{a}^\dagger$, $\hat{a}$

$$\begin{cases} \hat{x} = \sqrt{\frac{\hbar}{2m\omega}}(\hat{a}+\hat{a}^\dagger) \\ \hat{p} = -i\sqrt{\frac{m\hbar\omega}{2}}(\hat{a}-\hat{a}^\dagger). \end{cases}$$

Since the creation and annihilation operators act on a generic eigenstate as step up and step down operators

$$\hat{a}^\dagger |n\rangle = \sqrt{n+1}\,|n+1\rangle \qquad \hat{a}|n\rangle = \sqrt{n}\,|n-1\rangle$$

the term $d\psi_n/dx$ can be written as

$$\frac{d\psi_n}{dx} = \frac{i}{\hbar}\hat{p}\psi_n = -\frac{i}{\hbar}\left(\langle n|\hat{p}|n-1\rangle\psi_{n-1} + \langle n|\hat{p}|n+1\rangle\psi_{n+1}\right)$$

meaning that the oscillator momentum has non zero matrix elements for transitions between n and $n\pm 1$. Also, the following relations hold

$$\langle n|\hat{p}|n-1\rangle = i\sqrt{\frac{m\hbar\omega}{2}}\sqrt{n}$$

$$\langle n|\hat{p}|n+1\rangle = -i\sqrt{\frac{m\hbar\omega}{2}}\sqrt{n+1}$$

$$\langle n|\hat{x}|n-1\rangle = \sqrt{\frac{\hbar}{2m\omega}}\sqrt{n}$$

$$\langle n|\hat{x}|n+1\rangle = \sqrt{\frac{\hbar}{2m\omega}}\sqrt{n+1}$$

which are all symmetric matrix elements. Moreover, from the relation between $\hat{p}$ and $\hat{x}$, we obtain

$$\langle n|\hat{p}|n-1\rangle = im\omega\langle n|\hat{x}|n-1\rangle$$

$$\langle n|\hat{p}|n+1\rangle = -im\omega\langle n|\hat{x}|n+1\rangle.$$

Therefore, for the derivative of the distribution function, we get

$$\frac{dP(x)}{dx} = -\frac{2iA}{\hbar}\sum_{n=0}^{+\infty} e^{-\frac{\varepsilon_n}{kT}}\left(\langle n|\hat{p}|n-1\rangle\psi_{n-1}(x)\psi_n(x) + \langle n|\hat{p}|n+1\rangle\psi_{n+1}(x)\psi_n(x)\right) =$$

$$\frac{2mA\omega}{\hbar}\sum_{n=0}^{+\infty} e^{-\frac{\varepsilon_n}{kT}}\left(\langle n|\hat{x}|n-1\rangle\psi_{n-1}(x)\psi_n(x) - \langle n|\hat{x}|n+1\rangle\psi_{n+1}(x)\psi_n(x)\right) =$$

$$\frac{2mA\omega}{\hbar}\sum_{n=0}^{+\infty} e^{-\frac{\varepsilon_n}{kT}}\left(e^{-\frac{\hbar\omega}{kT}}\langle n|\hat{x}|n+1\rangle\psi_n(x)\psi_{n+1}(x) - \langle n|\hat{x}|n+1\rangle\psi_{n+1}(x)\psi_n(x)\right) =$$

$$-\frac{2mA\omega}{\hbar}(1 - e^{-\frac{\hbar\omega}{kT}})\sum_{n=0}^{+\infty} e^{-\frac{\varepsilon_n}{kT}}\langle n|\hat{x}|n+1\rangle\psi_{n+1}(x)\psi_n(x)$$

where we have used the matrix elements of $\hat{p}$ and $\hat{x}$ and moved to the new variable $n' = n - 1$. In this new variable, the sum goes from $n' = -1$ to $n' = +\infty$, but the matrix element $\langle -1|\hat{x}|0\rangle$ is zero, so that $n' = n = 0, \dots +\infty$. Finally, we have used the fact that the matrix elements are symmetric. A similar calculation can be done for $xP(x)$

$$xP(x) = A\sum_{n=0}^{+\infty} e^{-\frac{\varepsilon_n}{kT}}\left(\langle n|\hat{x}|n-1\rangle\psi_{n-1}(x)\psi_n(x) + \langle n|\hat{x}|n+1\rangle\psi_{n+1}(x)\psi_n(x)\right) =$$

$$A(1 + e^{-\frac{\hbar\omega}{kT}})\sum_{n=0}^{+\infty} e^{-\frac{\varepsilon_n}{kT}}\langle n|\hat{x}|n+1\rangle\psi_{n+1}(x)\psi_n(x).$$

We are now ready to compute the ratio of the quantities obtained previously

$$\frac{1}{xP(x)}\frac{dP(x)}{dx} = -\frac{2m\omega}{\hbar}\tanh\left(\frac{\hbar\omega}{2kT}\right).$$

The solution for the differential equation

$$\frac{dP}{P} = -\frac{2m\omega}{\hbar}x\tanh\left(\frac{\hbar\omega}{2kT}\right)dx$$

is

$$P(x) = Ae^{-\frac{m\omega}{\hbar}x^2\tanh\left(\frac{\hbar\omega}{2kT}\right)}.$$

Integrating in x between $-\infty$ and $+\infty$ we are able to find the normalization. Therefore, the probability density function for the position is

$$P(x) = \sqrt{\frac{m\omega}{\pi\hbar}\tanh\left(\frac{\hbar\omega}{2kT}\right)}\, e^{-\frac{m\omega}{\hbar}x^2\tanh\left(\frac{\hbar\omega}{2kT}\right)}.$$

For $kT \gg \hbar\omega$ we obtain

$$P(x) = \sqrt{\frac{m\omega^2}{2\pi kT}}\, e^{-\frac{m\omega^2}{2kT}x^2}$$

that is the classical distribution. For $\hbar\omega \gg kT$ we get

$$P(x) = \sqrt{\frac{m\omega}{\pi\hbar}}\, e^{-\frac{m\omega}{\hbar}x^2}$$

which is the square of the wave function in the ground state, i.e. $\psi_0(x) = \left(\frac{m\omega}{\pi\hbar}\right)^{1/4} e^{-\frac{m\omega}{2\hbar}x^2}$. If we set $\beta = 1/kT$, the average energy becomes

$$\langle \hat{H} \rangle = \frac{Tr(\hat{H}e^{-\beta\hat{H}})}{Tr(e^{-\beta\hat{H}})} = -\frac{d}{d\beta}\ln Tr(e^{-\beta\hat{H}}) =$$

$$\frac{d}{d\beta}\ln\left(2\sinh\left(\frac{\beta\hbar\omega}{2}\right)\right) = \frac{1}{2}\hbar\omega\coth\left(\frac{\beta\hbar\omega}{2}\right).$$

We note that in the classical limit where $\beta\hbar\omega \ll 1$, we can expand

$$\coth\left(\frac{\beta\hbar\omega}{2}\right) \approx \frac{2}{\beta\hbar\omega}$$

and find $\langle \hat{H} \rangle \approx kT$, that is consistent with the equipartition theorem of the energy (see Problem 7.20) assigning to each degree of freedom a contribution equal to $kT/2$. For the quantum mechanical case, we first plug the value of $\langle \hat{H} \rangle$ in $P(x)$

$$P(x) = \sqrt{\frac{m\omega^2}{2\pi\langle \hat{H} \rangle}}\, e^{-\frac{m\omega^2 x^2}{2\langle \hat{H} \rangle}}.$$

Then, we note from the Hamiltonian that the momentum and position have the same quadratic functional form and, instead of x, we can set $p/(m\omega)$ to find the probability density function for the momentum

$$P(p) = \sqrt{\frac{1}{2\pi m\langle \hat{H} \rangle}}\, e^{-\frac{p^2}{2m\langle \hat{H} \rangle}}$$

where the normalization is properly found by normalizing $P(p)$ to unity. It is now possible to compute the following average quantities

$$\left\langle \frac{1}{2}m\omega^2 \hat{x}^2 \right\rangle = \sqrt{\frac{m\omega^2}{2\pi\langle\hat{H}\rangle}} \int_{-\infty}^{+\infty} \left(\frac{1}{2}m\omega^2 x^2\right) e^{-\frac{m\omega^2 x^2}{2\langle\hat{H}\rangle}} dx = \frac{1}{2}\langle\hat{H}\rangle$$

$$\left\langle \frac{\hat{p}^2}{2m} \right\rangle = \sqrt{\frac{1}{2\pi m\langle\hat{H}\rangle}} \int_{-\infty}^{+\infty} \left(\frac{p^2}{2m}\right) e^{-\frac{p^2}{2m\langle\hat{H}\rangle}} dp = \frac{1}{2}\langle\hat{H}\rangle$$

where we see that the averages of the momentum and position operators give the same contribution to the average energy.

Problem 7.22.
The potential energy of a one dimensional classical oscillator can be written as the sum of harmonic and anharmonic contributions

$$V(x) = \lambda x^2 - \gamma x^3 - \alpha x^4$$

with λ, γ, $\alpha > 0$ and γ and α small with respect to λ. The oscillator is in thermal equilibrium at temperature T. Use perturbation theory in γ with $\alpha = 0$ to show that the leading anharmonic contribution to the average value of x is given by

$$\langle x \rangle_{An} = \frac{3\gamma kT}{4\lambda^2}.$$

Then, in the general case when both γ and α are different from zero, show that the leading anharmonic correction to the specific heat C is given by the following expression

$$C_{An} = \frac{3}{2}k^2 T \left(\frac{\alpha}{\lambda^2} + \frac{5\gamma^2}{4\lambda^3} \right)$$

where we have used the subscript An to indicate that we are dealing with the anharmonic part.

Solution
For the classical harmonic oscillator, the average value of x is obviously zero. In the anharmonic case, due to the presence of the cubic terms in the potential, such value becomes different from zero. The average position is

$$\langle x \rangle = \frac{\int x e^{-\beta H} dx dp}{\int e^{-\beta H} dx dp}$$

with

$$H = \frac{p^2}{2m} + V(x) = \frac{p^2}{2m} + \lambda x^2 - \gamma x^3 - \alpha x^4.$$

The first anharmonic contribution in perturbation theory comes from the Taylor expansion of the exponential function with respect to γ and α

$$e^{-\beta(\lambda x^2 - \gamma x^3 - \alpha x^4)} = e^{-\beta\lambda x^2}\left(1 + \beta\gamma x^3 + \beta\alpha x^4 + \beta^2\frac{\gamma^2 x^6}{2} + \ldots\right).$$

In the ratio defining $\langle x \rangle$, the integrals in dp cancel out

$$\langle x \rangle = \frac{\int_{-\infty}^{+\infty} x e^{-\beta\lambda x^2}\left(1 + \beta\gamma x^3 + \beta\alpha x^4 + \beta^2\frac{\gamma^2 x^6}{2} + \ldots\right)dx}{\int_{-\infty}^{+\infty} e^{-\beta\lambda x^2}\left(1 + \beta\gamma x^3 + \beta\alpha x^4 + \beta^2\frac{\gamma^2 x^6}{2} + \ldots\right)dx}.$$

When $\alpha = 0$, the average $\langle x \rangle$ may be written as

$$\langle x \rangle = \frac{\int_{-\infty}^{+\infty} x e^{-\beta\lambda x^2}\left(1 + \beta\gamma x^3\right)dx}{\int_{-\infty}^{+\infty} e^{-\beta\lambda x^2}\,dx} + \mathcal{O}(\gamma^2)$$

and the first non zero correction to the harmonic contribution is

$$\langle x \rangle_{An} = \frac{\int_{-\infty}^{+\infty} e^{-\beta\lambda x^2}\beta\gamma x^4\,dx}{\int_{-\infty}^{+\infty} e^{-\beta\lambda x^2}\,dx}.$$

If we use the integrals

$$\int_{-\infty}^{+\infty} e^{-ax^2}\,dx = \sqrt{\frac{\pi}{a}}$$

$$\int_{-\infty}^{+\infty} x^4 e^{-ax^2}\,dx = \frac{1}{a^2}\lim_{s \to 1}\frac{d^2}{ds^2}\int_{-\infty}^{+\infty} e^{-sax^2}\,dx = \frac{3}{4}\sqrt{\frac{\pi}{a^5}}$$

with $a = \beta\lambda$, we find

$$\langle x \rangle_{An} = \frac{3\gamma kT}{4\lambda^2}.$$

To obtain the second result, we set $\alpha \neq 0$ and we expand the partition function

$$Q(\beta) = \frac{1}{h}\int_{-\infty}^{+\infty} e^{-\beta\frac{p^2}{2m}}\,dp\int_{-\infty}^{+\infty} e^{-\beta\lambda x^2}\left(1 + \beta\gamma x^3 + \beta\alpha x^4 + \beta^2\frac{\gamma^2 x^6}{2} + \ldots\right)dx$$

where it is important to consider the Taylor expansion of the exponentials up to β^2. Together with the above integrals, we also use

$$\int_{-\infty}^{+\infty} x^6 e^{-ax^2}\,dx = -\frac{1}{a^3}\lim_{s \to 1}\frac{d^3}{ds^3}\int_{-\infty}^{+\infty} e^{-sax^2}\,dx = \frac{15}{8}\sqrt{\frac{\pi}{a^7}}$$

and we find the following result

$$Q(\beta) = \frac{\pi}{h\beta}\left(\frac{2m}{\lambda}\right)^{1/2}(1 + B\beta^{-1})$$

where we have set $B = \frac{3}{4}\frac{\alpha}{\lambda^2} + \frac{15}{16}\frac{\gamma^2}{\lambda^3}$. The first term is the partition function of the harmonic oscillator, i.e. $Q_H(\beta) = \frac{\pi}{\hbar\beta}\left(\frac{2m}{\lambda}\right)^{1/2}$. The second term contains the corrections of the anharmonic potential. We then recall the definition of the specific heat, $C = \frac{dU}{dT}$, and the relation between the average energy and the partition function, $U = -\frac{d\ln Q(\beta)}{d\beta}$. For $\ln Q(\beta)$ we find

$$\ln Q(\beta) = \ln Q_H(\beta) + \ln(1 + B\beta^{-1})$$

where the first term is the harmonic part, while the second term is the anharmonic correction. This second term may be expanded as

$$\ln(1 + B\beta^{-1}) \approx B/\beta$$

and we obtain the anharmonic correction to the energy

$$U_{An} \approx \frac{B}{\beta^2} = Bk^2T^2.$$

The specific heat is

$$C_{An} \approx 2Bk^2T = \frac{3}{2}k^2T\left(\frac{\alpha}{\lambda^2} + \frac{5}{4}\frac{\gamma^2}{\lambda^3}\right)$$

that is the desired result.

Problem 7.23.
Study the magnetization of a three dimensional system with N identical dipoles, each one with magnetic moment μ and momentum of inertia m_I, in presence of a constant magnetic field $H = (0, 0, H)$ (directed along the z direction) at temperature T. These dipoles may be considered as distinguishable and localized in space. Write down the total partition function and concentrate on the terms related to the coupling with the magnetic field. Identify the magnetization in the limit of high and low temperatures and compute the susceptibility when $H \to 0$.

Solution
The physical picture of this problem is that the thermal fluctuations disrupt the ordered situation in which the magnetic dipoles are oriented along the direction of the magnetic field. In this way, we expect that when $T \to 0$ the system exhibits a magnetization $M \neq 0$, while in the limit of high temperatures the dipoles orientate in some non coherent way with resulting zero total magnetization $M = 0$.

The Hamiltonian $\mathcal{H}$ of the system is characterized by the kinetic terms due to the rotational freedom of the dipoles, plus the coupling with the magnetic field

$$\mathcal{H} = \sum_{i=1}^{N}\left(\frac{p_{\theta_i}^2}{2m_I} + \frac{p_{\phi_i}^2}{2m_I\sin^2\theta_i} - \mu_i \cdot H\right) = \mathcal{H}_{rot} - \sum_{i=1}^{N}\mu H\cos\theta_i$$

where

$$\mathscr{H}_{rot} = \sum_{i=1}^{N} \left(\frac{p_{\theta_i}^2}{2m_I} + \frac{p_{\phi_i}^2}{2m_I \sin^2 \theta_i} \right)$$

is the rotational part of the Hamiltonian, with (θ_i, ϕ_i) the angles determining the position of the i-th dipole in the phase space and $(p_{\theta_i}, p_{\phi_i})$ the associated momenta. The partition function for the system is

$$Q_N(T,H,N) = \frac{1}{h^{2N}} \int \left(\prod_{i=1}^{N} dp_{\theta_i} dp_{\phi_i} d\theta_i d\phi_i \right) e^{-\beta \sum_{i=1}^{N} \left(\frac{p_{\theta_i}^2}{2m_I} + \frac{p_{\phi_i}^2}{2m_I \sin^2 \theta_i} - \mu H \cos \theta_i \right)}$$

and satisfies $Q_N(T,H,N) = Q_1^N(T,H)$, where

$$Q_1(T,H) = \frac{1}{h^2} \int_{-\infty}^{+\infty} dp_\theta \int_{-\infty}^{+\infty} dp_\phi \int_0^\pi d\theta \int_0^{2\pi} d\phi \ e^{-\beta \left(\frac{p_\theta^2}{2m_I} + \frac{p_\phi^2}{2m_I \sin^2 \theta} - \mu H \cos \theta \right)}.$$

The Gaussian integrals in p_θ and p_ϕ lead to

$$Q_1(T,H) = \frac{2\pi m_I kT}{h^2} \int_0^{2\pi} d\phi \int_0^\pi e^{\beta \mu H \cos \theta} \sin \theta \, d\theta$$

due to the well known formula $\int_{-\infty}^{+\infty} e^{-ax^2} dx = \sqrt{\frac{\pi}{a}}$. We recognize in

$$Q_1^{(M)}(T,H) = \int_0^{2\pi} d\phi \int_0^\pi e^{\beta \mu H \cos \theta} \sin \theta \, d\theta$$

the contribution to the partition function due to the coupling with the magnetic field. In this way, we can easily compute the average value of the total magnetization

$$M = \mu \left\langle \sum_{i=1}^{N} \cos \theta_i \right\rangle = NkT \left(\frac{\partial \ln Q_1^{(M)}}{\partial H} \right)_T.$$

Therefore, to compute the magnetization, it is necessary only the single dipole partition function

$$Q_1^{(M)}(T,H) = \int_0^{2\pi} d\phi \int_0^\pi e^{\frac{\mu H \cos \theta}{kT}} \sin \theta \, d\theta = \frac{4\pi kT}{\mu H} \sinh \left(\frac{\mu H}{kT} \right).$$

Given the magnetization, we can define the average magnetic moment for the single dipole as

$$\bar{\mu} = \frac{M}{N} = kT \left(\frac{\partial \ln Q_1^{(M)}}{\partial H} \right)_T = kT \left(\frac{\partial}{\partial H} \ln \left(\frac{4\pi kT}{\mu H} \sinh \left(\frac{\mu H}{kT} \right) \right) \right)_T =$$

$$\mu \left[\coth \left(\frac{\mu H}{kT} \right) - \frac{kT}{\mu H} \right] = \mu L \left(\frac{\mu H}{kT} \right)$$

where $L(\mu H/kT)$ is the Langevin function such that

$$L \left(\frac{\mu H}{kT} \right) \approx \begin{cases} 1 & \frac{\mu H}{kT} \gg 1 \quad \text{low temperatures} \\ \frac{\mu H}{3kT} & \frac{\mu H}{kT} \ll 1 \quad \text{high temperatures.} \end{cases}$$

If N is the number of dipoles, the magnetization is

$$M = N\bar{\mu} = \begin{cases} N\mu & \text{low temperatures} \\ \frac{N\mu^2 H}{3kT} & \text{high temperatures.} \end{cases}$$

For high temperatures $M \to 0$ if $H \to 0$. Also, we can define the susceptibility χ as the variation of the magnetization with respect to the magnetic field. In the limit of small H and for high temperatures it is found that

$$\lim_{H \to 0} \chi = \lim_{H \to 0} \left(\frac{\partial M}{\partial H} \right)_{T,N} \approx \frac{N_0 \mu^2}{3kT} = \frac{C}{T}$$

where the constant $C = N_0\mu^2/3k$ is the Curie constant. The quantum mechanical analogue of this situation is discussed in Problem 7.24.

Problem 7.24.
When we treat the problem of paramagnetism (see Problem 7.23) from the point of view of Quantum Mechanics, the starting point is the relationship between the magnetic moment of a dipole and its total angular momentum operator $\hat{J} = (\hat{J}_x, \hat{J}_y, \hat{J}_z)$

$$\hat{\mu} = g\mu_B \frac{\hat{J}}{\hbar}$$

with μ_B Bohr magneton and g the Landè degeneracy factor. Let us consider a system of N localized dipoles, each one with a fixed total angular momentum equal to J. These dipoles are placed in a constant magnetic field $H = (0,0,H)$ (directed along the z direction). It follows that the number of allowed orientation of the magnetic moment in the direction of the applied field is limited to the eigenvalues of $\hat{J}_z$. Compute the partition function and, in the limit of high temperatures, determine the magnetization and the susceptibility.

Solution

The Hamiltonian of the system is written as

$$\mathscr{H} = -\sum_{i=1}^{N} \hat{\mu}_i \cdot H.$$

We know that the magnetic field is directed along the z direction. Therefore, we rewrite the Hamiltonian as

$$\mathscr{H} = -\sum_{i=1}^{N} H g \mu_B \frac{\hat{J}_{zi}}{\hbar}.$$

The major difference with respect to the classical case (treated in Problem 7.23) arises from the fact that the magnetic moment in the direction of the applied magnetic field does not have arbitrary values. The eigenvalues for $\hat{J}_{zi}/\hbar$ are $-J \le m \le J$. This means that the partition function of the single dipole is

$$Q_1(T,H) = \sum_{m=-J}^{J} e^{\frac{g \mu_B H m}{kT}}$$

i.e. the integral over all the possible orientation angles is here replaced by a discrete sum over all the possible projection values of $\hat{J}$ along z. If we set $x = g\mu_B H/kT$ and $z = e^x$, we obtain

$$Q_1(T,H) = \sum_{m=-J}^{J} e^{mx} = \sum_{m=-J}^{J} z^m = \sum_{m=-J}^{-1} z^m + \sum_{m=0}^{J} z^m =$$

$$\sum_{m=1}^{J} \frac{1}{z^m} + \sum_{m=0}^{J} z^m = \sum_{m'=0}^{J-1} \frac{1}{z^{m'+1}} + \sum_{m=0}^{J} z^m = \frac{1}{z} \frac{1 - \frac{1}{z^J}}{1 - \frac{1}{z}} + \frac{1 - z^{J+1}}{1 - z} =$$

$$\frac{1 - z^{2J+1}}{z^J(1-z)} = \frac{z^{-(J+\frac{1}{2})} - z^{(J+\frac{1}{2})}}{z^{-\frac{1}{2}} - z^{\frac{1}{2}}} = \frac{\sinh[(J+\frac{1}{2})x]}{\sinh(\frac{x}{2})} = \frac{\sinh[(J+\frac{1}{2})\frac{g\mu_B H}{kT}]}{\sinh\left(\frac{g\mu_B H}{2kT}\right)}$$

where we have used

$$\sum_{m=0}^{J} z^m = \sum_{m=0}^{+\infty} z^m - \sum_{m=J+1}^{+\infty} z^m = \frac{1}{1-z} - z^{J+1} \sum_{m'=0}^{+\infty} z^{m'} = \frac{1 - z^{J+1}}{1-z}.$$

The magnetization of the single dipole is given by

$$\bar{\mu} = kT \left(\frac{\partial \ln Q_1}{\partial H}\right)_T = g\mu_B \left[\left(J + \frac{1}{2}\right) \coth\left(J + \frac{1}{2}\right)x - \frac{1}{2}\coth\left(\frac{x}{2}\right)\right] = g\mu_B B_J(x)$$

where we have used the Brillouin function $B_J(x)$

$$B_J(x) = \left[\left(J + \frac{1}{2}\right)\coth\left(J + \frac{1}{2}\right)x - \frac{1}{2}\coth\left(\frac{x}{2}\right)\right]$$

whose low and high temperature limits are

$$B_J(x) \approx \begin{cases} J & x = \frac{g\mu_B H}{kT} \gg 1 & \text{low temperature} \\ \frac{1}{3}J(J+1)\frac{g\mu_B H}{kT} & x = \frac{g\mu_B H}{kT} \ll 1 & \text{high temperature.} \end{cases}$$

If we introduce the number of dipoles N, the total magnetization at high temperatures is

$$M = \frac{Ng^2\mu_B^2 J(J+1)H}{3kT}.$$

For the susceptibility $\chi = \left(\frac{\partial M}{\partial H}\right)_{T,N}$ we obtain the Curie law

$$\lim_{H\to 0} \chi = \lim_{H\to 0} \left(\frac{\partial M}{\partial H}\right)_{T,N} = \frac{Ng^2\mu_B^2 J(J+1)}{3kT} = \frac{C}{T}$$

with $C = \frac{Ng^2\mu_B^2 J(J+1)}{3k}$.

Problem 7.25.
Determine the canonical partition function for a quantum oscillator with the potential

$$\hat{V} = \frac{\hat{x}^2}{4} + \alpha\hat{x}^4.$$

To determine the energy spectrum, treat α as a small parameter and use the first order perturbative results. Finally, verify the classical limit at high temperatures. For simplicity, use $\hbar = \omega = 1$, $m = 1/2$.

Solution
The Hamiltonian of the system is

$$\hat{H} = \hat{H}_0 + \delta\hat{H} = \hat{p}^2 + \frac{\hat{x}^2}{4} + \alpha\hat{x}^4$$

where $\hat{H}_0 = \hat{p}^2 + \frac{\hat{x}^2}{4}$ is the unperturbed Hamiltonian for the harmonic oscillator. We first have to determine the quantum mechanical correction to the energy eigenvalues of $\hat{H}_0$. To this end, it is convenient to use the creation and annihilation operators and write

$$\hat{H}_0 = \hat{a}^\dagger\hat{a} + \frac{1}{2}\mathbb{1}.$$

The relations between these operators and the position and momentum operators ($\hat{x}$ and $\hat{p}$) are given by

$$\hat{x} = (\hat{a}^\dagger + \hat{a}) \qquad \hat{p} = -i\frac{(\hat{a} - \hat{a}^\dagger)}{2}.$$

Also, the commutation rules for $\hat{a}$ and $\hat{a}^+$ are written as

$$[\hat{a}, \hat{a}^\dagger] = 1.$$

We recall that the action of $\hat{a}$ and $\hat{a}^+$ on the eigenstates is such that

$$\hat{a}|n\rangle = \sqrt{n}|n-1\rangle$$

$$\hat{a}^\dagger|n\rangle = \sqrt{n+1}|n+1\rangle$$

$$\hat{a}^\dagger\hat{a}|n\rangle = n|n\rangle$$

$$\hat{a}\hat{a}^\dagger|n\rangle = (n+1)|n\rangle.$$

In perturbation theory, the first order correction to the n-th eigenvalue (due to $\hat{x}^4$) is given by

$$\langle n|\hat{x}^4|n\rangle = \langle n|(\hat{a}+\hat{a}^\dagger)^4|n\rangle.$$

In principle, we can expand the binomial with $\hat{a}$ and $\hat{a}^+$, but the whole calculation may be done in a more elegant and efficient way. We first observe that we only need to consider those terms that possess two $\hat{a}$ and two $\hat{a}^\dagger$. The number of these terms is equal to the number of couples available out of 4 elements, i.e. $\binom{4}{2} = 6$ in total: 1) $\hat{a}^\dagger\hat{a}^\dagger\hat{a}\hat{a}$, 2) $\hat{a}^\dagger\hat{a}\hat{a}^\dagger\hat{a}$, 3) $\hat{a}^\dagger\hat{a}\hat{a}\hat{a}^\dagger$, 4) $\hat{a}\hat{a}^\dagger\hat{a}^\dagger\hat{a}$, 5) $\hat{a}\hat{a}^\dagger\hat{a}\hat{a}^\dagger$, 6) $\hat{a}\hat{a}\hat{a}^\dagger\hat{a}^\dagger$. We also note that we can combine $\hat{a}\hat{a}\hat{a}^\dagger\hat{a}^\dagger$ and $\hat{a}^\dagger\hat{a}^\dagger\hat{a}\hat{a}$ in such a way that

$$\hat{a}\hat{a}\hat{a}^\dagger\hat{a}^\dagger + \hat{a}^\dagger\hat{a}^\dagger\hat{a}\hat{a} = \hat{a}(\hat{a}^\dagger\hat{a}+1)\hat{a}^\dagger + \hat{a}^\dagger(\hat{a}\hat{a}^\dagger-1)\hat{a} = \hat{a}\hat{a}^\dagger\hat{a}\hat{a}^\dagger + \hat{a}^\dagger\hat{a}\hat{a}^\dagger\hat{a} + 1$$

where we have used the identity $[\hat{a},\hat{a}^\dagger] = \hat{a}\hat{a}^\dagger - \hat{a}^\dagger\hat{a} = 1$. To summarize, the whole correction is such that

$$\langle n|\hat{x}^4|n\rangle = \langle n|2\hat{a}^\dagger\hat{a}\hat{a}^\dagger\hat{a} + \hat{a}^\dagger\hat{a}\hat{a}\hat{a}^\dagger + 2\hat{a}\hat{a}^\dagger\hat{a}\hat{a}^\dagger + \hat{a}\hat{a}^\dagger\hat{a}^\dagger\hat{a} + 1 |n\rangle = 6\left(n^2+n+\frac{1}{2}\right).$$

The resulting partition function of the oscillator is

$$Q_1(T) = \sum_{n=0}^{+\infty} e^{-\frac{[(n+\frac{1}{2})+6\alpha(n^2+n+\frac{1}{2})]}{kT}} \approx e^{-\frac{1}{2kT}} \sum_{n=0}^{+\infty}\left(1-6\alpha\frac{(n^2+n+\frac{1}{2})}{kT}\right)e^{-\frac{n}{kT}} =$$

$$\sqrt{x}\sum_{n=0}^{+\infty}\left(1-6\alpha\frac{(n^2+n+\frac{1}{2})}{kT}\right)x^n$$

where we have set $x = e^{-1/kT}$. We recall some useful formulae

$$\sum_{n=0}^{+\infty} x^n = \frac{1}{1-x}$$

$$\sum_{n=0}^{+\infty} nx^n = x\frac{d}{dx}\left(\sum_{n=0}^{+\infty} x^n\right) = \frac{x}{(1-x)^2}$$

$$\sum_{n=0}^{+\infty} n^2x^n = x^2\frac{d^2}{dx^2}\left(\sum_{n=0}^{+\infty} x^n\right) + \sum_{n=0}^{+\infty} nx^n = x^2\frac{d^2}{dx^2}\left(\sum_{n=0}^{+\infty} x^n\right) + x\frac{d}{dx}\left(\sum_{n=0}^{+\infty} x^n\right) = \frac{x(x+1)}{(1-x)^3}$$

needed to evaluate the partition function

$$Q_1(T) = \sqrt{x} \left[\frac{1}{1-x} - \frac{3\alpha}{kT} \frac{(x+1)^2}{(1-x)^3} \right].$$

In the limit of high temperatures, we find

$$x \approx 1 - \frac{1}{kT} \qquad \sqrt{x} \approx 1 - \frac{1}{2kT}$$

so that the partition function becomes

$$Q_1(T) \approx kT - 12\alpha(kT)^2$$

where we have neglected terms proportional to α and αT. We can also check that the expansion of $Q_1(T,V)$ for high temperatures (i.e. the classical limit) is recovered with the classical formalism

$$Q_1(T) = \frac{1}{2\pi} \int_{-\infty}^{+\infty} e^{-\frac{p^2}{kT}} dp \int_{-\infty}^{+\infty} e^{-\frac{\frac{1}{2}x^2 + \alpha x^4}{kT}} dx \approx$$

$$\frac{1}{2\pi} \int_{-\infty}^{+\infty} e^{-\frac{p^2}{kT}} dp \int_{-\infty}^{+\infty} e^{-\frac{x^2}{4kT}} \left(1 - \alpha \frac{x^4}{kT} \right) dx = kT - 12\alpha(kT)^2.$$

The term $\frac{1}{2\pi}$ in front of the integral is due to the normalization of the phase space in the system of units where $h = 2\pi$. Also, to compute the integrals, we have used the following identities

$$\int_{-\infty}^{+\infty} e^{-ax^2} dx = \sqrt{\frac{\pi}{a}}$$

$$\int_{-\infty}^{+\infty} x^4 e^{-ax^2} dx = \frac{1}{a^2} \lim_{\beta \to 1} \frac{d^2}{d\beta^2} \int_{-\infty}^{+\infty} e^{-a\beta x^2} dx = \frac{3}{4} \sqrt{\frac{\pi}{a^5}}.$$

Problem 7.26.

In a very crude approximation, a biatomic molecule may be modelled with two beads connected by an undeformable rod (see Fig. 7.5). Therefore, we have five degrees of freedom, i.e. the motion of the center of mass (three degrees) and the angular variables (two degrees). The partition function of this system is the product of the one of the center of mass (that is equal to the one of a free particle) and the contribution due to the angular part. Concentrate on the angular part and determine the internal energy U and the specific heat C due to the rotational contributions for a gas of biatomic heteronuclear molecules. Repeat the calculation for homonuclear molecules. Discuss both the classical and quantum mechanical cases.

Solution
We choose as generalized coordinates $q = (\theta, \phi)$ and associated momenta $p = (p_\theta, p_\phi)$ (see Fig. 7.5). The Hamiltonian of the rigid rotator takes the form

$$H = \frac{L^2}{2I} = \frac{p_\theta^2}{2I} + \frac{p_\phi^2}{2I \sin^2 \theta}$$

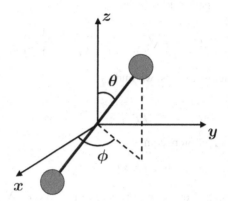

Fig. 7.5 We show a very crude approximation for a biatomic molecule: two beads (the two atoms) connected by an undeformable rod. Such approximation allows us to determine the thermodynamic properties and the specific heat deriving from the rotational degrees of freedom (see Problem 7.26)

with I the momentum of inertia. The partition function is

$$Q_N(T,N) = Q_1^N(T) = \frac{(2IkT)^N}{\hbar^{2N}}$$

because

$$Q_1(T) = \frac{1}{h^2} \int_{-\infty}^{+\infty} dp_\theta \int_{-\infty}^{+\infty} dp_\phi \int_0^\pi d\theta \int_0^{2\pi} d\phi\, e^{-\frac{1}{kT}\left(\frac{p_\theta^2}{2I} + \frac{p_\phi^2}{2I\sin^2\theta}\right)} = \frac{2IkT}{\hbar^2}$$

where we have first done the integral over the momenta and then the one over the angular coordinates. The specific heat is given by

$$C = \left(\frac{\partial U}{\partial T}\right)_N = -\left(\frac{\partial}{\partial T}\left(\frac{\partial \ln Q_N}{\partial \beta}\right)_N\right)_N = Nk.$$

For the case of homonuclear molecules, the integration over ϕ in the partition function is between 0 and π. In fact, when $\theta \to \pi - \theta$, a rigid rotation of π for the angle ϕ leaves the system unchanged since the two atoms are identical. In other words, there is a relation between the partition functions for heteronuclear and homonuclear molecules, that is

$$Q_{hete} = 2Q_{homo}.$$

Therefore, the energy U and the specific heat are unchanged because $U = -\left(\frac{\partial \ln Q_N}{\partial \beta}\right)_N$. As for the quantum mechanical case, the Hamiltonian becomes an operator $\hat{H}$ that is diagonal with respect to the basis given by the spherical harmonics $Y_{l,m}(\theta,\phi) =$

$\langle \theta, \phi | l, m \rangle$

$$\hat{H} |l,m\rangle = \frac{\hat{L}^2}{2I} |l,m\rangle = \varepsilon_l |l,m\rangle$$

where the energies are

$$\varepsilon_l = \frac{l(l+1)\hbar^2}{2I}$$

with degeneracy $2l+1$, because the quantum number m does not appear explicitly in the Hamiltonian. The resulting partition function for the single molecule is

$$Q_1(T) = \sum_{l=0}^{+\infty} (2l+1) e^{-\frac{l(l+1)\hbar^2}{2IkT}} = \sum_{l=0}^{+\infty} (2l+1) e^{-\frac{l(l+1)\Theta}{T}}$$

where $\Theta = \hbar^2/2Ik$. For $T \gg \Theta$ we approximate the sum with an integral and obtain

$$Q_1(T) \approx \frac{T}{\Theta} \int_0^{+\infty} e^{-\xi} d\xi = \frac{T}{\Theta}$$

from which

$$C = \left(\frac{\partial U}{\partial T}\right)_N = -\left(\frac{\partial}{\partial T} \left(\frac{\partial \ln Q_N}{\partial \beta}\right)_N\right)_N = Nk$$

that is the classical limit. For $T \ll \Theta$ we keep only the first terms of the sum

$$Q_1(T) \approx 1 + 3e^{-\frac{2\Theta}{T}} + 5e^{-\frac{6\Theta}{T}} + \cdots$$

leading to

$$C \approx 12Nk \left(\frac{\Theta}{T}\right)^2 e^{-\frac{2\Theta}{T}}$$

that goes exponentially to zero. For the case of homonuclear molecules, the wave function has to satisfy

$$|\psi(x)|^2 = |\psi(-x)|^2$$

which means that the physical content (i.e. the probability density function) of the wave function does not change when interchanging the atoms. This is nothing but a parity symmetry. In other words, the wave function must be symmetric (even) or antisymmetric (odd) under the interchange of the two atoms, $\psi(x) = \pm\psi(-x)$. The angular dependence of the wave function is given by the spherical harmonics $Y_{l,m}(\theta,\phi)$ which, under the interchange of the atoms, behave like

$$Y_{l,m}(\theta,\phi) \rightarrow (-1)^l Y_{l,m}(\theta,\phi).$$

The resulting wave function is symmetric (even) for l even and antisymmetric (odd) for l odd. These two cases are associated with two different partition functions

$$Q_1^{odd}(T) = \sum_{l=1,3,\ldots}^{+\infty} (2l+1) e^{-\frac{l(l+1)\hbar^2}{2IkT}}$$

$$Q_1^{even}(T) = \sum_{l=0,2,\ldots}^{+\infty} (2l+1)e^{-\frac{l(l+1)\hbar^2}{2IkT}}.$$

For high temperatures, $T \gg \Theta$, we obtain

$$Q_1^{odd}(T) \approx Q_1^{even}(T) \approx \frac{1}{2}Q_1(T) = \frac{T}{2\Theta}$$

that is the classical result. For low temperatures, $T \ll \Theta$, we get

$$Q_1^{odd}(T) \approx 3e^{-\frac{2\Theta}{T}} + 7e^{-\frac{12\Theta}{T}} + \cdots = e^{-\frac{2\Theta}{T}}(3 + 7e^{-\frac{10\Theta}{T}} + \cdots)$$

$$Q_1^{even}(T) \approx 1 + 5e^{-\frac{6\Theta}{T}} + \ldots$$

from which

$$C^{odd} \approx \frac{700}{3}\left(\frac{\Theta}{T}\right)^2 kNe^{-\frac{10\Theta}{T}}$$

$$C^{even} \approx 180\left(\frac{\Theta}{T}\right)^2 kNe^{-\frac{6\Theta}{T}}.$$

Problem 7.27.
Determine the probability distribution function in the phase space for a relativistic particle in a volume V and with energy $\varepsilon(p) = \sqrt{m^2c^4 + p^2c^2}$, where p is the absolute value of the momentum, m the mass, and c the speed of light. Give the final result in terms of the modified Bessel functions

$$K_v(z) = \int_0^{+\infty} e^{-z\cosh t}\cosh(vt)\,dt$$

$$K_v(z) \approx \frac{(v-1)!}{2}\left(\frac{z}{2}\right)^{-v} \qquad z \ll 1, v > 0.$$

Check what happens in the limit $\frac{mc^2}{kT} \to 0$.

Solution
In general, the probability distribution function $\rho(p,q)$ is

$$\rho(p,q) = Ae^{-\varepsilon(p,q)/kT}.$$

Such distribution must be normalized in order to be considered a proper probability distribution function

$$\int \rho(p,q)\,d^3p\,d^3q = A\int e^{-\varepsilon(p,q)/kT}\,d^3p\,d^3q = 1.$$

From this normalization condition we determine the constant A. For this specific case, $\varepsilon(p,q) = \varepsilon(p)$ is dependent only on the absolute value of the momentum and not on the spatial coordinates. Therefore, we have to set

$$4\pi AV \int_0^{+\infty} e^{-\varepsilon(p)/kT} p^2 dp = 1.$$

We then use $\varepsilon(p) = \sqrt{m^2c^4 + p^2c^2}$ and set $\sinh x = \xi = \frac{p}{mc}$. The above integral becomes

$$\frac{1}{A} = 4\pi V \int_0^{+\infty} e^{-\frac{1}{kT}\sqrt{m^2c^4+p^2c^2}} p^2 dp = 4\pi V(mc)^3 \int_0^{+\infty} e^{-\frac{mc^2}{kT}\sqrt{1+\xi^2}} \xi^2 d\xi =$$

$$4\pi V(mc)^3 \int_0^{+\infty} e^{-\frac{mc^2}{kT}\cosh x} \sinh^2 x \cosh x \, dx =$$

$$4\pi V(mc)^3 \left(\int_0^{+\infty} e^{-\frac{mc^2}{kT}\cosh x} \cosh^3 x \, dx - \int_0^{+\infty} e^{-\frac{mc^2}{kT}\cosh x} \cosh x \, dx \right).$$

These integrals may be expressed in terms of the modified Bessel functions K_ν given in the text. Therefore, the normalization is such that

$$\frac{1}{A} = \frac{4\pi(kT)^3}{c^3} V z^3 \left(\frac{d^2 K_1(z)}{dz^2} - K_1(z) \right)$$

where $z = \frac{mc^2}{kT}$. When $z \ll 1$ ($\frac{mc^2}{kT} \ll 1$), we use the asymptotic property given in the text to find

$$\frac{1}{A} \approx \frac{4\pi(kT)^3}{c^3} V z^3 \left(\frac{2}{z^3} - \frac{1}{z} \right) \approx \frac{8\pi(kT)^3}{c^3} V.$$

This is the same result we obtain when we consider the Hamiltonian $H = pc$ and impose the normalization

$$1 = 4\pi AV \int_0^{+\infty} e^{-pc/kT} p^2 dp = \frac{4\pi}{c^3} AV(kT)^3 \int_0^{+\infty} x^2 e^{-x} dx = \frac{8\pi}{c^3} AV(kT)^3$$

where $\int_0^{+\infty} x^2 e^{-x} dx = 2$.

Problem 7.28.
A gas of N indistinguishable and non interacting particles is placed in a volume V and is in thermal equilibrium at temperature T. The Hamiltonian of the single particle can be written as

$$H = ap^b = a(p_x^2 + p_y^2 + p_z^2)^{b/2} \qquad a, b > 0$$

where p is the absolute value of the momentum and (p_x, p_y, p_z) its components. Compute the average internal energy U, the pressure P, and the chemical potential μ. Finally, verify the limit

$$a = \frac{1}{2m} \qquad b = 2$$

with m a constant with physical dimensions of a mass.

Solution

The particles are non interacting and, therefore, the total partition function Q_N is the product of the partition functions of the single particles

$$Q_N(T,V,N) = \frac{Q_1^N(T,V)}{N!}$$

where $N!$ takes into account the indistinguishability of the particles. For Q_1, we can write down the integral in the phase space

$$Q_1(T,V) = \int \frac{d^3p\,d^3q}{h^3} e^{-\beta H} = \int \frac{d^3p\,d^3q}{h^3} e^{-\beta a p^b} = \frac{4\pi V}{h^3} \int_0^{+\infty} e^{-\beta a p^b} p^2\,dp.$$

For the integral in the momentum space, we have used spherical polar coordinates and integrated over the whole solid angle 4π. Introducing the new variable $x = \beta a p^b$, we easily obtain

$$p = \left(\frac{x}{a\beta}\right)^{\frac{1}{b}} \qquad dp = \frac{1}{ab\beta}\left(\frac{x}{a\beta}\right)^{\frac{1-b}{b}} dx$$

and the integral for Q_1 becomes

$$Q_1(T,V) = \frac{4\pi V}{bh^3}\left(\frac{1}{a\beta}\right)^{\frac{3}{b}} \int_0^{+\infty} x^{\frac{(3-b)}{b}} e^{-x}\,dx.$$

We can use the definition of the Euler Gamma function

$$\int_0^{+\infty} x^{\gamma-1} e^{-x}\,dx = \Gamma(\gamma)$$

to simplify the integral as

$$Q_1(T,V) = \frac{4\pi V}{bh^3}\left(\frac{1}{a\beta}\right)^{\frac{3}{b}} \Gamma\left(\frac{3}{b}\right).$$

The average internal energy U is

$$U = -\left(\frac{\partial \ln Q_N}{\partial \beta}\right)_{V,N} \approx -N\left(\frac{\partial \ln Q_1}{\partial \beta}\right)_V$$

and, since Q_1 is written as $A\beta^{-3/b}$ with A independent of β, we find that

$$U = \frac{3NkT}{b}.$$

For the pressure, we can write

$$P = \frac{N}{\beta}\left(\frac{\partial \ln Q_1}{\partial V}\right)_T = \frac{NkT}{V}.$$

Finally, for the chemical potential, we know that

$$\mu = -\frac{1}{\beta}\left(\frac{\partial \ln Q_N}{\partial N}\right)_{T,V}$$

and, using the Stirling approximation ($\ln N! \approx N\ln N - N$), we obtain

$$\mu \approx -\frac{1}{\beta}\left(\frac{\partial (N\ln Q_1 - N\ln N + N)}{\partial N}\right)_{T,V} = \frac{1}{\beta}\ln\left(\frac{N}{Q_1}\right) = \frac{1}{\beta}\ln\left(\frac{Nbh^3(a\beta)^{3/b}}{4\pi V\Gamma(3/b)}\right).$$

We note that when $a = \frac{1}{2m}$ and $b = 2$ we obtain, as expected, the results for an ideal classical gas with Hamiltonian $H = \frac{p^2}{2m}$.

Problem 7.29.

Consider N distinguishable and non interacting particles. The single particle energy spectrum is $\varepsilon_n = n\varepsilon$, with $n = 0, 1, 2, ..., +\infty$ and degeneracy $g_n = n+1$ ($\varepsilon > 0$ is a constant). Compute the canonical partition function Q_N, the internal average energy U, the energy fluctuations

$$\langle(\Delta E)^2\rangle = \langle E^2\rangle - \langle E\rangle^2$$

and the specific heat C.

Solution
The particles are non interacting and distinguishable, that implies

$$Q_N(\beta, N) = Q_1^N(\beta)$$

where $Q_1(\beta)$ is the partition function of the single particle

$$Q_1(\beta) = \sum_{n=0}^{+\infty}(n+1)e^{-\beta\varepsilon n}$$

that we write as the sum of two terms

$$Q_1(\beta) = \sum_{n=0}^{+\infty}ne^{-\beta\varepsilon n} + \sum_{n=0}^{+\infty}e^{-\beta\varepsilon n}.$$

The first term on the right can be written as the derivative of the second term

$$Q_1(\beta) = -\frac{d}{d(\beta\varepsilon)}\sum_{n=0}^{+\infty}e^{-\beta\varepsilon n} + \sum_{n=0}^{+\infty}e^{-\beta\varepsilon n}$$

so that we simply have to compute a single sum

$$\sum_{n=0}^{+\infty}e^{-\beta\varepsilon n} = \frac{1}{1-e^{-\beta\varepsilon}}.$$

The final result is

$$Q_1(\beta) = -\frac{d}{d(\beta\varepsilon)}\left(\frac{1}{1-e^{-\beta\varepsilon}}\right) + \frac{1}{1-e^{-\beta\varepsilon}} = \frac{e^{-\beta\varepsilon}}{(1-e^{-\beta\varepsilon})^2} + \frac{1}{1-e^{-\beta\varepsilon}} =$$

$$= \frac{1}{(1-e^{-\beta\varepsilon})^2}.$$

The energy is evaluated as

$$U = -\left(\frac{\partial \ln Q_N}{\partial \beta}\right)_N = -N\frac{d \ln Q_1(\beta)}{d\beta} = \frac{2N\varepsilon}{e^{\beta\varepsilon}-1}.$$

As for $\langle(\Delta E)^2\rangle$, we observe that

$$\langle(\Delta E)^2\rangle = \frac{1}{Q_N}\left(\frac{\partial^2 Q_N}{\partial \beta^2}\right)_N - \left(\frac{1}{Q_N}\frac{\partial Q_N}{\partial \beta}\right)_N^2 = \frac{\partial}{\partial \beta}\left(\frac{1}{Q_N}\frac{\partial Q_N}{\partial \beta}\right)_N = -\left(\frac{\partial U}{\partial \beta}\right)_N$$

and obtain

$$\langle(\Delta E)^2\rangle = \frac{2N\varepsilon^2 e^{\beta\varepsilon}}{(e^{\beta\varepsilon}-1)^2}.$$

The value of the specific heat is related to $\langle(\Delta E)^2\rangle$ in the following way

$$C = \left(\frac{\partial U}{\partial T}\right)_N = -\frac{1}{kT^2}\left(\frac{\partial U}{\partial \beta}\right)_N = \frac{\langle(\Delta E)^2\rangle}{kT^2} = \frac{2N\varepsilon^2 e^{\beta\varepsilon}}{(e^{\beta\varepsilon}-1)^2 kT^2}.$$

Problem 7.30.
An ideal classical gas is in thermal equilibrium at temperature T and is formed by N independent indistinguishable molecules in a spherical container of radius R. A force is directed towards the center of the sphere, with the following potential

$$V(r) = \alpha r \qquad \alpha > 0.$$

Find the pressure P and the particles density close to the surface of container.

Solution
Let us start from the Hamiltonian of the single particle

$$H = \frac{p^2}{2m} + V(r)$$

and compute the total partition function

$$Q_N(T,R,N) = \frac{1}{N!}Q_1^N(T,R)$$

where

$$Q_1(T,R) = \frac{1}{h^3} \int e^{-\beta H} d^3 p d^3 q = \frac{1}{h^3} \left[\int_{-\infty}^{+\infty} e^{-\beta p^2/2m} dp \right]^3 4\pi \int_0^R r^2 e^{-\beta \alpha r} dr =$$

$$\frac{1}{h^3} \left(\frac{2\pi m}{\beta} \right)^{3/2} 4\pi \int_0^R r^2 e^{-\beta \alpha r} dr.$$

We now set $\beta \alpha r = x$, $dr = \frac{1}{\beta \alpha} dx$, so that we have

$$Q_1(T,R) = \frac{1}{h^3} \left(\frac{2\pi m}{\beta} \right)^{3/2} \frac{4\pi}{\beta^3 \alpha^3} \int_0^{\beta \alpha R} x^2 e^{-x} dx.$$

The integral in x can be done

$$\int_0^{\beta \alpha R} x^2 e^{-x} dx = \left[e^{-x}(-x^2 - 2x - 2) \right]_0^{\beta \alpha R}$$

from which we obtain the partition function

$$Q_1(T,R) = \frac{1}{h^3} \left(\frac{2\pi m}{\beta} \right)^{3/2} \frac{4\pi}{\beta^3 \alpha^3} \left(-(\alpha \beta R)^2 e^{-\alpha \beta R} - 2(\alpha \beta R) e^{-\alpha \beta R} - 2e^{-\alpha \beta R} + 2 \right).$$

Once we know the partition function, we can compute the pressure as

$$P = -\left(\frac{\partial F}{\partial V} \right)_{T,N} = \frac{N}{\beta} \frac{1}{4\pi R^2} \left(\frac{\partial \ln Q_1}{\partial R} \right)_T$$

where

$$\left(\frac{\partial \ln Q_1}{\partial R} \right)_T = \frac{R^2 (\alpha \beta)^3 e^{-\alpha \beta R}}{\left(-(\alpha \beta R)^2 e^{-\alpha \beta R} - 2(\alpha \beta R) e^{-\alpha \beta R} - 2e^{-\alpha \beta R} + 2 \right)}$$

so that

$$P = \frac{NkT}{4\pi} \frac{(\alpha \beta)^3 e^{-\alpha \beta R}}{\left(-(\alpha \beta R)^2 e^{-\alpha \beta R} - 2(\alpha \beta R) e^{-\alpha \beta R} - 2e^{-\alpha \beta R} + 2 \right)}.$$

From the equation of state for the ideal gas $P = n(R)kT$, we extract the density $n(R)$ close to the surface of the container

$$n(R) = \frac{N}{4\pi} \frac{(\alpha \beta)^3 e^{-\alpha \beta R}}{\left(-(\alpha \beta R)^2 e^{-\alpha \beta R} - 2(\alpha \beta R) e^{-\alpha \beta R} - 2e^{-\alpha \beta R} + 2 \right)}.$$

Problem 7.31.

An ideal classical gas is formed by N indistinguishable non interacting molecules. Each one of the molecules is localized in space and has an electric dipole equal to $\tilde{d}$. The whole gas is in thermal equilibrium at temperature T and is under the effect of a constant electric field with intensity E directed along the z axis. The Hamiltonian of the single dipole is

$$H = \frac{1}{2I}p_\theta^2 + \frac{1}{2I\sin^2\theta}p_\phi^2 - \tilde{d}E\cos\theta$$

that is the rotational energy plus the coupling with the electric field. In the above expression, I is the momentum of inertia of the molecule, (θ,ϕ) its angles in spherical polar coordinates, and (p_θ,p_ϕ) the associated momenta. Prove that the partition function of the single dipole can be written as

$$Q_1(T,E) = \frac{2I\sinh(\beta\tilde{d}E)}{\hbar^2\beta^2\tilde{d}E}.$$

Then, defining the polarization as $P = \frac{N}{V}\langle\tilde{d}\cos\theta\rangle$, show that

$$P = \frac{N}{V}\left(\tilde{d}\coth(\beta\tilde{d}E) - \frac{1}{\beta E}\right)$$

where $\langle...\rangle$ is the average over the single particle statistics. Finally, in the limit of weak field ($\beta\tilde{d}E \to 0$), show that the dielectric constant ε, defined by

$$\varepsilon E = \varepsilon_0 E + P$$

is equal to $\varepsilon = \varepsilon_0 + \frac{N\beta\tilde{d}^2}{3V}$.

Solution
The partition function of the single dipole can be written as

$$Q_1(T,E) = \frac{1}{4\pi^2\hbar^2}\int_0^\pi d\theta \int_0^{2\pi} d\phi \int_{-\infty}^{+\infty} dp_\theta \int_{-\infty}^{+\infty} dp_\phi\, e^{-\beta\left(\frac{p_\theta^2}{2I} + \frac{p_\phi^2}{2I\sin^2\theta} - \tilde{d}E\cos\theta\right)}.$$

The integral in $d\phi$ produces 2π. Also, we can perform the Gaussian integrals in dp_θ and dp_ϕ which produce $\sqrt{2\pi I/\beta}$ and $\sqrt{2\pi I\sin^2\theta/\beta}$ respectively. To summarize, we obtain

$$Q_1(T,E) = \frac{I}{\hbar^2\beta}\int_0^\pi \sin\theta\, e^{\beta\tilde{d}E\cos\theta}\, d\theta$$

which we can write, by setting $x = \cos\theta$, as

$$Q_1(T,E) = \frac{I}{\hbar^2\beta}\int_{-1}^{+1} e^{\beta\tilde{d}Ex}\, dx = \frac{I}{\hbar^2\beta^2\tilde{d}E}(e^{\beta\tilde{d}E} - e^{-\beta\tilde{d}E})$$

that is

$$Q_1(T,E) = \frac{2I\sinh(\beta\tilde{d}E)}{\hbar^2\beta^2\tilde{d}E}.$$

As for the second point, we note that

$$\langle \tilde{d}\cos\theta \rangle = \tilde{d}\frac{\int e^{-\beta H}\cos\theta\, d\phi\, d\theta\, dp_\phi\, dp_\theta}{\int e^{-\beta H} d\phi\, d\theta\, dp_\phi\, dp_\theta} = \frac{1}{\beta}\left(\frac{\partial\ln Q_1}{\partial E}\right)_T,$$

and we obtain

$$\langle \tilde{d}\cos\theta \rangle = \tilde{d}\coth(\beta\tilde{d}E) - \frac{1}{\beta E}.$$

In the limit $x \to 0$, the function $\coth x$ is expanded as $\coth x \approx \frac{1}{x} + \frac{x}{3} + \dots$. Therefore, it is possible to verify that, in the limit $\beta\tilde{d}E \to 0$, we get

$$P \approx \frac{N}{V}\left(\frac{1}{\beta E} + \frac{\beta\tilde{d}^2 E}{3} - \frac{1}{\beta E}\right) = \frac{N\beta\tilde{d}^2 E}{3V}$$

and, hence

$$\varepsilon_0 E + P = \varepsilon_0 E + \frac{N\beta\tilde{d}^2 E}{3V} = \left(\varepsilon_0 + \frac{N\beta\tilde{d}^2}{3V}\right)E = \varepsilon E$$

with $\varepsilon = \varepsilon_0 + \frac{N\beta\tilde{d}^2}{3V}$.

Problem 7.32.

Consider an ideal gas of N non interacting indistinguishable particles placed in a volume V. The single particle Hamiltonian is $H = p^2/2m$, with p the absolute value of the momentum and m the mass of each particle. Prove the following relations

$$\frac{S(T,V,N)}{Nk} = \ln\left(\frac{Q_1(T,V)}{N}\right) + T\left(\frac{\partial\ln Q_1(T,V)}{\partial T}\right)_V + 1$$

$$\frac{S(T,P,N)}{Nk} = \ln\left(\frac{Q_1(T,P,N)}{N}\right) + T\left(\frac{\partial\ln Q_1(T,P,N)}{\partial T}\right)_{P,N}$$

where S is the entropy, P the pressure, and Q_1 is the canonical partition function of the single particle.

Solution

Since we are dealing with non interacting particles, the total partition function of the system Q_N can be written in terms of the single particle partition function Q_1 as

$$Q_N(T,V,N) = \frac{Q_1^N(T,V)}{N!}$$

where $N!$ accounts for the indistinguishability of the particles. From the definition of the free energy F, we also know that $ST = U - F$, where U is the internal energy. The relation between the free energy and the canonical partition function is

$$F(T,V,N) = -kT\ln Q_N(T,V,N) = -kT\left(\ln Q_1^N(T,V) - \ln N!\right)$$

where, using the Stirling approximation for the factorials, we find

$$F(T,V,N) = -kT\left(N\ln Q_1(T,V) - N\ln N + N\right)$$

that implies

$$F(T,V,N) = -NkT\ln\left(\frac{Q_1(T,V)}{N}\right) - NkT.$$

The average energy in terms of the partition function is

$$U = -\left(\frac{\partial \ln Q_N}{\partial \beta}\right)_{V,N} = kT^2\left(\frac{\partial \ln Q_N}{\partial T}\right)_{V,N}$$

and, using $Q_N(T,V,N) = Q_1^N(T,V)/N!$, we easily obtain

$$U = NkT^2\left(\frac{\partial \ln Q_1(T,V)}{\partial T}\right)_V.$$

Combining all these results, we get

$$\frac{S(T,V,N)}{Nk} = \ln\left(\frac{Q_1(T,V)}{N}\right) + T\left(\frac{\partial \ln Q_1(T,V)}{\partial T}\right)_V + 1.$$

Furthermore, for an ideal gas of massive particles (see Problems 7.1 and 7.19), we have

$$Q_1(T,V) = V\left(\frac{2\pi mkT}{h^2}\right)^{3/2}$$

and $PV = NkT$. Combining these results, we can rewrite the partition function of the single particle in terms of T and P

$$Q_1(T,P,N) = \frac{TNk}{P}\left(\frac{2\pi mkT}{h^2}\right)^{3/2}$$

and therefore

$$\left(\frac{\partial \ln Q_1(T,V)}{\partial T}\right)_V = \left(\frac{\partial \ln Q_1(T,P,N)}{\partial T}\right)_{P,N} - \frac{1}{T}.$$

The previous equation for the entropy is then equivalent to

$$\frac{S(T,P,N)}{Nk} = \ln\left(\frac{Q_1(T,P,N)}{N}\right) + T\left(\frac{\partial \ln Q_1(T,P,N)}{\partial T}\right)_{P,N}$$

that is the desired result.

Problem 7.33.
Consider a statistical system with a fixed number of particles and prove the Gibbs-

Helmholtz equation

$$H = \Phi + TS = -T^2 \left[\frac{\partial(\Phi/T)}{\partial T} \right]_P$$

with H the enthalpy and Φ the Gibbs potential. Finally, substituting in H and $\Phi + TS$ the corresponding expressions in terms of partition functions, verify that we obtain an identity.

Solution

Let us recall the definitions of the thermodynamic potentials

$$H \equiv H(S,P) = U + PV$$

$$F \equiv F(T,V) = U - TS$$

$$\Phi \equiv \Phi(T,P) = F + PV.$$

From the second equation we obtain $U = F + TS$. Substituting this result in the Gibbs potential, we find

$$\Phi = F + PV = U - TS + PV = H - TS.$$

We also know that

$$d\Phi = -SdT + VdP$$

from which

$$S = -\left(\frac{\partial \Phi}{\partial T} \right)_P$$

and

$$H = \Phi + TS = \Phi - T\left(\frac{\partial \Phi}{\partial T} \right)_P = -T^2 \left[\frac{\partial(\Phi/T)}{\partial T} \right]_P.$$

In the context of the canonical ensemble we can write

$$F = -kT \ln Q_N$$

$$S = -\left(\frac{\partial F}{\partial T} \right)_V = k \left[\frac{\partial(T \ln Q_N)}{\partial T} \right]_V = k \ln Q_N + kT \left(\frac{\partial \ln Q_N}{\partial T} \right)_V,$$

$$P = -\left(\frac{\partial F}{\partial V} \right)_T = kT \left(\frac{\partial \ln Q_N}{\partial V} \right)_T$$

$$U = kT^2 \left(\frac{\partial \ln Q_N}{\partial T} \right)_V.$$

Therefore, we obtain

$$H = U + PV = kT^2 \left(\frac{\partial \ln Q_N}{\partial T} \right)_V + VkT \left(\frac{\partial \ln Q_N}{\partial V} \right)_T$$

$$\Phi = F + PV = -kT \ln Q_N + VkT \left(\frac{\partial \ln Q_N}{\partial V} \right)_T$$

from which we see that the relation $H = \Phi + TS$ is verified.

Problem 7.34.

A three dimensional gas is in thermal equilibrium at temperature T and is characterized by N indistinguishable and non interacting ultrarelativistic particles. The Hamiltonian of the single particle is

$$H = pc$$

with p the absolute value of the momentum and c the speed of light. Write down the canonical partition function for this system. Also, find the density of states $g(E,V,N)$, verifying explicitly that

$$\left(\frac{\partial S}{\partial E} \right)_{V,N} \Bigg|_{E=U} = \frac{1}{T}$$

where $S = k \ln g$ is the entropy and U is the average energy such that $U = \langle E \rangle$.

Solution

The particles are non interacting and the canonical partition function Q_N is written as

$$Q_N(T,V,N) = \frac{Q_1^N(T,V)}{N!}$$

where $Q_1(T,V)$ is the partition function of the single particle expressed as

$$Q_1(T,V) = \int \frac{d^3 p \, d^3 q}{h^3} e^{-\beta pc} = \frac{4\pi V}{h^3} \int_0^{+\infty} p^2 e^{-\beta pc} dp.$$

We can solve the integral exactly to find

$$Q_N(T,V,N) = \frac{1}{N!} \left[8\pi V \left(\frac{kT}{hc} \right)^3 \right]^N.$$

The density of states is the inverse Laplace transform of the partition function

$$g(E,V,N) = \frac{1}{2\pi i} \int_{\beta'-i\infty}^{\beta'+i\infty} Q_N(T,V,N) e^{\beta E} d\beta =$$

$$\frac{1}{2\pi} \int_{-\infty}^{+\infty} Q_N \left(\frac{1}{k(\beta'+i\beta'')}, V, N \right) e^{(\beta'+i\beta'')E} d\beta''$$

where $\beta' > 0$. The variable $\beta = \beta' + i\beta''$ is now treated as a complex variable and the integration path is parallel and to the right of the imaginary axis, i.e. along the straight line $Re(\beta) = \beta' = \text{const}$. We can make use of the general result

$$\frac{1}{2\pi i} \int_{\beta'-i\infty}^{\beta'+i\infty} \frac{e^{\beta x}}{\beta^{\alpha+1}} d\beta = \frac{x^\alpha}{\alpha!} \quad x \geq 0$$

and set $\alpha = 3N - 1$, $x = E$ to obtain

$$g(E,V,N) = \frac{1}{N!(3N-1)!} \left(\frac{8\pi V}{h^3 c^3}\right)^N E^{3N-1}.$$

In the thermodynamic limit, $N \gg 1$, and we can use $3N$ instead of $3N - 1$. Thus, the entropy (see also Problem 6.19) is

$$S = 3Nk \ln E + \text{const.}$$

where the constant does not depend on E. This means that

$$\left(\frac{\partial S}{\partial E}\right)_{V,N} = \frac{3Nk}{E}.$$

At the same time, we can compute the average energy from the canonical partition function as

$$U = -\left(\frac{\partial \ln Q_N}{\partial \beta}\right)_{V,N} = 3NkT$$

which can be substituted in the previous expression to get

$$\left(\frac{\partial S}{\partial E}\right)_{V,N}\Bigg|_{E=U} = \frac{1}{T}.$$

Problem 7.35.
Characterize the energy fluctuations for a statistical system composed of a fixed number of N distinguishable particles with two energy levels, ε_1 and ε_2, with degeneracy g_1 and g_2 respectively.

Solution

From the canonical ensemble we know that the fluctuations of the energy are given by

$$\langle (\Delta E)^2 \rangle = kT^2 C$$

where C is the specific heat that can be computed from the partition function

$$Q_N(T,N) = \left(\sum_{r=1}^{2} g_r e^{-\beta \varepsilon_r}\right)^N = \left(g_1 e^{-\beta \varepsilon_1} + g_2 e^{-\beta \varepsilon_2}\right)^N.$$

The average energy is computed as

$$U = -\left(\frac{\partial \ln Q_N}{\partial \beta}\right)_N = N\frac{g_1\varepsilon_1 + g_2\varepsilon_2 e^{-\beta\Delta}}{g_1 + g_2 e^{-\beta\Delta}}$$

with $\Delta = \varepsilon_2 - \varepsilon_1$. We therefore find the specific heat

$$C = \left(\frac{\partial U}{\partial T}\right)_N = -\frac{1}{kT^2}\left(\frac{\partial U}{\partial \beta}\right)_N = \frac{N}{kT^2}\frac{g_1 g_2 \Delta^2 e^{-\beta\Delta}}{(g_1 + g_2 e^{-\beta\Delta})^2}$$

leading to

$$\langle(\Delta E)^2\rangle = N\frac{g_1 g_2 \Delta^2 e^{-\beta\Delta}}{(g_1 + g_2 e^{-\beta\Delta})^2}$$

that is the desired result.

Problem 7.36.
An ideal gas with N non interacting indistinguishable particles with mass m is located in a spherical container with radius R. The system is subject to an external force with potential

$$V(r) = \alpha r^3 \qquad \alpha > 0$$

with r the distance from the center of the sphere. Find the partition function for the system in thermal equilibrium at temperature T and the pressure as a function of the distance from the center. At some point, a small hole with area σ is pierced on the surface of the container, perpendicular to some given direction (say x). In the limit of external weak field ($\alpha \to 0$), find the pressure on the surface of the container and give an estimate for the x momentum component transfered in the unit time through the small hole at constant T.

Solution
The partition function of the single particle is

$$Q_1(T) = \int \frac{d^3q\, d^3p}{h^3} e^{-\beta\frac{p^2}{2m} - \beta\alpha r^3}.$$

With the help of spherical polar coordinates $d^3q = 4\pi r^2 dr$, and the substitution $x = r^3$, we rewrite $Q_1(T)$ as

$$Q_1(T) = \frac{4\pi}{3}\int e^{-\beta\frac{p^2}{2m}} d^3p \int_0^{R^3} e^{-\beta\alpha x} dx = \frac{4\pi}{3}\lambda^{-3}\int_0^{R^3} e^{-\beta\alpha x} dx$$

where we use the characteristic length $\lambda = \frac{h}{\sqrt{2\pi m kT}}$ coming from the Gaussian integral over momenta. The final result is

$$Q_1(T) = \frac{4\pi kT}{3\alpha}\lambda^{-3}(1 - e^{-\beta\alpha R^3})$$

that, combined with the expression for the total partition function

$$Q_N(T,N) = \frac{Q_1^N(T)}{N!}$$

is the answer to the first question. As for the second point, we can look at the local chemical potential which is given by

$$\mu_{tot} = \mu_{id} + \alpha r^3$$

where

$$\mu_{id} = kT \ln(n\lambda^3)$$

is the chemical potential of an ideal free gas with density n and Hamiltonian $H = p^2/2m$. Due to the presence of the potential $V(r)$, the system develops an inhomogeneous density $n(r)$ depending on the radial position r. When moving from $r = 0$ to a generic r, we can impose the condition that μ_{tot} stays constant and find

$$kT \ln(n(r)\lambda^3) + \alpha r^3 = kT \ln(n(0)\lambda^3)$$

that is

$$\ln\left(\frac{n(r)}{n(0)}\right) = -\beta \alpha r^3$$

from which

$$n(r) = n(0)\, e^{-\beta \alpha r^3}.$$

In order to find $n(0)$, we impose that the integral of $n(r)$ over all the domain is exactly equal to N

$$4\pi \int_0^R r^2 n(r)dr = 4\pi n(0) \int_0^R r^2 e^{-\beta \alpha r^3} dr = \frac{4\pi kT\, n(0)}{3\alpha}\left(1 - e^{-\beta \alpha R^3}\right) = N$$

and we find

$$n(0) = \frac{3\alpha N}{4\pi kT}\frac{1}{1 - e^{-\beta \alpha R^3}}.$$

To compute the pressure, we use the ideal gas law $P(r) = n(r)kT$ to find

$$P(r) = \frac{3\alpha N}{4\pi}\frac{e^{-\beta \alpha r^3}}{1 - e^{-\beta \alpha R^3}}.$$

When $\alpha \to 0$, the numerator is getting close to 1 while the denominator to $\beta \alpha R^3 = \alpha R^3/kT$. We know that $4\pi R^3/3$ is the volume of the sphere and we find

$$P = \frac{NkT}{V}.$$

As for the effusion problem, the transferred momentum M can be found using simple considerations (see also the section on kinetic physics, Problem 9.3). If we consider that the surface of the hole is perpendicular to the x axis and we impose that the transferred momentum obeys a Maxwellian distribution function

$$f(v) = n \left(\frac{m}{2\pi kT} \right)^{3/2} e^{-\frac{1}{2}\beta m v_x^2 - \frac{1}{2}\beta m v_y^2 - \frac{1}{2}\beta m v_z^2}$$

we can write

$$M = \sigma m \int_{v_x > 0} f(v) v_x^2 d^3 v =$$

$$\sigma m n \left(\frac{m}{2\pi kT} \right)^{3/2} \left(\int_{-\infty}^{+\infty} e^{-\frac{1}{2}\beta m s^2} ds \right)^2 \left(\int_0^{+\infty} e^{-\frac{1}{2}\beta m v_x^2} v_x^2 dv_x \right) = \frac{1}{2} \sigma n kT$$

where, in the limit $\alpha \to 0$, the density $n = N/V$ is homogeneous and does not dependent on the radial position.

Problem 7.37.
A statistical system is characterized by N distinguishable and non interacting atoms in thermal equilibrium with a reservoir at temperature T. Each atom can occupy the energy levels $E_n = (n+1)\varepsilon$ ($\varepsilon > 0$, $n = 0, 1, 2, ..., +\infty$) and the degeneracy of the n-th level is equal to $g_n = \lambda^n$, with $\lambda > 1$. Compute the canonical partition function, the specific heat, and analyze the result at low temperatures. Is there any temperature T_c above which the canonical description is not well posed?

Solution
The total partition function is

$$Q_N(T,N) = Q_1^N(T)$$

where Q_1 is the partition function for the single atom. A direct calculation can be performed with the result

$$Q_1(T) = \sum_{n=0}^{+\infty} \lambda^n e^{-\beta(n+1)\varepsilon} = e^{-\beta\varepsilon} \sum_{n=0}^{+\infty} e^{-(\beta\varepsilon - \ln\lambda)n}$$

that is

$$Q_1(T) = \frac{1}{e^{\beta\varepsilon} - \lambda}.$$

The previous calculation is meaningless if we have

$$e^{\ln\lambda - \beta\varepsilon} \geq 1$$

that is $T \geq T_c = \frac{\varepsilon}{k \ln\lambda}$, where this T_c is a characteristic temperature above which our statistical treatment is not well posed. When $T < T_c$, we can define and compute the average energy as

$$U = -\left(\frac{\partial \ln Q_N}{\partial \beta} \right)_N = \frac{N\varepsilon}{1 - \lambda e^{-\beta\varepsilon}}$$

and the specific heat as

$$C = \left(\frac{\partial U}{\partial T} \right)_N = \frac{N\lambda\varepsilon^2 e^{-\beta\varepsilon}}{kT^2(1 - \lambda e^{-\beta\varepsilon})^2}.$$

In the limit of low temperatures the specific heat goes to zero

$$\lim_{T \to 0} C = 0.$$

Problem 7.38.
Consider an ideal gas formed by atoms of type A and atoms of type B. These atoms can bound to each other and form the molecule AB, according to the reaction

$$A + B \leftrightarrow AB$$

taking place at temperature T in a volume V. If N_A, N_B, N_{AB} are the numbers ($\gg 1$) of particles for the species entering the reaction, show that

$$\frac{N_{AB}}{N_A N_B} = \frac{f_{AB}}{f_A f_B}$$

where f_X ($X = A, B, AB$) is the single particle partition function. Treat all the atoms as indistinguishable.

Solution
The equilibrium condition is given by the equality of the chemical potentials

$$\mu_A + \mu_B = \mu_{AB}$$

where μ_X is the chemical potential of the system X. From the canonical ensemble we know that

$$\mu_X = -kT \left(\frac{\partial \ln Q_X}{\partial N_X} \right)_{T,V}$$

where $Q_X(T,V,N)$ is the partition function. We find

$$\left(\frac{\partial \ln Q_A}{\partial N_A} \right)_{T,V} + \left(\frac{\partial \ln Q_B}{\partial N_B} \right)_{T,V} = \left(\frac{\partial \ln Q_{AB}}{\partial N_{AB}} \right)_{T,V}.$$

We recall that

$$Q_X(T,V,N_X) = \frac{f_X^{N_X}(T,V)}{N_X!}$$

and we use the Stirling approximation for the factorials

$$\ln \left(\frac{f_A}{N_A} \right) + \ln \left(\frac{f_B}{N_B} \right) = \ln \left(\frac{f_{AB}}{N_{AB}} \right)$$

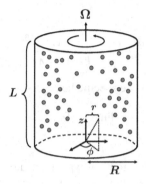

Fig. 7.6 An ideal classical gas formed by N molecules with mass m is placed inside a cylinder with radius R and height L. The cylinder is rotating with angular velocity Ω around its symmetry axis. The particles density is inhomogeneous in space and is studied in Problem 7.39

leading to

$$\frac{N_{AB}}{N_A N_B} = \frac{f_{AB}}{f_A f_B}$$

which is the desired result.

Problem 7.39.
Let us consider an ideal classical gas formed by $N \gg 1$ indistinguishable particles with mass m placed in a cylinder of radius R and height L rotating around its axis with angular velocity Ω. The resulting Hamiltonian of the single particle is

$$H = \frac{1}{2m}p^2 - \frac{1}{2}m\Omega^2 r^2.$$

with p the absolute value of the momentum and r the distance from the rotation axis. The whole system is in equilibrium at temperature T. Using the canonical ensemble, compute the gas density in the cylinder and discuss the limit $\Omega \to 0$. Finally, determine the pressure on the surface of the container.

Solution
We use cylindrical coordinates (r, z, ϕ) and start from the Hamiltonian of the single particle H (see Fig. 7.6). The presence of the term $\frac{1}{2}m\Omega^2 r^2$ in the Hamiltonian produces an inhomogeneous density of particles throughout the cylinder. The local density $n(r, z, \phi)$ can be written as

$$n(r, z, \phi) = N\frac{\int e^{-\beta H} d^3 p}{Q_1(T, R)}$$

with $Q_1(T, R)$ the partition function of the single particle. We remark that $n(r, z, \phi)$ is normalized in such a way that

$$N = \int_0^L dz \int_0^{2\pi} d\phi \int_0^R n(r, z, \phi)\, r\, dr.$$

The single particle partition function is

$$Q_1(T,R) = \frac{1}{h^3} \int e^{-\beta \frac{p^2}{2m}} d^3p \int_0^L dz \int_0^{2\pi} d\phi \int_0^R r e^{\beta \frac{m\Omega^2 r^2}{2}} dr =$$

$$\left(\frac{2\pi m}{h^2\beta}\right)^{3/2} 2\pi L \int_0^R r e^{\beta m\Omega^2 r^2/2} dr = \left(\frac{2\pi m}{h^2\beta}\right)^{3/2} \frac{2\pi L}{\beta m\Omega^2} \left(e^{\frac{\beta m\Omega^2 R^2}{2}} - 1\right).$$

From the previous expression we find

$$n(r,z,\phi) = n(r) = \frac{Nm\Omega^2\beta}{2\pi L} \frac{e^{\beta m\Omega^2 r^2/2}}{e^{\beta m\Omega^2 R^2/2} - 1}.$$

The dependence of the density on the local coordinates is only in the radial distance r, as it should be expected because the term $m\Omega^2 r^2/2$ (that is responsible for the inhomogeneity) is only dependent on r. The same result can be obtained by imposing a constant local chemical potential in the cylinder (similarly to what we have done in Problem 7.36). The local chemical potential is defined by

$$\mu_{TOT} = \mu_{id} - \frac{m\Omega^2 r^2}{2}.$$

In the above expression, μ_{id} is the chemical potential of an ideal gas with density n and Hamiltonian $H = p^2/2m$

$$\mu_{id} = kT \ln(n\lambda^3)$$

where

$$\lambda = \frac{h}{\sqrt{2\pi mkT}}$$

is the thermal length scale. The density n must depend on r in order to guarantee the constancy of μ_{TOT}, i.e.

$$kT \ln(n(0)\lambda^3) = kT \ln(n(r)\lambda^3) - \frac{m\Omega^2 r^2}{2}$$

from which we find

$$n(r) = n(0) e^{\frac{\beta m\Omega^2 r^2}{2}}.$$

The normalization $n(0)$ is found by imposing $\int n(r) \, r dr \, d\phi \, dz = N$, yielding the very same result previously found. The limit $\Omega \to 0$ corresponds to the case where the density becomes homogeneous. This can be seen by Taylor expanding the exponential functions in the density for $\Omega \to 0$

$$n(r) \approx \frac{Nm\Omega^2\beta}{2\pi L} \frac{\left(1 + \frac{\beta m\Omega^2 r^2}{2} + \mathcal{O}(\Omega^2)\right)}{\left(\frac{\beta m\Omega^2 R^2}{2} + \mathcal{O}(\Omega^2)\right)}$$

to find

$$n(r) \approx \frac{N}{\pi R^2 L} = \frac{N}{V}$$

with $V = \pi R^2 L$ the volume of the cylinder. As for the pressure, it can be derived directly from the total free energy $F = -kT \ln Q_N$, with the total partition function written as $Q_N(T,R,N) = \frac{Q_1^N(T,R)}{N!}$, and the definition

$$P = -\left(\frac{\partial F}{\partial V}\right)_{T,N}.$$

Differentiation with respect to V is connected to differentiation with respect to R as $\frac{\partial}{\partial V} = \frac{1}{2\pi RL} \frac{\partial}{\partial R}$, so that we find

$$P = \frac{Nm\Omega^2}{2\pi L} \frac{e^{\frac{\beta m\Omega^2 R^2}{2}}}{\left(e^{\frac{\beta m\Omega^2 R^2}{2}} - 1\right)}$$

that is the equation for the ideal gas, $P = n(R)kT$, with the local density in R

$$n(R) = \frac{Nm\Omega^2\beta}{2\pi L} \frac{e^{\beta m\Omega^2 R^2/2}}{e^{\beta m\Omega^2 R^2/2} - 1}.$$

Problem 7.40.
Using the first law of thermodynamics, write the chemical potential in terms of energy derivatives. Repeat this computation writing it in terms of entropy derivatives. Using the Sackur-Tetrode formula for the entropy (see Problem 6.26), show that these two formulae for the chemical potential lead to the same result that is found using the formalism of the canonical ensemble.

Solution
The first law of thermodynamics, including a chemical potential, can be written as

$$dU = TdS - PdV + \mu dN.$$

Differentiating with respect to N, keeping S and V constant, we get

$$\mu = \left(\frac{\partial U}{\partial N}\right)_{S,V}.$$

Let us now rewrite our first equation as

$$dS = \frac{dU}{T} + \frac{P}{T}dV - \frac{\mu}{T}dN.$$

At the same time, differentiating with respect to N, keeping V and U constant, we get

$$-\frac{\mu}{T} = \left(\frac{\partial S}{\partial N}\right)_{U,V}.$$

To compute these derivatives we will need the Sackur-Tetrode formula for an ideal gas in a box of volume V

$$S(U,V,N) = Nk\left\{\frac{5}{2} - \ln\left[\left(\frac{3\pi\hbar^2}{m}\right)^{3/2}\frac{N^{\frac{5}{2}}}{VU^{\frac{3}{2}}}\right]\right\}$$

and the associated energy

$$U(S,V,N) = \left(\frac{3\pi\hbar^2}{m}\right)\frac{N^{\frac{5}{3}}}{V^{\frac{2}{3}}}e^{\frac{2S}{3Nk}-\frac{5}{3}}.$$

An explicit computation gives

$$-\frac{\mu}{T} = \left(\frac{\partial S}{\partial N}\right)_{U,V} = k\ln\left[\left(\frac{m}{3\pi\hbar^2}\right)^{3/2}\frac{VU^{3/2}}{N^{\frac{5}{2}}}\right].$$

From the first law of thermodynamics we find

$$T = \left(\frac{\partial U}{\partial S}\right)_{V,N} = \left(\frac{3\pi\hbar^2}{m}\right)\frac{N^{\frac{5}{3}}}{V^{\frac{2}{3}}}e^{\frac{2S}{3Nk}-\frac{5}{3}}\frac{2}{3Nk} = \frac{2U}{3Nk}$$

from which we find $U = \frac{3}{2}NkT$ which can be plugged back in $-\frac{\mu}{T}$ to recover the chemical potential written as

$$\mu = -kT\ln\left[\frac{V}{N}\left(\frac{mkT}{2\pi\hbar^2}\right)^{\frac{3}{2}}\right].$$

As for the expression of μ in terms of energy derivatives, we get

$$\mu = \left(\frac{\partial U}{\partial N}\right)_{S,V} = \left(\frac{5\pi\hbar^2}{m}\right)\left(\frac{N}{V}\right)^{\frac{2}{3}}e^{\frac{2S}{3Nk}-\frac{5}{3}} - \frac{2S}{3N^2k}\left(\frac{3\pi\hbar^2}{m}\right)\frac{N^{\frac{5}{3}}}{V^{\frac{2}{3}}}e^{\frac{2S}{3Nk}-\frac{5}{3}} =$$

$$e^{\frac{2S}{3Nk}-\frac{5}{3}}\frac{\pi\hbar^2}{mV^{\frac{2}{3}}}\left(5N^{\frac{2}{3}} - \frac{2SN^{-\frac{1}{3}}}{k}\right) = \frac{1}{3}N^{-\frac{5}{3}}U\left(5N^{\frac{2}{3}} - \frac{2SN^{-\frac{1}{3}}}{k}\right) =$$

$$\frac{2U}{3N}\ln\left[\left(\frac{3\pi\hbar^2}{m}\right)^{\frac{3}{2}}\frac{N^{\frac{5}{2}}}{VU^{\frac{3}{2}}}\right] = -kT\ln\left[\frac{V}{N}\left(\frac{mkT}{2\pi\hbar^2}\right)^{\frac{3}{2}}\right]$$

which coincides with the expression of μ in terms of entropy derivatives. It is now left to check these formulae against the result that can be found using the canonical ensemble formalism. The single particle partition function is

$$Q_1(T,V) = \frac{1}{h^3} \int e^{-\beta \frac{p^2}{2m}} d^3p \, d^3q = \frac{V}{h^3} \left(\int_{-\infty}^{+\infty} e^{-\beta \frac{p^2}{2m}} dp \right)^3 = \frac{V}{h^3} (2\pi mkT)^{\frac{3}{2}}.$$

The N particles partition function is readily obtained

$$Q_N(T,V,N) = \frac{Q_1(T,V)}{N!}$$

from which the free energy $F = -kT \ln Q_N$ follows. At last, using the Stirling approximation ($N! \approx N^N e^{-N}$), we get

$$\mu = \left(\frac{\partial F}{\partial N} \right)_{T,V} \approx -kT \ln \left(\frac{Q_1}{N} \right) = -kT \ln \left[\frac{V}{N} \left(\frac{mkT}{2\pi \hbar^2} \right)^{\frac{3}{2}} \right].$$

8

Grand Canonical Ensemble

Problem 8.1.
A gas is in contact with a surface. On the surface we find N_0 localized and distinguishable sites adsorbing N ($N \leq N_0$) molecules of the gas (each site can adsorb zero or one molecule of the gas). Find the grand canonical partition function of the system, and determine the chemical potential as a function of the average number of particles $\langle N \rangle$ which are adsorbed by the surface. You can think that the canonical partition function of an adsorbed molecule is a function only of the temperature, $Q(T)$, and that all the adsorbed molecules are non interacting.

Solution
Due to the independence of the molecules, the canonical partition function for N adsorbed molecules is the product of N single molecule partition functions. Moreover, being the sites localized (distinguishable), we have to determine all the possible ways to select N sites out of the N_0 available. This provides the following canonical partition function for the N adsorbed molecules

$$Q_N(T,N) = \frac{N_0!}{N!(N_0-N)!} Q^N(T).$$

We now determine the grand canonical partition function summing over all the possible values of N, i.e. $N = 0, 1, 2, ..., N_0$. The result is

$$\mathscr{Q}(T,z) = \sum_{N=0}^{N_0} z^N Q_N(T,N) = \sum_{N=0}^{N_0} \frac{N_0!}{N!(N_0-N)!} z^N Q^N(T)$$

with z the fugacity. The above expression can be directly summed using the binomial formula

$$\mathscr{Q}(T,z) = \sum_{N=0}^{N_0} \frac{N_0!}{N!(N_0-N)!} z^N Q^N(T) =$$
$$\sum_{N=0}^{N_0} \frac{N_0!}{N!(N_0-N)!} (zQ(T))^N 1^{N_0-N} = (zQ(T)+1)^{N_0}.$$

Cini M., Fucito F., Sbragaglia M.: Solved Problems in Quantum and Statistical Mechanics.
DOI 10.1007/978-88-470-2315-4_8, © Springer-Verlag Italia 2012

The average number of adsorbed molecules is

$$\langle N \rangle = z \left(\frac{\partial \ln \mathscr{Q}}{\partial z} \right)_T = N_0 \frac{zQ(T)}{zQ(T)+1} = \frac{N_0}{z^{-1}Q^{-1}(T)+1}$$

that implies

$$1 + \frac{1}{zQ(T)} = \frac{N_0}{\langle N \rangle}$$

or, alternatively

$$zQ(T) = e^{\beta \mu} Q(T) = \frac{\langle N \rangle}{N_0 - \langle N \rangle}.$$

Therefore, the chemical potential can be given as a function of T, $Q(T)$, $\langle N \rangle$ and N_0

$$\mu = kT \ln \left(\frac{\langle N \rangle}{Q(T)(N_0 - \langle N \rangle)} \right).$$

Problem 8.2.
Let us consider a gas of N molecules in a volume V at temperature T with mass m and fugacity

$$z(T,V,N) = \left(\frac{N}{V} \right) \lambda^3$$

with $\lambda = \frac{h}{\sqrt{2\pi mkT}}$ and h the Planck constant. Some of these molecules are bound to some independent attraction sites (each site cannot have more than one molecule), and the total number of these sites is N_0. The associated canonical partition function, $Q(T)$, of the site-molecule system is made of the bound state formed from the site and the molecule (see also Fig. 8.1). Using the grand canonical ensemble, compute the average number of molecules for each site, together with the associated probability to have zero and one molecule respectively. Comment on the limit where the molecules density goes to zero and the temperature is high.

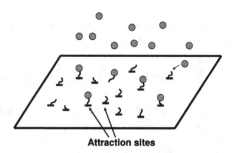

Attraction sites

Fig. 8.1 A molecular gas is in contact with a wall where N_0 independent attraction sites are localized. The average number of adsorbed molecules can be computed using the grand canonical ensemble (see Problem 8.2)

Solution

The grand canonical partition function for the single site is

$$\mathcal{Q}_{site}(T,z) = 1 + zQ(T)$$

and the probabilities to find 0 or 1 molecule in the site are evaluated as

$$P_0 = \frac{1}{1+zQ(T)} \qquad P_1 = \frac{zQ(T)}{1+zQ(T)}.$$

From the probability, we compute the average number of molecules in each site

$$\langle N \rangle_{site} = \sum_{n=0}^{1} nP_n = \frac{zQ(T)}{1+zQ(T)} = \frac{1}{z^{-1}Q^{-1}(T)+1}$$

meaning that the total average number of molecules adsorbed by the N_0 sites is

$$\langle N \rangle_{sites} = N_0 \langle N \rangle_{site} = \frac{N_0}{z^{-1}Q^{-1}(T)+1}.$$

The same result is obtained by considering the grand canonical partition function of the N_0 sites (see Problem 8.1). Considering the fugacity $z = \left(\frac{N}{V}\right)\lambda^3$, where $\lambda = h(2\pi mkT)^{-1/2}$, we see that only in the limit of small density and high temperature (i.e. $z \ll 1$) all the sites are empty

$$\lim_{z \to 0} \langle N \rangle_{sites} = 0.$$

Problem 8.3.

Consider a solid-gas (s-g) system in equilibrium. Compute the variation of the pressure with respect to the temperature using the Clausius-Clapeyron equation

$$\left(\frac{dP}{dT}\right)_c = \frac{\Delta h}{T\Delta v}$$

with $\Delta v = v_g - v_s$ and $\Delta h = h_g - h_s$ the variations of the specific volumes and enthalpies respectively (i.e. volume/enthalpy for a single particle). Consider that the specific volume of the solid is small, so that $v_s \ll v_g$, and assume that the gas is well approximated by an ideal one. Then, define the specific heat at the coexistence of the two phases, $c_c = T\left(\frac{\partial s}{\partial T}\right)_c$, and prove the following relation

$$c_c = c_P - T\left(\frac{\partial v}{\partial T}\right)_P \left(\frac{dP}{dT}\right)_c$$

with c_P the specific heat at constant pressure. In the above expression, the subscript c means that we evaluate the different quantities on the coexistence curve for the solid-gas equilibrium. To answer the first question, approximate the solid as a collection of $N_s \gg 1$ distinguishable three dimensional quantum harmonic oscillators. To answer the second question, choose the equilibrium pressure as a function of the temperature, $P = P(T)$, justifying this choice.

Solution

If we assume that the change in the volume is mainly due to the gas, we find that $\Delta v \approx v_g = \frac{kT}{P}$ and

$$\left(\frac{dP}{dT} \right)_c \approx \frac{P \Delta h}{kT^2}.$$

We now need to compute both the solid and gas enthalpy from which we can extract Δh. The grand canonical partition function of the gas is given by

$$\mathscr{Q}(T, V_g, z_g) = \sum_{N=0}^{+\infty} z_g^N Q_N(T, V_g, N) = \sum_{N=0}^{+\infty} \frac{(z_g V_g f(T))^N}{N!} = e^{z_g V_g f(T)}$$

where $Q_N(T, V_g, N) = \frac{(V_g f(T))^N}{N!}$ is the N particles canonical partition function (see Problem 7.1) and

$$f(T) = h^{-3} (2\pi mkT)^{\frac{3}{2}} \qquad z_g = e^{\mu_g / kT}$$

where μ_g is the chemical potential of the gas. The pressure is found by taking the logarithm of the grand canonical partition function

$$P = \frac{kT}{V_g} \ln \mathscr{Q} = z_g kT f(T)$$

and for the average number (N_g) and average energy (U_g) we get

$$N_g = z_g \left(\frac{\partial \ln \mathscr{Q}}{\partial z_g} \right)_{T, V_g} = z_g V_g f(T)$$

$$U_g = -\left(\frac{\partial \ln \mathscr{Q}}{\partial \beta} \right)_{V_g, z_g} = z_g V_g kT^2 f'(T).$$

As for the solid, we can model it as a system of three dimensional quantum harmonic oscillators which are distinguishable and localized in a volume V_s. The grand canonical partition function is

$$\mathscr{Q}(T, z_s) = \sum_{N=0}^{+\infty} (z_s \phi(T))^N = \sum_{N=0}^{+\infty} \left(z_s \left(\sum_{n=0}^{+\infty} e^{-\beta \hbar \omega (n + \frac{1}{2})} \right)^3 \right)^N = (1 - z_s \phi(T))^{-1}$$

where

$$\phi(T) = \left(2 \sinh \left(\frac{\hbar \omega}{2kT} \right) \right)^{-3}.$$

The resulting pressure (P_s), average number (N_s), and average energy (U_s) are

$$P_s = \frac{kT}{V_s} \ln \mathscr{Q} = -\frac{kT}{V_s} \ln(1 - z_s \phi(T))$$

$$N_s = z_s \left(\frac{\partial \ln \mathscr{Q}}{\partial z_s} \right)_T = \frac{z_s \phi(T)}{1 - z_s \phi(T)}$$

$$U_s = -\left(\frac{\partial \ln \mathscr{Q}}{\partial \beta} \right)_{z_s} = \frac{z_s kT^2 \phi'(T)}{1 - z_s \phi(T)}.$$

Since z_s and T are intensive, the pressure of the solid goes to zero when $V_s \to +\infty$. Moreover, the equation for N_g is equivalent to

$$z_g = N_g / (V_g f(T))$$

while, from the equation for N_s, we have

$$z_s \phi(T) = \frac{N_s}{N_s + 1} \approx 1 - \frac{1}{N_s}$$

so that $z_s = \phi^{-1}(T)$ in the limit $N_s \gg 1$. The equilibrium condition implies the equality of fugacities $(z_s = z_g)$

$$\frac{N_g}{V_g} = \frac{P}{kT} = \frac{f(T)}{\phi(T)}.$$

This justifies the choice $P = P(T)$ to be used later. Let us now determine the variation of the specific enthalpy. We know that

$$U_g = z_g V_g kT^2 f'(T) = N_g kT^2 f'(T)/f(T)$$

from which we extract the enthalpy for the gas

$$H_g = U_g + PV_g = \frac{N_g kT^2 f'(T)}{f(T)} + N_g kT$$

where we have used the ideal gas law. In a similar way, the enthalpy for the solid reads

$$H_s = U_s - kT \ln(1 - z_s \phi(T)) = N_s kT^2 \frac{\phi'(T)}{\phi(T)} - kT \ln(1 - z_s \phi(T)).$$

Therefore, we can evaluate $\Delta h = \frac{H_g}{N_g} - \frac{H_s}{N_s}$ which is the specific enthalpy variation. As we noted above, the choice $P = P(T)$ is justified by the condition of thermodynamic equilibrium and, when looking at the entropy density s as a function of the pressure P and temperature T, we have

$$c_c = T\left(\frac{\partial s}{\partial T}\right)_c = T\left(\frac{\partial s}{\partial T}\right)_P + T\left(\frac{\partial s}{\partial P}\right)_T\left(\frac{dP}{dT}\right)_c = c_P - T\left(\frac{\partial v}{\partial T}\right)_P\left(\frac{dP}{dT}\right)_c$$

where we have used the Maxwell relation

$$\left(\frac{\partial s}{\partial P}\right)_T = -\left(\frac{\partial v}{\partial T}\right)_P.$$

The latter can be derived from the differential expression for the chemical potential $d\mu = -s dT + v dP$, that implies

$$\left(\frac{\partial \mu}{\partial T}\right)_P = -s \qquad \left(\frac{\partial \mu}{\partial P}\right)_T = v.$$

If we differentiate the first expression with respect to P, and the second with respect to T, we find the aforementioned Maxwell relation. To compute c_c, we first need to compute c_P

$$c_P = \left(\frac{\partial h}{\partial T}\right)_P$$

with $h = h_s + h_g$. Since h_s and h_g have been computed before, c_P is known. Moreover, $\left(\frac{\partial v}{\partial T}\right)_P \approx \left(\frac{\partial v_g}{\partial T}\right)_P = k/P$, because the ideal gas law is valid. Finally, using the Clausius-Clapeyron equation for $\left(\frac{dP}{dT}\right)_c$, we know all the terms entering the definition of c_c.

Problem 8.4.

Consider a three dimensional classical gas of independent and indistinguishable particles with a single particle Hamiltonian $H(p,q) = F(p)$, where p and q are the momentum and position of the single particle. Prove that, in the grand canonical ensemble, the probability to have N molecules

$$P(N) = \frac{z^N Q_N(T,V,N)}{\mathscr{Q}(T,V,z)}$$

is a Poisson distribution. In the above expression, $Q_N(T,V,N)$, z and $\mathscr{Q}(T,V,z)$ are the canonical partition function of N particles, the fugacity, and the grand canonical partition function respectively. Determine the fluctuations $\langle (\Delta N)^2 \rangle$, verifying that the correct result for the Poisson distribution is recovered, i.e. that the fluctuations coincide with the average value.

Solution
The definition of the grand canonical partition function is

$$\mathscr{Q}(T,V,z) = \sum_{N=0}^{+\infty} z^N Q_N(T,V,N)$$

with $Q_N(T,V,N)$ the canonical partition function for N particles

$$Q_N(T,V,N) = \frac{1}{N!h^{3N}} \int e^{-\beta H_N(p_1,p_2,\dots,q_1,q_2,\dots)} d^{3N}p \, d^{3N}q =$$

$$\frac{1}{N!h^{3N}} \left(\int e^{-\beta F(p)} d^3p \, d^3q \right)^N = \frac{1}{N!} Q_1^N(T,V) = \frac{1}{N!}(Vf(T))^N$$

where H_N is the total Hamiltonian (the sum of the single particle Hamiltonians), and $f(T)$ is a function of the temperature coming from the integral in the momentum space: since the exact dependence on the temperature does not play a relevant role in what follows, we will keep it in a very general form. The average number of particles is

$$\langle N \rangle = \sum_{N=0}^{+\infty} \frac{z^N Q_N(T,V,N)}{\mathcal{Q}(T,V,z)} N = z \left(\frac{\partial \ln \mathcal{Q}}{\partial z} \right)_{T,V} = z \left(\frac{\partial}{\partial z} \ln \left(\sum_{N=0}^{+\infty} \frac{(zVf(T))^N}{N!} \right) \right)_{T,V} =$$

$$z \left(\frac{\partial}{\partial z}(zVf(T)) \right)_{T,V} = zVf(T)$$

where we have used that

$$\sum_{N=0}^{+\infty} \frac{(zVf(T))^N}{N!} = e^{zVf(T)}.$$

The above expression reveals the probability $P(N)$ to find a number N of particles

$$P(N) = \frac{z^N Q_N(T,V,N)}{\mathcal{Q}(T,V,z)} = \frac{(zVf(T))^N}{N!} e^{-zVf(T)} = \frac{\langle N \rangle^N}{N!} e^{-\langle N \rangle}$$

which is exactly a Poisson distribution. The fluctuations in the grand canonical ensemble are

$$\langle (\Delta N)^2 \rangle = \langle N^2 \rangle - \langle N \rangle^2 = z \left(\frac{\partial \langle N \rangle}{\partial z} \right)_{T,V} = zVf(T) = \langle N \rangle.$$

It is instructive to show that we obtain the same result when evaluating $\langle (\Delta N)^2 \rangle$ with the probability function previously found

$$\langle (\Delta N)^2 \rangle = \langle N^2 \rangle - \langle N \rangle^2 = \sum_{N=0}^{+\infty} N^2 \frac{e^{-\langle N \rangle} \langle N \rangle^N}{N!} - \langle N \rangle^2 =$$

$$\sum_{N=1}^{+\infty} N^2 \frac{e^{-\langle N \rangle} \langle N \rangle^N}{N!} - \langle N \rangle^2 = \sum_{N=1}^{+\infty} N \frac{e^{-\langle N \rangle} \langle N \rangle^N}{(N-1)!} - \langle N \rangle^2 =$$

$$\sum_{N=1}^{+\infty} (N-1) \frac{e^{-\langle N \rangle} \langle N \rangle^N}{(N-1)!} + \sum_{N=1}^{+\infty} \frac{e^{-\langle N \rangle} \langle N \rangle^N}{(N-1)!} - \langle N \rangle^2 =$$

$$\sum_{N=2}^{+\infty} (N-1) \frac{e^{-\langle N \rangle} \langle N \rangle^N}{(N-1)!} + e^{-\langle N \rangle} \langle N \rangle \sum_{N=1}^{+\infty} \frac{\langle N \rangle^{N-1}}{(N-1)!} - \langle N \rangle^2 =$$

$$e^{-\langle N \rangle} \langle N \rangle^2 \sum_{N=2}^{+\infty} \frac{\langle N \rangle^{N-2}}{(N-2)!} + \langle N \rangle - \langle N \rangle^2 = \langle N \rangle^2 + \langle N \rangle - \langle N \rangle^2 = \langle N \rangle$$

where we have used the series expansion of the exponential function

$$e^A = \sum_{N=0}^{+\infty} \frac{A^N}{N!} = \sum_{N=1}^{+\infty} \frac{A^{N-1}}{(N-1)!} = \sum_{N=2}^{+\infty} \frac{A^{N-2}}{(N-2)!}.$$

Problem 8.5.
Consider a reaction involving proton (p), electron (e), and hydrogen (H) gases

$$p + e \leftrightarrow H$$

taking place in a volume V and in thermal equilibrium at temperature T. Let us also assume that the gases may be treated as ideal classical ones of non interacting particles, taking into account only the spin degeneracy. Find the electron density n_e as a function of the hydrogen density n_H and the temperature T, assuming the condition of zero total charge. As for the energy spectrum of the hydrogen atom, consider only the ground state with energy $-E_0$. Also, you can ignore the electron mass as compared to the one of the proton, so that the hydrogen mass is well approximated by the proton mass.

Solution
The grand canonical partition function for a classical ideal gas with independent particles is

$$\mathcal{Q}(T,V,z) = \sum_{N=0}^{+\infty} \frac{e^{\beta \mu N} Q_1^N}{N!} = e^{z Q_1}$$

where Q_1 is the partition function of the single particle and $z = e^{\beta \mu}$ its fugacity. We then use the result for the ideal classical gas (see also Problem 7.1) and add the spin degeneracy (g) as a multiplicative factor

$$Q_1(T,V) = gV \left(\frac{2\pi m k T}{h^2} \right)^{3/2}.$$

The average number is

$$\langle N \rangle = z \left(\frac{\partial \ln \mathcal{Q}}{\partial z} \right)_{T,V} = z Q_1 = z g V \left(\frac{2\pi m k T}{h^2} \right)^{3/2}$$

so that we find

$$n = \frac{\langle N \rangle}{V} = \frac{z Q_1}{V} = g e^{\beta \mu} \left(\frac{2\pi m}{\beta h^2} \right)^{3/2}.$$

This leads to the following result for the electron $(g = 2)$, proton $(g = 2)$, and hydrogen $(g = 4)$ densities

$$n_e = 2 e^{\beta \mu_e} \left(\frac{2\pi m_e}{\beta h^2} \right)^{3/2}$$

$$n_p = 2e^{\beta \mu_p} \left(\frac{2\pi m_p}{\beta h^2} \right)^{3/2}$$

$$n_H = 4e^{\beta \mu_H} e^{\beta E_0} \left(\frac{2\pi m_p}{\beta h^2} \right)^{3/2}$$

where we have written the hydrogen energy as the one of the ground state $(-E_0)$ plus the kinetic energy of the proton ($\frac{p_H^2}{2m_H} \approx \frac{p_H^2}{2m_p}$). The equilibrium condition imposes

$$\mu_H = \mu_e + \mu_p$$

so that, using the previous expressions, we can write

$$n_e = 2\lambda_e^{3/2} e^{\beta(\mu_H - \mu_p)}$$

where $\lambda_e = \frac{2\pi m_e}{\beta h^2}$. The ratio $\frac{n_H}{n_p}$ is

$$\frac{n_H}{n_p} = 2e^{\beta E_0} e^{\beta(\mu_H - \mu_p)}$$

and, hence, $e^{\beta(\mu_H - \mu_p)} = \frac{n_H}{2n_p} e^{-\beta E_0}$. Plugging this result back in the relation for n_e, we get

$$n_e = \lambda_e^{3/2} \frac{n_H}{n_p} e^{-\beta E_0}.$$

The condition that the total charge is zero implies $n_e = n_p$. This allows to write the previous expression as

$$n_e^2 = n_H \lambda_e^{3/2} e^{-\beta E_0}$$

from which we extract the desired result

$$n_e = \sqrt{n_H} \lambda_e^{3/4} e^{-\beta E_0/2}.$$

Problem 8.6.

An ideal gas of N non interacting molecules with magnetic moment μ and mass m is immersed in a magnetic field $B = (0,0,B)$, so that the single particle Hamiltonian is

$$H = \frac{p^2}{2m} - s\mu B$$

with p the absolute value of the momentum and $s = \pm 1$ depending on the particle we consider, i.e. with the momentum parallel or antiparallel to B. These two kinds of molecules have densities n_+ and n_- respectively. What is the ratio n_-/n_+ as a function of B at equilibrium?

Solution

The problem can be solved by imposing the constancy of the chemical potentials. Starting from the Hamiltonian of the single particle, we can write down the two chemical potentials for the molecules with magnetic moment parallel (+) or antiparallel $(-)$ to B

$$\mu_+ = \mu_+^{id} - \mu B$$

$$\mu_- = \mu_-^{id} + \mu B$$

where $\mu_\pm^{id}$ is the chemical potential of an ideal free gas with density $n_\pm$ and Hamiltonian $H = p^2/2m$

$$\mu_\pm^{id} = kT \ln(n_\pm \lambda^3)$$

with $\lambda = \dfrac{h}{\sqrt{2\pi m k T}}$ a thermal length scale. The densities $n_\pm$ must show a dependence on B in order to guarantee the equality of the chemical potentials, otherwise non equilibrium fluctuations appear (see also Problems 7.36 and 7.39 based on similar ideas). This leads to

$$kT \ln(n_-(B)\lambda^3) + \mu B = kT \ln(n_+(B)\lambda^3) - \mu B$$

from which we can extract directly the ratio n_-/n_+ as a function of B and the temperature T

$$\frac{n_-}{n_+} = e^{-\frac{2\mu B}{kT}}.$$

Problem 8.7.
Consider a reaction involving oxygen (O), hydrogen (H), and water (H_2O) molecules

$$O + 2H \leftrightarrow H_2O$$

taking place in a volume V and in thermal equilibrium at temperature T. Consider that the hydrogen density (n_H) is constantly twice the one of the oxygen (n_O). Also, all the gases involved may be considered as ideal ones with independent particles. Compute the water density n_{H_2O} as a function of n_O and T. Assume that the masses of water and oxygen molecules are such that $m_{H_2O} = 18 m_H$ and $m_O = 16 m_H$.

Solution
Chemical equilibrium requires the equality of the chemical potentials. This condition reads

$$\mu_O + 2\mu_H = \mu_{H_2O}.$$

In general, the chemical potential μ is defined as

$$\mu = \left(\frac{\partial F}{\partial N}\right)_{T,V}$$

where the free energy F may be defined in terms of the canonical partition function Q_N

$$F = -kT \ln Q_N \approx -NkT \left(\ln\left(\frac{Q_1}{N}\right) + 1\right)$$

$$Q_N(T,V,N) = \frac{Q_1^N(T,V)}{N!}$$

where we used the Stirling approximation for N and where Q_1 is the single particle canonical partition function

$$Q_1(T,V) = V \left(\frac{2\pi mkT}{h^2} \right)^{3/2} .$$

The result for the chemical potential is

$$\mu = \left(\frac{\partial F}{\partial N} \right)_{T,V} = -kT \ln \left(\frac{Q_1}{N} \right) .$$

Alternatively, we can use the grand canonical ensemble for the partition function

$$\mathscr{Q}(T,V,z) = \sum_{N=0}^{+\infty} z^N \frac{Q_1^N(T,V)}{N!} = e^{zQ_1(T,V)}$$

where $z = e^{\mu/kT}$ is the fugacity. Recalling that

$$\langle N \rangle = z \left(\frac{\partial \ln \mathscr{Q}}{\partial z} \right)_{T,V} = zQ_1 = e^{\mu/kT} Q_1$$

we find for μ the very same result obtained before, with the only difference that the number of particles in the canonical ensemble has to be replaced with the averaged number of particles. When we balance the chemical potentials, we obtain

$$\ln \left(\frac{Q_{1O}}{N_O} \right) + 2\ln \left(\frac{Q_{1H}}{N_H} \right) = \ln \left(\frac{Q_{1H_2O}}{N_{H_2O}} \right)$$

from which we get

$$\frac{N_{H_2O}}{N_O N_H^2} = \frac{Q_{1H_2O}}{Q_{1H}^2 Q_{1O}} .$$

With the conditions on the masses given in the text, we find

$$n_{H_2O} = 4 \left(\frac{9}{8} \right)^{3/2} \left(\frac{h^2}{2\pi m_H kT} \right)^3 n_O^3 .$$

9

Kinetic Physics

Problem 9.1.
Determine the number of particle-wall collisions per unit area for an ideal quantum gas composed of N independent particles inside a cubic container of volume V. Treat the general case with a single particle energy $\varepsilon = \varepsilon(p)$, i.e. only dependent on the absolute value of the momentum. Also, determine the pressure and discuss the corresponding classical limit. Finally, discuss the case of a photon gas with a single particle energy

$$\varepsilon = cp \qquad c = \text{const.}$$

and a chemical potential $\mu = 0$. For the general case of a quantum gas, the number of particles per unit volume with the momentum between p and $p + dp$ is

$$\frac{dN_p}{dV} = \frac{g}{(2\pi\hbar)^3} \frac{1}{e^{\frac{(\varepsilon-\mu)}{kT}} \pm 1} d^3p = \frac{2}{(2\pi\hbar)^3} \frac{1}{e^{\frac{(\varepsilon-\mu)}{kT}} \pm 1} d\phi \, \sin\theta \, d\theta \, p^2 \, dp$$

where g is the spin degeneracy, and ± 1 refers to fermions/bosons respectively. In the above expression, p, θ and ϕ refer to spherical polar coordinates in the momentum space.

Solution
The number of collisions per unit area and time is provided by all those particles whose velocity v (with absolute value v) is sufficiently high to cover the distance between their position and the walls of the container. Let us choose a wall, and set the z axis perpendicular to the wall. The number of collisions per unit time (dt) and area (dA) is given by those particles contained in the volume (see also Fig. 9.1).

$$dV = dAv_z dt = v\cos\theta \, dt.$$

The velocities are in general not directed along z, and that is the reason why we have to consider only their projection $v\cos\theta$. We also have to set $v_z > 0$ or, alternatively, $\cos\theta > 0$, that implies $0 \le \theta < \pi/2$, because we are interested only in those particles that move towards the wall and not in the opposite direction. For a given θ and

Cini M., Fucito F., Sbragaglia M.: Solved Problems in Quantum and Statistical Mechanics.
DOI 10.1007/978-88-470-2315-4_9, © Springer-Verlag Italia 2012

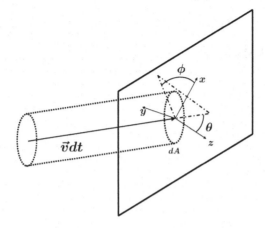

Fig. 9.1 The effusion of gas molecules through an opening of area dA in the walls of a vessel containing the gas. The case of a quantum gas is discussed in Problem 9.1

momentum p, the total number of particles hitting the wall per unit time and area is obtained by multiplying $\frac{dN_p}{dV}$ by the quantity $v\cos\theta$

$$dN_{hit} = \frac{g}{(2\pi\hbar)^3}\langle n_\varepsilon\rangle \, v\cos\theta \, d\phi \, \sin\theta \, d\theta \, p^2 \, dp$$

where we have defined

$$\langle n_\varepsilon\rangle = \frac{1}{e^{\frac{(\varepsilon-\mu)}{kT}}\pm 1}$$

as the mean occupation number for the energy level ε (see problems on the quantum gases in the next chapters). From the kinetic point of view, the pressure is given by the momentum transferred to the walls. If p_i and p_f are the z components of the momentum before and after the collisions, the total transferred momentum is $\Delta p = p_i - p_f = p_z - (-p_z) = 2p_z = 2p\cos\theta$, if we assume that the collision is elastic, i.e. the net effect of collision is to change sign to the normal component of the momentum. Therefore, the pressure P is given by the following expression

$$P = 2\int p\cos\theta \, dN_{hit} \, d^3p = \frac{2g}{(2\pi\hbar)^3}\int_0^{2\pi}d\phi\int_0^{\frac{\pi}{2}}\sin\theta\cos^2\theta d\theta\int_0^{+\infty}vp^3\frac{1}{e^{\frac{(\varepsilon-\mu)}{kT}}\pm 1}dp.$$

As for the classical limit, it corresponds to the case $e^{\frac{(\mu-\varepsilon)}{kT}}\ll 1$. If this condition is well verified for all ε of physical interest, the probability function reduces to the standard Maxwell-Boltzmann factor of classical statistical mechanics. Also, for the average number of particles $\langle n_\varepsilon\rangle$, the condition that $e^{\frac{(\mu-\varepsilon)}{kT}}\ll 1$ may be read as $\langle n_\varepsilon\rangle\ll 1$, i.e. the probability that the energy level ε is occupied is very small. In this limit, the pressure may be approximated by

$$P \approx \frac{2g}{(2\pi\hbar)^3} \int_0^{2\pi} d\phi \int_0^{\frac{\pi}{2}} \sin\theta \cos^2\theta d\theta \int_0^{+\infty} vp^3 e^{\frac{(-\varepsilon+\mu)}{kT}} \left(1 \mp e^{\frac{(-\varepsilon+\mu)}{kT}}\right) dp.$$

Let us now come to the last question of the problem, i.e. the photon gas with $\mu = 0$ (see also Problems 10.10-10.14 on black body radiation in the chapter on Bose-Einstein gases). In particular, the condition $\mu = 0$ may be read as the mathematical condition to have thermodynamic equilibrium: if we consider the free energy F, it reads

$$0 = \left(\frac{\partial F}{\partial N}\right)_{T,V} = \mu.$$

For a photon gas, we have to consider a spin degeneracy $g = 2$ to take into account the multiplicity of transverse modes and the fact that the longitudinal modes do not appear in the radiation. If we now use $\varepsilon = pc$ and $v = c$ in the previous formula, we obtain

$$P = \frac{4}{(2\pi\hbar)^3} \int_0^{2\pi} d\phi \int_0^{\frac{\pi}{2}} \sin\theta \cos^2\theta d\theta \int_0^{+\infty} cp^3 \frac{1}{e^{\frac{cp}{kT}} - 1} dp.$$

If we use the known result

$$\int_0^{+\infty} \frac{x^3}{e^x - 1} dx = \frac{\pi^4}{15}$$

we can solve exactly the integral on the right hand side

$$P = \frac{8\pi}{3h^3 c^3} (kT)^4 \int_0^{+\infty} \frac{x^3}{e^x - 1} dx = \frac{8\pi^5}{45h^3 c^3} (kT)^4.$$

Problem 9.2.
An ideal three dimensional classical gas composed of N molecules with mass m is in thermal equilibrium at temperature T in a container with volume V. Determine the relation between the temperature T and:

- the average absolute value of the velocity for the gas;
- the average squared velocity of the gas.

Solution

Let us start by computing the *Maxwell distribution*. We consider a single molecule with coordinates $q = (q_x, q_y, q_z)$ and momentum $p = (p_x, p_y, p_z)$ in the three dimensional space. The molecule is in contact with a reservoir at temperature T, and we can use the canonical distribution to compute the probability that the molecule occupies a cell with volume $d^3q d^3p$ in the phase space

$$P(q,p)d^3q d^3p \propto e^{-\beta \frac{p^2}{2m}} d^3q d^3p$$

or, equivalently, using the velocity v instead of the momentum p, we find

$$P'(q,v)d^3q d^3p \propto e^{-\beta \frac{mv^2}{2}} d^3q d^3v.$$

We now look for the number of molecules per unit volume with velocity between v and $v + dv$. This number is proportional to the probability multiplied by the total number N and divided by the volume element d^3q

$$f(v)d^3v \propto \frac{NP'd^3q d^3v}{d^3q}$$

that is

$$f(v)d^3v = Ce^{-\beta \frac{mv^2}{2}} d^3v$$

where C is a normalization constant. The previous distribution function is the Maxwell distribution function for the velocity. The constant C is determined by imposing that the integral of the distribution function is equal to the total density $n = N/V$

$$C \int e^{-\beta \frac{mv^2}{2}} d^3v = C \int_{-\infty}^{+\infty} e^{-\beta \frac{mv_x^2}{2}} dv_x \int_{-\infty}^{+\infty} e^{-\beta \frac{mv_y^2}{2}} dv_y \int_{-\infty}^{+\infty} e^{-\beta \frac{mv_z^2}{2}} dv_z = n$$

leading to $C = n \left(\frac{\beta m}{2\pi} \right)^{3/2}$. Also, we can determine the probability distribution $g(v_x)$ associated with one component of the velocity, for example in the x direction. This is found by integrating in dv_y and dv_z the previous expression

$$g(v_x)dv_x = dv_x \int f(v)dv_y dv_z = n \left(\frac{\beta m}{2\pi} \right)^{1/2} e^{-\beta mv_x^2} dv_x.$$

Using the last equation, we can compute the average value $\langle v_x \rangle \equiv \bar{v}_x$ in the x direction

$$\bar{v}_x = \frac{1}{n} \int_{-\infty}^{+\infty} g(v_x)v_x dv_x = 0$$

that is obvious, due to the symmetry of the distribution function. The result is different from zero if we consider the absolute value of the velocity. We first use spherical polar coordinates, and then obtain the probability $F(v)$ that the molecule has the absolute value of the velocity between v and $v + dv$ by integrating out the angular variables

$$F(v)dv = 4\pi n \left(\frac{\beta m}{2\pi}\right)^{3/2} e^{-\beta \frac{mv^2}{2}} v^2 dv.$$

With this new distribution function, we can compute $\bar{v} = \langle v \rangle$

$$\bar{v} = \frac{1}{n}\int_0^{+\infty} F(v)v\,dv = \left(\frac{8kT}{\pi m}\right)^{1/2}$$

that is the answer to the first question posed by the text. Finally, using the probability $F(v)$, we have

$$\langle v^2 \rangle = \frac{1}{n}\int_0^{+\infty} F(v)v^2 dv$$

from which we obtain the relation

$$\frac{1}{2}m\langle v^2 \rangle = \frac{3}{2}kT.$$

Problem 9.3.
An ideal three dimensional classical gas composed of N particles with unitary mass is in thermal equilibrium at temperature T in a container of volume V. A small hole with area A is present on the surface of the container. Compute:

- the effusion rate as a function of the average value of the absolute velocity;
- the momentum transferred as a function of the gas pressure.

Solution
The effusion is defined as the number of particles exiting the hole per unit time and area. If we assume the area of the hole to be perpendicular to the x axis, this number is given by

$$R = \int_{v_x>0} f(v)v_x d^3v$$

and knowing that $f(v)$ is a Maxwellian distribution function

$$f(v) = n\left(\frac{\beta}{2\pi}\right)^{3/2} e^{-\beta \frac{v^2}{2}} \qquad n = \frac{N}{V}$$

we get

$$R = \frac{n}{2}\left(\frac{\beta}{2\pi}\right)^{1/2}\int_0^{+\infty} e^{-\beta \frac{v_x^2}{2}} d(v_x^2) = \frac{n}{(2\pi\beta)^{1/2}}.$$

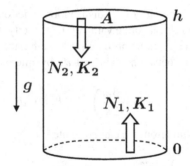

Fig. 9.2 A schematic view of the isothermal atmosphere under the effect of gravity. The barometric formula for the pressure is deduced with simple kinetic considerations in Problem 9.4

We also recall the average absolute value of the velocity $\bar{v} = \langle v \rangle = \left(\frac{8kT}{\pi}\right)^{1/2}$ (see Problem 9.2 and use $m = 1$), from which we obtain

$$R = \frac{n\bar{v}}{4}.$$

This effusion may be regarded as an efficient method to separate molecules with different masses because the average velocity depends on the mass itself and, for fixed temperature, it is larger for lighter molecules. As for the momentum transferred through the hole, we get

$$M = \int_{v_x>0} v_x^2 f(v)d^3v = n\left(\frac{\beta}{2\pi}\right)^{1/2}\int_0^{+\infty} e^{-\beta\frac{v_x^2}{2}}v_x^2 dv_x = \frac{nkT}{2}$$

from which $M = \frac{P}{2}$, because $P = nkT$.

Problem 9.4.
Using simple kinetic considerations, derive the barometric formula for the pressure in the atmosphere. Assume that we can treat the atmosphere as an ideal gas at equilibrium, composed of particles with mass m. For the sake of simplicity, treat the problem as one dimensional along the vertical coordinate z, with a local Maxwellian distribution function of the velocity v_z

$$f(v_z) = n(z)\left(\frac{m}{2\pi kT(z)}\right)^{1/2} e^{-\frac{mv_z^2}{2kT(z)}}$$

with a space dependent density $(n(z))$ and temperature $(T(z))$.

Solution
Let us consider a column of air with height h (see also Fig. 9.2) and base A. The density of the air is not constant and it varies with the height, due to the effect of

the gravitational field. The standard derivation of the barometric formula is done by computing the difference in pressure at the ends of the cylinder (similarly to what we have done in Problem 6.14). However, this is unnecessary if a Boltzmann microscopic approach is adopted. Here one considers a vertical column of gas, isolated from all external disturbances and in equilibrium, i.e. with no mass, momentum, heat transfers. Consider the two regions of the column at $z = 0$ and $z = h$. Let n_1 and T_1 be the air density and the temperature at height $z = 0$, and n_2 and T_2 those at height $z = h$. As suggested by the text, we can assume that the Maxwellian distribution for the velocities is valid. Let us compute the number of molecules that move downwards through A at $z = h$ per unit time and area. To do that, we start from the Maxwell distribution at height h

$$f_2(v_z)\,dv_z = n_2 \left(\frac{m}{2\pi kT_2}\right)^{1/2} e^{-\frac{mv_z^2}{2kT_2}}\,dv_z.$$

The number of molecules moving per unit time and area through A is obtained by integrating $f_2(v_z)$ times the velocity v_z (gas effusion)

$$N_2 = n_2 \left(\frac{m}{2\pi kT_2}\right)^{1/2} \int_0^{+\infty} v_z e^{-\frac{mv_z^2}{2kT_2}}\,dv_z = n_2 \left(\frac{kT_2}{2\pi m}\right)^{1/2}.$$

On the contrary, molecules leaving the lower region at $z = 0$ lose kinetic energy in moving to the upper region, and only those whose velocity is such that

$$\frac{1}{2}mv_z^2 \geq mgh \rightarrow v_z \geq \sqrt{2gh}$$

will reach the height h. It follows that

$$N_1 = n_1 \left(\frac{m}{2\pi kT_1}\right)^{1/2} \int_{\sqrt{2gh}}^{+\infty} v_z e^{-\frac{mv_z^2}{2kT_1}}\,dv_z = n_1 \left(\frac{kT_1}{2\pi m}\right)^{1/2} e^{-\frac{mgh}{kT_1}}.$$

If the gas is in equilibrium, a mass transfer through A at $z = h$ is not possible, so that $N_1 = N_2$, and

$$n_1 \sqrt{T_1}\, e^{-\frac{mgh}{kT_1}} = n_2 \sqrt{T_2}.$$

Let us now consider the transport of the energy through A per unit time and area. For the molecules moving downwards we have

$$K_2 = \langle \frac{1}{2}mv_z^2 \times v_z \rangle = n_2 \left(\frac{m}{2\pi kT_2}\right)^{1/2} \frac{1}{2}m \int_0^{+\infty} v_z^3 e^{-\frac{mv_z^2}{2kT_2}}\,dv_z = n_2 \frac{(kT_2)^{3/2}}{(2\pi m)^{1/2}}.$$

Each of the molecules moving upwards will only succeed in carrying an energy $\frac{1}{2}mv_z^2 - mgh$ to the height h, so that the net transfer of the energy upwards is

$$K_1 = n_1 \left(\frac{m}{2\pi kT_1}\right)^{1/2} \int_{\sqrt{2gh}}^{+\infty} \left(\frac{1}{2}mv_z^2 - mgh\right) v_z e^{-\frac{mv_z^2}{2kT_1}}\,dv_z.$$

If we set $\xi^2 = v_z^2 - 2gh$ we end up with the following integral

$$K_1 = n_1 \left(\frac{m}{2\pi kT_1}\right)^{1/2} e^{-\frac{mgh}{kT_1}} \frac{1}{2} m \int_0^{+\infty} \xi^3 e^{-\frac{m\xi^2}{2kT_1}} d\xi = n_1 \frac{(kT_1)^{3/2}}{(2\pi m)^{1/2}} e^{-\frac{mgh}{kT_1}}.$$

If there is not a net transfer of the energy (thermal equilibrium), we need to set

$$n_1 T_1^{3/2} e^{-\frac{mgh}{kT_1}} = n_2 T_2^{3/2}.$$

If we use this equation with the one obtained previously for the density, we find $T_1 = T_2$, that is the condition that the atmosphere is isothermal. We now call T the atmosphere temperature and write the equation relating n_1 and n_2 as

$$n_2 = n_1 e^{-\frac{mgh}{kT}}$$

from which, using the equation of state for the ideal gas, we find the barometric formula for the atmosphere

$$p(h) = p(0) e^{-\frac{mgh}{kT}}.$$

Problem 9.5.
Consider an ideal classical gas in thermal equilibrium at temperature T. The gas is composed of N molecules in total: $N/2$ with mass m_1 (type 1) and $N/2$ with mass $m_2 = 4m_1$ (type 2). The gas is placed in a container with volume V. At time $t = 0$, a small hole with area A is produced on the container's wall, and the effusion process starts. Assuming that the whole process takes place at equilibrium with temperature T, compute the number of particles of type 2 that remain in the container when the number of particles of type 1 is reduced by a factor 2.

Solution
We denote by $N_i(t)$ the number of particles of type i as a function of time and write down the following differential equations ($i = 1, 2$)

$$\frac{dN_i(t)}{dt} = -AR_i$$

where R_i denotes the number of particles of i-th type exiting the hole per unit time and area, i.e. the i-th rate of effusion. From the general theory of effusion (see also Problem 9.3) we know that $R_i = \frac{n_i \langle v_i \rangle}{4}$, where n_i and v_i denote the density and absolute value of the velocity for the molecules of type i. Given this relation, the differential equations ($i = 1, 2$) become

$$\frac{dN_i(t)}{dt} = -\frac{AN_i \langle v_i \rangle}{4V}$$

which can be solved

$$N_i(t) = \frac{N}{2} e^{-A\langle v_i\rangle t/4V}.$$

At time t_1 the particles of the first type are reduced by a factor 2, which means that their number is equal to $N/4$

$$\frac{N}{2} e^{-A\langle v_1\rangle t_1/4V} = \frac{N}{4}$$

from which we get $t_1 = \frac{4V\ln 2}{A\langle v_1\rangle}$. It is now sufficient to plug t_1 in the expression for $N_2(t)$ to find the number of particles of type 2 present when $t = t_1$

$$N_2(t_1) = \frac{N}{2} e^{-\langle v_2\rangle \ln 2/\langle v_1\rangle} = \frac{N}{2} e^{-\ln 2/2}$$

because $\langle v_i\rangle = \left(\frac{8kT}{\pi m_i}\right)^{1/2}$ and $m_2 = 4m_1$ (see Problem 9.2). This means that when the lighter particles are reduced by a factor 0.5, the heavier ones are reduced by a factor ≈ 0.707.

Problem 9.6.
A box with volume V is separated in two equal parts by a tiny wall. In the left side of the box there is an ideal classical gas of N_0 molecules with mass m and pressure P_0, while the right side is empty. At time $t = 0$, a small hole with area A is produced on the wall. Determine the pressure $P(t)$ as a function of time that is exerted by the gas on the left side. Assume that the temperature stays constant during the effusion process. Discuss the limit $t \to +\infty$.

Solution
When the small hole is made on the wall, the number of particles changes in time. If we call $N(t)$ the number of particles in the left side, the right side is then filled with $N_0 - N(t)$ molecules. In the time interval between t and $t + dt$ the number of molecules moving from left to right is (see Problem 9.3)

$$R_{l\to r} = \frac{AN(t)\bar{v}}{4V}$$

with $\bar{v}$ the average value of the absolute velocity, $\bar{v} = \langle v\rangle = \left(\frac{8kT}{\pi m}\right)^{1/2}$ (see Problem 9.2). At the same time, the number of molecules moving from right to left is

$$R_{r\to l} = \frac{A(N_0 - N(t))\bar{v}}{4V}.$$

Therefore, we can write

$$\frac{dN(t)}{dt} = -\frac{AN(t)\bar{v}}{4V} + \frac{A(N_0 - N(t))\bar{v}}{4V} = -\frac{A\bar{v}}{4V}(2N(t) - N_0).$$

If we set $M(t) = 2N(t) - N_0$ and $dM = 2dN$, the resulting differential equation is

$$\frac{dM(t)}{dt} = -bM(t)$$

where we have used $b = A\bar{v}/2V$. We obtain $M(t) = Ce^{-bt}$, with C an integration constant computed from the condition that at time $t = 0$ we have $M = N_0$. This means that $C = N_0$ and

$$N(t) = \frac{N_0}{2}(1 + e^{-bt}).$$

Since the pressure is proportional to N for a fixed volume and temperature, we get

$$P(t) = \frac{P_0}{2}(1 + e^{-bt}).$$

In the limit $t \to +\infty$ the pressure on both sides is equal to $P_0/2$.

Problem 9.7.
A cylindrical container (whose height L is very large, $L \gg 1$, with respect to the radius of its base) with base located at $z = 0$, contains N particles with mass $m_1 = m$ and N particles with mass $m_2 = 4m$. The gas is ideal, non interacting, in thermal equilibrium at temperature T, and under the effect of a constant gravitational acceleration g. At some point, a small hole is opened in the lateral surface and, at later times, it is observed that the number of particles of mass m_1 is equal to the number of particles of mass m_2. Under the assumption that the effusion process takes place at equilibrium, determine the height h at which the hole was opened.

Solution
When the number of particles of mass m_1 and m_2 is the same, the effusion rate of both particles is the same, i.e. $R_1 = R_2$. When studying the effusion of a gas through a small hole, it is known (see Problem 9.3) that the number of particles per unit time and area is $R = \frac{n(h)\bar{v}}{4}$, where $n(h)$ is the local density at height h and $\bar{v}$ the average value of the absolute velocity of the gas, $\bar{v}_i = \langle v_i \rangle = \left(\frac{8kT}{\pi m_i}\right)^{1/2}$ (see Problem 9.2). As a consequence, the condition $R_1 = R_2$ implies

$$n_1(h)\bar{v}_1 = n_2(h)\bar{v}_2.$$

Using the previous expressions for $\bar{v}_i$, $i = 1, 2$, and recalling the relation between the particles masses, i.e. $m_2 = 4m_1$, we obtain

$$n_2(h) = 2n_1(h).$$

When studying the change of density for a gas under the effect of gravity, we obtain (see also Problem 9.4)

$$n_i(h) = C_i e^{-\frac{m_i gh}{kT}}$$

where C_i $(i = 1, 2)$ are normalization constants which can be found by imposing that the total number of particles for each mass is N. We find

$$C_i \int_0^L e^{-\frac{m_i g h}{kT}} \, dh \approx C_i \int_0^{+\infty} e^{-\frac{m_i g h}{kT}} \, dh = N$$

leading to $C_i = N m_i g / kT$. Therefore, we can write

$$n_1(h) = \frac{Nmg}{kT} e^{-\frac{mgh}{kT}} \qquad n_2(h) = \frac{4Nmg}{kT} e^{-\frac{4mgh}{kT}}$$

and the condition $n_2(h) = 2n_1(h)$ becomes

$$4e^{-\frac{4mgh}{kT}} = 2e^{-\frac{mgh}{kT}}$$

or, alternatively $h = \frac{kT \ln 2}{3mg}$.

Problem 9.8.

Let us consider a two dimensional classical system composed of N particles with mass m and single particle Hamiltonian

$$H = \frac{p^2}{2m} + V(r) = \frac{p^2}{2m} + \frac{1}{2}\mu^2 r^2 - \frac{1}{4}\lambda r^4$$

where p is the absolute value of the momentum and where $V(r)$ is a potential ($\mu > 0$ and $\lambda > 0$) due to external forces, with r the radial distance ($r > 0$) from a fixed center. The whole system is in thermal equilibrium at temperature T. Compute the rate of particles $dN(t)/dt$ escaping from 'the peak' of the potential energy barrier under the assumption that the non quadratic part of the potential is a small perturbation ($\lambda \ll 1$).

Solution

We need to find the peak of the potential, say located at $r = b$. From the condition $dH/dr = 0$ we find

$$\mu^2 r - \lambda r^3 = 0.$$

If we neglect the solution $r = 0$, we get $r = b = \mu/\sqrt{\lambda}$ (see also Fig. 9.3, where we plot $V(r)$ with $\lambda = \mu = 1$).

The number of particles exiting a small hole per unit time and area (see Problems 9.2 and 9.3) is

$$R = \rho(b)\frac{\langle v \rangle}{4} = \rho(b)\sqrt{\frac{kT}{2\pi m}}$$

where $\rho(b)$ is the local density at $r = b$. Multiplying this result by $2\pi b$ (the perimeter of the circle with radius b), we get the total number of particles escaping from the peak of the potential per unit time

$$\frac{dN(t)}{dt} = \rho(b)b\sqrt{\frac{2\pi kT}{m}}.$$

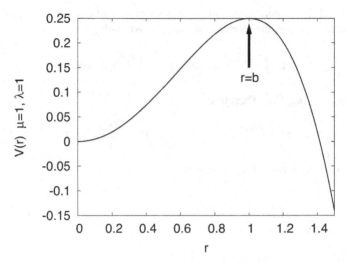

Fig. 9.3 We plot the potential energy barrier $V(r) = \frac{1}{2}\mu^2 r^2 - \frac{1}{4}\lambda r^4$ for $\lambda = \mu = 1$. In Problem 9.8 we compute the number of particles per unit time escaping from 'the peak' of the potential energy barrier located at $r = b = \mu/\sqrt{\lambda}$

To solve the problem, we finally need to compute $\rho(b)$, i.e. the particles density in the peak of the potential. From the assumption of equilibrium (see also Problems 7.12, 7.36 and 7.39), we know that

$$\rho(r) = ce^{-\frac{V(r)}{kT}} = ce^{-\left(\frac{\mu^2 r^2}{2kT} - \frac{\lambda r^4}{4kT}\right)}$$

and we have to determine the constant c by imposing that the integral of $\rho(r)$ over the space gives the total number of particles. We obtain

$$N = 2\pi \int_0^{+\infty} r\rho(r)\,dr = 2\pi c \int_0^{+\infty} re^{-\left(\frac{\mu^2 r^2}{2kT} - \frac{\lambda r^4}{4kT}\right)}\,dr.$$

As reported in the text, we can treat λ as a small parameter, so that we can Taylor expand the exponential function in the integrand

$$N \approx 2\pi c \int_0^{+\infty} r\left(1 + \frac{\lambda r^4}{4kT}\right)e^{-\frac{\mu^2 r^2}{2kT}}\,dr.$$

If we set $\mu^2 r^2/2kT = y$, we get

$$N = \frac{2\pi kTc}{\mu^2} \int_0^{+\infty}\left(1 + \frac{\lambda kTy^2}{\mu^4}\right)e^{-y}\,dy$$

from which

$$c = \frac{N\mu^2}{2\pi kT} \frac{1}{\left(1 + \frac{2\lambda kT}{\mu^4}\right)}.$$

Finally, we have to plug this value of c in the equation for $\rho(b)$, consider that $b = \mu/\sqrt{\lambda}$, and plug $\rho(b)$ in the equation for $dN(t)/dt$. The final result is

$$\frac{dN(t)}{dt} = \frac{N\mu^3}{\sqrt{2\pi m\lambda kT}\left(1 + \frac{2\lambda kT}{\mu^4}\right)} e^{-\frac{\mu^4}{4kT\lambda}}.$$

Problem 9.9.
Let us consider a cubic container with volume V and negligible mass which is on a smooth surface without friction. In the container we find N molecules of an ideal gas (each particle has mass m) in thermal equilibrium at temperature T. At time $t = 0$ a small hole with area A is opened on one side of the container. This hole stays opened for a small time interval Δt. Determine the velocity of the container as a function of $T, m, V, \Delta t$. Assume that in the time interval Δt the pressure and the temperature of the container stay constant.

Solution
During the effusion, the momentum transferred (see also Problem 9.3) per unit time through A is equal to $\Pi = \frac{AP}{2}$, where P is the pressure inside the container. The velocity of the container is $v = \frac{AP\Delta t}{2M}$, where M is the total mass of the gas, that we approximate with Nm (some particles will exit, so it won't be exactly equal). Using this information and the equation of state for the ideal gas ($PV = NkT$), we find

$$v = \frac{AkT\Delta t}{2mV}.$$

10

Bose-Einstein Gases

Problem 10.1.
Consider a three dimensional gas of bosons with spin 0 and single particle energy
given by

$$\varepsilon = \frac{p^2}{2m}$$

where p is the absolute value of the momentum and m the mass of the particle. Write
down the equation determining the critical temperature (T_c) for the Bose-Einstein
condensation in the ground state.

Solution
We start by writing down the logarithm of the grand canonical partition function for
the quantum gas obeying the Bose-Einstein statistics

$$\ln \mathscr{Q} = -\int \ln(1 - ze^{-\beta\varepsilon})\frac{d^3p\,d^3q}{h^3} = -\frac{4\pi V}{h^3}\int_0^{+\infty} \ln(1 - ze^{-\beta\varepsilon})p^2 dp$$

where $\varepsilon = \frac{p^2}{2m}$. For the average density of particles $\langle N \rangle/V$, we find

$$\frac{\langle N \rangle}{V} = \frac{z}{V}\left(\frac{\partial \ln \mathscr{Q}}{\partial z}\right)_{T,V} = \frac{1}{V}\int \frac{1}{z^{-1}e^{\beta\varepsilon} - 1}\frac{d^3p\,d^3q}{h^3} = \frac{4\pi}{h^3}\int_0^{+\infty} \frac{1}{z^{-1}e^{\beta\varepsilon} - 1}p^2 dp.$$

The above integral may be written in terms of the Bose-Einstein functions $g_\alpha(z)$

$$g_\alpha(z) = \frac{1}{\Gamma(\alpha)}\int_0^{+\infty} \frac{x^{\alpha-1}dx}{z^{-1}e^x - 1}$$

as

$$\frac{\langle N \rangle}{V} = \left(\frac{2\pi mkT}{h^2}\right)^{3/2} g_{3/2}(z).$$

For a given $\langle N \rangle$, the fugacity z depends on the temperature T. At the same time,
for a given T, z cannot be above the maximum value $z_{max} = e^{\beta\varepsilon}\big|_{p=0} = 1$, to ensure

Cini M., Fucito F., Sbragaglia M.: Solved Problems in Quantum and Statistical Mechanics.
DOI 10.1007/978-88-470-2315-4_10, © Springer-Verlag Italia 2012

positive occupation numbers. The equation for the critical temperature is found by setting $z = 1$ and $T = T_c$ ($\beta = \beta_c = \frac{1}{kT_c}$) in the equation for $\langle N \rangle$

$$\frac{\langle N \rangle}{V} = \left(\frac{2\pi m k T_c}{h^2} \right)^{3/2} g_{3/2}(1).$$

Making use of the Riemann zeta function such that $\zeta(3/2) = g_{3/2}(1)$, we find

$$T_c = \frac{h^2}{2\pi mk} \left(\frac{\langle N \rangle}{V \zeta(3/2)} \right)^{2/3}.$$

Problem 10.2.
Using the continuous approximation, discuss the phenomenon of the Bose-Einstein condensation and the existence of a critical temperature for one ($d = 1$) and two dimensional ($d = 2$) gases placed in a cubic d dimensional volume of edge L with a single particle energy $\varepsilon = \frac{p^2}{2m}$, where p is the absolute value of the momentum and m its mass.

Solution
In order to reveal the existence of a critical temperature, we need to consider the equation determining the particles average density. The dimensionless volume element in the phase space is given by $\frac{gL^d d^d p}{(2\pi\hbar)^d}$, where g is the spin degeneracy ($g = 2S + 1$) and d the space dimensionality ($d = 1, 2$). For the case with $d = 2$ we find

$$\frac{gL^2 d^2 p}{(2\pi\hbar)^2} = \frac{gL^2 2\pi p dp}{(2\pi\hbar)^2} = \frac{gL^2 m d\varepsilon}{2\pi\hbar^2}$$

while, for $d = 1$

$$\frac{gL dp}{2\pi\hbar} = \frac{gL\sqrt{m} d\varepsilon}{2\sqrt{2\varepsilon}\pi\hbar}.$$

The corresponding average densities are given by

$$\frac{\langle N \rangle}{L^2} = \frac{gm}{2\pi\hbar^2} \int_0^{+\infty} \frac{d\varepsilon}{z^{-1} e^{\beta\varepsilon} - 1} = \frac{gm}{2\pi\hbar^2} kT \int_0^{+\infty} \frac{dx}{z^{-1} e^x - 1} = \frac{gm}{2\pi\hbar^2} kT g_1(z)$$

and

$$\frac{\langle N \rangle}{L} = \frac{g\sqrt{m}}{2\sqrt{2}\pi\hbar} \int_0^{+\infty} \frac{1}{\sqrt{\varepsilon}} \frac{d\varepsilon}{z^{-1} e^{\beta\varepsilon} - 1} = \frac{g\sqrt{mkT}}{2\sqrt{2}\pi\hbar} \int_0^{+\infty} \frac{1}{\sqrt{x}} \frac{dx}{z^{-1} e^x - 1} = \frac{g\sqrt{mkT}}{2\sqrt{2}\sqrt{\pi}\hbar} g_{\frac{1}{2}}(z)$$

for $d = 2$ and $d = 1$ respectively. Since the average number of particles must be a positive quantity, we must have $0 \leq z \leq 1$. When we reach the critical value $z = 1$, the Bose-Einstein condensation sets in for the particles in the ground state $\varepsilon = 0$. Therefore, the equation for the critical temperature T_c is obtained by setting $z = 1$ and $T = T_c$ in the equation for $\frac{\langle N \rangle}{L}$, $\frac{\langle N \rangle}{L^2}$. For the case $d = 2$ we can solve the integral exactly because

$$g_1(z) = -\ln(1-z).$$

Moreover, we know that when $z \approx 1$ we find $g_n(z) \approx (\ln z)^{-(1-n)}$ $(0 < n < 1)$. This means that for the $d = 1$ case (where $n = 1/2$) we have

$$g_{\frac{1}{2}}(z) \approx (\ln z)^{-\frac{1}{2}}.$$

Therefore, the integrals defining $\frac{\langle N \rangle}{L}$, $\frac{\langle N \rangle}{L^2}$ for $d = 1$ and $d = 2$ diverge when $z \to 1$ and the corresponding critical temperatures are zero, i.e. the Bose-Einstein condensation, contrary to what happens in the three dimensional gas (see Problem 10.1), does not take place.

Problem 10.3.
Consider a gas composed of $\langle N \rangle$ particles obeying the Bose-Einstein statistics. The gas is in a d dimensional container with volume V. The single particle energy is $\varepsilon = p^b$, where p is the absolute value of the momentum and b a positive constant. Suppose that $\langle N \rangle / V$ is fixed and determine the conditions for which the Bose-Einstein condensation takes place.

Solution
Similarly to Problem 10.1, we write the logarithm of the grand canonical partition function for the quantum gas obeying the Bose-Einstein statistics in d dimensions

$$\ln \mathscr{Q} = -\int \ln(1 - ze^{-\beta\varepsilon}) \frac{d^d p d^d q}{h^d} = -\frac{\Omega_d V}{h^3} \int_0^{+\infty} \ln(1 - ze^{-\beta\varepsilon}) p^{d-1} dp$$

and find the average density of particles $\langle N \rangle / V$ as

$$\frac{\langle N \rangle}{V} = \frac{z}{V} \left(\frac{\partial \ln \mathscr{Q}}{\partial z} \right)_{T,V} = \frac{1}{V} \int \frac{1}{z^{-1} e^{\beta\varepsilon} - 1} \frac{d^d p d^d q}{h^d} = \frac{\Omega_d}{h^d} \int_0^{+\infty} \frac{1}{z^{-1} e^{\beta\varepsilon} - 1} p^{d-1} dp.$$

In the above expression, Ω_d is the d dimensional solid angle. The equation for the critical temperature (T_c) is found by setting $z = 1$, $\beta = \beta_c = \frac{1}{kT_c}$ in the equation for $\langle N \rangle$. Using the new variable $x = \beta\varepsilon = p^b$, the existence of a non zero T_c is then closely related to the behaviour of the integral

$$I = \int_0^{+\infty} \frac{x^{d/b-1} dx}{e^x - 1}.$$

In particular, it is important to examine its behaviour in a neighborhood of the origin $x = 0$, since for large x the integrand is exponentially small and the integral is surely convergent. We find that the integral converges only if $d/b > 1$, while diverges when $d/b \leq 1$. When the integral converges $(d/b > 1)$ there is a finite critical temperature T_c at which the ground state starts to be occupied by a significant fraction of the particles. At such critical temperature, we find

$$\frac{\langle N \rangle}{V} = \frac{\Omega_d (kT_c)^{d/b}}{bh^d} \int_0^{+\infty} \frac{x^{d/b-1} dx}{e^x - 1} = \frac{\Omega_d (kT_c)^{d/b}}{bh^d} \Gamma(d/b) g_{d/b}(1)$$

with

$$g_{d/b}(z) = \frac{1}{\Gamma(d/b)} \int_0^{+\infty} \frac{x^{d/b-1} dx}{z^{-1} e^x - 1}.$$

The final result is

$$kT_c = \left(\frac{\langle N \rangle}{V} \frac{bh^d}{\Omega_d \Gamma(d/b) \zeta(d/b)} \right)^{b/d}$$

where $\zeta(d/b)$ is the Riemann zeta function such that $\zeta(d/b) = g_{d/b}(1)$.

Problem 10.4.
A Bose-Einstein gas is characterized by particles of mass m and spin $S = 0$ in a volume V. The single particle energy is

$$\varepsilon = \frac{p^2}{2m} + n\Delta$$

where $\frac{p^2}{2m}$ is the kinetic term, $\Delta > 0$ a constant, and n an integer number equal to $n = 0, 1$. Determine the equation for the critical temperature T_c of the Bose-Einstein condensation. In the limit $\Delta \gg kT$, is the critical temperature increasing or decreasing with respect to the case $\varepsilon = \frac{p^2}{2m}$ (see Problem 10.1)?

Solution
We start from the equation determining the average number of particles

$$\langle N \rangle = \sum_{n=0,1} \int \frac{1}{z^{-1} e^{\beta \frac{p^2}{2m} + \beta n\Delta} - 1} \frac{d^3 p\, d^3 q}{h^3} = \sum_{n=0,1} \frac{4\pi V}{h^3} \int_0^{+\infty} \frac{1}{z^{-1} e^{\beta \frac{p^2}{2m} + \beta n\Delta} - 1} p^2 dp$$

where we have considered that the whole energy spectrum is made up of a continuous part ($\frac{p^2}{2m}$) and a discrete one ($n\Delta$). Furthermore

$$\langle N \rangle = \sum_{n=0,1} \frac{V(2\pi mkT)^{3/2}}{h^3} \frac{1}{\Gamma(3/2)} \int_0^{+\infty} \frac{1}{z^{-1} e^{x+\beta n\Delta} - 1} x^{1/2} dx.$$

Using the Bose-Einstein functions

$$g_\alpha(z) = \frac{1}{\Gamma(\alpha)} \int_0^{+\infty} \frac{x^{\alpha-1} dx}{z^{-1} e^x - 1}$$

we get

$$\frac{\langle N \rangle}{V} = \frac{1}{\lambda^3} g_{3/2}(z) + \frac{1}{\lambda^3} g_{3/2}(z e^{-\beta \Delta})$$

with $\lambda = \sqrt{\frac{h^2}{2\pi mkT}}$ a thermal length scale. The equation defining the critical temperature T_c is obtained by imposing that $z = e^{\beta E_0} = 1$ when $T = T_c$, where $E_0 = 0$ is the ground state energy. Therefore, we find

$$\frac{\langle N \rangle}{V} = \frac{1}{\lambda_c^3} g_{3/2}(1) + \frac{1}{\lambda_c^3} g_{3/2}(e^{-\beta_c \Delta}).$$

In the limit $\Delta \gg kT_c$ we can use $g_{3/2}(e^{-\beta_c \Delta}) \approx e^{-\beta_c \Delta}$ (because $g_{3/2}(x) \approx x$ when $x \ll 1$) and we get

$$\frac{\langle N \rangle}{V} = \frac{1}{\lambda_c^3} g_{3/2}(1) + \frac{1}{\lambda_c^3} e^{-\beta_c \Delta}$$

or, equivalently

$$\lambda_c^3 \frac{\langle N \rangle}{V} = \left(\frac{h^2}{2\pi mkT_c} \right)^{3/2} \frac{\langle N \rangle}{V} = g_{3/2}(1) + e^{-\beta_c \Delta} = \zeta(3/2) + e^{-\beta_c \Delta}.$$

In Fig. 10.1 we have considered the case $k = 1$, $\left(\frac{h^2}{2\pi mk} \right)^{3/2} \frac{\langle N \rangle}{V} = 0.05$ and reported the functions $y = 0.05 T_c^{-3/2}$ and $y = \zeta(3/2) + e^{-\Delta/T_c}$ for $\Delta = 0.1$. From the intersection we find the critical temperature. The intersection with the curve $y = \zeta(3/2)$, instead, gives the critical temperature for the case with $\varepsilon = \frac{p^2}{2m}$ (see Problem 10.1). We see that the critical temperature is decreased by the discrete spectrum when compared to the case with $\varepsilon = \frac{p^2}{2m}$.

Problem 10.5.
Consider a two dimensional Bose-Einstein gas in a domain of area A. This gas is characterized by a fixed average number $\langle N \rangle$ of ultrarelativistic particles with spin 0 and single particle energy

$$\varepsilon = cp$$

with p the absolute value of the momentum and c the speed of light. Prove that, unlike the non relativistic case, we find a critical temperature T_c for the Bose-Einstein condensation. Also, below T_c, determine the way the particles density changes in the condensed phase as a function of the temperature T.

Solution
Let us start from the average number of particles

$$\langle N \rangle = \frac{1}{4\pi^2 \hbar^2} \int \frac{1}{z^{-1} e^{\beta cp} - 1} d^2q \, d^2p$$

where we can use spherical polar coordinates in the momentum space and perform exactly the integral in the spatial coordinates

$$\langle N \rangle = \frac{A}{2\pi \hbar^2} \int_0^{+\infty} \frac{p}{z^{-1} e^{\beta cp} - 1} dp.$$

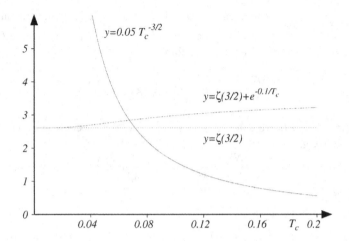

Fig. 10.1 The critical temperature T_c for the Bose-Einstein condensation of the gas described in Problem 10.4 is found from the intersection of the two curves $y = \left(\frac{h^2}{2\pi m k T_c}\right)^{3/2} \frac{\langle N \rangle}{V}$ and $y = g_{3/2}(1) + e^{-\beta_c \Delta}$. We have considered the case $k = 1$, $\left(\frac{h^2}{2\pi m k}\right)^{3/2} \frac{\langle N \rangle}{V} = 0.05$ and reported the functions $y = 0.05 T_c^{-3/2}$, $y = \zeta(3/2) + e^{-0.1/T_c}$. The intersection with the curve $y = \zeta(3/2)$ corresponds to the critical temperature of Problem 10.1

If we set $\beta c p = x$, we can write the density $n = \frac{\langle N \rangle}{A}$ as

$$n = \frac{\langle N \rangle}{A} = \frac{1}{2\pi\beta^2 c^2 \hbar^2} \int_0^{+\infty} \frac{zxe^{-x}}{1 - ze^{-x}} dx.$$

We now expand the denominator as a geometric series

$$n = \frac{1}{2\pi\beta^2 c^2 \hbar^2} \int_0^{+\infty} \left(zxe^{-x} \sum_{k=0}^{+\infty} z^k e^{-kx} \right) dx = \frac{1}{2\pi\beta^2 c^2 \hbar^2} \sum_{k=1}^{+\infty} z^k \int_0^{+\infty} xe^{-kx} dx$$

where we have interchanged the series with the integral. We then solve the integral and obtain

$$n = \frac{1}{2\pi\beta^2 c^2 \hbar^2} \sum_{k=1}^{+\infty} \frac{z^k}{k^2}.$$

The equation for the critical temperature T_c is found by setting $z = 1$ and $T = T_c$ ($\beta = \beta_c = \frac{1}{kT_c}$) in the equation for n

$$n = \frac{1}{2\pi\beta_c^2 c^2 \hbar^2} \sum_{k=1}^{+\infty} \frac{1}{k^2} = \frac{\pi}{12\beta_c^2 c^2 \hbar^2}$$

where we have used $\sum_{k=1}^{+\infty} \frac{1}{k^2} = \frac{\pi^2}{6}$. The above equation, for a given density, determines the critical temperature

$$\beta_c = \frac{1}{kT_c} = \sqrt{\frac{\pi}{3n}} \frac{1}{2c\hbar}.$$

When β increases above β_c, a significant fraction (say n_0) of the total number of bosons condenses in the ground state, so that the total density is

$$n = n_0 + \frac{\pi}{12\beta^2 c^2 \hbar^2} = n_0 + n \left(\frac{\beta_c}{\beta} \right)^2$$

and, hence

$$n_0 = n \left(1 - \left(\frac{\beta_c}{\beta} \right)^2 \right).$$

Problem 10.6.
Let us consider a collection of $\langle N \rangle$ bosonic quantum harmonic oscillators in two dimensions, with spin 0 and frequency ω. Identify the critical value of the chemical potential μ_c for which the density of states has a divergence in the ground state. Then, give an estimate for the chemical potential μ in the limit $z \to 0$. Finally, discuss the existence of a critical temperature for the phenomenon of the Bose-Einstein condensation.

Solution
In the most general case, the expression for the average number of states characterizing the system is

$$\langle N \rangle = \sum_{\text{states}} \frac{1}{z^{-1} e^{\beta \varepsilon_{\text{state}}} - 1}$$

with $z = e^{\beta\mu}$ and μ the chemical potential. In our case, we can write down this number using the known expression for the energy levels of a two dimensional quantum harmonic oscillator (see for example Problem 5.8), i.e. $E_{n_x,n_y} = \hbar\omega(n_x + n_y + 1)$, with $n_{x,y}$ non negative integers. The result for $\langle N \rangle$ is

$$\langle N \rangle = \sum_{n_x,n_y=0}^{+\infty} \frac{1}{z^{-1} e^{\beta \hbar \omega (n_x + n_y + 1)} - 1}.$$

If we set $n = n_x + n_y$, the degeneracy of a given n is $g(n) = n + 1$. In this way, for a generic function of the sum $f(n) = f(n_x + n_y)$, we have

$$\sum_{n_x,n_y} f(n_x + n_y) = \sum_n g(n) f(n).$$

The summation over n can be used to determine the average number as

$$\langle N \rangle = \sum_{n_x,n_y=0}^{+\infty} \frac{1}{z^{-1}e^{\beta\hbar\omega(n_x+n_y+1)} - 1} = \sum_{n=0}^{+\infty} \frac{n+1}{z^{-1}e^{\beta\hbar\omega(n+1)} - 1}.$$

Particles in the ground state, $n = 0$, have a critical value of the chemical potential. In such a case, the denominator becomes zero when $e^{\beta\hbar\omega} = z_c$, i.e. $\mu_c = \hbar\omega$. In order to answer the second question, we note that when $z \to 0$ we can approximate $\langle N \rangle$ as

$$\langle N \rangle = \sum_{n_x,n_y=0}^{+\infty} \frac{1}{z^{-1}e^{\beta\hbar\omega(n_x+n_y+1)} - 1} \approx ze^{-\beta\hbar\omega} \sum_{n_x,n_y=0}^{+\infty} e^{-\beta\hbar\omega(n_x+n_y)} =$$

$$ze^{-\beta\hbar\omega} \left(\sum_{n=0}^{+\infty} e^{-n\beta\hbar\omega} \right)^2 = \frac{ze^{-\beta\hbar\omega}}{(1 - e^{-\beta\hbar\omega})^2} = \frac{z}{4\sinh^2\left(\frac{\beta\hbar\omega}{2}\right)}$$

leading to

$$\mu = \frac{1}{\beta} \ln\left(4\langle N \rangle \sinh^2\left(\frac{\beta\hbar\omega}{2}\right) \right).$$

The critical temperature, if any, is found by imposing simultaneously that in the limit $\mu \to \mu_c$, $N_0 = 0$, $T = T_c = \frac{1}{k\beta_c}$ and that the sum defining $\langle N \rangle$ is convergent (N_0 is the average number of bosons in the condensed state). We have

$$\langle N \rangle = N_0 + \sum_{n=1}^{+\infty} \frac{n+1}{e^{n\beta\hbar\omega} - 1} = \sum_{n=1}^{+\infty} \frac{n+1}{e^{n\beta_c\hbar\omega} - 1}.$$

The sum converges and, for a fixed $\langle N \rangle$, the previous equation defines implicitly the critical temperature.

Problem 10.7.
Make use of the Clausius-Clapeyron equation for the Bose-Einstein gas treated in Problem 10.1 and find the relationship between the pressure and the temperature in presence of a Bose-Einstein condensation. Suppose that the condensed phase has a negligible specific volume. Make use of the fact that the latent heat is $L = \frac{5}{2}kT\frac{\zeta(\frac{5}{2})}{\zeta(\frac{3}{2})}$, where ζ is the Riemann zeta function.

Solution
The Clausius-Clapeyron equation (see also Problem 8.3) is given by

$$\frac{dP}{dT} = \frac{L}{T\Delta v}$$

where Δv is the variation of the specific volume and L the latent heat. For a Bose gas, when both phases coexist, the gaseous one has a specific volume v_g, whereas the condensed one has a negligible specific volume $v_c \approx 0$. This means that

$$\Delta v = v_g = \left(\frac{V}{N}\right)_g = \frac{\lambda^3}{\zeta\left(\frac{3}{2}\right)}$$

where $\lambda = \sqrt{\frac{2\pi\hbar^2}{mkT}}$. We also know that $L = \frac{5}{2}kT\frac{\zeta\left(\frac{5}{2}\right)}{\zeta\left(\frac{3}{2}\right)}$. Therefore, we can write the Clausius-Clapeyron equation as

$$\frac{dP}{dT} = \frac{L}{T\Delta v} = \frac{5}{2}kT\frac{\zeta\left(\frac{5}{2}\right)}{\zeta\left(\frac{3}{2}\right)}\frac{1}{Tv_g} = \frac{5}{2}k\zeta\left(\frac{5}{2}\right)\frac{1}{\lambda^3} = \frac{5}{2}\frac{P}{T}.$$

that implies $PT^{-\frac{5}{2}} = $ const.

Problem 10.8.
An average number $\langle N \rangle$ of bosons (spin $S = 1$) in three dimensions is subject to a constant magnetic field with intensity B directed along the z axis. The single particle energy is

$$\varepsilon = \frac{p^2}{2m} - \tau s_z B$$

where τ is the magnetic moment, p the absolute value of the momentum, and $s_z = -1, 0, +1$. Using the formalism of quantum gases, determine the average occupation numbers associated with the three different values of the spin. Also, write down the magnetization M and its approximation in the limit of small B. Finally, in the limit of high temperatures and small densities (the classical limit), define the susceptibility

$$\chi = \left(\frac{\partial M}{\partial B}\right)_{T,\langle N \rangle}$$

and prove Curie law

$$\lim_{B\to 0} \chi \approx \frac{\langle N \rangle}{T}.$$

Solution
We start by writing down the total average number of particles

$$\langle N \rangle = \sum_{states} \frac{1}{e^{\beta(E_{state}-\mu)} - 1}$$

where $\sum_{states}$ indicates all the possible states available for the system, each one with energy E_{state}. We note that the energy has a kinetic term ($\frac{p^2}{2m}$) plus the potential energy of the dipoles arising from the presence of the magnetic field ($\tau s_z B$). When computing $\sum_{states}$, the kinetic term represents a continuous spectrum and contributes with the integral

$$\int \frac{d^3p\, d^3q}{h^3}$$

while we have to consider a discrete sum (because $s_z = -1, 0, +1$) for the magnetic term. The final result is

$$\langle N \rangle = \sum_{s_z = -1, 0, +1} \int \frac{d^3 p\, d^3 q}{h^3} \frac{1}{e^{\beta \left(\frac{p^2}{2m} - \tau B s_z - \mu \right)} - 1}.$$

We can think that this total number is composed of three terms, i.e. $\langle N_0 \rangle$, $\langle N_+ \rangle$ and $\langle N_- \rangle$, identifying the average occupation numbers for the states with different spin projection along the z axis

$$\langle N \rangle = \langle N_0 \rangle + \langle N_+ \rangle + \langle N_- \rangle$$

from which

$$\langle N_{s_z} \rangle = \frac{V}{h^3} \int \frac{d^3 p}{e^{\beta \left(\frac{p^2}{2m} - \tau B s_z - \mu \right)} - 1} \qquad s_z = \pm 1, 0.$$

If we set $z = e^{\beta \mu}$, $x = \beta \frac{p^2}{2m}$ and $\lambda = \frac{h}{\sqrt{2\pi m k T}}$, we have

$$\langle N_{s_z} \rangle = \frac{V}{\lambda^3} g_{3/2}(z e^{\beta \tau s_z B}) \qquad g_{3/2}(z) = \frac{1}{\Gamma(3/2)} \int_0^{+\infty} \frac{x^{1/2} dx}{z^{-1} e^x - 1}.$$

The magnetization is defined as

$$M = \tau \sum_{s_z} s_z \langle N_{s_z} \rangle = \tau (\langle N_+ \rangle - \langle N_- \rangle)$$

that is

$$M = \frac{\tau V}{\lambda^3} \left[g_{3/2}(z e^{\beta \tau B}) - g_{3/2}(z e^{-\beta \tau B}) \right].$$

In the limit of small B, we can expand the exponentials to obtain

$$g_{3/2}(z e^{\pm \beta \tau B}) \approx g_{3/2}(z(1 \pm \beta \tau B))$$

and a Taylor series of $g_{3/2}(z(1 \pm \beta \tau B))$ for small $z \beta \tau B$ yields

$$g_{3/2}(z(1 \pm \beta \tau B)) = g_{3/2}(z) \pm z \beta \tau B \frac{d g_{3/2}(z)}{dz} + \mathcal{O}(B^2).$$

We can also use the properties of the Bose-Einstein functions

$$z \frac{d g_{3/2}(z)}{dz} = g_{1/2}(z)$$

to get the magnetization when $B \to 0$

$$M = \frac{2\tau^2 V}{k T \lambda^3} B g_{1/2}(z) + \mathcal{O}(B^2).$$

Moreover, for high temperatures and low densities, we have $z \to 0$ (classical limit) and we can approximate $g_{3/2}(z) \approx z$. Therefore, when $B \to 0$, the total average number of particles in the classical limit is

$$\langle N \rangle = \langle N_0 \rangle + \langle N_+ \rangle + \langle N_- \rangle = \frac{3V}{\lambda^3} g_{3/2}(z) \approx \frac{3zV}{\lambda^3}$$

from which we extract z as a function of the density

$$z = \frac{\lambda^3}{3} \frac{\langle N \rangle}{V}$$

which can be substituted in the magnetization

$$M = \frac{2\tau^2 V}{kT\lambda^3} B g_{1/2}(z) + \mathscr{O}(B^2) \approx \frac{2\tau^2 V}{kT\lambda^3} Bz + \mathscr{O}(B^2) = \frac{2\tau^2 \langle N \rangle}{3kT} B + \mathscr{O}(B^2)$$

and, hence

$$\lim_{B \to 0} \chi = \lim_{B \to 0} \left(\frac{\partial M}{\partial B} \right)_{T, \langle N \rangle} = \frac{2\tau^2 \langle N \rangle}{3kT}$$

that is Curie law (see also Problems 6.13, 7.23 and 7.24).

Problem 10.9.
Let us consider a system of $\langle N \rangle$ non relativistic bosons with mass m and spin 0 in a cylindrical volume of base A and height L. The gas is under the effect of a constant gravitational field with acceleration g. Show that the critical temperature for the Bose-Einstein condensation is well approximated by

$$T_c \approx T_c^{(0)} \left[1 + \frac{8}{9} \frac{1}{\zeta(3/2)} \left(\frac{\pi m g L}{k T_c^{(0)}} \right)^{1/2} \right]$$

where $T_c^{(0)}$ is the critical temperature in absence of an external field (see Problem 10.1). Assume that $mgL \ll kT_c^{(0)}$. To solve the problem, set $\omega = z e^{-\beta mgh}$, where h is the vertical coordinate. Then, with $\alpha = -\ln(\omega)$, make use of the following expansion valid for $\omega \approx 1$

$$g_{3/2}(\omega) \approx \Gamma\left(-\frac{1}{2}\right) \alpha^{1/2} + \zeta\left(\frac{3}{2}\right) + \cdots$$

with

$$\Gamma\left(\frac{3}{2}\right) = \frac{\sqrt{\pi}}{2} \qquad \Gamma\left(-\frac{1}{2}\right) = -2\sqrt{\pi}.$$

Solution
We write down the equation for the total average number of bosons

$$\langle N \rangle = \frac{4\pi A}{h^3} \int_0^L dh \int_0^{+\infty} \frac{p^2}{z^{-1} e^{\beta mgh} e^{\beta p^2/2m} - 1} dp$$

and, by setting $\omega = ze^{-\beta mgh}$ and $x = \beta \frac{p^2}{2m}$, we can write

$$\langle N \rangle = 2\pi A \left(\frac{2mkT}{h^2} \right)^{3/2} \Gamma \left(\frac{3}{2} \right) \frac{1}{\beta mg} \int_{ze^{-\beta mgL}}^{z} \frac{g_{3/2}(\omega)}{\omega} d\omega$$

where

$$\int_{0}^{+\infty} \frac{x^{1/2}}{\omega^{-1} e^x - 1} dx = \Gamma \left(\frac{3}{2} \right) g_{3/2}(\omega).$$

We now make use of the expansion suggested by the text for $g_{3/2}(\omega)$, where $\alpha = -\ln(\omega)$, valid for values of ω close to 1 (small α)

$$g_{3/2}(\omega) \approx \Gamma \left(-\frac{1}{2} \right) \alpha^{1/2} + \zeta \left(\frac{3}{2} \right) + \cdots.$$

We know that $\Gamma \left(\frac{3}{2} \right) = \frac{\sqrt{\pi}}{2}$ and $\Gamma \left(-\frac{1}{2} \right) = -2\sqrt{\pi}$, and close to the transition (where $z \approx 1$ and $\alpha \approx \beta mgh$) we can write

$$g_{3/2}(\omega) \approx -2\sqrt{\pi}(\beta mgh)^{1/2} + \zeta \left(\frac{3}{2} \right)$$

which can be plugged into the equation for the average number $\langle N \rangle$

$$\langle N \rangle = A \left(\frac{2\pi mkT}{h^2} \right)^{3/2} \left[L\zeta \left(\frac{3}{2} \right) - 2\sqrt{\pi}(\beta mg)^{1/2} \int_{0}^{L} h^{1/2} dh \right].$$

We then factorize the term $L\zeta \left(\frac{3}{2} \right)$ and define $V = AL$ to obtain

$$\langle N \rangle = V \left(\frac{2\pi mkT}{h^2} \right)^{3/2} \zeta \left(\frac{3}{2} \right) \left[1 - \frac{4}{3} \frac{1}{\zeta \left(\frac{3}{2} \right)} \sqrt{\frac{\pi mgL}{kT}} \right].$$

This equation, for fixed $\langle N \rangle / V$, implicitly defines the critical temperature T_c for the Bose-Einstein condensation. To compute it, we recall that we have to assume $mgL \ll kT_c^{(0)}$. To give a zeroth order approximation for T_c, we can neglect the second term inside the square bracket, obtaining the known result

$$T_c^{(0)} = \frac{h^2}{2\pi mk} \left(\frac{\langle N \rangle}{V\zeta \left(\frac{3}{2} \right)} \right)^{2/3}$$

which coincides with the critical temperature for a gas of free bosons without any external field (see Problem 10.1). As a first order approximation, we can plug the zeroth order result obtained previously into the equation for $\langle N \rangle$ and get

$$T_c \approx \frac{h^2}{2\pi mk} \left(\frac{\langle N \rangle}{V\zeta \left(\frac{3}{2} \right)} \right)^{2/3} \frac{1}{\left[1 - \frac{4}{3} \frac{1}{\zeta \left(\frac{3}{2} \right)} \sqrt{\frac{\pi mgL}{kT_c^{(0)}}} \right]^{2/3}}$$

where, if we expand the denominator with $\frac{1}{(1-x)^\alpha} \approx 1 + \alpha x \ (x \ll 1)$, we finally obtain

$$T_c \approx T_c^{(0)} \left[1 + \frac{8}{9} \frac{1}{\zeta(3/2)} \left(\frac{\pi mgL}{kT_c^{(0)}} \right)^{1/2} \right]$$

that is the desired result.

Problem 10.10.
Determine the maximum of the Planck distribution (for the three dimensional case) as a function of the frequency and the wavelength. Show that this is possible if we maximize the function $x^a/(e^x - 1)$ for $a = 3$ and $a = 5$ respectively. This means solving the equation $x = a(1 - e^{-x})$, which can be done in an iterative way $x_n = a(1 - e^{-x_{n-1}})$, starting from $x_1 = 1$ (stop after 5 iterations). Verify Wien law, $\lambda_{max} T =$ const., and comment on the fact that we find two different constants in the two approaches.

We know that the sun produces the largest amount of radiation around the wavelength $\approx 5 \times 10^{-5}$ cm. Using the results previously obtained, determine:

- the temperature of the sun;
- the amount of energy produced, knowing that the main mechanism of production of such energy is the transformation of hydrogen into helium, and that this reaction stops when 10% of the hydrogen has been converted. A good approximation is to take the whole mass of the hydrogen equal to the mass of the sun (use Einstein relation $E = \Delta M c^2$);
- the lifetime of the sun.

You can use the following numerical constants: $h = 6.625 \times 10^{-27}$ erg s; $c = 3 \times 10^{10}$ cm s^{-1}; $\sigma_B = 5.67 \times 10^{-5}$ erg cm^{-2} s^{-1} K^{-4}; $R_{sun} = 7 \times 10^{10}$ cm; $M_{sun} = 2 \times 10^{33}$ g; $k = 1.38 \times 10^{-16}$ erg K^{-1}.

Solution
The dimensionless volume element in the phase space for a free photon gas is given by

$$2 \frac{V}{h^3} dp_x dp_y dp_z = \frac{8\pi V}{h^3} p^2 dp = \frac{8\pi V}{c^3} v^2 dv$$

with $p = \frac{hv}{c}$. The energy density per unit frequency u_v is obtained by multiplying the dimensionless volume element by the Bose-Einstein distribution and by the energy, and dividing by the volume V

$$u_v dv = \frac{8\pi h}{c^3} \frac{v^3 dv}{e^{\frac{hv}{kT}} - 1}.$$

Using the relation $\lambda v = c$, we also obtain the energy density per unit wavelength u_λ

$$u_\lambda d\lambda = \frac{8\pi hc}{\lambda^5} \frac{d\lambda}{e^{\frac{hc}{\lambda kT}} - 1}.$$

Local maxima are found by imposing the following conditions

$$\frac{du_v}{dv} = \frac{8\pi h}{c^3} \left(\frac{kT}{h}\right)^2 \frac{d}{dx}\left(\frac{x^3}{e^x - 1}\right) = 0$$

$$\frac{du_\lambda}{d\lambda} = \frac{d\lambda'}{d\lambda}\frac{du_{\lambda'}}{d\lambda'} = 8\pi hc \left(\frac{kT}{h}\right)^4 \frac{d\lambda'}{d\lambda}\frac{d}{dx}\left(\frac{x^5}{e^x - 1}\right) = 0$$

with $\lambda' = 1/\lambda$ and $x = \frac{h v}{kT}$ in the first case, and $x = \frac{hc}{\lambda kT}$ in the second. These are the functions given in the text of the problem. Using an iterative procedure, $x_n = a(1 - e^{-x_{n-1}})$, we get $x_1 = 1, x_2 = 1.89636, x_3 = 2.54966, x_4 = 2.76568, x_5 = 2.8112$ (for $a = 3$) and $x_1 = 1, x_2 = 3.1606, x_3 = 4.788, x_4 = 4.9584, x_5 = 4.9649$ (for $a = 5$), from which

$$\frac{h v_{max}}{kT} = 2.82144$$

$$\frac{hc}{\lambda_{max}kT} = 4.96511.$$

The last relation proves Wien law. Using $v_{max} = c/\lambda_{max}$ in the first relation, we note that the two constants are not equal. The reason is that both distributions, u_v and u_λ, are given per unit frequency and wavelength respectively. It then follows from the relation $d\lambda = -\lambda^2 dv/c$ that a unitary interval of frequencies is not corresponding to a unitary interval of wavelengths, i.e. they are not directly proportional.

Wien law proves that the spectrum of our radiation is dependent only on $\lambda_{max}T$, a fact that is now used to answer the other questions of the problem. Using the numerical constants given by the text, we obtain

$$T_{sun} = \frac{1}{4.96}\frac{hc}{k\lambda} \approx 6000K.$$

Let us now compute the total energy emitted by the sun through the mechanism of transformation of hydrogen into Helium. From the text, we know that such reaction stops when the 10% of hydrogen has been converted. Therefore, the total energy produced is

$$E_{TOT} = \Delta Mc^2 = 0.1 \times 2 \times 10^{33} \times 9 \times 10^{20}\text{erg} \approx 2 \times 10^{53}\text{erg}.$$

If we assume the sun behaves like a black body, we can use Stefan-Boltzmann law to estimate the release of energy per unit time and area. If we multiply by the total surface $4\pi R_{sun}^2$, we obtain the energy released in the unit time. Consequently, an estimate for the lifetime of the sun naturally emerges

$$t_{sun} = \frac{E_{TOT}}{\sigma_B T^4 4\pi R_{sun}^2} \approx \frac{2 \times 10^{53}\text{erg}}{4 \times 10^{33}\text{erg s}^{-1}} \approx 0.5 \times 10^{20}\text{s} \approx 1.5 \times 10^{12}\text{years}.$$

Problem 10.11.
At the temperature $T_\gamma = 3000K$ in a volume V_γ, the electromagnetic radiation de-coupled from the plasma originating after the Big Bang. Knowing that the present temperature is $T_{now} = 3K$, what is the ratio between the present volume and V_γ if we assume an adiabatic expansion? Has the peak of the Planck distribution moved? How much these values change if we consider a d dimensional universe? Let us suppose that for $T < T_\gamma$ the electrons can only interact electromagnetically. Is it reasonable to consider them not to be interacting?

Solution
We first recall the infinitesimal volume element for a d dimensional sphere of radius R

$$dV_d = \frac{d\pi^{d/2}}{(d/2)!} R^{d-1} dR.$$

Then, we consider the dimensionless volume element in the phase space for the photon gas

$$\frac{gV}{(2\pi\hbar)^d} d^d p = \frac{gV}{(2\pi\hbar)^d} \frac{d\pi^{\frac{d}{2}}}{(\frac{d}{2})!} p^{d-1} dp = \frac{gVd}{\pi^{\frac{d}{2}}(2c)^d (\frac{d}{2})!} \omega^{d-1} d\omega$$

where V is the volume, $\omega = \frac{pc}{\hbar}$, and $g = d - 1$ accounts for the independent polar-izations in d dimensions. For the average energy we find

$$U = \frac{gVd\hbar}{\pi^{\frac{d}{2}}(2c)^d (\frac{d}{2})!} \int_0^{+\infty} \frac{\omega^d d\omega}{e^{\frac{\hbar\omega}{kT}} - 1} = \frac{gVd}{\pi^{\frac{d}{2}}(2\hbar c)^d (\frac{d}{2})!} (kT)^{d+1} \int_0^{+\infty} \frac{x^d dx}{e^x - 1}.$$

A photon gas has a zero chemical potential, that implies $\Phi = \mu N = PV + F = 0$. This leads to $PV = -F$. The term PV is related to the grand canonical partition function

$$-F = PV = kT \ln \mathcal{Z} = -kT \frac{gVd}{\pi^{\frac{d}{2}}(2c)^d (\frac{d}{2})!} \int_0^{+\infty} \omega^{d-1} \ln(1 - e^{\frac{-\hbar\omega}{kT}}) d\omega =$$

$$kT \frac{gV}{\pi^{\frac{d}{2}}(2c)^d (\frac{d}{2})!} \frac{\hbar}{kT} \int_0^{+\infty} \frac{\omega^d d\omega}{e^{\frac{\hbar\omega}{kT}} - 1} = \frac{gV}{\pi^{\frac{d}{2}}(2c)^d (\frac{d}{2})!\hbar^d} (kT)^{d+1} \int_0^{+\infty} \frac{x^d dx}{e^x - 1} = \frac{U}{d}$$

where we have used the integration by parts. Finally, the entropy is

$$S = \frac{U - F}{T} = \frac{d+1}{d} \frac{U}{T} \propto VT^d$$

meaning that, for a fixed d, the adiabatic transformation ($S = $ const.) leads to

$$VT^d = \text{const.}$$

Using this result with $d = 3$, we find the ratio of the volumes at the temperatures $T_\gamma = 3000K$ and $T_{now} = 3K$ as

$$\frac{V_{T_{now}}}{V_{T_\gamma}} = \left(\frac{T_\gamma}{T_{now}}\right)^3 \approx 10^9.$$

As for the peak of Plank distribution, we set $x = \frac{\hbar\omega}{kT}$, consider the function $f(x) = \frac{x^{d-1}}{e^x-1}$, and find its maximum x_M (see also Problem 10.10). After differentiation, we obtain

$$\frac{df(x)}{dx}\bigg|_{x=x_M} = \frac{(d-1)x_M^{d-2}}{e^{x_M}-1} - \frac{e^{x_M}x_M^{d-1}}{(e^{x_M}-1)^2} = 0$$

leading to

$$(d-1)e^{x_M} - (d-1) - x_M e^{x_M} = 0.$$

In general, the maximum of the distribution verifies $x_M = M(d)$ with $M(d)$ a generic function of d. Therefore, given the temperature, the ω_p for the peak is such that

$$\frac{\hbar\omega_p}{kT} = M(d).$$

For a fixed d, when moving from T_γ to T_{now}, we can estimate the change in frequency

$$\frac{\omega_p(T_{now})}{\omega_p(T_\gamma)} = \frac{T_{now}}{T_\gamma} \approx 10^{-3}$$

valid for every d. Let us now answer the last question. We introduce the specific volume $v = V/\langle N \rangle$, i.e. the volume for a single particle. The average distance among the particles is thus $r \approx v^{\frac{1}{3}} = n^{-\frac{1}{3}}$, where n is the average density which, for a photon gas, reads

$$n = \frac{1}{\pi^2 c^3} \int_0^{+\infty} \frac{\omega^2 d\omega}{e^{\hbar\omega/kT}-1} = \frac{2\zeta(3)(kT)^3}{\pi^2\hbar^3 c^3}.$$

We then get

$$\frac{e^2/r}{kT} \approx \frac{e^2 n^{\frac{1}{3}}}{kT} = \frac{e^2}{\hbar c}\left(\frac{2\zeta(3)}{\pi^2}\right)^{1/3} \approx 0.62 \times \frac{e^2}{\hbar c} = 0.62 \times \frac{1}{137} \approx 0.0045.$$

This justifies the approximation in the text of the problem.

Problem 10.12.
An ideal black body is able to absorb all the incident radiation and emit it according to the Planck formula. At equilibrium, the emissivity of the black body is equal to its absorption. Assume that both the earth and the sun behave like ideal black bodies in equilibrium. Knowing that:

- the earth's temperature is $T_E \approx 14\,C = 287$ K;
- its distance from the sun is $D = 1.5 \times 10^{13}$cm;
- the radius of the sun is $R_S = 6.96 \times 10^{10}$cm;

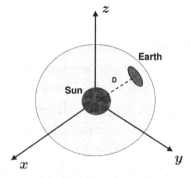

Fig. 10.2 A schematic view of the problem of two black bodies (the earth and the sun) in thermodynamic equilibrium and separated by a very large distance (see Problem 10.12)

give an estimate for the sun's temperature. The earth's surface may be well approximated by a two dimensional area on the surface of the sphere with radius D (see Fig. 10.2).

Solution
To determine the radiation emitted by the sun, we need to compute the average energy density of a black body with a volume V. We get

$$\frac{U}{V} = \frac{\hbar}{\pi^2 c^3} \int_0^{+\infty} \frac{\omega^3 d\omega}{e^{\frac{\hbar\omega}{kT}} - 1} = \frac{k^4 T^4}{\pi^2 \hbar^3 c^3} \int_0^{+\infty} \frac{x^3 dx}{e^x - 1} = \frac{\pi^2 k^4}{15\hbar^3 c^3} T^4.$$

Using the formula for the effusion through a small hole, we find the energy emitted per unit area and time as

$$R_S = \frac{c}{4}\frac{U}{V} = \sigma T_S^4$$

which is the Stefan-Boltzmann law. This emission is clearly isotropic and the total emissivity is given by the product of the radiation flux per unit area and the total spherical surface of the sun, i.e. $P_S = 4\pi R_S^2 \sigma T_S^4$. Only a fraction of this total emitted radiation reaches the earth: this fraction is the ratio $\pi R_E^2/(4\pi D^2)$ (see Fig. 10.2) which represents the solid angle of emission in the direction of the earth. From the information of the text, we know that all the radiation absorbed is emitted by the earth. This implies the equation

$$(4\pi R_S^2)(\sigma T_S^4)\frac{\pi R_E^2}{4\pi D^2} = (4\pi R_E^2)(\sigma T_E^4)$$

which gives

$$T_S = \sqrt{\frac{2D}{R_S}} T_E \approx 5960 K.$$

Problem 10.13.
Using the formalism of the ideal quantum gases with a continuous spectrum in the energy, prove the pressure-energy relation

$$PV = \frac{\alpha}{3}U$$

when the single particle energy is given by $\varepsilon = p^\alpha$, with p the absolute value of the momentum and $\alpha > 0$. Using this result, determine the relation between the energy and the temperature for an ultrarelativistic gas ($\varepsilon = p$ taking the speed of light $c = 1$) with zero chemical potential. To this end, make use of the first law of thermodynamics $dU = TdS - PdV$, where $U(T,V) = u(T)V$ ($u(T)$ is the energy density) and use the Maxwell relations derived from the thermodynamic potential suited to describe this case.

Solution
A generic quantum gas is described by the following occupation number

$$\langle n_\varepsilon \rangle = \frac{1}{z^{-1}e^{\frac{\varepsilon}{kT}} + a}$$

with $a = \pm 1$ ($a = -1$ means Bose-Einstein and $a = 1$ means Fermi-Dirac). It follows that

$$\frac{PV}{kT} = \ln \mathscr{Q} = \frac{gV}{a} \int_0^{+\infty} \ln \left[1 + aze^{-\frac{\varepsilon}{kT}} \right] \frac{4\pi p^2 dp}{h^3} =$$

$$\frac{4\pi gV}{ah^3} \left[\frac{p^3}{3} \ln \left[1 + aze^{-\frac{\varepsilon}{kT}} \right]_0^{+\infty} + \int_0^{+\infty} \frac{p^3}{3} \frac{aze^{-\frac{\varepsilon}{kT}}}{1 + aze^{-\frac{\varepsilon}{kT}}} \frac{1}{kT} \frac{d\varepsilon}{dp} dp \right]$$

where $\mathscr{Q} = \mathscr{Q}(T,V,z)$ is the grand canonical partition function and g the spin degeneracy. The boundary term in the expression for PV/kT is equal to zero and we find

$$PV = \frac{4\pi gV}{h^3} \int_0^{+\infty} \frac{p^3}{3} \frac{1}{z^{-1}e^{\frac{\varepsilon}{kT}} + a} \frac{d\varepsilon}{dp} dp = \frac{4\pi gV}{3h^3} \int_0^{+\infty} \frac{1}{z^{-1}e^{\frac{\varepsilon}{kT}} + a} \left(p \frac{d\varepsilon}{dp} \right) p^2 dp.$$

The average number of states and the average energy are

$$\langle N \rangle = \frac{4\pi gV}{h^3} \int_0^{+\infty} \frac{1}{z^{-1}e^{\frac{\varepsilon}{kT}} + a} p^2 dp.$$

$$U = \frac{4\pi gV}{h^3} \int_0^{+\infty} \frac{1}{z^{-1}e^{\frac{\varepsilon}{kT}} + a} \varepsilon p^2 dp$$

from which

$$P = \frac{4\pi g}{3h^3} \int_0^{+\infty} \frac{1}{z^{-1}e^{\frac{\varepsilon}{kT}} + a} \left(p \frac{d\varepsilon}{dp} \right) p^2 dp = \frac{4\pi g\alpha}{3h^3} \int_0^{+\infty} \frac{1}{z^{-1}e^{\frac{\varepsilon}{kT}} + a} p^\alpha p^2 dp = \frac{\alpha U}{3V}.$$

Let us now look at the relation between the average energy and the temperature for an ultrarelativistic gas with $\varepsilon = p$ and zero chemical potential. From the first law of thermodynamics, we know that $dU = TdS - PdV$. If we differentiate this expression with respect to V at constant temperature, we get

$$\left(\frac{\partial U}{\partial V}\right)_T = T\left(\frac{\partial S}{\partial V}\right)_T - P = T\left(\frac{\partial P}{\partial T}\right)_V - P$$

where we have used the Maxwell relation

$$\left(\frac{\partial S}{\partial V}\right)_T = \left(\frac{\partial P}{\partial T}\right)_V$$

coming from the differential expression of the free energy $dF = -SdT - PdV$. If we now assume that $U = Vu(T)$ and that $P = \frac{u}{3}$, we end up with the following differential equation

$$u = \frac{T}{3}\frac{du}{dT} - \frac{u}{3}$$

that is simplified to yield $T\frac{du}{dT} = 4u$. The solution is $u(T) = AT^4$ with A a normalization constant. This result may be somehow useful when computing the average energy for a a photon gas, in the sense that it predicts the correct power law in the temperature associated with the energy density. Unfortunately, the use of a classical approach would lead to an infinite normalization constant, $A \rightarrow +\infty$. In this case, the use of Quantum Mechanics is needed to obtain a finite A (see Problems 10.10, 10.11 and 10.12).

Problem 10.14.
Compute the density of the states $a(\varepsilon)$, the entropy S, the free energy F, the enthalpy H, and the Gibbs potential Φ, for a photon gas in three dimensions using the grand canonical formalism for the quantum gases.

Solution
We first determine the density of states $a(\varepsilon)$ and then use it for the computation of the partition function. The single particle energy is $\varepsilon = pc$, with p the absolute value of the momentum and c the speed of light. Our gas has $z = 1$ and the spin degeneracy is $g = 2$ to take into account the two different directions of the transverse modes and the fact that the longitudinal modes do not appear in the radiation. The density of states $a(\varepsilon)$ is determined from

$$2\frac{Vd^3p}{h^3} = \frac{8\pi V p^2 dp}{h^3} = a(\varepsilon)d\varepsilon$$

where we have used spherical polar coordinates in the momentum and integrated over the angular variables, because the energy depends only on the absolute value of the momentum. If we use $p = \varepsilon/c$ we get

$$a(\varepsilon)d\varepsilon = \frac{8\pi V}{h^3 c^3}\varepsilon^2 d\varepsilon.$$

The pressure is given by

$$
\frac{PV}{kT} = \ln \mathscr{Q} = -\frac{8\pi V}{h^3 c^3} \int_0^{+\infty} \ln\left(1 - e^{-\frac{\varepsilon}{kT}}\right) \varepsilon^2 d\varepsilon = \frac{8\pi V}{3kTh^3 c^3} \int_0^{+\infty} \frac{\varepsilon^3}{e^{\frac{\varepsilon}{kT}} - 1} d\varepsilon =
$$

$$
\frac{8\pi V}{3h^3 c^3}(kT)^3 \int_0^{+\infty} \frac{x^3}{e^x - 1} dx = \frac{8\pi^5 V}{45h^3 c^3}(kT)^3 = C(kT)^3
$$

where we have integrated by parts and used the following definition for the constant C

$$
C = \frac{8\pi^5 V}{45h^3 c^3}.
$$

The average energy can be computed taking the derivative of the logarithm of the partition function

$$
U = kT^2 \left(\frac{\partial \ln \mathscr{Q}}{\partial T}\right)_V = 3C(kT)^4
$$

so that we can write down the pressure in terms of the energy (see also Problem 10.13) as

$$
PV = \frac{U}{3}.
$$

Moreover, for the entropy we find

$$
S = k \left(\frac{\partial (T \ln \mathscr{Q})}{\partial T}\right)_V = 4Ck^4 T^3.
$$

We can now compute the free energy and the enthalpy

$$
F = U - TS = -C(kT)^4 \qquad H = U + PV = 4C(kT)^4.
$$

As a check, using the results previously obtained, we compute

$$
\Phi = F + PV = 0.
$$

This is the correct result because, given an average number of particles N, the Gibbs potential is related to the chemical potential μ, $\Phi = \mu N$, and $\mu = 0$ in our case because we are dealing with a photon gas.

Problem 10.15.

An average number $\langle N \rangle$ of bosons with spin $S = 0$ is confined in a two dimensional domain with surface A. The gas is ultrarelativistic with a single particle energy $\varepsilon = pc$, where p is the absolute value of the momentum and c the speed of light. Compute the number of states and determine the correction to the equation of state for the ideal gas at high temperatures. We recall that the pressure is such that $PA = U/d$ (see Problem 10.13), with U the average energy and d the space dimensionality. Also, the Gamma function

$$
\Gamma(n) = \int_0^{+\infty} e^{-t} t^{n-1} dt = (n-1)!
$$

may be useful.

Solution

The average number of particles is

$$\langle N \rangle = \frac{2\pi A}{h^2} \int_0^{+\infty} \frac{p\,dp}{e^{\beta(pc-\mu)} - 1}$$

and, at high temperatures, we can set $z \ll 1$ with $z = e^{\beta\mu}$. We first write the integral by setting $x = \beta pc$

$$\langle N \rangle = \frac{2\pi A}{h^2(\beta c)^2} \int_0^{+\infty} \frac{ze^{-x}x\,dx}{1 - ze^{-x}}$$

and we Taylor expand in z by keeping only the first two terms of the expansion

$$\langle N \rangle \approx \frac{2\pi A}{(\hbar\beta c)^2} \left[z \int_0^{+\infty} e^{-x}x\,dx + z^2 \int_0^{+\infty} e^{-2x}x\,dx \right]$$

from which, using the values of the Gamma function, we get

$$z\left(1 + \frac{z}{4}\right) = n'$$

with $n' = \frac{\langle N \rangle(\hbar\beta c)^2}{2\pi A}$. Solving for z and approximating $\sqrt{1+n'} \approx 1 + \frac{1}{2}n' - \frac{1}{8}(n')^2$ we find $z \approx n'\left(1 - \frac{n'}{4}\right)$. We now need the average energy to evaluate the equation of state ($PA = \frac{U}{2}$). Again, making use of the Gamma function, we find

$$U = \frac{2\pi Ac}{h^2} \int_0^{+\infty} \frac{p^2 dp}{e^{\beta(pc-\mu)} - 1} \approx \frac{2\pi Ac}{h^2(\beta c)^3} \left(z \int_0^{+\infty} x^2 e^{-x} dx + z^2 \int_0^{+\infty} x^2 e^{-2x} dx \right) = $$

$$\frac{4\pi Ac}{h^2(\beta c)^3} z\left(1 + \frac{z}{8}\right)$$

where we can plug $z = n'\left(1 - \frac{n'}{4}\right)$ to obtain

$$U \approx \frac{4\pi Ac}{h^2(\beta c)^3} n'\left(1 - \frac{1}{8}n'\right) + \mathcal{O}((n')^3).$$

From $PA = \frac{U}{2}$ we obtain the pressure

$$P = \frac{\langle N \rangle kT}{A}\left(1 - \frac{1}{16}\frac{\langle N \rangle(\hbar\beta c)^2}{A\pi}\right)$$

showing a negative first order correction to the classical equation of state. A similar problem (see Problem 11.13) can be considered for a Fermi-Dirac gas and shows a positive correction to the ideal equation of state.

11

Fermi-Dirac Gases

Problem 11.1.
An atomic nucleus of Helium consists of a gas of 0.18 nucleons in a volume of 1 fm^3 (1 fm $= 10^{-13}$cm). In this system, we can find two kinds of nucleons (protons and neutrons) with spin $S = 1/2$ and their masses, $m_n \approx 1$GeV, may be considered equal. Compute the Fermi energy for the system and comment about its nature at the ambient temperature. Assume that the particles are independent with a single particle energy

$$\varepsilon = \frac{p^2}{2m_n}$$

where p is the absolute value of the momentum. Also, make use of: $h = 7 \times 10^{-24}$GeVs, 10^{-14}cm $\approx 7 \times 10^{-25}$s, $k = 8.61 \times 10^{-5}$eV/K

Solution
We first need to find the Fermi energy ε_F as a function of the particles density. To this end, we impose that at zero temperature the gas occupies all the energy levels up to ε_F, and the average number of particles is

$$\langle N \rangle = \frac{4\pi g V}{h^3} \int_0^{p_F} p^2 dp = \frac{4\pi g V p_F^3}{3h^3}$$

with g the spin degeneracy factor and $p_F = \sqrt{2m_n\varepsilon_F}$ the Fermi momentum. Therefore, the Fermi energy for the system is

$$\varepsilon_F = \left(\frac{3\langle N \rangle}{4\pi g V} \right)^{2/3} \frac{h^2}{2m_n}$$

where one has to use the degeneracy factor $g = 4$ because each nucleon has the z component of the spin $S_z = \pm\frac{\hbar}{2}$ (see also Problem 3.9). We obtain $\varepsilon_F \approx 25$M eV. At the ambient temperature we know that $kT \approx 0.0259$eV. It is verified that $\varepsilon_F \gg kT$ and, therefore, the system is fully degenerate.

Cini M., Fucito F., Sbragaglia M.: Solved Problems in Quantum and Statistical Mechanics. DOI 10.1007/978-88-470-2315-4_11, © Springer-Verlag Italia 2012

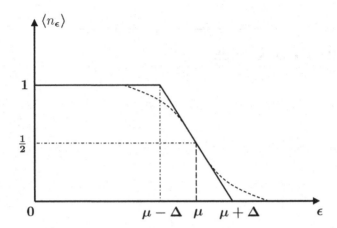

Fig. 11.1 A possible approximation for the Fermi-Dirac occupation number ($\langle n_\varepsilon \rangle$): in the point where $\varepsilon = \mu$ we draw the tangent to $\langle n_\varepsilon \rangle$. The resulting thermodynamic properties for the two dimensional gas are discussed in Problem 11.2

Problem 11.2.
Consider a Fermi gas with spin S and energy $\varepsilon = p^2/2m$, where p is the absolute value of the momentum and m the associated mass. The gas is placed on a two dimensional surface A at finite temperature T. Let us approximate the Fermi-Dirac occupation number ($\langle n_\varepsilon \rangle$) in the following way: in the point where $\varepsilon = \mu$ draw the tangent to $\langle n_\varepsilon \rangle$ as shown in Fig. 11.1. Write down the expression for $\langle n_\varepsilon \rangle$ obtained in this way and determine the average energy. Finally, compute the specific heat C_A and comment on the result.

Solution
For this system, the density of states $a(\varepsilon)$ is such that

$$(2S+1) \int \frac{d^2 p \, d^2 q}{h^2} = \int a(\varepsilon) d\varepsilon = C \int d\varepsilon$$

where $C = (2S+1)mA/(2\pi\hbar^2)$ is a constant. To proceed with the calculations, we need to determine the form of $\langle n_\varepsilon \rangle$. We start by computing the tangent to $\langle n_\varepsilon \rangle$ where $\varepsilon = \mu$. For the Fermi-Dirac statistics we can use

$$\langle n_\varepsilon \rangle = \frac{1}{e^{\beta(\varepsilon - \mu)} + 1}$$

so that

$$\langle n_\varepsilon \rangle |_{\varepsilon = \mu} = \frac{1}{2}$$

while, for the derivative, we have

$$\frac{d\langle n_\varepsilon \rangle}{d\varepsilon} = -\frac{\beta e^{\beta(\varepsilon-\mu)}}{\left(e^{\beta(\varepsilon-\mu)}+1\right)^2}$$

from which

$$\frac{d\langle n_\varepsilon \rangle}{d\varepsilon}\bigg|_{\varepsilon=\mu} = -\frac{\beta}{4}.$$

Therefore, we have to write the equation for a straight line $r(\varepsilon)$ touching the point $(\mu, \frac{1}{2})$ and with angular coefficient $-\frac{\beta}{4}$. Such equation is

$$r(\varepsilon) = \frac{1}{2} + \frac{\beta}{4}(\mu - \varepsilon).$$

The desired form for $\langle n_\varepsilon \rangle$ is

$$\langle n_\varepsilon \rangle = \begin{cases} 1 & 0 \le \varepsilon \le \mu - \Delta \\ r(\varepsilon) & \mu - \Delta \le \varepsilon \le \mu + \Delta \end{cases}$$

with $\Delta = 2kT$. For simplicity, we can write $r(\varepsilon)$ as

$$r(\varepsilon) = \frac{1}{2} + \frac{\beta}{4}(\mu - \varepsilon) = \frac{1}{2}\left(1 - \frac{x}{\Delta}\right)$$

with $x = \varepsilon - \mu$. The average energy is

$$U = C \int \langle n_\varepsilon \rangle \varepsilon d\varepsilon = C \int_0^{\mu - \Delta} \varepsilon d\varepsilon + \frac{C}{2} \int_{-\Delta}^{\Delta} \left(1 - \frac{x}{\Delta}\right)(\mu + x)dx$$

where we can expand the integrand of the second integral as

$$\left(1 - \frac{x}{\Delta}\right)(\mu + x) = \mu + x - \frac{\mu x}{\Delta} - \frac{x^2}{\Delta}$$

so that

$$U = \frac{C}{2}(\mu - \Delta)^2 + \frac{C}{2} \int_{-\Delta}^{\Delta} \left[\mu + x - \frac{\mu x}{\Delta} - \frac{x^2}{\Delta}\right] dx = \frac{C}{2}(\mu - \Delta)^2 + C\mu\Delta - \frac{C}{3}\Delta^2.$$

The final result is

$$U = \frac{C}{2}\mu^2 + \frac{C}{6}\Delta^2$$

from which

$$C_A = \left(\frac{\partial U}{\partial T}\right)_{A,N} = \frac{4k^2 CT}{3} = \frac{4mA(2S+1)k^2 T}{6\pi\hbar^2}.$$

The result is linear in the temperature, and this is in agreement with the specific heat of the Fermi gas only for low temperatures, where we know that $C_A = \frac{mA\pi k^2 T}{3\hbar^2}$ (see the exact solution analyzed in Problem 11.3) when $S = 1/2$. In the approximate case of this problem, when $S = 1/2$, we find $C_A = \frac{4mAk^2 T}{3\pi\hbar^2}$. Due to the approximations in $\langle n_\varepsilon \rangle$, the constant in front of the temperature is not the same as that of the exact solution.

Problem 11.3.
A Fermi gas with $\langle N \rangle$ particles of spin $S = 1/2$ and mass m is placed in a two dimensional domain of area A at finite temperature T. Determine:

- the Fermi energy ε_F as a function of the density;
- the chemical potential μ as a function of T and ε_F;
- the limit $\lim_{T \to 0} \mu$, verifying that it is equal to ε_F;
- the specific heat at constant area C_A in the low temperature limit.

Solution
The Fermi energy can be computed from the total number of particles at zero temperature

$$\langle N \rangle = \frac{2}{4\pi^2\hbar^2} \int d^2q \int d^2p$$

where the domain of integration is such that $\frac{p^2}{2m} \leq \varepsilon_F$. The integral in the position coordinates gives the area of the region A. For what the momentum is concerned, using polar coordinates, we get

$$n = \frac{\langle N \rangle}{A} = \frac{1}{\pi\hbar^2} \int_0^{\sqrt{2m\varepsilon_F}} p\,dp = \frac{m\varepsilon_F}{\pi\hbar^2}$$

and we easily find $\varepsilon_F = \frac{n\pi\hbar^2}{m}$. When $T \neq 0$, the average number of particles is written as

$$n = \frac{1}{\pi\hbar^2} \int_0^{+\infty} \frac{1}{e^{\beta(\varepsilon-\mu)} + 1} p\,dp$$

where we have used $\varepsilon = p^2/2m$. We can set $x = \beta\varepsilon$, and obtain

$$n = \frac{m}{\beta\pi\hbar^2} \int_0^{+\infty} \frac{1}{e^{-\beta\mu}e^x + 1}dx.$$

The integral can be done exactly, with the result

$$n = \frac{m}{\beta\pi\hbar^2} \int_0^{+\infty} \frac{1}{e^{-\beta\mu}e^x + 1}dx = \frac{m}{\beta\pi\hbar^2} \ln\left(\frac{e^x}{e^{-\beta\mu}e^x + 1}\right)\Bigg|_0^{+\infty} = \frac{m}{\beta\pi\hbar^2} \ln(1 + e^{\beta\mu}).$$

Using the Fermi energy $\varepsilon_F = \frac{n\pi\hbar^2}{m}$, we obtain $\ln(1 + e^{\beta\mu}) = \beta\varepsilon_F$ or, equivalently

$$\mu = \frac{1}{\beta} \ln(e^{\beta\varepsilon_F} - 1).$$

It is now immediate to show that in the limit of low temperatures ($\beta \to +\infty$) the chemical potential equals the Fermi energy

$$\lim_{\beta \to +\infty} \frac{1}{\beta} \ln(e^{\beta \varepsilon_F} - 1) = \varepsilon_F.$$

As for the last point, we need to study the internal energy U for low temperatures. We first write down U as

$$U = \frac{mA}{\pi \hbar^2} \int_0^{+\infty} \frac{\varepsilon}{e^{\beta(\varepsilon - \mu)} + 1} d\varepsilon$$

and note that it is an integral of type

$$I = \int_0^{+\infty} \frac{f(\varepsilon)}{e^{\beta(\varepsilon - \mu)} + 1} d\varepsilon$$

with $f(\varepsilon) = \varepsilon$ for our case. To solve the integral I, we change variables by setting $x = \beta(\varepsilon - \mu)$, and obtain

$$I = \frac{1}{\beta} \int_{-\beta\mu}^{+\infty} \frac{f(\mu + x/\beta)}{e^x + 1} dx$$

that we write as the sum of two integrals

$$I = \frac{1}{\beta} \left(\int_{-\beta\mu}^0 \frac{f(\mu + x/\beta)}{e^x + 1} dx + \int_0^{+\infty} \frac{f(\mu + x/\beta)}{e^x + 1} dx \right).$$

In the first integral, we change x into $-x$ to get

$$I = \frac{1}{\beta} \left(\int_0^{\beta\mu} \frac{f(\mu - x/\beta)}{e^{-x} + 1} dx + \int_0^{+\infty} \frac{f(\mu + x/\beta)}{e^x + 1} dx \right).$$

We also note that $1/(e^{-x} + 1) = 1 - 1/(e^x + 1)$, so that we can write

$$I = \frac{1}{\beta} \left[\int_0^{\beta\mu} f(\mu - x/\beta) dx + \int_0^{+\infty} \frac{f(\mu + x/\beta)}{e^x + 1} dx - \int_0^{\beta\mu} \frac{f(\mu - x/\beta)}{e^x + 1} dx \right].$$

In the limit $T \to 0$, i.e. $\beta \to +\infty$, we find

$$I = \mu f(\mu) + \frac{2f'(\mu)}{\beta^2} \int_0^{+\infty} \frac{x}{e^x + 1} dx$$

because, when $\beta \to +\infty$, we have used

$$f(\mu + x/\beta) - f(\mu - x/\beta) \approx \frac{2x}{\beta} f'(\mu).$$

If we now set $f(\varepsilon) = \varepsilon$, and we note that the relevant integral involved is

$$\int_0^{+\infty} \frac{x}{e^x + 1} dx = \frac{\pi^2}{12}$$

we can determine the behaviour of U for low temperatures as

$$\lim_{T \to 0} U \approx \frac{mA}{\pi \hbar^2} \left[\mu^2 + \frac{k^2 T^2 \pi^2}{6} \right]$$

from which, when $T \to 0$, we find

$$C_A = \left(\frac{\partial U}{\partial T} \right)_{A,N} = \frac{mA \pi k^2 T}{3\hbar^2}.$$

Problem 11.4.
At a finite temperature T, we want to describe a Fermi gas in three dimensions with an average number of particles N, spin S, and with a single particle energy given by

$$\varepsilon = \frac{p^2}{2m}$$

where p is the absolute value of the momentum and m the associated mass. If we define the fugacity as $z = e^{\beta \mu}$, show that

$$\frac{1}{z} \left(\frac{\partial z}{\partial T} \right)_P = -\frac{5}{2T} \frac{f_{5/2}(z)}{f_{3/2}(z)}$$

with $f_\alpha(z)$ the Fermi function of order α. Then, define the specific volume as $v = \frac{V}{N}$ and, in the limit of low temperatures, determine the ratio $\gamma \equiv \frac{C_P - C_V}{C_V}$, where C_P and C_V are the specific heats at constant pressure and specific volume. You can make use of the following formulae

$$f_\alpha(z) = \frac{1}{\Gamma(\alpha)} \int_0^{+\infty} \frac{x^{\alpha-1} dx}{z^{-1} e^x + 1} \qquad z \frac{d f_\alpha(z)}{dz} = f_{\alpha-1}(z)$$

$$S = Nk \left[\frac{5}{2} \frac{f_{5/2}(z)}{f_{3/2}(z)} - \ln z \right] \qquad \text{(Entropy)}$$

$$f_{1/2}(z) \approx \frac{2}{\pi^{1/2}} (\ln z)^{1/2} \left(1 - \frac{1}{24} \pi^2 (\ln z)^{-2} \right) \qquad T \to 0$$

$$f_{3/2}(z) \approx \frac{4}{3\pi^{1/2}} (\ln z)^{3/2} \left(1 + \frac{1}{8} \pi^2 (\ln z)^{-2} \right) \qquad T \to 0$$

$$f_{5/2}(z) \approx \frac{8}{15\pi^{1/2}} (\ln z)^{5/2} \left(1 + \frac{5}{8} \pi^2 (\ln z)^{-2} \right) \qquad T \to 0.$$

Solution
We start by considering the relation between the pressure and the internal energy for a three dimensional quantum gas with a single particle energy $\varepsilon = \frac{p^2}{2m}$ (see Problem 10.13)

$$P = \frac{2}{3}\frac{U}{V}.$$

The average energy U is then expressed in terms of the function $f_{5/2}(z)$

$$U = \frac{4\pi g V}{2mh^3} \int_0^{+\infty} \frac{1}{z^{-1}e^{\beta \frac{p^2}{2m}}+1} p^4 dp = \frac{3}{2}kT\frac{gV}{\lambda^3}f_{5/2}(z)$$

where $g = 2S+1$ is the degeneracy factor and $\lambda = \sqrt{\frac{h^2}{2\pi mkT}}$ a thermal length scale. The previous result is equivalent to

$$U = c_1 V T^{5/2} f_{5/2}(z)$$

with c_1 a constant independent of the temperature. If we go back to the equation for the pressure P, we get

$$P = \frac{2c_1}{3}T^{5/2}f_{5/2}(z).$$

We can now differentiate both sides with respect to the temperature keeping P constant

$$0 = \frac{5}{2}T^{3/2}f_{5/2}(z) + T^{5/2}\left(z\frac{df_{5/2}(z)}{dz}\right)\left(\frac{\partial z}{\partial T}\right)_P \frac{1}{z}$$

from which, using the property $z\frac{df_{5/2}(z)}{dz} = f_{3/2}(z)$, we obtain

$$\frac{1}{z}\left(\frac{\partial z}{\partial T}\right)_P = -\frac{5}{2T}\frac{f_{5/2}(z)}{f_{3/2}(z)}$$

that is the desired result. We can proceed in a similar way for the observable N and obtain

$$\frac{N}{V} = c_2 T^{3/2}f_{3/2}(z)$$

where, again, c_2 is a numerical constant. Differentiating both sides with respect to the temperature and keeping $v = \frac{V}{N}$ constant, we obtain

$$0 = \frac{3}{2}T^{1/2}f_{3/2}(z) + T^{3/2}\left(z\frac{df_{3/2}(z)}{dz}\right)\left(\frac{\partial z}{\partial T}\right)_v \frac{1}{z}$$

from which, using that $z\frac{df_{3/2}(z)}{dz} = f_{1/2}(z)$, we get

$$\frac{1}{z}\left(\frac{\partial z}{\partial T}\right)_v = -\frac{3}{2T}\frac{f_{3/2}(z)}{f_{1/2}(z)}.$$

We then compute the ratio of the specific heats at constant pressure and specific volume, C_P and C_V. We note that the entropy S is a function of N and z, that means

$$\frac{C_P}{C_V} = \frac{\left(\frac{T\partial S}{\partial T}\right)_{P,N}}{\left(\frac{T\partial S}{\partial T}\right)_{V,N}} = \frac{\left(\frac{T\partial S}{\partial z}\right)\left(\frac{\partial z}{\partial T}\right)_P}{\left(\frac{T\partial S}{\partial z}\right)\left(\frac{\partial z}{\partial T}\right)_V} = \frac{\left(\frac{\partial z}{\partial T}\right)_P}{\left(\frac{\partial z}{\partial T}\right)_V} = \frac{5}{3}\frac{f_{5/2}(z)f_{1/2}(z)}{(f_{3/2}(z))^2}$$

where we have used the results obtained previously for $\left(\frac{\partial z}{\partial T}\right)_P$ and $\left(\frac{\partial z}{\partial T}\right)_V$. In the low temperature limit, we can use the expansions given in the text and assume that $\ln z = \beta \mu \approx \beta \varepsilon_F$

$$\frac{C_P}{C_V} = \frac{\left(1 + \frac{5}{8}\pi^2\left(\frac{kT}{\varepsilon_F}\right)^2\right)\left(1 - \frac{1}{24}\pi^2\left(\frac{kT}{\varepsilon_F}\right)^2\right)}{\left(1 + \frac{1}{8}\pi^2\left(\frac{kT}{\varepsilon_F}\right)^2\right)^2} \approx 1 + \frac{\pi^2}{3}\left(\frac{kT}{\varepsilon_F}\right)^2.$$

Consistently, we find

$$\gamma = \frac{C_P - C_V}{C_V} \approx \frac{\pi^2}{3}\left(\frac{kT}{\varepsilon_F}\right)^2.$$

Problem 11.5.

An average number $\langle N \rangle$ of fermions with spin $1/2$ is confined in a one dimensional segment of length L. We have the following dispersion relation between the energy ε and the momentum p

$$\begin{cases} \varepsilon = \varepsilon_0 \sin^2\left(\frac{|p|a}{\hbar}\right) & 0 \le |p| \le \frac{\pi\hbar}{2a} \\ \varepsilon = \varepsilon_0[3 - \sin^2\left(\frac{|p|a}{\hbar}\right)] & \frac{\pi\hbar}{2a} < |p| \le \frac{\pi\hbar}{a}. \end{cases}$$

In the above expressions, $\varepsilon_0 = \frac{\hbar^2}{2ma^2}$ with a and $\hbar$ constants. In the limit of a fully degenerate gas, compute the average energy for the following cases:

- $L = \frac{\langle N \rangle a}{2}$;
- $L = \langle N \rangle a$;
- $L = \frac{3\langle N \rangle a}{2}$.

Solution

We first need to determine the Fermi momentum as a function of $\langle N \rangle / L$. To do that, we first compute the average particles density

$$\frac{\langle N \rangle}{L} = \frac{2}{2\pi\hbar}\int_{-p_F}^{+p_F} dp$$

from which we extract the value of the Fermi momentum

$$p_F = \frac{N}{L}\frac{\pi\hbar}{2}.$$

When $L = \frac{\langle N \rangle a}{2}$, we have $p_F = \frac{\pi \hbar}{a}$, so that

$$U = \frac{2L\varepsilon_0}{\pi \hbar} \left(\int_0^{\frac{\pi \hbar}{2a}} \sin^2 \left(\frac{|p|a}{\hbar} \right) dp + \int_{\frac{\pi \hbar}{2a}}^{\frac{\pi \hbar}{a}} \left[3 - \sin^2 \left(\frac{|p|a}{\hbar} \right) \right] dp \right)$$

leading to

$$U = \frac{3L\varepsilon_0}{a}.$$

In the second case, when $L = \langle N \rangle a$, we find $p_F = \frac{\pi \hbar}{2a}$ and

$$U = \frac{2L\varepsilon_0}{\pi \hbar} \int_0^{\frac{\pi \hbar}{2a}} \sin^2 \left(\frac{|p|a}{\hbar} \right) dp$$

that means

$$U = \frac{L\varepsilon_0}{2a}.$$

Finally, in the last case, we get $p_F = \frac{\pi \hbar}{3a}$ and

$$U = \frac{2L\varepsilon_0}{\pi \hbar} \int_0^{\frac{\pi \hbar}{3a}} \sin^2 \left(\frac{|p|a}{\hbar} \right) dp$$

leading to

$$U = \frac{L\varepsilon_0}{3a} \left[1 - \frac{3\sqrt{3}}{4\pi} \right].$$

Problem 11.6.
Consider a one dimensional fully degenerate Fermi gas with spin $S = 1/2$. The gas
is confined in a one dimensional segment of length L ($0 \leq x \leq L$). The dispersion
relation between the energy and the momentum (see also Problem 11.5) has the
following structure

$$\begin{cases} \varepsilon = \varepsilon_0 \sin^2 \left(\frac{|p|a}{\hbar} \right) & 0 \leq |p| \leq \frac{\pi \hbar}{2a} \\ \varepsilon = \varepsilon_0 \left[1 + \ln \left(\frac{|p|a}{\hbar} \right) \right] & \frac{\pi \hbar}{2a} \leq |p| \leq \frac{\pi \hbar}{a}. \end{cases}$$

Compute the average energy when the first energy band ($0 \leq |p| \leq \frac{\pi \hbar}{2a}$) is filled and
the corresponding number of states. Repeat the calculation when both energy bands
are filled.

Solution
When the first energy band is completely filled, all the states in the momentum space
are occupied up to $|p| = \frac{\pi \hbar}{2a}$. In this situation, the average number of states is given
by

$$\langle N \rangle = 2 \int_0^L dq \int_{|p| \leq \frac{\pi \hbar}{2a}} \frac{dp}{h} = \frac{2L}{h} \frac{\pi \hbar}{a} = \frac{L}{a}$$

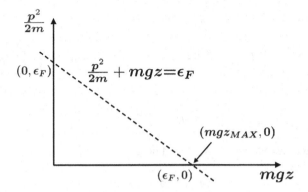

Fig. 11.2 We construct the allowed continuous energy spectrum for a fully degenerate Fermi gas in a cylindrical container under the effect of gravity g. The kinetic energy ($\frac{p^2}{2m}$) and the potential energy (mgz) have to be chosen in such a way that the total energy (kinetic plus potential) is below the Fermi energy ε_F. In Problem 11.7 we compute the Fermi energy and the internal energy

and the average energy is

$$U = 4 \int_0^L dq \int_0^{\frac{\pi\hbar}{2a}} \varepsilon_0 \sin^2\left(\frac{pa}{\hbar}\right) \frac{dp}{h} = \frac{L\varepsilon_0}{2a}.$$

When both energy bands are filled, the corresponding average number of states is

$$\langle N \rangle = 2 \int_0^L dq \int_{|p| \leq \frac{\pi\hbar}{a}} \frac{dp}{h} = \frac{4L}{h} \frac{\pi\hbar}{a} = \frac{2L}{a}$$

while, for the average energy, we have

$$U = 4 \int_0^L dq \int_0^{\frac{\pi\hbar}{2a}} \varepsilon_0 \sin^2\left(\frac{pa}{\hbar}\right) \frac{dp}{h} + 4 \int_0^L dq \int_{\frac{\pi\hbar}{2a}}^{\frac{\pi\hbar}{a}} \varepsilon_0 \left[1 + \ln\left(\frac{pa}{\hbar}\right)\right] \frac{dp}{h} =$$

$$\frac{L\varepsilon_0}{a}\left[\frac{1}{2} + \ln\pi + \ln 2\right].$$

Problem 11.7.
A fully degenerate Fermi gas with spin $S = 1/2$ and $\langle N \rangle$ particles is placed in a cylindrical container with base A and height H. The gas is under the effect of a constant gravitational acceleration g acting along the negative z direction. The maximum height allowed for the gas coincides with H. Compute the Fermi energy as a function of $\langle N \rangle / A$. Finally, compute the average energy of the system.

Solution
The energy is written as

$$\varepsilon = \frac{p^2}{2m} + mgz.$$

The gas is fully degenerate, and we have $\varepsilon \leq \varepsilon_F$, so that

$$mgz \leq \varepsilon_F - \frac{p^2}{2m}.$$

The maximum height (z_{MAX}) reached by the particles is obtained by minimizing the kinetic term $\frac{p^2}{2m}$ (i.e. setting it equal to 0) and we get $mgz_{MAX} = \varepsilon_F$. From the information of the text, we know that this maximum height is exactly the height of the container. This implies a relationship between the Fermi energy ε_F and H

$$H = \frac{\varepsilon_F}{mg}.$$

The average number of states is

$$\langle N \rangle = 2 \int \frac{d^3 p d^3 q}{h^3} = \frac{2A}{h^3} \int_0^{\frac{\varepsilon_F}{mg}} dz \int d^3 p$$

where the integral in the momentum space is over the domain defined by

$$\frac{p^2}{2m} + mgz \leq \varepsilon_F.$$

Therefore, for a given z, we need to integrate in the spherical region of radius $p = (2m(\varepsilon_F - mgz))^{1/2}$, and the integral is

$$\langle N \rangle = \frac{2A}{h^3} \int_0^{\frac{\varepsilon_F}{mg}} \frac{4\pi}{3} (2m(\varepsilon_F - mgz))^{3/2} dz =$$

$$\frac{8\pi}{3} \frac{A}{h^3} (2m\varepsilon_F)^{3/2} \frac{\varepsilon_F}{mg} \int_0^1 t^{3/2} dt = \frac{32}{15} \frac{(2m)^{1/2}}{g} \frac{\pi A}{h^3} \varepsilon_F^{5/2}$$

from which we extract the relation between ε_F and $\langle N \rangle /A$

$$\varepsilon_F = \left(\frac{15 h^3 g \langle N \rangle}{32 \pi \sqrt{2mA}} \right)^{2/5}.$$

For the average energy, we can write

$$U = \frac{8\pi A}{h^3} \int_0^{\frac{\varepsilon_F}{mg}} dz \int_0^{\sqrt{2m(\varepsilon_F - mgz)}} \left(\frac{p^2}{2m} + mgz \right) p^2 dp$$

that reduces to

$$U = \frac{8\pi A}{h^3} \frac{(2m)^{5/2}}{10m} \varepsilon_F^{5/2} \frac{\varepsilon_F}{mg} \int_0^1 (1 - \phi)^{5/2} d\phi +$$

$$\frac{8\pi A}{h^3} \frac{mg}{3} (2m)^{3/2} \varepsilon_F^{3/2} \left(\frac{\varepsilon_F}{mg} \right)^2 \int_0^1 \phi(1 - \phi)^{3/2} d\phi$$

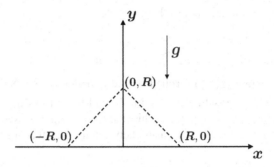

Fig. 11.3 A fully degenerate Fermi gas is confined in a triangular region with base $2R$ and height R. The gas is under the effect of gravity g. In Problem 11.8 we determine the average number of fermions, assuming that the maximum height reached by the gas coincides with the height of the triangle

where we can compute the two relevant integrals as

$$\int_0^1 (1-\phi)^{5/2}d\phi = \frac{2}{7} \qquad \int_0^1 \phi(1-\phi)^{3/2}d\phi = \frac{4}{35}.$$

Problem 11.8.
Let us consider a Fermi gas with spin $1/2$ and mass m confined in the two dimensional domain sketched in Fig. 11.3: in the (x,y) plane, it is a triangle with base $2R$ and height R. This gas is at temperature $T = 0$ and is under the effect of a constant gravitational field g. Compute the average number $\langle N \rangle$ of fermions under the assumption that the maximum height reached by the gas is exactly the height of the triangle.

Solution
The single particle energy is

$$\varepsilon = \frac{p^2}{2m} + mgy$$

where p is the absolute value of the momentum. The gas is fully degenerate ($T = 0$), and we have $\varepsilon \leq \varepsilon_F$, defining the allowed energy domain

$$\frac{p^2}{2m} + mgy \leq \varepsilon_F$$

with ε_F the Fermi energy. Due to the symmetry of the problem, we can work with half of the domain (the one with $x \geq 0$) and multiply the final result by two. We therefore have

$$\langle N \rangle = \frac{2\pi}{\pi^2\hbar^2} \int_0^{\sqrt{2m(\varepsilon_F - mgy)}} p\,dp \int_0^{y_M} dy \int_0^{R-y} dx.$$

We finally need to compute the maximum height y_M reached by the gas (similar ideas are discussed in Problem 11.7). In particular, y_M is reached when $p = 0$ and $\varepsilon = \varepsilon_F$, so that $y_M = \frac{\varepsilon_F}{mg}$. From the information of the text we learn that $y_M = R$ and $\varepsilon_F = mgR$. Therefore, the integral is

$$\langle N \rangle = \frac{2\pi}{\pi^2 \hbar^2} \int_0^{\sqrt{2m(\varepsilon_F - mgy)}} p \, dp \int_0^{y_M = R} dy \int_0^{R-y} dx = \frac{2\pi}{\pi^2 \hbar^2} m^2 g \int_0^R (R-y)^2 dy$$

with the final result

$$\langle N \rangle = \frac{2}{3\pi\hbar^2} m^2 g R^3.$$

Problem 11.9.

Compute the average number of particles for a fully degenerate relativistic gas with spin $S = 1/2$ in a cubic container of edge L. The gas is under the effect of a constant gravitational acceleration g acting along one direction (say $-x$). Consider explicitly the cases:

- $v \ll c$ with negligible rest energy;
- $v \approx c$ (ultrarelativistic),

where v is the velocity of the particles and c the speed of light. Assume that $L > \frac{\varepsilon_F}{mg}$, with m the mass of each particle and ε_F the Fermi energy.

Solution
The general expression for the momentum is

$$p = \frac{mv}{\sqrt{1 - \frac{v^2}{c^2}}}$$

with associated kinetic energy $\varepsilon_k(p) = c\sqrt{p^2 + m^2 c^2}$. For small velocities and negligible rest energy, we have $\varepsilon_k(p) \approx \frac{p^2}{2m}$. For $v \approx c$, instead, we get $\varepsilon_k(p) = pc$. In our case, we also have a constant acceleration g. Therefore, the single particle energy in the two cases is

$$\varepsilon = \frac{p^2}{2m} + mgx$$

$$\varepsilon = cp + mgx.$$

Moreover, the gas is fully degenerate and the distribution function in the energy space is 0 for $\varepsilon \geq \varepsilon_F$ and 1 for $\varepsilon < \varepsilon_F$, with ε_F the Fermi energy. The total average number of particles is found by integrating in the energy space up to $\varepsilon = \varepsilon_F$. In Fig. 11.4 we have sketched the allowed regions of integration in the two cases. The average number of particles for $v \ll c$ is

$$\langle N \rangle = \frac{4\pi(2S+1)}{(2\pi\hbar)^3} L^2 \int_0^{\sqrt{2m\varepsilon_F}} p^2 \, dp \int_0^{-\frac{p^2}{2m^2 g} + \frac{\varepsilon_F}{mg}} dx = \left(\frac{L^2}{\pi^2 \hbar^3}\right) \frac{4\sqrt{2}}{15} \frac{\varepsilon_F^{\frac{5}{2}} \sqrt{m}}{g}$$

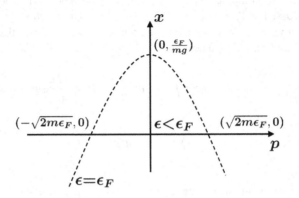

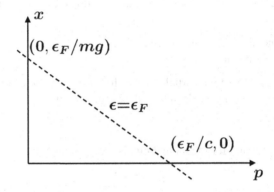

Fig. 11.4 We construct the allowed continuous energy spectrum for a fully degenerate relativistic Fermi gas under the effect of a constant gravitational acceleration (g). In Problem 12.17 we treat the following cases: 1) $v \ll c$ and negligible rest energy (top) 2) $v \approx c$ (bottom), where v is the velocity of the particles and c the speed of light

while, for $v \approx c$, we get

$$\langle N \rangle = \left(\frac{L^2}{\pi^2 \hbar^3} \right) \int_0^{\frac{\epsilon_F}{c}} p^2 dp \int_0^{\frac{-pc+\epsilon_F}{mg}} dx = \left(\frac{L^2}{\pi^2 \hbar^3} \right) \frac{\epsilon_F^4}{12mgc^3}.$$

Problem 11.10.
Determine the average number of particles $\langle N \rangle$ as a function of the Fermi energy for a fully degenerate Fermi gas (with spin $S = 1/2$) in two dimensions with the following single particle energy

$$\varepsilon = (p_x^2 + p_y^2)^s \qquad s > 0$$

where p_x and p_y are the two components of the momentum and s is a positive integer. Finally, compute the average energy U as a function of $\langle N \rangle$ and the area of the domain occupied by the gas.

Solution

The gas is fully degenerate so that the distribution in the energy space is a Heaviside function around the Fermi energy

$$\varepsilon_F = p_F^{2s}$$

with p_F the absolute value of the Fermi momentum. In the momentum space, the energy levels are occupied up to p_F and we can write the following expression for the total average number

$$\langle N \rangle = \frac{2}{(2\pi\hbar)^2} \int dx\,dy \int dp_x\,dp_y = \frac{A}{\pi\hbar^2} \int_0^{p_F} p\,dp = \frac{A}{2\pi\hbar^2} p_F^2$$

with A the area of the region occupied by the gas. The average energy is

$$U = \frac{2}{(2\pi\hbar)^2} \int dx\,dy \int (p_x^2 + p_y^2)^s dp_x\,dp_y = \frac{A}{\pi\hbar^2} \int_0^{p_F} p^{2s+1} dp = \frac{A p_F^{2s+2}}{\pi\hbar^2 (2s+2)}.$$

We can easily determine p_F from the first equation and plug it back in the second one to get

$$U = \frac{(2\langle N \rangle)^{s+1}}{2s+2} \left(\frac{\pi\hbar^2}{A} \right)^s.$$

Problem 11.11.

A fully degenerate Fermi gas (with spin $S = 1/2$) is characterized by $\langle N \rangle$ non inter-acting electrons confined in 2 dimensions within a circular region of radius R. The single particle energy is

$$\varepsilon = \frac{p^2}{2m} + \alpha r$$

where r is the distance from the center of the circular region, p the absolute value of the momentum, and $\alpha > 0$ a constant. Determine the Fermi energy ε_F and the average energy when:

- the potential energy term is small, i.e. $\alpha R \ll \varepsilon_F$;
- the potential energy at the border of the disc is larger than the Fermi energy, i.e. $\alpha R > \varepsilon_F$.

Solution

The allowed domain of integration is obtained by imposing $\varepsilon \leq \varepsilon_F$, that is

$$\varepsilon = \frac{p^2}{2m} + \alpha r \leq \varepsilon_F.$$

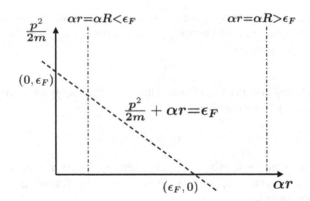

Fig. 11.5 We construct the allowed continuous energy spectrum for a fully degenerate Fermi gas confined in a two dimensional circular region with radius R. The gas is under the effect of a central force with potential $V(r) = \alpha r$. In Problem 11.11 we compute the Fermi energy ε_F and the average energy U when $\alpha R \ll \varepsilon_F$ and $\alpha R > \varepsilon_F$.

Furthermore, we have to consider that the radial distance of the particles is $r \le R$. When $\alpha R < \varepsilon_F$ (and in particular when $\alpha R \ll \varepsilon_F$) we can use the integral (see also Fig. 11.5) over the whole circular region $0 \le r \le R$, and $0 \le p \le \sqrt{2m(\varepsilon_F - \alpha r)}$ with the certainty that the argument of the square root never changes its sign. The average number of fermions is given by the following integral

$$\langle N \rangle = \frac{2}{4\pi^2\hbar^2} \int \int_D d^2p\, d^2q = \frac{2}{\hbar^2} \int_0^{\sqrt{2m(\varepsilon_F - \alpha r)}} p\, dp \int_0^R r\, dr = \frac{2m}{\hbar^2}\left(\varepsilon_F \frac{R^2}{2} - \alpha \frac{R^3}{3} \right)$$

where we have used polar coordinates in both the momentum ($d^2p = 2\pi p\, dp$) and position ($d^2q = 2\pi r\, dr$) space. If we consider the condition $\alpha R \ll \varepsilon_F$, the above result is well approximated by

$$\langle N \rangle = \frac{2m}{\hbar^2}\left(\varepsilon_F \frac{R^2}{2} - \alpha \frac{R^3}{3} \right) \approx \frac{2m}{\hbar^2} \varepsilon_F \frac{R^2}{2}$$

from which $\varepsilon_F = \frac{\hbar^2 \langle N \rangle}{mR^2}$. The average energy is

$$U = \frac{2}{\hbar^2} \int_0^{\sqrt{2m(\varepsilon_F - \alpha r)}} p\, dp \int_0^R r\left(\frac{p^2}{2m} + \alpha r \right) dr = \frac{2}{\hbar^2} m \left(\frac{\varepsilon_F^2 R^2}{4} - \frac{\alpha^2 R^4}{8} \right) \approx \frac{m}{\hbar^2} \frac{\varepsilon_F^2 R^2}{2}.$$

In the second case ($\alpha R > \varepsilon_F$), we cannot vary r between 0 and R (see Fig. 11.5). In the ($\alpha r, \frac{p^2}{2m}$) plane, we need to span the triangular region $0 \le \alpha r \le \varepsilon_F, 0 \le \frac{p^2}{2m} \le \varepsilon_F$, with area $\varepsilon_F^2/2$. Therefore, the average number is given by

$$\langle N \rangle = \frac{2}{4\pi^2\hbar^2} \int \int_D d^2p\, d^2q = \frac{2}{\hbar^2} \int_0^{p_F} p\, dp \int_0^{\frac{1}{\alpha}(\varepsilon_F - p^2/2m)} r\, dr$$

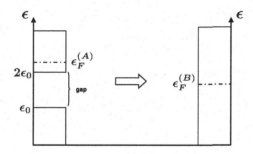

Fig. 11.6 The allowed continuous energy spectrum for a fully degenerate Fermi gas is characterized by a forbidden gap of energy: the particles have energies below ε_0 or above $2\varepsilon_0$. The Fermi energy changes when the energy gap disappears. Details are reported in Problem 11.12

where $p_F = \sqrt{2m\varepsilon_F}$. A calculation leads to

$$\langle N \rangle = \frac{1}{\alpha^2 \hbar^2} \int_0^{p_F} p \left(\varepsilon_F - \frac{p^2}{2m} \right)^2 dp = \frac{m\varepsilon_F^3}{3\alpha^2 \hbar^2}$$

from which $\varepsilon_F = \left(\frac{3\alpha^2 \hbar^2 \langle N \rangle}{m} \right)^{1/3}$. For the average energy, we have the following integral

$$U = \frac{2}{\hbar^2} \int_0^{p_F} p \, dp \int_0^{\frac{1}{\alpha}(\varepsilon_F - p^2/2m)} \left(\frac{p^2}{2m} + \alpha r \right) r \, dr$$

leading to $U = \frac{m\varepsilon_F^4}{4\alpha^2 \hbar^2}$.

Problem 11.12.
Let us consider a fully degenerate Fermi gas with spin $1/2$ placed in a volume V. The single particle energy for this gas is $\varepsilon = \frac{p^2}{2m}$, with p the absolute value of the momentum and m the mass of the particle. The average number of particles is

$$\langle N \rangle = \frac{16\pi V (2m\varepsilon_0)^{3/2}}{3h^3}.$$

Furthermore, there is a forbidden gap of energy: the particles may have an energy below ε_0 or above $2\varepsilon_0$, without the possibility to fill in between (see Fig. 11.6). Compute the variation in the internal energy due to the disappearance of the energy gap.

Solution
Let us start by computing the Fermi energy for the system without energy gap (we call this situation 'case B'). We have

$$\langle N \rangle = \frac{8\pi V}{h^3} \int_0^{p_F^{(B)}} p^2 \, dp = \frac{8\pi V (2m\varepsilon_F^{(B)})^{3/2}}{3h^3}$$

where we have used that

$$\frac{gV}{h^3} d^3 p = \frac{4\pi V}{h^3} (2m)^{\frac{3}{2}} \varepsilon^{\frac{1}{2}} d\varepsilon$$

with $g = 2$ the spin degeneracy. The relation between ε_F and ε_0 is given by

$$\langle N \rangle = \frac{8\pi V}{h^3} \int_0^{p_F^{(B)}} p^2 dp = \frac{8\pi V (2m\varepsilon_F^{(B)})^{3/2}}{3h^3} = \frac{16\pi V (2m\varepsilon_0)^{3/2}}{3h^3}$$

leading to

$$\varepsilon_F^{(B)} = 2^{2/3} \varepsilon_0$$

from which we learn that the Fermi level for the system without the band is larger than ε_0. The expression for the internal energy is

$$U^{(B)} = \frac{8\pi V}{h^3} \int_0^{p_F^{(B)}} \frac{p^4}{2m} dp$$

that is

$$U^{(B)} = \frac{16\pi V m (2m)^{1/2}}{5h^3} 2^{5/3} \varepsilon_0^{5/2}.$$

Let us now consider the situation with the energy gap (we call this situation 'case A'). From the above considerations, we already know that the Fermi level has to be located above $2\varepsilon_0$ in order to accommodate all the particles. Therefore, we obtain

$$\langle N \rangle = \frac{8\pi V}{h^3} \left(\int_0^{\sqrt{2m\varepsilon_0}} p^2 dp + \int_{\sqrt{2}\sqrt{2m\varepsilon_0}}^{\varepsilon_F^{(A)}} p^2 dp \right) = \frac{4\pi V (2m)^{3/2}}{h^3} \left(\int_0^{\varepsilon_0} \sqrt{\varepsilon} d\varepsilon + \right.$$

$$\left. \int_{2\varepsilon_0}^{\varepsilon_F^{(A)}} \sqrt{\varepsilon} d\varepsilon \right) = \frac{8\pi V (2m)^{3/2}}{3h^3} (\varepsilon_0^{3/2} + (\varepsilon_F^{(A)})^{3/2} - (2\varepsilon_0)^{3/2})$$

from which we get $\varepsilon_F^{(A)}$ as a function of ε_0

$$\varepsilon_F^{(A)} = \varepsilon_0 (1 + 2^{3/2})^{2/3}.$$

For the internal energy we have

$$U^{(A)} = \frac{16\pi V m (2m)^{1/2}}{5h^3} \varepsilon_0^{5/2} (1 + (1 + 2^{3/2})^{5/3} - 2^{5/2})$$

and, for the difference $\Delta U = U^{(A)} - U^{(B)}$, we obtain the following result

$$\Delta U = \frac{16\pi V m (2m)^{1/2}}{5h^3} \varepsilon_0^{5/2} (1 + (1 + 2^{3/2})^{5/3} - 2^{5/2} - 2^{5/3})$$

that is the variation in the internal energy due to the disappearance of the gap.

Problem 11.13.

Determine the first order correction, for $z \ll 1$, to the equation of state for a gas of ultrarelativistic three dimensional fermions with spin $S = 1/2$. In the previous expression, $z = e^{\beta \mu}$ is the fugacity of the gas and μ its chemical potential.

Solution

Starting from the energy of the ultrarelativistic particle

$$\varepsilon = cp$$

with c the speed of light and p the absolute value of the momentum, the expressions for the pressure P and particles density n are such that

$$\frac{P}{kT} = \frac{8\pi}{h^3} \int_0^{+\infty} \ln(1 + ze^{-\beta cp}) p^2 dp$$

and

$$n = \frac{8\pi}{h^3} \int_0^{+\infty} \frac{ze^{-\beta cp}}{1 + ze^{-\beta cp}} p^2 dp.$$

Since $z \ll 1$, we can expand the logarithm defining P

$$\frac{P}{kT} = \frac{8\pi}{h^3} \int_0^{+\infty} \left(ze^{-\beta cp} - \frac{z^2 e^{-2\beta cp}}{2} + ...\right) p^2 dp.$$

To simplify matters, let us introduce

$$A = \frac{8\pi}{h^3} \int_0^{+\infty} e^{-\beta cp} p^2 dp \qquad B = \frac{8\pi}{h^3} \int_0^{+\infty} e^{-2\beta cp} p^2 dp$$

in such a way that

$$\frac{P}{kT} = zA - \frac{1}{2} z^2 B.$$

Also, in the expression for n, we can expand the denominator up to the second order in z

$$n = zA - z^2 B.$$

Our objective is now to find z as a function of n and plug this result in the equation for P. We write

$$z = \frac{1}{A}(n + z^2 B)$$

and solve this equation in an iterative way. When z is small, the first approximation is $z = z_0 = 0$. If we use this information in the above equation we have the first order approximation $z = z_1 = n/A$. Finally, the next order (i.e. z_2) is found by substituting z_1

$$z = z_2 = \frac{n}{A}\left(1 + \frac{nB}{A^2}\right).$$

Using this expression in the equation for P, we get

$$\frac{P}{kT} = n\left(1 + \frac{nB}{2A^2}\right)$$

where, when evaluating z^2, we have neglected terms like n^3 and n^4. Finally, we need to compute the integrals A and B. Recalling the definition of the Gamma function

$$\Gamma(n) = \int_0^{+\infty} e^{-t} t^{n-1} dt = (n-1)!$$

we find

$$A = \frac{16\pi}{h^3 c^3 \beta^3} \qquad B = \frac{2\pi}{h^3 c^3 \beta^3}$$

so that

$$\frac{P}{kT} = n\left(1 + \frac{n}{16A}\right)$$

showing a positive first order correction to the classical equation of state. A similar problem (see Problem 10.15) can be considered for a Bose-Einstein gas and shows a negative correction to the ideal equation of state.

Problem 11.14.
An average number N of fermions is placed in a volume V at temperature $T = 0$. The single particle energy is $\varepsilon = \frac{p^2}{2m}$, with m the mass of the particle and p the absolute value of the momentum. Give an estimate for the isothermal compressibility

$$\kappa_T = -\frac{1}{V}\left(\frac{\partial V}{\partial P}\right)_{T,N}.$$

Solution
When $T = 0$, we know that $F = U - TS = U$. Moreover, for a quantum three dimensional gas, the following pressure-energy relation

$$PV = \frac{2}{3}U$$

holds (see also Problems 10.13, 12.16). Therefore, we can write down the equation for the pressure as

$$P = -\left(\frac{\partial F}{\partial V}\right)_{T,N} = -\left(\frac{\partial U}{\partial V}\right)_{T,N} = -\frac{\partial}{\partial V}\left(\frac{3}{2}PV\right)_{T,N} = -\frac{3}{2}\left[V\left(\frac{\partial P}{\partial V}\right)_{T,N} + P\right].$$

Simplifying this relation, we find

$$-V\left(\frac{\partial P}{\partial V}\right)_{T,N} = \frac{5}{3}P = \frac{10}{9}\frac{U}{V} = \frac{2}{3}\frac{N}{V}\varepsilon_F = \frac{\hbar^2}{3m}\left(\frac{6\pi^2}{g}\right)^{\frac{2}{3}}\left(\frac{N}{V}\right)^{\frac{5}{3}}$$

where g is the degeneracy factor due to the spin of the particles, and where we have used the relation between the average energy and the Fermi energy

$$U = \frac{3}{5}N\varepsilon_F = \frac{3}{5}\frac{\hbar^2}{2m}\left(\frac{6\pi^2}{g}\right)^{\frac{2}{3}}\left(\frac{N}{V}\right)^{\frac{5}{3}}.$$

This can be easily proved from the basic equations determining the average number N and average energy U for the fully degenerate Fermi gas

$$N = \frac{4\pi Vg}{h^3}\int_0^{p_F} p^2 dp = \frac{4\pi Vg}{h^3}\frac{p_F^3}{3}$$

$$U = \frac{4\pi Vg}{h^3}\int_0^{p_F}\frac{p^4}{2m}dp = \frac{4\pi Vg}{h^3}\frac{p_F^5}{10m}$$

with $p_F = \sqrt{2m\varepsilon_F}$. From the first we get

$$\varepsilon_F = \frac{\hbar^2}{2m}\left(\frac{6\pi^2 N}{gV}\right)^{\frac{2}{3}}$$

while, from the second

$$U = \frac{4\pi Vg}{h^3}\frac{p_F^5}{10m} = \frac{3}{5}N\frac{p_F^2}{2m} = \frac{3}{5}N\varepsilon_F.$$

The requested compressibility is given by

$$\kappa_T = -\frac{1}{V}\left(\frac{\partial V}{\partial P}\right)_{T,N} = \left(\frac{\hbar^2}{3m}\left(\frac{6\pi^2}{g}\right)^{\frac{2}{3}}\left(\frac{N}{V}\right)^{\frac{5}{3}}\right)^{-1}.$$

Problem 11.15.
A gas is composed of an average number $\langle N \rangle$ of non interacting fermions with spin $1/2$ in equilibrium at temperature $T = 0$. The gas is placed on a two dimensional disc with radius R and is under the effect of a constant radial force. The energy of the single particle is

$$\varepsilon = \frac{p^2}{2m} - br$$

with p the absolute value of the momentum, m the mass of the particle, and $b > 0$ a constant. Compute the Fermi energy ε_F, the average internal energy U, the free energy F, and the pressure P exerted at the border of the disc.

Solution
We start by writing down the average number of particles in the system

$$\langle N \rangle = \frac{2}{h^2}\int_D d^2p\, d^2q$$

where D is a suitable domain of integration such that the energy ε is below the Fermi energy, i.e. $\frac{p^2}{2m} - br \leq \varepsilon_F$. If we set $\varepsilon_k = p^2/2m$ and we use polar coordinates in both the momentum $(d^2p = 2\pi p\,dp)$ and position $(d^2q = 2\pi r\,dr)$ space, we get

$$\langle N \rangle = \frac{2m}{\hbar^2} \int_D r\,dr\,d\varepsilon = \frac{2m}{\hbar^2} \int_0^R r\,dr \int_0^{\varepsilon_F + br} d\varepsilon_k = \frac{m\varepsilon_F R^2}{\hbar^2} + \frac{2mbR^3}{3\hbar^2}$$

from which it is easily obtained that $\varepsilon_F = \frac{\langle N \rangle \hbar^2}{mR^2} - \frac{2bR}{3}$. For the average energy, we find

$$U = \frac{2m}{\hbar^2} \int_D (\varepsilon_k - br)\,r\,dr\,d\varepsilon_k = \frac{mR^2}{2\hbar^2}\left(\varepsilon_F^2 - \frac{b^2 R^2}{2}\right).$$

The computation of the free energy is very simple because we are dealing with a fully degenerate gas with $T = 0$, that means $F = (U - TS)|_{T=0} = U$. The pressure is obtained using the appropriate derivative of the free energy with respect to the area $A = \pi R^2$

$$P = -\left(\frac{\partial F}{\partial A}\right)_{T=0,\langle N \rangle} = -\left(\frac{\partial U}{\partial A}\right)_{T=0,\langle N \rangle}$$

and, since the infinitesimal change of the area is written as $dA = 2\pi R\,dR$, we get

$$P = -\frac{1}{2\pi R}\left(\frac{\partial F}{\partial R}\right)_{T=0,\langle N \rangle} = -\frac{1}{2\pi R}\left(\frac{\partial U}{\partial R}\right)_{T=0,\langle N \rangle} = \frac{m}{2\pi\hbar^2}(bR + \varepsilon_F)^2.$$

Problem 11.16.
A container in d dimensions with volume $2V$ is separated in two equal parts, A and B, by a wall allowing for the exchange of particles. A fully degenerate Fermi gas with spin $S = 1/2$, mass m, and with a magnetic moment τ directed along the z direction, is placed in both regions. The gas has a kinetic energy

$$\varepsilon_k = \frac{p^2}{2m}$$

with p the absolute value of the momentum in d dimensions. At some point, a weak magnetic field H directed along the z direction is switched on in the region A. Under the assumption that the density is the same in both regions, determine the direction of the density flux of the particles as a function of d. Is there any dimension d_{eq} where the system is in chemical equilibrium?

Solution
We recall that the infinitesimal volume element for the sphere with radius R in d dimensions is

$$dV_d = \frac{d\pi^{\frac{d}{2}}}{(\frac{d}{2})!} R^{d-1} dR.$$

Therefore, we can write down the number of states in an infinitesimal cell of the momentum space as

$$\frac{gV}{(2\pi\hbar)^d} d^d p = \frac{gV}{(2\pi\hbar)^d} \frac{d\pi^{\frac{d}{2}}}{\left(\frac{d}{2}\right)!} p^{d-1} dp = \frac{2V}{(2\pi\hbar)^d} \frac{d(2m\pi)^{\frac{d}{2}}}{2\left(\frac{d}{2}\right)!} \varepsilon_k^{\frac{d}{2}-1} d\varepsilon_k$$

with $g = 2$ the spin degeneracy. The gas in the region A has also the interaction energy with the magnetic field $\pm\tau H$, where the sign depends on the orientation of the spin. This means that the spin degeneracy is removed and we have two groups of particles: those with spin parallel to H and energy

$$\varepsilon_+ = \varepsilon_k - \tau H = \frac{p^2}{2m} - \tau H$$

and those with spin antiparallel to H and energy

$$\varepsilon_- = \varepsilon_k + \tau H = \frac{p^2}{2m} + \tau H.$$

Since the gas is completely degenerate, we find $\varepsilon_\pm \leq \mu_A$ in the region A, with μ_A the Fermi energy in such a region. The average number of particles in this region is

$$\langle N_A \rangle = C \left[\int_0^{(\mu_A + \tau H)} \varepsilon_k^{\frac{d}{2}-1} d\varepsilon_k + \int_0^{(\mu_A - \tau H)} \varepsilon_k^{\frac{d}{2}-1} d\varepsilon_k \right] =$$

$$\frac{2C}{d} \left[(\mu_A + \tau H)^{\frac{d}{2}} + (\mu_A - \tau H)^{\frac{d}{2}} \right]$$

where we have used the definition

$$C = \frac{V}{2(2\pi\hbar)^d} \frac{d(2m\pi)^{\frac{d}{2}}}{\left(\frac{d}{2}\right)!}.$$

In the region B, the spin degeneracy is not removed because the magnetic field is not present, and we have

$$\langle N_B \rangle = \frac{4C}{d} \mu_B^{\frac{d}{2}}$$

with μ_B the Fermi energy in the region B. Both gases have the same density and the volume in both regions is the same. Therefore, we can set $\langle N_A \rangle = \langle N_B \rangle$ with the result

$$(\mu_A + \tau H)^{\frac{d}{2}} + (\mu_A - \tau H)^{\frac{d}{2}} = 2\mu_B^{\frac{d}{2}}.$$

If we factorize μ_A in the right hand side and set $x = \tau H/\mu_A$, we can expand the resulting binomial function in the limit of small H ($x \ll 1$) using

$$(1+x)^\alpha = 1 + \alpha x + \frac{\alpha}{2}(\alpha - 1)x^2 + \dots$$

so that

$$\left(\frac{\mu_B}{\mu_A}\right)^{\frac{d}{2}} = 1 + \frac{1}{8}d(d-2)\left(\frac{\tau H}{\mu_A}\right)^2.$$

For $d < 2$, we have $\mu_B < \mu_A$ and the particles flow from A to B (the system wants to minimize the chemical potential). For $d = d_{eq} = 2$ we have $\mu_B = \mu_A$ and there is no net flow of particles, i.e. the system is in chemical equilibrium. Finally, when $d > 2$, we have $\mu_B > \mu_A$ with a resulting flow from B to A.

Problem 11.17.

A three dimensional volume is separated in two parts by a rigid and impenetrable wall. The first part contains a Fermi gas composed of particles with spin $1/2$, while the second one a Fermi gas of particles with spin $3/2$. In both cases the single particle energy is $\varepsilon = \frac{p^2}{2m}$, with p the absolute value of the momentum and m the associated mass. Determine the density ratio at the mechanical equilibrium in the limit of zero temperature.

Solution

We need to impose the condition of the mechanical equilibrium between the two parts. If the single particle energy is $\varepsilon = \frac{p^2}{2m}$, the relation between the pressure (P) and the average internal energy (U) is $P = \frac{2U}{3V}$ (see also Problems 10.13 and 11.5). The mechanical equilibrium requires

$$P_1 = P_2$$

that implies

$$\frac{U_1}{V_1} = \frac{U_2}{V_2}.$$

Let us now compute the internal energy for the gas of particles with spin $1/2$. When the gas is fully degenerate (zero temperature) it reads

$$U_1 = g_1 \frac{4\pi V_1}{h^3} \int_0^{p_{1F}} \frac{p^4}{2m} dp$$

where p_{1F} is the Fermi momentum, and the factor $g_1 = 2S+1 = 2$ corresponds to the spin degeneracy. We easily obtain

$$\frac{U_1}{V_1} = \frac{4\pi}{5mh^3} p_{1F}^5.$$

The computation of the energy for the gas with spin $3/2$ goes along the same lines, with the only exception that we have a different spin degeneracy $g_2 = 2S+1 = 4$ and a different Fermi momentum p_{2F}. Therefore, we find

$$\frac{U_2}{V_2} = \frac{8\pi}{5mh^3} p_{2F}^5.$$

The final step is to replace the Fermi momentum with some function of the average density. For a fully degenerate Fermi gas, the particles density is

$$n_i = \frac{\langle N_i \rangle}{V_i} = g_i \frac{4\pi}{h^3} \int_0^{p_{iF}} p^2 \, dp \quad i = 1, 2$$

so that

$$p_{1F} = \left(n_1 \frac{3h^3}{8\pi} \right)^{1/3} \quad p_{2F} = \left(n_1 \frac{3h^3}{16\pi} \right)^{1/3}.$$

If we plug the expressions for p_{1F} and p_{2F} in the equation defining the internal energy, we find that $\frac{U_1}{V_1} = \frac{U_2}{V_2}$ is equivalent to

$$\frac{(3n_1)^{5/3} h^2}{10m(8\pi)^{2/3}} = \frac{(3n_2)^{5/3} h^2}{10m(16\pi)^{2/3}}$$

that implies $\frac{n_1}{n_2} = 2^{-2/5}$.

Problem 11.18.
An electron gas is at equilibrium at temperature $T = 0$. The single particle energy is $\varepsilon = \frac{p^2}{2m}$, with m the mass of the electron and p the absolute value of the momentum. The electrons occupy a container with volume V which, in turn, occupies a larger container of volume $V + \Delta V$, with $\Delta V \ll V$. The gas container is initially isolated from the larger one by some walls. At some point, the walls are removed, and the electron gas reaches a new equilibrium state in the larger volume. The total average energy is unchanged. What is the temperature of the new equilibrium state? Given the condition $\Delta V \ll V$, we assume that the temperature is small. Compare the result with the classical counterpart.

Solution
For a Fermi gas at $T = 0$ in a volume V, all the energy levels are occupied up to the Fermi energy $\varepsilon_F(V)$. The function giving the occupation number is a Heaviside theta function that is 1 for energies $\varepsilon \leq \varepsilon_F(V)$ and 0 otherwise. The total average energy is

$$U_{T=0} = \frac{8\pi V}{h^3} \int_0^{p_F} \frac{p^2}{2m} p^2 dp = \frac{4\pi V}{5mh^3} p_F^5 = \frac{3}{5} \langle N \rangle \varepsilon_F(V)$$

where $p_F = \sqrt{2m\varepsilon_F(V)}$ is the Fermi momentum. Given the average number

$$\langle N \rangle = \frac{8\pi V}{h^3} \int_0^{p_F} p^2 dp = \frac{8\pi V}{3h^3} p_F^3$$

we can write

$$\varepsilon_F(V) = \frac{p_F^2}{2m} = \frac{1}{2m} \left(\frac{3h^3}{8\pi} \frac{\langle N \rangle}{V} \right)^{2/3}.$$

When $T \neq 0$ in the volume $V + \Delta V$, the energy is always connected to the logarithm of the grand canonical partition function as

$$U_{T \neq 0} = - \left(\frac{\partial \ln \mathscr{Z}}{\partial \beta} \right)_{V,z} = kT^2 \left(\frac{\partial \ln \mathscr{Z}}{\partial T} \right)_{V,z} = \frac{3kTV}{\lambda^3} f_{5/2}(z) =$$

$$\frac{3}{2} \langle N \rangle kT \frac{f_{5/2}(z)}{f_{3/2}(z)} = \frac{3}{5} \langle N \rangle \varepsilon_F (V + \Delta V) \left[1 + \frac{5\pi^2}{12} \left(\frac{kT}{\varepsilon_F(V + \Delta V)} \right)^2 + \ldots \right]$$

where $\lambda = h/(2\pi m kT)^{1/2}$, and where we have used the Fermi-Dirac functions

$$f_\alpha(z) = \frac{1}{\Gamma(\alpha)} \int_0^{+\infty} \frac{x^{\alpha-1} dx}{z^{-1} e^x + 1}$$

with their asymptotic expansions (valid when $T \to 0$)

$$f_{3/2}(z) = \frac{4}{3\sqrt{\pi}} (\ln z)^{3/2} \left[1 + \frac{\pi^2}{8} (\ln z)^{-2} + \ldots \right]$$

$$f_{5/2}(z) = \frac{8}{15\sqrt{\pi}} (\ln z)^{5/2} \left[1 + \frac{5\pi^2}{8} (\ln z)^{-2} + \ldots \right]$$

$$\ln z \approx \frac{\varepsilon_F (V + \Delta V)}{kT} \left[1 - \frac{\pi^2}{12} \left(\frac{kT}{\varepsilon_F(V + \Delta V)} \right)^2 + \ldots \right].$$

In the new equilibrium state, we know that the energy is the same as before, so that

$$\frac{3}{5} \langle N \rangle \varepsilon_F (V + \Delta V) \left[1 + \frac{5\pi^2}{12} \left(\frac{kT}{\varepsilon_F(V + \Delta V)} \right)^2 + \ldots \right] = \frac{3}{5} \langle N \rangle \varepsilon_F(V).$$

The Fermi energy for the volume $V + \Delta V$ can be expanded as

$$\varepsilon_F (V + \Delta V) = \frac{1}{2m} \left(\frac{3h^3}{8\pi} \frac{\langle N \rangle}{V + \Delta V} \right)^{2/3} = \varepsilon_F(V) \left(1 - \frac{2}{3} \frac{\Delta V}{V} \right) + \ldots$$

which can be substituted in the previous equation to obtain

$$T = \frac{\varepsilon_F(V)}{k\pi} \sqrt{\frac{8}{5} \frac{\Delta V}{V}} + \ldots.$$

In the classical case, making use of the equipartition theorem, we assign to each degree of freedom an energy contribution equal to $kT/2$ (see also Problems 7.18, 7.19, 7.20). For N particles in three dimensions with a single particle energy $\varepsilon = \frac{p_x^2 + p_y^2 + p_z^2}{2m}$, the total energy would be $U = \frac{3}{2} NkT$. Consequently, for a fixed energy and fixed number of particles, the temperatures in the volumes V and $V + \Delta V$ must be the same.

12

Fluctuations and Complements

Problem 12.1.

Characterize the fluctuations of the energy in the grand canonical ensemble and prove that

$$\langle (\Delta E)^2 \rangle = \langle (\Delta E)^2 \rangle_{can} + \langle (\Delta N)^2 \rangle \left(\frac{\partial U}{\partial N} \right)^2_{T,V}$$

where $U = \langle E \rangle$ is the average energy, N the average number of particles, $\langle (\Delta N)^2 \rangle$ its fluctuations, and $\langle (\Delta E)^2 \rangle_{can}$ the energy fluctuations as obtained from the canonical ensemble.

Solution

If the grand canonical partition function $\mathscr{Q} = \mathscr{Q}(T,V,z)$ is known, we can define the average energy

$$U = \langle E \rangle = - \left(\frac{\partial \ln \mathscr{Q}}{\partial \beta} \right)_{V,z} = - \left(\frac{\partial \ln \mathscr{Q}}{\partial \beta} \right)_{V,\alpha}$$

with $\alpha = -\ln z$, and the associated fluctuations are

$$\langle (\Delta E)^2 \rangle = \langle E^2 \rangle - \langle E \rangle^2 = - \left(\frac{\partial U}{\partial \beta} \right)_{V,z} = kT^2 \left(\frac{\partial U}{\partial T} \right)_{V,z}.$$

The average number of particles is

$$N = - \left(\frac{\partial \ln \mathscr{Q}}{\partial \alpha} \right)_{\beta,V}$$

and its fluctuations are

$$\langle (\Delta N)^2 \rangle = \langle N^2 \rangle - \langle N \rangle^2 = kT \left(\frac{\partial N}{\partial \mu} \right)_{T,V}.$$

Cini M., Fucito F., Sbragaglia M.: Solved Problems in Quantum and Statistical Mechanics.
DOI 10.1007/978-88-470-2315-4_12, © Springer-Verlag Italia 2012

From the previous expressions for U and N, we obtain

$$\left(\frac{\partial N}{\partial \beta}\right)_{V,\alpha} = -\left(\frac{\partial}{\partial \beta}\left(\frac{\partial \ln \mathscr{Q}}{\partial \alpha}\right)_{\beta,V}\right)_{V,\alpha} = -\left(\frac{\partial}{\partial \alpha}\left(\frac{\partial \ln \mathscr{Q}}{\partial \beta}\right)_{V,\alpha}\right)_{\beta,V} = \left(\frac{\partial U}{\partial \alpha}\right)_{\beta,V}$$

or, equivalently

$$\left(\frac{\partial N}{\partial T}\right)_{V,z} = \frac{1}{T}\left(\frac{\partial U}{\partial \mu}\right)_{T,V}.$$

We can now consider $U = U(T,V,N)$ with N a function of T,V,z

$$dU(T,V,N(T,V,z)) = \left(\frac{\partial U}{\partial T}\right)_{V,N} dT + \left(\frac{\partial U}{\partial V}\right)_{T,N} dV + \left(\frac{\partial U}{\partial N}\right)_{T,V} dN(T,V,z)$$

$$dN(T,V,z) = \left(\frac{\partial N}{\partial T}\right)_{V,z} dT + \left(\frac{\partial N}{\partial V}\right)_{T,z} dV + \left(\frac{\partial N}{\partial z}\right)_{T,V} dz$$

so that

$$dU = \left(\left(\frac{\partial U}{\partial T}\right)_{V,N} + \left(\frac{\partial U}{\partial N}\right)_{T,V}\left(\frac{\partial N}{\partial T}\right)_{V,z}\right) dT +$$

$$\left(\left(\frac{\partial U}{\partial V}\right)_{T,N} + \left(\frac{\partial U}{\partial N}\right)_{T,V}\left(\frac{\partial N}{\partial V}\right)_{T,z}\right) dV + \left(\frac{\partial U}{\partial N}\right)_{T,V}\left(\frac{\partial N}{\partial z}\right)_{T,V} dz.$$

If we compute $\left(\frac{\partial U}{\partial T}\right)_{V,z}$, we obtain

$$\left(\frac{\partial U}{\partial T}\right)_{V,z} = \left(\frac{\partial U}{\partial T}\right)_{V,N} + \left(\frac{\partial U}{\partial N}\right)_{T,V}\left(\frac{\partial N}{\partial T}\right)_{V,z}.$$

The fluctuations become

$$\langle(\Delta E)^2\rangle = kT^2\left(\frac{\partial U}{\partial T}\right)_{V,z} = kT^2\left(\frac{\partial U}{\partial T}\right)_{V,N} + kT^2\left(\frac{\partial U}{\partial N}\right)_{T,V}\left(\frac{\partial N}{\partial T}\right)_{V,z} =$$

$$\langle(\Delta E)^2\rangle_{can} + kT\left(\frac{\partial U}{\partial N}\right)_{T,V}\left(\frac{\partial U}{\partial \mu}\right)_{T,V}$$

where we have used the formula for the fluctuations of the energy in the canonical ensemble

$$\langle(\Delta E)^2\rangle_{can} = kT^2\left(\frac{\partial U}{\partial T}\right)_{V,N}.$$

It is also noted that

$$\left(\frac{\partial U}{\partial \mu}\right)_{T,V} = \left(\frac{\partial U}{\partial N}\right)_{T,V}\left(\frac{\partial N}{\partial \mu}\right)_{T,V}$$

and, hence

$$\langle (\Delta E)^2 \rangle = \langle (\Delta E)^2 \rangle_{can} + \left(\frac{\partial U}{\partial N} \right)^2_{T,V} \langle (\Delta N)^2 \rangle$$

that is the desired result.

Problem 12.2.
A system is in equilibrium with a reservoir at temperature T' and pressure P'. Prove that, in a fluctuation with respect to the equilibrium state at constant pressure, the increase or decrease of the entropy only depends upon the sign of the thermal expansion coefficient

$$\alpha = \frac{1}{V} \left(\frac{\partial V}{\partial T} \right)_P.$$

For simplicity, no exchange of particles is allowed. To solve the problem, relate the entropy variation to α and C_P (the specific heat at constant pressure) and finally show that $C_P \geq 0$. Also, consider that the total energy and volume of the system and the reservoir is conserved.

Solution
The quantity we are interested in is the variation of the entropy during the expansion, i.e. $(\partial S/\partial V)_P$. Following the suggestion of the text, we can use the method of the Jacobians and relate $\left(\frac{\partial S}{\partial V} \right)_P$ to α and C_P

$$\left(\frac{\partial S}{\partial V} \right)_P = \frac{\partial(S,P)}{\partial(V,P)} = \frac{\frac{\partial(S,P)}{\partial(P,T)}}{\frac{\partial(V,P)}{\partial(P,T)}} = \frac{\left(\frac{\partial S}{\partial T} \right)_P}{\left(\frac{\partial V}{\partial T} \right)_P} = \frac{C_P}{\alpha TV}.$$

Therefore, it is sufficient to show that $C_P \geq 0$. Our system is initially at equilibrium with the reservoir at temperature T' and pressure P'. The total entropy, i.e. the one of the system (s) plus the one of the reservoir (r), attains its maximum value at the equilibrium. Therefore, a fluctuation can only reduce it

$$\Delta S_{tot} = \Delta S_s + \Delta S_r \leq 0.$$

From the first law of thermodynamics we know that

$$\Delta S_s = \int_i^f \frac{1}{T_s} dU_s + \int_i^f \frac{P_s}{T_s} dV_s$$

and

$$\Delta S_r = \int_i^f \frac{1}{T_r} dU_r + \int_i^f \frac{P_r}{T_r} dV_r$$

where i and f stand for the initial and final state. Also, the temperature and the pressure of the reservoir do not change appreciably and stay equal to their equilibrium values (the reservoir is so large that the expansion of the system does not perturb its pressure and temperature). Therefore, we have

$$\Delta S_r = \int_i^f \frac{1}{T_r} dU_r + \int_i^f \frac{P_r}{T_r} dV_r = \frac{1}{T'} \int_i^f dU_r + \frac{P'}{T'} \int_i^f dV_r = \frac{1}{T'} \Delta U_r + \frac{P'}{T'} \Delta V_r.$$

On the contrary, for the system we have to use

$$\Delta S_s = \int_i^f \frac{1}{T_s} dU_s + \int_i^f \frac{P_s}{T_s} dV_s$$

where we are not allowed to bring out of the integral T_s and P_s because, in principle, those parameters vary. Moreover, if the total volume and energy are conserved

$$\Delta V_r = -\Delta V_s \qquad \Delta U_r = -\Delta U_s.$$

Therefore, we can write

$$\Delta S_s + \Delta S_r = \Delta S_s - \frac{1}{T'} \Delta U_s - \frac{P'}{T'} \Delta V_s \leq 0$$

or, equivalently

$$\Delta U_s + P' \Delta V_s - T' \Delta S_s \geq 0.$$

Let us now drop the subscript s

$$\Delta U + P' \Delta V - T' \Delta S \geq 0$$

knowing that we are referring to the properties of the system hereafter. We can now expand $\Delta U = \Delta U(S, V)$ up to second order

$$\Delta U = \left(\frac{\partial U}{\partial S}\right)_V \Delta S + \left(\frac{\partial U}{\partial V}\right)_S \Delta V +$$
$$\frac{1}{2} \left(\left(\frac{\partial^2 U}{\partial S^2}\right)_V (\Delta S)^2 + \left(\frac{\partial^2 U}{\partial V^2}\right)_S (\Delta V)^2 + 2 \left(\frac{\partial}{\partial S}\left(\frac{\partial U}{\partial V}\right)_S\right)_V \Delta S \Delta V \right)$$

and we can substitute this in the previous expression, with $T = (\partial U / \partial S)_V$ and $P = -(\partial U / \partial V)_S$

$$\left(\frac{\partial^2 U}{\partial S^2}\right)_V (\Delta S)^2 + \left(\frac{\partial^2 U}{\partial V^2}\right)_S (\Delta V)^2 + 2 \left(\frac{\partial}{\partial S}\left(\frac{\partial U}{\partial V}\right)_S\right)_V \Delta S \Delta V =$$
$$\left(\frac{\partial T}{\partial S}\right)_V (\Delta S)^2 + \left(\frac{\partial T}{\partial V}\right)_S \Delta S \Delta V - \left(\frac{\partial P}{\partial V}\right)_S (\Delta V)^2 - \left(\frac{\partial P}{\partial S}\right)_V \Delta S \Delta V \geq 0.$$

We note that ΔT and ΔP can be written as

$$\Delta T = \left(\frac{\partial T}{\partial S}\right)_V \Delta S + \left(\frac{\partial T}{\partial V}\right)_S \Delta V$$

$$\Delta P = \left(\frac{\partial P}{\partial V}\right)_S \Delta V + \left(\frac{\partial P}{\partial S}\right)_V \Delta S$$

that implies

$$\Delta T \Delta S - \Delta P \Delta V \geq 0.$$

This final expression is a relationship between T, P, S and V. If we choose T and P as independent variables and expand S and V at first order, we have

$$\Delta T \left[\left(\frac{\partial S}{\partial T} \right)_P \Delta T + \left(\frac{\partial S}{\partial P} \right)_T \Delta P \right] - \Delta P \left[\left(\frac{\partial V}{\partial T} \right)_P \Delta T + \left(\frac{\partial V}{\partial P} \right)_T \Delta P \right] =$$
$$\left(\frac{\partial S}{\partial T} \right)_P (\Delta T)^2 - \left(\frac{\partial V}{\partial P} \right)_T (\Delta P)^2 + \left[\left(\frac{\partial S}{\partial P} \right)_T - \left(\frac{\partial V}{\partial T} \right)_P \right] \Delta T \Delta P \geq 0.$$

If we keep P constant ($\Delta P = 0$), we find that

$$\left(\frac{\partial S}{\partial T} \right)_P (\Delta T)^2 = T C_P (\Delta T)^2 \geq 0$$

that implies $C_P \geq 0$ because both T and $(\Delta T)^2$ are positive.

Problem 12.3.
Unlike an ideal gas, which cools down during an adiabatic expansion, a one dimensional rubber band (with spring constant K and rest position $x_0 = 0$) is increasing its temperature T when elongated in an adiabatic way. Determine the probability for the fluctuations from the rest position (Δx) and compute $\langle (\Delta x)^2 \rangle$ and $\langle \Delta x \Delta T \rangle$.

Solution
As already discussed in Problem 6.6, the first law of thermodynamics reads

$$T dS = dE - K x dx.$$

Let us now derive the probability for the fluctuations. In the case of a fluid with pressure P and volume V, the first law is written as $T dS = dE + P dV$, and the probability of the fluctuations reads

$$p \propto e^{-\frac{1}{2kT}(\Delta S \Delta T - \Delta P \Delta V)}.$$

When comparing the first law of thermodynamics for the fluid and the one for the rubber band, we see that the elastic force $-Kx$ plays the role of the pressure P, while the elongation Δx plays the role of the variation of the volume ΔV. This means that we can follow the same derivation for the probability of the fluctuations for a fluid and replace P with $-Kx$ and ΔV with Δx. Therefore, we find

$$p \propto e^{-\frac{1}{2kT}(\Delta S \Delta T + K(\Delta x)^2)}.$$

To compute $\langle (\Delta x)^2 \rangle$ and $\langle \Delta x \Delta T \rangle$ we need to expand ΔS in terms of Δx and ΔT

$$\Delta S = \left(\frac{\partial S}{\partial T} \right)_x \Delta T + \left(\frac{\partial S}{\partial x} \right)_T \Delta x = \frac{C_x}{T} \Delta T + \left(\frac{\partial S}{\partial x} \right)_T \Delta x$$

where we have introduced the specific heat at constant elongation

$$C_x = T \left(\frac{\partial S}{\partial T} \right)_x.$$

As already discussed in Problem 6.6, the entropy stays constant during an isothermal elongation

$$\left(\frac{\partial S}{\partial x} \right)_T = 0$$

and the probability of the fluctuations becomes

$$p \propto e^{-\frac{C_x}{2kT^2}(\Delta T)^2 - \frac{K}{2kT}(\Delta x)^2}$$

from which we see that Δx and ΔT are statistically independent Gaussian variables with the property

$$\langle (\Delta x)^2 \rangle = \frac{kT}{K}$$

$$\langle \Delta x \Delta T \rangle = 0.$$

Problem 12.4.

Consider a gas with a fixed number of particles. Using the probability of the fluctuations from the thermodynamic equilibrium, prove that:

- $\langle \Delta T \Delta P \rangle = \frac{T^2}{C_V} \left(\frac{\partial P}{\partial T} \right)_V$;
- $\langle \Delta V \Delta P \rangle = -T$;
- $\langle \Delta T \Delta S \rangle = T$;
- $\langle \Delta V \Delta S \rangle = T \left(\frac{\partial V}{\partial T} \right)_P$.

To simplify matters, assume $k = 1$.

Solution

As we know from the theory, the probability of the fluctuations from the thermodynamic equilibrium is

$$p \propto e^{\frac{1}{2T}[\Delta V \Delta P - \Delta S \Delta T]}$$

where we can use the differential expressions for $P = P(T,V)$ and $S = S(T,V)$

$$\Delta P = \left(\frac{\partial P}{\partial V} \right)_T \Delta V + \left(\frac{\partial P}{\partial T} \right)_V \Delta T$$

$$\Delta S = \left(\frac{\partial S}{\partial V} \right)_T \Delta V + \left(\frac{\partial S}{\partial T} \right)_V \Delta T$$

to obtain

$$[\Delta V \Delta P - \Delta S \Delta T] = \left(\frac{\partial P}{\partial V}\right)_T (\Delta V)^2 + \left(\frac{\partial P}{\partial T}\right)_V \Delta T \Delta V -$$

$$\left(\frac{\partial S}{\partial V}\right)_T \Delta V \Delta T - \left(\frac{\partial S}{\partial T}\right)_V (\Delta T)^2 = \left(\frac{\partial P}{\partial V}\right)_T (\Delta V)^2 - \left(\frac{\partial S}{\partial T}\right)_V (\Delta T)^2.$$

Then, using the Maxwell relation $\left(\frac{\partial S}{\partial V}\right)_T = \left(\frac{\partial P}{\partial T}\right)_V$, we can simplify the formula for the probability as

$$p \propto e^{\frac{1}{2T}\left[\left(\frac{\partial P}{\partial V}\right)_T (\Delta V)^2 - \left(\frac{\partial S}{\partial T}\right)_V (\Delta T)^2\right]}$$

from which we extract $\langle(\Delta V)^2\rangle$ and $\langle(\Delta T)^2\rangle$, the standard deviations of V and T respectively

$$\langle(\Delta V)^2\rangle = -T\left(\frac{\partial V}{\partial P}\right)_T \qquad \langle(\Delta T)^2\rangle = T\left(\frac{\partial T}{\partial S}\right)_V \qquad \langle\Delta V \Delta T\rangle = 0.$$

It is now immediate to show that

$$\langle\Delta P \Delta T\rangle = \left(\frac{\partial P}{\partial V}\right)_T \langle\Delta V \Delta T\rangle + \left(\frac{\partial P}{\partial T}\right)_V \langle(\Delta T)^2\rangle =$$

$$T\left(\frac{\partial P}{\partial T}\right)_V \left(\frac{\partial T}{\partial S}\right)_V = \frac{T^2}{C_V}\left(\frac{\partial P}{\partial T}\right)_V,$$

where we have used $P = P(T,V)$. Similarly, we can obtain

$$\langle\Delta P \Delta V\rangle = \left(\frac{\partial P}{\partial V}\right)_T \langle(\Delta V)^2\rangle + \left(\frac{\partial P}{\partial T}\right)_V \langle\Delta V \Delta T\rangle = -T\left(\frac{\partial P}{\partial V}\right)_T \left(\frac{\partial V}{\partial P}\right)_T = -T.$$

Then, we can consider $S = S(T,V)$ and obtain

$$\langle\Delta T \Delta S\rangle = \left(\frac{\partial S}{\partial V}\right)_T \langle\Delta V \Delta T\rangle + \left(\frac{\partial S}{\partial T}\right)_V \langle(\Delta T)^2\rangle = T\left(\frac{\partial S}{\partial T}\right)_V \left(\frac{\partial T}{\partial S}\right)_V = T$$

$$\langle\Delta V \Delta S\rangle = \left(\frac{\partial S}{\partial V}\right)_T \langle(\Delta V)^2\rangle + \left(\frac{\partial S}{\partial T}\right)_V \langle\Delta T \Delta V\rangle =$$

$$-T\left(\frac{\partial S}{\partial V}\right)_T \left(\frac{\partial V}{\partial P}\right)_T = -T\left(\frac{\partial S}{\partial P}\right)_T = T\left(\frac{\partial V}{\partial T}\right)_P,$$

where we have used the Maxwell relation $-\left(\frac{\partial S}{\partial P}\right)_T = \left(\frac{\partial V}{\partial T}\right)_P$.

Problem 12.5.
Consider a system with internal energy E, volume V, temperature T, and with a fixed number N of particles. Using the probability of the fluctuations, compute the

energy fluctuations showing that they are

$$\langle (\Delta E)^2 \rangle = C_V T^2 + k_T T V \left(\frac{\partial E}{\partial V} \right)_T^2$$

where C_V is the specific heat at constant volume and $\kappa_T = -\frac{1}{V} \left(\frac{\partial V}{\partial P} \right)_T$ the isothermal compressibility. Compare this result with the one obtained in Problem 12.1. To solve the problem, take the differential of the energy, square it and take the average. To simplify matters, assume $k = 1$.

Solution
Following the suggestion in the text, we take $E = E(T,V)$ and compute

$$\Delta E = \left(\frac{\partial E}{\partial V} \right)_T \Delta V + \left(\frac{\partial E}{\partial T} \right)_V \Delta T = \left(\frac{\partial E}{\partial V} \right)_T \Delta V + C_V \Delta T.$$

Squaring and taking the average we get

$$\langle (\Delta E)^2 \rangle = \left(\frac{\partial E}{\partial V} \right)_T^2 \langle (\Delta V)^2 \rangle + C_V^2 \langle (\Delta T)^2 \rangle + 2 \langle (\Delta V)(\Delta T) \rangle C_V \left(\frac{\partial E}{\partial V} \right)_T.$$

The formula for the probability of the fluctuations is

$$p \propto e^{\frac{1}{2T}[\Delta V \Delta P - \Delta S \Delta T]}.$$

Substituting in it $P = P(T,V)$, $S = S(T,V)$ we find (see Problem 12.4)

$$\langle \Delta V \Delta T \rangle = 0 \quad \langle (\Delta T)^2 \rangle = \frac{T^2}{C_V} \quad \langle (\Delta V)^2 \rangle = -T \left(\frac{\partial V}{\partial P} \right)_T = V T \kappa_T.$$

Plugging this back in $\langle (\Delta E)^2 \rangle$, we get the desired result

$$\langle (\Delta E)^2 \rangle = C_V T^2 + k_T T V \left(\frac{\partial E}{\partial V} \right)_T^2.$$

All these expressions are obtained with the assumption that the number of particles N is fixed. In particular, the equation for $\langle (\Delta V)^2 \rangle$ may also be used to derive an expression for the fluctuations in the specific volume $v = V/N$

$$\langle (\Delta v)^2 \rangle = \frac{V T \kappa_T}{N^2}$$

or also the fluctuations in the density $n = N/V$

$$\langle (\Delta n)^2 \rangle = \frac{\langle (\Delta v)^2 \rangle}{v^4} = \frac{T \kappa_T N^2}{V^3}.$$

Furthermore, considering a system with a fixed volume V and a variable number of particles N, we can write

$$\langle (\Delta N)^2 \rangle = \langle (\Delta n)^2 \rangle V^2 = \frac{T \kappa_T N^2}{V}.$$

Substituting this back in the previous result for $\langle (\Delta E)^2 \rangle$, we get

$$\langle (\Delta E)^2 \rangle = C_V T^2 + \langle (\Delta N)^2 \rangle \frac{V^2}{N^2} \left(\frac{\partial E}{\partial V} \right)_T^2.$$

By the same token, at a fixed temperature T, we can connect the derivative of the energy with respect to V in a system with fixed N, to that of the energy with respect to N in a system with fixed V, i.e.

$$\left(\frac{\partial E}{\partial V} \right)_T^2 = \left(\frac{\partial E}{\partial N} \right)_T^2 \frac{N^2}{V^2}.$$

In fact N, V are extensive functions (homogeneous functions of order one) and $N = V f(P, T)$, i.e. N is proportional to the volume (the ideal gas gives an explicit example). The final result is

$$\langle (\Delta E)^2 \rangle = C_V T^2 + \langle (\Delta N)^2 \rangle \left(\frac{\partial E}{\partial N} \right)_T^2$$

that is in agreement with the result of Problem 12.1.

Problem 12.6.
A system is in thermal equilibrium at temperature T and is composed of N particles ($N = 1, 2$). They can be found in three energy levels $E = n\varepsilon, n = 0, 1, 2$. Determine the grand canonical partition function for particles obeying:

- Fermi-Dirac Statistics (FD);
- Bose-Einstein statistics (BE);
- Maxwell-Boltzmann statistics for indistinguishable particles (MB).

Solution
The grand canonical partition function is

$$\mathscr{Q}(T, z) = \sum_{N=1}^{2} z^N Q_N(T, N) = z Q_1(T, 1) + z^2 Q_2(T, 2)$$

where the canonical partition function has been used

$$Q_N(T, N) = \widetilde{\sum}_{\{n_E\}} g\{n_E\} e^{-\beta \Sigma_E n_E E}$$

with $\{n_E\} = (n_0, n_1, n_2)$ the set of occupation numbers for the energy levels and $g\{n_E\}$ the associated degeneracy. With $\widetilde{\sum}$ we mean that the set $\{n_E\}$ is satisfying

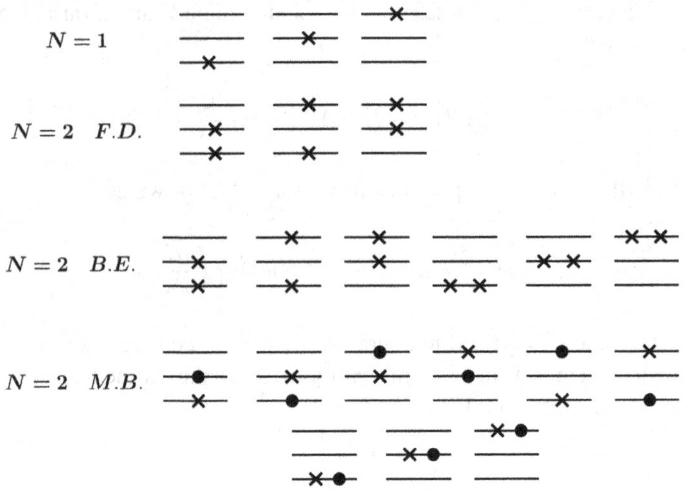

Fig. 12.1 The possible arrangements of $N = 1, 2$ particles in 3 energy levels. We treat explicitly the Fermi-Dirac (FD), Bose-Einstein (BE) and Maxwell-Boltzmann (MB) cases. The resulting grand canonical partition functions are discussed in Problem 12.6

the constraint $N = \sum_E n_E$. Regarding the degeneracy coefficient, for the Fermi-Dirac case we have

$$\begin{cases} g^{(FD)}\{n_E\} = 1 & n_E = 0, 1 \\ g^{(FD)}\{n_E\} = 0 & n_E = 2, 3, 4, \ldots \end{cases}$$

For the Bose-Einstein case we have

$$g^{(BE)}\{n_E\} = 1.$$

Finally, the Maxwell-Boltzmann case for indistinguishable particles leads to

$$g^{(MB)}\{n_E\} = \frac{1}{\prod_E n_E!}.$$

For the case of a single particle, the canonical partition function is independent of the statistics and is equal to

$$Q_1^{(FD)}(T, 1) = Q_1^{(BE)}(T, 1) = Q_1^{(MB)}(T, 1) = 1 + e^{-\beta \varepsilon} + e^{-2\beta \varepsilon}.$$

For the case of two particles, we find three different results, dependent on the statistics. In Fig. 12.1 we report all the resulting configurations. To be noted that in the FD and BE statistics, particles (denoted by the crosses) are indistinguishable. On the other hand, in the MB statistics, we first consider the particles as distinguishable (crosses and filled symbols), and then divide by the factor 2! to account for their indistinguishability. The final result is:

- $Q_2^{(FD)}(T,2) = e^{-\beta\varepsilon} + e^{-2\beta\varepsilon} + e^{-3\beta\varepsilon};$
- $Q_2^{(BE)}(T,2) = 1 + e^{-\beta\varepsilon} + 2e^{-2\beta\varepsilon} + e^{-3\beta\varepsilon} + e^{-4\beta\varepsilon};$
- $Q_2^{(MB)}(T,2) = \frac{1}{2} + e^{-\beta\varepsilon} + \frac{3}{2}e^{-2\beta\varepsilon} + e^{-3\beta\varepsilon} + \frac{1}{2}e^{-4\beta\varepsilon}.$

Problem 12.7.

Consider a system with a single particle and two energy levels (0 and ε). Write down the canonical partition function Q_1. Then, consider the same system with two particles and write down the partition function when:

- the particles are treated as classical and distinguishable (the Maxwell-Boltzmann case);
- the particles obey Bose-Einstein statistics;
- the particles obey Fermi-Dirac statistics.

What is the relation between these partition functions and Q_1? Repeat the calculation for the case with three particles and three energy levels (0, ε_1 and ε_2). Using these examples, determine the number of terms generated in the general case of n particles in m energy levels. Characterize the limit where the Maxwell-Boltzmann case gives the same result as the quantum (Bose-Einstein and Fermi-Dirac) cases.

Solution
The single particle partition function is very simple and independent of the statistics

$$Q_1(T) = 1 + e^{-\beta\varepsilon}.$$

In the case of two particles with the Maxwell Boltzmann (MB) statistics, we find

$$Q_2^{(MB)}(T) = 1 + e^{-\beta\varepsilon} + e^{-\beta\varepsilon} + e^{-2\beta\varepsilon} = (1 + e^{-\beta\varepsilon})^2 = (Q_1(T))^2.$$

In the quantum cases, i.e. Bose-Einstein (BE) and Fermi-Dirac (FD), we have indistinguishable particles. In the Bose-Einstein case, a generic energy level can be occupied by any number of bosons

$$Q_2^{(BE)}(T) = 1 + e^{-\beta\varepsilon} + e^{-2\beta\varepsilon}.$$

In the Fermi-Dirac case, due to the Pauli exclusion principle, a generic energy level is occupied by at most one particle

$$Q_2^{(FD)}(T) = e^{-\beta\varepsilon}.$$

The case with three particles and three energy levels provides the following result for the Maxwell-Boltzmann statistics

$$Q_3^{(MB)}(T) = 1 + e^{-3\beta\varepsilon_1} + e^{-3\beta\varepsilon_2} + 3e^{-\beta\varepsilon_1} + 3e^{-\beta\varepsilon_2} + 6e^{-\beta(\varepsilon_1+\varepsilon_2)} + 3e^{-\beta(2\varepsilon_1+\varepsilon_2)} +$$
$$3e^{-\beta(\varepsilon_1+2\varepsilon_2)} + 3e^{-2\beta\varepsilon_1} + 3e^{-2\beta\varepsilon_2} = (1 + e^{-\beta\varepsilon_1} + e^{-\beta\varepsilon_2})^3 = Q_1^3(T)$$

where $Q_1(T) = 1 + e^{-\beta\varepsilon_1} + e^{-\beta\varepsilon_2}$ in this case. The total number of terms is $N_{MB} = 3 \times 3 \times 3 = 3^3 = 27$ because each particle can occupy the energy levels independently from the others. For the case of n particles in m energy levels, we find

$$N_{MB} = n^m$$

and the coefficients in front of the terms have the form $\frac{n!}{n_1!n_2!n_3!}$, where n_i is the occupation number of the i-th level. In the Bose-Einstein case, the case of three particles in three energy levels gives

$$Q_3^{(BE)}(T) = 1 + e^{-3\beta\varepsilon_1} + e^{-3\beta\varepsilon_2} + e^{-\beta\varepsilon_1} + e^{-\beta\varepsilon_2} + e^{-\beta(\varepsilon_1+\varepsilon_2)} + e^{-\beta(2\varepsilon_1+\varepsilon_2)} +$$
$$e^{-\beta(\varepsilon_1+2\varepsilon_2)} + e^{-2\beta\varepsilon_1} + e^{-2\beta\varepsilon_2}.$$

The total number of terms is 10. In general, when dealing with m energy levels and n particles, we find a total number of terms equal to

$$N_{BE} = \frac{(n+m-1)!}{(m-1)!n!}.$$

Finally, the case of three Fermi-Dirac particles in three energy levels, produces

$$Q_3^{(FD)}(T) = e^{-\beta(\varepsilon_1+\varepsilon_2)}.$$

With this quantum statistics, the case of n particles and m energy levels has a total number of terms equal to

$$N_{FD} = \frac{m!}{(m-n)!n!}.$$

Only in the limit of high temperatures (more energy levels are accessible) and low densities (not so many particles to accommodate), the behaviour of all physical systems tends asymptotically to what we expect on classical grounds. In such limit, the occupation number for each energy level is $n_i = 0, 1$, and all the Maxwell-Boltzmann terms, once divided by the Gibbs factor $n!$ (to account for the indistinguishability of the particles), are the same as those found in $Q^{(FD)}(T)$ or $Q^{(BE)}(T)$. To give an example, in the case of three particles in three energy levels, only the term $6e^{-\beta(\varepsilon_1+\varepsilon_2)}$ survives. Dividing by $3! = 6$, we get the quantum result of $Q_3^{(FD)}(T)$ and $Q_3^{(BE)}(T)$ (in which we have also neglected the contributions from states with more than one particle per energy level).

Problem 12.8.

A volume V is filled with N independent distinguishable particles of a given gas with constant energy E. Let us split the volume V in two subvolumes

$$V = V_1 + V_2$$

where V_1 and V_2 are filled with N_1 and N_2 particles, respectively. Let

$$p = \frac{V_1}{V} \qquad q = \frac{V_2}{V}$$

be the associated volume fractions such that $p+q=1$. Find the probability P_K that K particles are located in V_1 and compute $\langle K \rangle$ and $\langle (\Delta K)^2 \rangle = \langle (K - \langle K \rangle)^2 \rangle$, where $\langle ... \rangle$ means the average with respect to P_K. Finally, compute the fluctuations $\frac{\sqrt{\langle (\Delta K)^2 \rangle}}{\langle K \rangle}$.

Solution
The number of possible spatial configurations for a particle located in a volume V is proportional to V itself. If there is no spatial correlation (particles are independent), the probability that any one of the particles is found in a given region is totally independent of the positions of the other particles. It follows that the probability to find K particles in V_1 (and obviously $(N-K)$ particles in V_2) is proportional to $V_1^K V_2^{N-K}$

$$P_K = A \frac{N!}{K!(N-K)!} V_1^K V_2^{N-K}$$

where we have considered the factor $\frac{N!}{K!(N-K)!}$ to take into account all the possible ways to choose K distinguishable particles out of N. In the above expression, A is a normalization constant found by imposing that the sum of P_K over all the possible realizations of K is equal to 1. Using the binomial representation

$$\sum_{K=0}^{N} \frac{N!}{K!(N-K)!} V_1^K V_2^{N-K} = (V_1 + V_2)^N = V^N$$

one can write

$$\sum_{K=0}^{N} P_K = A \sum_{K=0}^{N} \frac{N!}{K!(N-K)!} V_1^K V_2^{N-K} = AV^N = 1$$

so that $A = \frac{1}{V^N}$. The resulting probability is

$$P_K = \frac{1}{V^N} \binom{N}{K} V_1^K V_2^{N-K} = \frac{1}{V^N} \binom{N}{K} V_1^K V_2^{N-K} = \binom{N}{K} p^K q^{N-K}.$$

The average number $\langle K \rangle$ is

$$\langle K \rangle = \sum_{K=0}^{N} K P_K = \sum_{K=0}^{N} K \binom{N}{K} p^K q^{N-K} = p \frac{d}{dp} \sum_{K=0}^{N} \binom{N}{K} p^K q^{N-K} =$$

$$p \frac{d}{dp} (p+q)^N = Np(p+q)^{N-1} = Np$$

where we have considered that $p + q = 1$. The fluctuations are given by

$$\langle (\Delta K)^2 \rangle = \langle (K - \langle K \rangle)^2 \rangle = \sum_{K=0}^{N} K^2 P_K - \left(\sum_{K=0}^{N} K P_K \right)^2 =$$

$$\left(p \frac{d}{dp} \right)^2 \sum_{K=0}^{N} \binom{N}{K} p^K q^{N-K} - (Np)^2 =$$

$$p \frac{d}{dp} (pN(p+q)^{N-1}) - (Np)^2 =$$

$$pN(p+q)^{N-1} + p^2 N(N-1)(p+q)^{N-2} - (Np)^2 = Npq.$$

The final result is

$$\frac{\sqrt{\langle (\Delta K)^2 \rangle}}{\langle K \rangle} = \frac{\sqrt{Npq}}{Np} = \sqrt{\frac{q}{p}} \frac{1}{\sqrt{N}}.$$

We remark that for large N the fluctuations are negligible for a finite value of p and q. Only when p gets very small they become large.

Problem 12.9.
Consider an ideal gas with $N \gg 1$ independent particles in a volume V. Using simple considerations, determine the probability P_K that $K \ll N$ (i.e. $p = \frac{V_0}{V} \ll 1$) particles are located in the subvolume $V_0 \ll V$. Assume that $N - K \gg 1$.

Solution
The probability that a single particle is in the volume V_0 is $p = \frac{V_0}{V}$. The resulting probability that K particles are in V_0 and (simultaneously) $N - K$ particles are in the volume $V - V_0$ is

$$P_K = C_{K,N} p^K (1-p)^{N-K} = \frac{N!}{K!(N-K)!} p^K (1-p)^{N-K}$$

where $C_{K,N}$ represents all the possible ways to select K particles out of N. It immediately follows that the average number of particles in V_0 is

$$\langle K \rangle = Np.$$

All these results are found in Problem 12.8. We now derive some asymptotic formula for the probability P_K. In the limit suggested by the text, we can use the Stirling approximation, $N! \approx \sqrt{2\pi N} \left(\frac{N}{e} \right)^N$, to get

$$\frac{N!}{(N-K)!} \approx \sqrt{2\pi N} \left(\frac{N}{e} \right)^N \frac{e^{N-K}}{\sqrt{2\pi(N-K)}(N-K)^{N-K}} =$$

$$\sqrt{\frac{N}{N-K}} e^{-K} \left(\frac{N}{N-K} \right)^N (N-K)^K =$$

$$\left(1 - \frac{K}{N} \right)^{-\frac{1}{2}} e^{-K} \left(1 - \frac{K}{N} \right)^{-N} N^K \left(1 - \frac{K}{N} \right)^K \approx$$

$$\left(1 + \frac{K}{2N} \right) e^{-K} (1+K) N^K \left(1 - \frac{K^2}{N} \right) \approx e^{-K} N^K.$$

In the same limit, since we know that $\langle K \rangle = Np$, we find

$$(1-p)^{N-K} \approx (1-p)^N = (1-p)^{\frac{\langle K \rangle}{p}} \approx e^{-\langle K \rangle}.$$

If we use all these results together, we find

$$P_K = N^K p^K \frac{e^{-\langle K \rangle}}{K!} = \frac{\langle K \rangle^K e^{-\langle K \rangle}}{K!}$$

that is a Poisson distribution with average value equal to $\langle K \rangle$.

Problem 12.10.

During a thermally induced emission, some electrons leave the surface of a metal. Let us assume that the electron emissions are statistically independent events and that the emission probability during a time interval dt is λdt, with λ a constant. If the process of emission starts at time $t = 0$, determine the probability of emission of n electrons in a time interval $t > 0$.

Solution
Let us call $P_n(t)$ the probability to emit n electrons in the time interval t. The composition rule for the probabilities gives

$$P_n(t+dt) = P_{n-1}(t)P_1 + P_n(t)(1-P_1)$$

$$P_0(t+dt) = P_0(t)(1-P_1)$$

where $P_1 = \lambda dt$ is the probability to emit one electron, as explained in the text. Clearly, $(1-P_1)$ is the probability not to emit the electron. If we expand

$$P_n(t+dt) \approx P_n(t) + \frac{dP_n(t)}{dt} dt$$

and we use $P_1 = \lambda dt$, we obtain the following differential equations

$$\frac{dP_n(t)}{dt} = \lambda [P_{n-1}(t) - P_n(t)] \quad n \neq 0$$

$$\frac{dP_0(t)}{dt} = -\lambda P_0(t).$$

These differential equations need boundary conditions which are $P_n(0) = 1$ for $n = 0$, and $P_n(0) = 0$ for $n \neq 0$, i.e. at time $t = 0$ we have probability one to find zero electrons and probability zero to find at least one electron. For P_0 we can solve immediately and find

$$P_0(t) = e^{-\lambda t}.$$

As for P_n with $n \neq 0$, we start by considering the solution of the differential equation

$$\frac{dy(t)}{dt} + p(t)y(t) = q(t)$$

that is given by

$$y(t) = e^{-\int_0^t p(x)dx} \left[\int_0^t q(s)e^{\int_0^s p(t')dt'} \, ds + \text{const.} \right].$$

If we set $y(t) = P_n(t)$, $q(t) = \lambda P_{n-1}(t)$ and $p(t) = \lambda$ we have

$$P_n(t) = e^{-\lambda t} \left[\lambda \int_0^t P_{n-1}(s)e^{\lambda s} \, ds + \text{const.} \right].$$

Using the boundary conditions, we can solve for $n = 0, 1, 2, 3, \ldots$ and find

$$P_0(t) = e^{-\lambda t}$$

$$P_1(t) = (\lambda t)e^{-\lambda t}$$

$$P_2(t) = \frac{(\lambda t)^2}{2}e^{-\lambda t}$$

$$P_3(t) = \frac{(\lambda t)^3}{3!}e^{-\lambda t}$$

from which we guess the following form of the solution

$$P_n(t) = \frac{(\lambda t)^n}{n!}e^{-\lambda t}.$$

It is straightforward to verify that this $P_n(t)$ is the solution of the differential equations given above.

Problem 12.11.
The Ising model is characterized by a number N of particles with spin $S = 1/2$ localized on a given lattice. The Hamiltonian of the system is given by the interaction energy between these spins

$$\mathcal{H} = -J \sum_{\langle ij \rangle} \sigma_i \sigma_j$$

where J is a coupling constant and where $\sigma_i = \pm 1$ are the values of the projections of the spin along the z axis. The sum $\sum_{\langle ij \rangle}$ is performed over all the values of i, j which are nearest neighbor. Using the canonical ensemble, compute the free energy and the specific heat in the case of a one dimensional chain of $N \gg 1$ spins with periodic boundary conditions ($\sigma_{N+1} \equiv \sigma_1$).

Solution
Starting from the Hamiltonian $\mathcal{H}$, we find the partition function

$$Q_N(T,N) = \sum_{\sigma_1 = \pm 1} \cdots \sum_{\sigma_N = \pm 1} e^{\tilde{\beta} \sum_{i=1}^N \sigma_i \sigma_{i+1}} = \sum_{\sigma_1 = \pm 1} \cdots \sum_{\sigma_N = \pm 1} \prod_{i=1}^N (\cosh \tilde{\beta} + \sigma_i \sigma_{i+1} \sinh \tilde{\beta})$$

with $\tilde{\beta} = J/kT$. In the above expression, we have used the following relation

$$e^{\tilde{\beta}\sigma_i\sigma_{i+1}} = \cosh\tilde{\beta} + \sigma_i\sigma_{i+1}\sinh\tilde{\beta}$$

because $\sigma_i\sigma_{i+1} = \pm 1$. If we expand the product, we obtain terms of type

$$(\cosh\tilde{\beta})^{N-k}(\sinh\tilde{\beta})^k(\sigma_{i_1}\sigma_{i_1+1})(\sigma_{i_2}\sigma_{i_2+1})\ldots(\sigma_{i_k}\sigma_{i_k+1}).$$

It is easy to see that all the mixed terms (i.e. those involving spins on different locations) give zero as a result when summed over all the possible realizations. Let us make an example and compute

$$Q_2(T,2) = \sum_{\sigma_1=\pm 1}\sum_{\sigma_2=\pm 1}(\cosh\tilde{\beta} + \sigma_1\sigma_2\sinh\tilde{\beta})(\cosh\tilde{\beta} + \sigma_2\sigma_3\sinh\tilde{\beta}) =$$

$$\sum_{\sigma_1=\pm 1}\sum_{\sigma_2=\pm 1}\left((\cosh\tilde{\beta})^2 + (\sinh\tilde{\beta})^2 + 2\sigma_1\sigma_2\sinh\tilde{\beta}\cosh\tilde{\beta}\right) =$$

$$4\left((\cosh\tilde{\beta})^2 + (\sinh\tilde{\beta})^2\right)$$

where we have used the periodic relation $\sigma_3 = \sigma_1$ and also

$$\sum_{\sigma_1=\pm 1}\sum_{\sigma_2=\pm 1}\sigma_1\sigma_2 = \sum_{\sigma_1=\pm 1}\sigma_1\sum_{\sigma_2=\pm 1}\sigma_2 = (1-1)(1-1) = 0.$$

Therefore, in the case of N spins, the partition function is

$$Q_N(T,N) = 2^N\left[(\cosh\tilde{\beta})^N + (\sinh\tilde{\beta})^N\right] \approx 2^N(\cosh\tilde{\beta})^N$$

because $\cosh\tilde{\beta} > \sinh\tilde{\beta}$ ($\tilde{\beta} \neq +\infty$) and $N \gg 1$. As for the free energy F, internal energy $U = F + TS$, and the specific heat $C = \left(\frac{\partial U}{\partial T}\right)_N$, we have

$$F = -NkT\ln(2\cosh\tilde{\beta})$$

$$U = F + TS = F - T\left(\frac{\partial F}{\partial T}\right)_N = -NJ\tanh\tilde{\beta}$$

$$C = \frac{Nk\tilde{\beta}^2}{(\cosh\tilde{\beta})^2}.$$

Problem 12.12.
Consider the Ising model discussed in Problem 12.11. In presence of an external magnetic field H, the Hamiltonian of the system is given by the interaction energy between the spins plus the coupling with the magnetic field

$$\mathcal{H} = -J\sum_{\langle ij\rangle}\sigma_i\sigma_j - H\sum_{i=1}^{N}\sigma_i$$

where J is a constant coupling and where $\sigma_i = \pm 1$ are the values of the projections of the spin along the z axis. The sum $\sum_{\langle ij \rangle}$ is performed over all the values of i, j which are nearest neighbor. Show that this model does not present a magnetization M in the limit $N \gg 1$, $H \to 0$ and $\beta < +\infty$.

Solution

We need to compute the magnetization. To this end, we start from the Hamiltonian

$$\mathcal{H} = -J \sum_{i=1}^{N} \sigma_i \sigma_{i+1} - H \sum_{i=1}^{N} \sigma_i$$

and the partition function

$$Q_N(T,H,N) = \sum_{\sigma_1 = \pm 1} \cdots \sum_{\sigma_N = \pm 1} e^{\frac{J}{kT} \sum_{i=1}^{N} \sigma_i \sigma_{i+1} + \frac{H}{kT} \sum_{i=1}^{N} \sigma_i} =$$

$$\sum_{\sigma_1 = \pm 1} \cdots \sum_{\sigma_N = \pm 1} e^{\frac{J}{kT} \sum_{i=1}^{N} \sigma_i \sigma_{i+1} + \frac{1}{2} \frac{H}{kT} \sum_{i=1}^{N} (\sigma_i + \sigma_{i+1})}$$

where we have used the periodic boundary condition $\sigma_{N+1} \equiv \sigma_1$. In this expression, the variables σ_i are just numbers assuming the values ± 1. Let us imagine to define the states $|s_i = \pm\rangle$ so that the i-th spin may be described by a complete set of eigenstates $|+\rangle, |-\rangle$ such that the 'spin' operator, $\hat{\sigma}_i$, has σ_i as eigenvalue

$$\hat{\sigma}_i |s_i\rangle = \sigma_i |s_i\rangle .$$

Therefore, we recognize in the partition function matrix elements of type

$$\langle s_i | e^{\frac{J}{kT} \hat{\sigma}_i \hat{\sigma}_{i+1} + \frac{1}{2} \frac{H}{kT} (\hat{\sigma}_i + \hat{\sigma}_{i+1})} |s_{i+1}\rangle = e^{\frac{J}{kT} \sigma_i \sigma_{i+1} + \frac{1}{2} \frac{H}{kT} (\sigma_i + \sigma_{i+1})}.$$

From the generic element, we immediately find out the components

$$\langle +| e^{\frac{J}{kT} \hat{\sigma}_i \hat{\sigma}_{i+1} + \frac{1}{2} \frac{H}{kT} (\hat{\sigma}_i + \hat{\sigma}_{i+1})} |+\rangle = e^{\frac{J}{kT} + \frac{H}{kT}} = e^{\beta(J+H)}$$

$$\langle -| e^{\frac{J}{kT} \hat{\sigma}_i \hat{\sigma}_{i+1} + \frac{1}{2} \frac{H}{kT} (\hat{\sigma}_i + \hat{\sigma}_{i+1})} |-\rangle = e^{\frac{J}{kT} - \frac{H}{kT}} = e^{\beta(J-H)}$$

$$\langle +| e^{\frac{J}{kT} \hat{\sigma}_i \hat{\sigma}_{i+1} + \frac{1}{2} \frac{H}{kT} (\hat{\sigma}_i + \hat{\sigma}_{i+1})} |-\rangle = e^{-\frac{J}{kT}} = e^{-\beta J}$$

$$\langle -| e^{\frac{J}{kT} \hat{\sigma}_i \hat{\sigma}_{i+1} + \frac{1}{2} \frac{H}{kT} (\hat{\sigma}_i + \hat{\sigma}_{i+1})} |+\rangle = e^{-\frac{J}{kT}} = e^{-\beta J}$$

giving us the representation of the well known transfer matrix $\hat{T}$

$$\hat{T} = \begin{pmatrix} e^{\beta(J+H)} & e^{-\beta J} \\ e^{-\beta J} & e^{\beta(J-H)} \end{pmatrix} .$$

Using these results and the completeness relation

$$\sum_{s_i = \pm} |s_i\rangle \langle s_i| = |+\rangle \langle +| + |-\rangle \langle -| = 1$$

we write down the partition function as

$$Q_N(T,H,N) = \sum_{s_1=\pm} \cdots \sum_{s_N=\pm} \langle s_1|\hat{T}|s_2\rangle \langle s_2|\hat{T}|s_3\rangle \ldots \langle s_N|\hat{T}|s_1\rangle =$$

$$\sum_{s_1=\pm} \langle s_1|\hat{T}^N|s_1\rangle = \langle +|\hat{T}^N|+\rangle + \langle -|\hat{T}^N|-\rangle = \lambda_1^N + \lambda_2^N$$

where λ_1, λ_2 are the eigenvalues of $\hat{T}$ given by

$$\lambda_{1,2} = e^{J\beta}\left(\cosh(\beta H) \pm \sqrt{\cosh^2(\beta H) - 2e^{-2J\beta}\sinh(2J\beta)}\right)$$

with the property $\lambda_1 > \lambda_2$ if $\beta < +\infty$. For large N, the specific magnetization is given by

$$M = \frac{1}{N}\left(\frac{\partial F}{\partial H}\right)_{T,N} = -\frac{1}{N\beta}\left(\frac{\partial \ln Q_N}{\partial H}\right)_{T,N} = -\frac{1}{N\beta}\left(\frac{\partial \ln(\lambda_1^N + \lambda_2^N)}{\partial H}\right)_{T,N} =$$

$$-\frac{1}{N\beta}\left(\frac{\partial}{\partial H}\ln\left[\lambda_1^N\left(1+\left(\frac{\lambda_2}{\lambda_1}\right)^N\right)\right]\right)_{T,N} \approx -\frac{1}{\beta}\left(\frac{\partial \ln \lambda_1}{\partial H}\right)_T =$$

$$-\frac{\sinh(\beta H) + \dfrac{\sinh(\beta H)\cosh(\beta H)}{\sqrt{\cosh^2(\beta H) - 2e^{-2J\beta}\sinh(2J\beta)}}}{\cosh(\beta H) + \sqrt{\cosh^2(\beta H) - 2e^{-2J\beta}\sinh(2J\beta)}} =$$

$$-\frac{\sinh(\beta H)}{\sqrt{\cosh^2(\beta H) - 2e^{-2J\beta}\sinh(2J\beta)}}$$

from which we see that M goes to zero in the limit $H \to 0$.

Problem 12.13.
Let us consider a generic ideal quantum gas of bosons in a volume V at temperature T. Show that the following relation for the entropy

$$S = k\sum_i\left[-\langle n_i\rangle \ln\langle n_i\rangle + (1 + \langle n_i\rangle)\ln(1 + \langle n_i\rangle)\right]$$

holds. In the above expression, $\langle n_i\rangle$ represents the occupation number for the i-th energy level. What does it happen in the case of fermions?

Solution
Let us start from the expression of the occupation number

$$\langle n_i\rangle = \frac{1}{z^{-1}e^{\beta\varepsilon_i} - 1}$$

from which we can write

$$z^{-1}e^{\beta\varepsilon_i} = \frac{1 + \langle n_i\rangle}{\langle n_i\rangle}$$

and, since $z = e^{\beta\mu}$, we get

$$-\beta\mu + \beta\varepsilon_i = \ln(1 + \langle n_i \rangle) - \ln\langle n_i \rangle.$$

From the logarithm of the grand canonical partition function we find the pressure

$$\ln \mathcal{Q} = \frac{PV}{kT} = -\sum_i \ln(1 - z e^{-\beta\varepsilon_i}) = -\sum_i \ln\left(\frac{1}{1 + \langle n_i \rangle}\right) = \sum_i \ln(1 + \langle n_i \rangle).$$

The entropy can be written as

$$S = \frac{1}{T}(-N\mu + PV + U)$$

where with N we mean the average number in the grand canonical ensemble. The average energy and average number are

$$U = -\left(\frac{\partial \ln \mathcal{Q}}{\partial \beta}\right)_{V,z}, \qquad N = z\left(\frac{\partial \ln \mathcal{Q}}{\partial z}\right)_{T,V}.$$

We first note that

$$\frac{U}{T} = -k\beta\left(\frac{\partial \ln \mathcal{Q}}{\partial \beta}\right)_{V,z} = k\sum_i \beta\varepsilon_i\langle n_i \rangle.$$

Then, for the quantity $\frac{\mu N}{T}$, we find

$$\frac{\mu N}{T} = \frac{\mu}{T}z\left(\frac{\partial \ln \mathcal{Q}}{\partial z}\right)_{T,V} = k\sum_i \beta\mu\langle n_i \rangle$$

so that

$$-\frac{\mu N}{T} + \frac{U}{T} = k\sum_i(-\beta\mu + \beta\varepsilon_i)\langle n_i \rangle = k\sum_i \langle n_i \rangle \ln(1 + \langle n_i \rangle) - k\sum_i \langle n_i \rangle \ln\langle n_i \rangle.$$

When we substitute this expression and

$$\frac{PV}{T} = k\sum_i \ln(1 + \langle n_i \rangle)$$

in the equation for the entropy, we find

$$S = k\sum_i [-\langle n_i \rangle \ln\langle n_i \rangle + (1 + \langle n_i \rangle) \ln(1 + \langle n_i \rangle)]$$

that is the desired result. Similar calculations can be done in the case of fermions with the occupation number

$$\langle n_i \rangle = \frac{1}{z^{-1}e^{\beta\varepsilon_i} + 1}$$

and the final result is

$$S = k \sum_i [-\langle n_i \rangle \ln \langle n_i \rangle - (1 - \langle n_i \rangle) \ln(1 - \langle n_i \rangle)].$$

Problem 12.14.
Write down the probability that an energy level is occupied by m particles in the following cases:

- a gas of bosonic (independent) particles with density $n = 10^{16}$ particles per cm^3 at temperature $T \approx 400$K;
- a photon gas at room temperature.

Before writing down the probability, determine whether the system can be treated with the classical or quantum statistics. Make use of the following constants $k = 10^{-16}$erg/K, $h = 10^{-27}$erg s, $m_e = 10^{-28}$g.

Solution
Let us start with the bosons gas. We first need to know if the gas is in a classical or quantum regime. To this end, we compute the thermal length scale

$$\lambda = \frac{h}{\sqrt{2\pi m_e kT}} \approx 2 \times 10^{-7} \text{cm}$$

and we see that we are in the regime where $n\lambda^3 \ll 1$: we can treat the gas with the classical statistics, where the occupation number of the energy level ε is

$$\langle n_\varepsilon \rangle \approx z e^{-\beta\varepsilon} \ll 1.$$

Summing $\langle n_\varepsilon \rangle$ over all the possible energies, we find the total average number of particles $\langle N \rangle$, i.e. $\langle N \rangle = \sum_\varepsilon \langle n_\varepsilon \rangle \approx \sum_\varepsilon z e^{-\beta\varepsilon}$, that implies

$$z = \frac{\langle N \rangle}{\sum_\varepsilon e^{-\beta\varepsilon}} \qquad \langle n_\varepsilon \rangle = \frac{\langle N \rangle e^{-\beta\varepsilon}}{\sum_\varepsilon e^{-\beta\varepsilon}}.$$

The probability to find a particle in the energy level ε is proportional to $\langle n_\varepsilon \rangle$, and the probability to have m particles is proportional to $\langle n_\varepsilon \rangle^m$, that is

$$P_\varepsilon(m) = \frac{A_1}{m!} \langle n_\varepsilon \rangle^m = \frac{1}{m!} \frac{\langle n_\varepsilon \rangle^m}{e^{\langle n_\varepsilon \rangle}}.$$

The factor $m!$ takes into account the (classical) indistinguishability of the particles and $A_1 = e^{-\langle N \rangle}$ sets the normalization of $P_\varepsilon(m)$ in such a way that $\sum_{m=0}^{+\infty} P_\varepsilon(m) = 1$. As for the case of the photon gas, we need to use the Bose-Einstein statistics leading to

$$P_\varepsilon(m) = A_2 \left(z e^{-\beta\varepsilon} \right)^m$$

with A_2 the normalization constant found with

$$\sum_{m=0}^{+\infty} P_\varepsilon(m) = A_2 \sum_{m=0}^{+\infty} \left(ze^{-\beta\varepsilon}\right)^m = \frac{A_2}{1 - ze^{-\beta\varepsilon}} = 1$$

leading to $A_2 = 1 - ze^{-\beta\varepsilon}$. Moreover, in the case of a photon gas, the chemical potential is zero and $z = 1$. The final result is

$$P_\varepsilon(m) = \left(e^{-\beta\varepsilon}\right)^m \left(1 - e^{-\beta\varepsilon}\right).$$

Problem 12.15.
In the primordial universe the particles formed a plasma at equilibrium. Among the various species of particles, there were electrons, positrons, photons interacting according to

$$\gamma \leftrightarrow e^+ + e^-$$

which was responsible for equilibrium. At that time the electrons were "hot", i.e. they were relativistic. Compute the density $n_{e\pm}$ in the two limits $kT \gg m_e c^2$ and $kT \ll m_e c^2 (v/c \ll 1)$ and show that in the latter $n_{e\pm}$ is negligible with respect to the photon density. Consider that, for the reaction $A \leftrightarrow B + C$ at equilibrium, the following relation among the chemical potentials holds: $\mu_A = \mu_B + \mu_C$. Furthermore, since the number density of the electrons and positrons is the same (for "hot" particles virtual annihilations dominate over other reactions), the charge density for the leptons is zero and an odd function of the chemical potentials, and $\mu_{e^+} + \mu_{e^-} = \mu_\gamma = 0$ leads to $\mu_{e^+} = \mu_{e^-} = 0$. Some useful constants are: $m_e = 10^{-28}$g, $c = 10^{10}$cm/s, $T > 4 \times 10^3$K, $k = 10^{-16}$erg/K. Moreover

$$\int_0^{+\infty} \frac{x^{n-1}}{e^x + 1} dx = (1 - 2^{1-n})\Gamma(n)\zeta(n)$$

$$\int_0^{+\infty} \frac{x^{n-1}}{e^x - 1} dx = \Gamma(n)\zeta(n).$$

Solution
The energy of a relativistic particle is $\varepsilon = \sqrt{c^2 p^2 + (m_e c^2)^2}$. Let us consider our two cases. When $kT \gg m_e c^2$, we have $\varepsilon/kT \approx pc/kT$ and this is the case of an ultrarelativistic particle of mass close to zero. The multiplicity factor is connected to the spin states and it is $g = 2$, corresponding to the two elicity states of a massless particle. Given that the chemical potential is zero, we get the average densities for $e^\pm$

$$n_{e\pm} = \frac{2}{(2\pi\hbar)^3} \int \frac{1}{e^{\beta cp} + 1} d^3 p = \frac{(kT)^3}{\pi^2 (\hbar c)^3} \int_0^{+\infty} \frac{x^2}{e^x + 1} dx = $$
$$\frac{(kT)^3}{\pi^2 (\hbar c)^3} \frac{3\Gamma(3)\zeta(3)}{4} = \frac{3(kT)^3 \zeta(3)}{2\pi^2 (\hbar c)^3}.$$

When $kT \ll m_ec^2 (v/c \ll 1)$, this is the standard relativistic case for which $\varepsilon \approx \frac{p^2}{2m_e} + m_ec^2$. Therefore

$$n_{e\pm} = \frac{2}{(2\pi\hbar)^3} \int \frac{1}{e^{\beta\frac{p^2}{2m_e}+\beta m_ec^2} + 1} d^3p \approx \frac{2}{(2\pi\hbar)^3} \int e^{-\beta\frac{p^2}{2m_e}-\beta m_ec^2} d^3p$$

$$= 2\left(\frac{2\pi m_e kT}{h^2}\right)^{\frac{3}{2}} e^{-\frac{m_ec^2}{kT}}.$$

The photon density is

$$n_\gamma = \frac{1}{\pi^2 c^3} \int_0^{+\infty} \frac{\omega^2 d\omega}{e^{\hbar\omega/kT} - 1} = \frac{2\zeta(3)(kT)^3}{\pi^2 \hbar^3 c^3}$$

while the argument of the exponential gives

$$\frac{m_ec^2}{kT} \approx 0.25 \times 10^5.$$

This makes the exponential very suppressed and therefore $n_{e\pm} \ll n_\gamma$.

Problem 12.16.
In the primordial universe the various species of particles were in thermal equilibrium due to the different subnuclear reactions. The cross section, σ, of one of these reactions times the particles density n gives $\ell \approx 1/n\sigma$ for the free mean path, i.e. the average distance between two collisions. Given that the cross section is the number of scattered particles per unit area and unit time, a particle which travels the unitary distance (1cm) collides against the other particles contained in the volume of base σ and height 1cm. This volume contains σn particles. When the universe becomes larger than ℓ, this specie of particles decouples from thermal equilibrium because the number of collisions is not enough to guarantee equilibrium. The Thermodynamics is that of a free expanding gas. Given that the universe expands linearly with a coefficient given by the Hubble constant $H \approx \sqrt{\rho G_N}$ (ρ is the energy density and G_N the Newton constant), the decoupling threshold is given by $\sigma n/H \approx 1$.

- Use the latter formula to compute the neutrinos decoupling temperature. Suppose the neutrinos have a zero chemical potential. Neutrinos are weakly interacting and their cross section is given by $\sigma \approx G_F^2 T^2$ where G_F is the Fermi constant. For the universe density, n, and energy density, ρ, neglect all constants and only retain the temperature dependence for a relativistic gas.
- The universe keeps expanding. For $kT \gg m_ec^2$ compute the entropy of the plasma which only contains electrons, positrons and photons (e^-, e^+, γ respectively). Consider these particles to have zero chemical potential.
- Compute the entropy of the photon gas for $kT \ll m_ec^2$ considering that all the electrons and positrons have annihilated yielding photons.
- Compute now $(VT^3)_{kT \gg m_ec^2}/(VT^3)_{kT \ll m_ec^2} = (T_{kT \gg m_ec^2}/T_{kT \ll m_ec^2})^3$ where, as a first approximation, we can take the volumes to be equal. This ratio is the

same between the temperatures of the photon background radiation and that of the neutrinos, ν. In fact, at $kT \approx m_e c^2$, the neutrinos were already decoupled and were evolving starting from $T_{kT \gg m_e c^2}$; γ's, instead, evolved starting from $T_{kT \ll m_e c^2}$. Find the temperature of the neutrinos at the present time, knowing that the ratio $T_{kT \gg m_e c^2}/T_{kT \ll m_e c^2}$ has kept constant from that epoch to present times.

Use the unit system for which $c = 1$, $k = 1$.
In this system [Energy]=[Mass]=[Temperature]=1/[Length]=1/[Time]. Furthermore $1\,\text{GeV} \approx 10^{-3}\text{erg}$, $1\,\text{GeV} \approx 10^{13}\text{K}$, $1\,\text{GeV} \approx 10^{-24}\text{g}$, $1\,\text{GeV} \approx 10^{-14}\text{cm}$, $1\,\text{GeV} = 6.6\,10^{-25}\text{s}$, $m_e = 0.5\text{MeV}$, $G_F \approx 10^{-5}\text{GeV}^{-2}$, $G_N = m_{Planck}^{-2} = 1/(10^{19}\text{GeV})^2$. Finally

$$\int_0^{+\infty} \frac{x^{n-1}}{e^x + 1}dx = (1 - 2^{1-n})\Gamma(n)\zeta(n)$$

$$\int_0^{+\infty} \frac{x^{n-1}}{e^x - 1}dx = \Gamma(n)\zeta(n).$$

Solution
Let us start by evaluating

$$1 \approx \frac{\sigma n}{H} = \frac{G_F^2 T^2 T^3}{\sqrt{G_N}T^2}$$

from which $T^3 = \frac{\sqrt{G_N}}{G_F^2} = 10^{-9}\text{GeV}^3 = (10^{10}\text{K})^3$. In fact, the relativistic particles density is such that $n \approx T^3$ and the energy density $\rho \approx T^4$ (see also Problem 12.15). This temperature is larger than $T \approx m_e = 0.5\text{MeV} = 5 \times 10^{-4}\text{GeV} = 5 \times 10^9\text{K}$. Let us compute the entropy for $T \gg m_e$. Since the chemical potential is zero, we find

$$0 = \Phi = \mu N = E - TS + PV$$

from which $S = \frac{V}{T}(\rho + P)$. For a relativistic gas $P = \rho/3$ holds from which $S = \frac{4}{3}\frac{V}{T}\rho$. For the electrons and positrons $e^\pm$, the energy density is

$$\rho_{e^\pm} = \frac{2}{(2\pi\hbar)^3} \int \frac{cp}{e^{\beta cp} + 1}d^3p = \frac{(kT)^4}{\pi^2(\hbar c)^3} \int_0^{+\infty} \frac{x^3}{e^x + 1}dx =$$

$$\frac{(kT)^4}{\pi^2(\hbar c)^3} \frac{7\Gamma(4)\zeta(4)}{8}$$

while for the photons we get

$$\rho_\gamma = \frac{2}{(2\pi\hbar)^3} \int \frac{cp}{e^{\beta cp} - 1}d^3p = \frac{(kT)^4}{\pi^2(\hbar c)^3} \int_0^{+\infty} \frac{x^3}{e^x - 1}dx =$$

$$\frac{(kT)^4}{\pi^2(\hbar c)^3}\Gamma(4)\zeta(4).$$

Therefore

$$S_{kT \gg m_e c^2} = \frac{4}{3}\frac{V}{T}(\rho_{e^+} + \rho_{e^-} + \rho_\gamma) = \frac{4}{3}\frac{V}{kT}\frac{11}{4}\frac{(kT)^4}{\pi^2 (\hbar c)^3}\Gamma(4)\zeta(4).$$

Analogously

$$S_{kT \ll m_e c^2} = \frac{4}{3}\frac{V}{T}\rho_\gamma = \frac{4}{3}\frac{V}{T}\frac{(kT)^4}{\pi^2(\hbar c)^3}\Gamma(4)\zeta(4).$$

Since the entropy stays constant, and in first approximation we can neglect volume changes, we find

$$\frac{(VT^3)_{kT \gg m_e c^2}}{(VT^3)_{kT \ll m_e c^2}} \approx \frac{T^3_{kT \gg m_e c^2}}{T^3_{kT \ll m_e c^2}} = \frac{4}{11} \Rightarrow \frac{T_{kT \gg m_e c^2}}{T_{kT \ll m_e c^2}} = \left(\frac{4}{11}\right)^{\frac{1}{3}}.$$

This ratio has stayed constant until modern times and therefore $T_\nu = (4/11)^{1/3} T_\gamma = 0.71 \times 3\mathrm{K} \approx 2\mathrm{K}$.

Problem 12.17.

A three dimensional container is separated in two parts by an adiabatic wall which moves without friction. One side is filled with a photon gas in equilibrium at temperature T. The other side contains a fully degenerate Fermi gas of $\langle N \rangle$ particles with spin $S = \frac{1}{2}$ and single particle energy $\varepsilon = pc$, with p the absolute value of the momentum and c the speed of light. Find the volume occupied by the fermions, V_F, under the assumption of mechanical equilibrium. We recall the relation between the pressure and the energy, $PV = U/d$ (see also Problem 10.13), with d the space dimensionality and V the volume occupied by the ultrarelativistic gas. Moreover, we recall that

$$\int_0^{+\infty} \frac{x^3}{e^x - 1}dx = \frac{\pi^4}{15}.$$

Solution

Let us start by analyzing the Fermi (F) gas. If we consider the single particle energy

$$\varepsilon = pc$$

we can evaluate the average number $\langle N \rangle$ of fermions

$$\langle N \rangle = 2\int_{\varepsilon \leq \varepsilon_F} \frac{d^3 p\, d^3 q}{h^3} = \frac{8\pi V_F}{h^3}\int_0^{p_F} p^2 dp = \frac{8\pi V_F}{h^3}\frac{p_F^3}{3} = \frac{8\pi V_F}{3h^3}\left(\frac{\varepsilon_F}{c}\right)^3$$

where we have used $p_F = \frac{\varepsilon_F}{c}$. The Fermi energy is

$$\varepsilon_F = \left(\frac{3\langle N \rangle h^3}{8\pi V_F}\right)^{1/3} c.$$

To determine the pressure, we use the information given in the text and look for the internal energy

$$U_F = 2 \int_{\varepsilon \leq \varepsilon_F} \frac{d^3 p \, d^3 q}{h^3} pc = \frac{8\pi V_F}{h^3} \int_0^{p_F} cp^3 dp = \frac{8\pi V_F c}{4h^3} \left(\frac{\varepsilon_F}{c} \right)^4$$

from which

$$P_F = \frac{1}{4} \left(\frac{\langle N \rangle}{V_F} \right)^{4/3} hc \left(\frac{3}{8\pi} \right)^{1/3}.$$

Let us then consider the photon (ph) gas. We can use, again, the relation $P_{ph} V_{ph} = U_{ph}/d$ with $d = 3$. For U_{ph} we write (see also Problems 10.10 and 10.11)

$$U_{ph} = 2 \frac{4\pi V_{ph}}{h^3} \int_0^{+\infty} \frac{1}{e^{\beta cp} - 1} p^2 (cp) dp$$

where, if we set $x = \beta cp$, we find

$$U_{ph} = \frac{8\pi V_{ph} c}{h^3} \frac{1}{(\beta c)^4} \int_0^{+\infty} \frac{x^3}{e^x - 1} dx.$$

Using the integral given in the text, we get

$$U_{ph} = \frac{8\pi}{h^3} V_{ph} \frac{k^4}{c^3} \frac{\pi^4 T^4}{15}$$

and the pressure is

$$P_{ph} = \frac{8\pi^5}{45} \frac{(kT)^4}{(hc)^3}.$$

The mechanical equilibrium requires $P_F = P_{ph}$ so that

$$V_F = \langle N \rangle \left(\frac{hc}{kT} \right)^3 \left(\frac{45}{32} \frac{1}{\pi^5} \right)^{3/4} \left(\frac{3}{8\pi} \right)^{\frac{1}{4}}.$$

Problem 12.18.

Our starting point is the first law of thermodynamics for a gas of N ($N \gg 1$) particles in a volume V.

1) Find the relation among the variations of the energy, dU, and volume, dV, in processes at constant T, V. Write $\left(\frac{\partial U}{\partial V} \right)_{T,N}$ in terms of the pressure P and a suitable derivative of the entropy.
2) Take the Sackur-Tetrode equation (see Problem 6.26) for the entropy and verify you obtain for $\left(\frac{\partial U}{\partial V} \right)_{T,N}$ the right result.
3) Compute the entropy for a Fermi and Bose-Einstein gas. Consider the bosons at a higher temperature with respect to their condensation point (what happens below

such temperature in these equations?). Give the final result in terms of

$$g_n(z) = \frac{1}{\Gamma(n)} \int_0^{+\infty} \frac{x^{n-1}dx}{z^{-1}e^x - 1}$$

$$f_n(z) = \frac{1}{\Gamma(n)} \int_0^{+\infty} \frac{x^{n-1}dx}{z^{-1}e^x + 1}$$

with a suitable n. Check these results in the classical limit against the Sackur-Tetrode equation.

4) Repeat the computation in 2) with the entropy computed in 3) writing $\left(\frac{\partial U}{\partial V}\right)_{T,N}$ explicitly in terms only of P and z (using the functions $f_n(z)$ and $g_n(z)$): what is changing? In the classical limit of high temperatures and low densities verify that you recover the result of the computation in 2).

For point 3) you can use the fact that, in the grand canonical ensemble, the entropy is given by

$$S = kT \left(\frac{\partial \ln \mathscr{Q}}{\partial T}\right)_{V,z} - Nk\ln z + k\ln \mathscr{Q}$$

where $\mathscr{Q}(T,V,z)$ is the grand canonical partition function and z the fugacity.

Solution
The first law of thermodynamics is

$$dU = TdS - PdV + \mu dN.$$

We can then expand the entropy as

$$dS = \left(\frac{\partial S}{\partial T}\right)_{V,N} dT + \left(\frac{\partial S}{\partial V}\right)_{T,N} dV + \left(\frac{\partial S}{\partial N}\right)_{T,V} dN$$

and plugging it back in the previous formula

$$dU = T\left[\left(\frac{\partial S}{\partial T}\right)_{V,N} dT + \left(\frac{\partial S}{\partial V}\right)_{T,N} dV + \left(\frac{\partial S}{\partial T}\right)_{T,V} dN\right] - PdV + \mu dN.$$

For constant T, N we get

$$\left(\frac{\partial U}{\partial V}\right)_{T,N} = \left[T\left(\frac{\partial S}{\partial V}\right)_{T,N} - P\right].$$

The Sackur-Tetrode entropy is a function of U, V, N

$$S(U,V,N) = Nk\left\{\frac{5}{2} - \ln\left[\left(\frac{3\pi\hbar^2}{m}\right)^{3/2} \frac{N^{\frac{5}{2}}}{VU^{\frac{3}{2}}}\right]\right\}$$

but, substituting the energy of a free gas ($U = \frac{3}{2}NkT$) we get

$$S(T,V,N) = Nk \left\{ \frac{5}{2} + \ln \left[\left(\frac{mkT}{2\pi\hbar^2} \right)^{\frac{3}{2}} \frac{V}{N} \right] \right\}.$$

We can now explicitly compute $\left(\frac{\partial S}{\partial V} \right)_{T,N}$

$$T \left(\frac{\partial S}{\partial V} \right)_{T,N} = \frac{NkT}{V} = P$$

where we used the equation of state of an ideal gas $PV = NkT$. Combining now the above expressions we find that

$$\left(\frac{\partial U}{\partial V} \right)_{T,N} = \left[T \left(\frac{\partial S}{\partial V} \right)_{T,N} - P \right] = [P - P] = 0.$$

This is the expected result for the classical gas since, in this case, the internal energy is not a function of the volume V. We must now repeat the previous computation in the quantum case

$$S = kT \left(\frac{\partial \ln \mathscr{Q}}{\partial T} \right)_{V,z} - Nk\ln z + k\ln \mathscr{Q}$$

where $\mathscr{Q}(T,V,z)$ is the grand canonical partition function and z the fugacity. We furthermore remember that for a quantum gas

$$\frac{PV}{kT} = \ln \mathscr{Q} = \frac{g_s V}{\lambda^3} \phi_{5/2}(z)$$

$$\frac{N}{V} = g_s \frac{\phi_{3/2}(z)}{\lambda^3}$$

are valid. In the above expressions, g_s is the spin degeneracy, $\phi = f, g$ are the Fermi-Dirac and Bose-Einstein functions respectively, and $\lambda = h/\sqrt{2\pi mkT}$ is the thermal length scale. The above equations are valid for temperatures above the point of the Bose-Einstein condensation, otherwise we must add to N a contribution representing the condensed phase. Finding S is now simple

$$S = \frac{5}{2} \frac{g_s kV \phi_{5/2}(z)}{\lambda^3} - Nk\ln z = S_1(T,V,z) + S_2(z,N)$$

where we have defined $S_1(T,V,z) = \frac{5}{2} \frac{g_s kV \phi_{5/2}(z)}{\lambda^3}$ and $S_2(z,N) = -Nk\ln z$. It is easy to compare this result with the classical result: in this limit $\phi_n(z) \approx z$ and using

$$\frac{N}{V} = g_s \frac{\phi_{3/2}(z)}{\lambda^3} \approx \frac{g_s z}{\lambda^3}$$

we eliminate z from S giving the result

$$S = \frac{5}{2}\frac{g_s k V \phi_{5/2}(z)}{\lambda^3} - Nk\ln z \approx \frac{5}{2}\frac{g_s k V z}{\lambda^3} - Nk\ln z = \frac{5}{2}Nk - Nk\ln z =$$

$$\frac{5}{2}Nk + Nk\ln\left(\frac{g_s}{\lambda^3}\frac{V}{N}\right) = Nk\left\{\frac{5}{2} + \ln\left[g_s\left(\frac{mkT}{2\pi\hbar^2}\right)^{\frac{3}{2}}\frac{V}{N}\right]\right\}.$$

The only difference lies in the spin degeneracy term. If we repeat the classical computation including g_s the two results are identical. Let us then start by evaluating

$$\left(\frac{\partial S}{\partial V}\right)_{T,N} = \left(\frac{\partial S_1}{\partial V}\right)_{T,N} + \left(\frac{\partial S_2}{\partial V}\right)_{T,N}$$

where we separated the contribution of S_1 and S_2. The entropy S_1 is a function of T, V, z, while in $\left(\frac{\partial U}{\partial V}\right)_{T,N}$ we must compute the entropy at constant T, N. We then need results at constant z and not at constant N. Using the formalism of the Jacobians we find

$$\left(\frac{\partial S_1}{\partial V}\right)_{T,N} = \frac{\partial(S_1,N)}{\partial(V,N)}\bigg|_T = \frac{\frac{\partial(S_1,N)}{\partial(V,z)}}{\frac{\partial(V,N)}{\partial(V,z)}}\bigg|_T = \frac{\left(\frac{\partial S_1}{\partial V}\right)_{T,z}\left(\frac{\partial N}{\partial z}\right)_{T,V} - \left(\frac{\partial S_1}{\partial z}\right)_{T,V}\left(\frac{\partial N}{\partial V}\right)_{T,z}}{\left(\frac{\partial N}{\partial z}\right)_{T,V}} =$$

$$\left(\frac{\partial S_1}{\partial V}\right)_{T,z} - \left(\frac{\partial S_1}{\partial z}\right)_{T,V}\left(\frac{\partial N}{\partial V}\right)_{T,z}\left(\frac{\partial z}{\partial N}\right)_{T,V} =$$

$$\left(\frac{\partial S_1}{\partial V}\right)_{T,z} + \left(\frac{\partial S_1}{\partial z}\right)_{T,V}\left(\frac{\partial z}{\partial V}\right)_{T,N}$$

where we used

$$\left(\frac{\partial N}{\partial V}\right)_{T,z}\left(\frac{\partial z}{\partial N}\right)_{T,V}\left(\frac{\partial V}{\partial z}\right)_{T,N} = -1.$$

The derivatives of S_1 with respect to V and z are easy to compute

$$\left(\frac{\partial S_1}{\partial V}\right)_{T,z} = \frac{5}{2}\frac{g_s k \phi_{5/2}(z)}{\lambda^3}$$

$$\left(\frac{\partial S_1}{\partial z}\right)_{T,V} = \frac{5}{2}\frac{g_s k V}{\lambda^3}\frac{d\phi_{5/2}(z)}{dz} = \frac{5}{2}\frac{k g_s V \phi_{3/2}(z)}{z\lambda^3}$$

remembering that

$$\frac{d\phi_n(z)}{dz} = \frac{\phi_{n-1}(z)}{z}.$$

Only the derivative of z with respect to V needs a little care. In fact, $g_s \phi_{3/2}(z) = \frac{\lambda^3 N}{V}$, from which we get

$$\left(\frac{\partial \phi_{3/2}(z)}{\partial V} \right)_{T,N} = -\frac{\lambda^3 N}{g_s V^2} = -\frac{\phi_{3/2}(z)}{V}$$

and

$$\left(\frac{\partial z}{\partial V} \right)_{T,N} = \left(\frac{\partial \phi_{3/2}(z)}{\partial V} \right)_{T,N} \left(\frac{dz}{d\phi_{3/2}(z)} \right) = -\frac{z\phi_{3/2}(z)}{V\phi_{1/2}(z)}.$$

The final result for $\left(\frac{\partial S_1}{\partial V} \right)_{T,N}$ is

$$\left(\frac{\partial S_1}{\partial V} \right)_{T,N} = \frac{5}{2} \frac{g_s k \phi_{5/2}(z)}{\lambda^3} - \frac{5}{2} \frac{k g_s \phi_{3/2}^2(z)}{\lambda^3 \phi_{1/2}(z)}.$$

The entropy $S_2(z,N)$ is a function of z and N and we get

$$\left(\frac{\partial S_2}{\partial V} \right)_{T,N} = -\frac{Nk}{z} \left(\frac{\partial z}{\partial V} \right)_{T,N} = \frac{Nk}{V} \frac{\phi_{3/2}(z)}{\phi_{1/2}(z)}.$$

All these results together give

$$T \left(\frac{\partial S}{\partial V} \right)_{T,N} = T \left(\frac{\partial S_1}{\partial V} \right)_{T,N} + T \left(\frac{\partial S_2}{\partial V} \right)_{T,N} =$$

$$\frac{5}{2} \frac{g_s kT \phi_{5/2}(z)}{\lambda^3} - \frac{5}{2} \frac{kT g_s \phi_{3/2}^2(z)}{\lambda^3 \phi_{1/2}(z)} + \frac{NkT}{V} \frac{\phi_{3/2}(z)}{\phi_{1/2}(z)}$$

and, reporting in terms of z and the pressure P, we get

$$T \left(\frac{\partial S}{\partial V} \right)_{T,N} = \left[\frac{5}{2} - \frac{3}{2} \frac{\phi_{3/2}^2(z)}{\phi_{5/2}(z)\phi_{1/2}(z)} \right] P$$

where we have used

$$P = \frac{g_s kT \phi_{5/2}(z)}{\lambda^3}.$$

The final result is

$$\left(\frac{\partial U}{\partial V} \right)_{T,N} = \left[T \left(\frac{\partial S}{\partial V} \right)_{T,N} - P \right] = \frac{3}{2} \left[1 - \frac{\phi_{3/2}^2(z)}{\phi_{5/2}(z)\phi_{1/2}(z)} \right] P.$$

In the classical limit $\phi_n(z) \approx z$ and

$$\left(\frac{\partial U}{\partial V} \right)_{T,N} = \frac{3}{2} \left[1 - \frac{\phi_{3/2}^2(z)}{\phi_{5/2}(z)\phi_{1/2}(z)} \right] P \approx \frac{3}{2} \left[1 - \frac{z^2}{z^2} \right] P = 0.$$

Index

Adiabatic compressibility, 196
Adiabatic transformation, 200, 329, 367
Angular momentum, 8, 113, 115–117, 121, 122, 131, 133, 134, 154, 174, 261
Anharmonic potential, 93, 257, 263

Black body radiation, 327, 329, 330, 332, 333
Bohr-Sommerfeld quantization, 21, 184, 185
Boltzmann formula, 28, 203, 204, 207, 208, 212–214, 216, 217, 219, 221, 223, 231
Bose-Einstein Condensation, 315–319, 325
Bose-Einstein functions, 34, 315, 318
Bose-Einstein gas, 29, 315–318, 321–323, 325, 332, 334, 371, 373

Canonical Ensemble, 28, 227–229, 231, 232, 234–236, 238, 239, 241, 243, 244, 246, 248–251, 253, 257, 259, 261, 263, 265, 268, 269, 271–273, 275, 276, 278–280, 282–284, 363, 369, 373, 378, 379
Canonical transformation, 238
Central forces, 145, 146, 148, 150, 151, 156, 157, 160, 180, 232, 272, 280, 311, 351, 357
Chemical potential, 194, 202, 269, 280, 283, 284, 289, 296–298, 321, 358, 384
Chemical reaction, 283, 296, 298, 384
Classical limit, 323, 355, 361, 371, 373, 383, 388
Clausius-Clapeyron equation, 291, 322
Clebsh-Gordan coefficients, 10, 126
Coherent State, 110, 111
Complex plane, 13, 17, 183, 279
Creation and annihilation operators, 166, 177, 179, 253, 263
Critical temperature, 236, 315–319, 321, 322, 325

Curie law, 259, 261, 323

De Broglie equation, 19, 40
Delta function, 92, 96, 99, 101
Density flux, 3, 58, 65, 79, 82, 160, 358
Density of states, 321, 327, 333, 338
Discrete spectra, 48, 53, 78, 118, 122, 135, 164, 165, 169, 203, 204, 212, 213, 227, 228, 234–236, 248
Dispersion relations, 344, 345

Effusion, 30, 301, 303, 305, 308–311, 313, 330
Electric field, 107, 172, 178, 179, 182, 183, 273
Enthalpy, 23, 198, 225, 276, 291, 333, 388
Entropy, 23, 194–196, 198, 200–205, 207, 208, 211–214, 216, 217, 221, 223, 225, 231, 235, 275, 329, 333, 342, 365, 367, 368, 388
Equipartition theorem, 214, 216, 221, 249–251, 361
Extensive function, 25, 193, 194, 201, 369

Fermi-Dirac functions, 35, 342
Fermi-Dirac gas, 29, 332, 337, 338, 340, 342, 344–346, 348–351, 353, 355–358, 360, 361, 371, 373, 385, 387
First law of Thermodynamics, 194, 198, 201, 202, 205, 210, 365, 367
Fluctuations, 31, 243, 244, 271, 279, 363, 365, 367–369, 374
Free energy, 23, 193, 200, 202, 225, 227–229, 231, 232, 234–236, 238, 239, 241, 243, 244, 246, 248–251, 253, 257, 259, 261, 263, 265, 268, 269, 271–273, 275, 276, 278–280, 282–284, 286, 298, 356, 378, 379, 388
Fugacity, 29, 289, 290, 321, 342

Gamma function, 32, 221, 269, 315, 317, 318, 323, 325, 334, 342, 355, 361, 384, 385, 388

Gibbs potential, 23, 193, 211, 225, 276, 333, 388

Grand Canonical Ensemble, 29, 289–291, 294, 296–298, 363, 369, 371, 381, 384

Grand potential, 23, 291, 315, 317, 355, 361

Gravity, 210, 244, 246, 306, 310, 325, 346, 348, 349

Harmonic oscillator, 5, 43, 51, 55, 77, 86, 93, 95, 103, 107, 110, 111, 145, 151, 153, 166, 171, 177–179, 184, 214, 221, 232, 253, 257, 263, 321, 367

Heisenberg representation, 51, 86

Heisenberg uncertainty principle, 40, 44, 60, 101, 103, 113, 116

Helium atom, 157

High/Low temperature limit, 204, 227, 229, 234, 236, 241, 243, 259, 261, 265, 282, 290, 323, 334, 338, 340, 342, 373, 388

Hubble constant, 385

Hydrogen atom, 12, 13, 133, 150, 154, 159, 161, 180, 182, 296

Hypergeometric series, 16, 79, 89, 146, 151

Ideal/Classical gas, 210, 227, 229, 272, 273, 275, 280, 284, 291, 303, 305, 306, 308–310, 371, 373

Indicial equation, 14, 16, 79, 89, 145, 146, 151

Internal energy, 194, 195, 198, 201–205, 207, 208, 212–214, 216, 217, 221, 225, 227–229, 231, 232, 235, 236, 244, 246, 248–251, 253, 257, 265, 269, 271, 273, 275, 276, 278, 279, 282, 291, 327, 329, 330, 332–334, 338, 340, 342, 344–346, 350, 351, 353, 356, 357, 360, 361, 363, 365, 369, 378, 381, 385, 387, 388

Irregular singularities, 14, 17, 89, 146, 151

Ising model, 208, 378, 379

Isothermal compressibility, 196, 356, 369

Jacobi Identity, 40, 75, 110

Jacobian, 25, 195, 365, 388

Magnetic field, 118, 139, 141, 201, 205, 208, 228, 259, 261, 297, 323, 358, 378, 379

Magnetic moment, 118, 122, 139, 141, 205, 208, 228, 259, 261, 297, 323, 358, 378, 379

Maxwell distribution, 227, 280, 303, 305, 306, 308–310

Maxwell relations, 24, 195, 196, 198, 200, 332, 368

Microcanonical Ensemble, 28, 203, 204, 207, 208, 212–214, 216–219, 221, 223

Momentum of inertia, 116, 163, 172, 218, 259, 265, 273

Multidimensional sphere, 33, 250, 358, 387

Neutrinos, 385

Newton constant, 385

Occupation number, 203, 212, 297, 301, 315–319, 321–323, 325, 332, 334, 338, 340, 342, 346, 348–351, 353, 355, 357, 358, 361, 371, 373, 381, 383, 384

One dimensional Schrödinger equation, 3, 44, 56, 58, 60, 62, 65, 66, 69, 73, 89, 92, 96, 99, 128, 168, 229

Particles emission, 183, 377

Pauli matrices, 10, 48, 71, 118, 120, 121, 124, 141, 173

Pauli principle, 159, 373

Perturbation theory, 21, 163–169, 171–174, 177–180, 182, 257, 263

Photon gas, 327, 329, 383, 384

Plane wave, 44, 73, 75, 79, 92, 139

Positrons, 384, 385

Potential well, 56, 58, 60, 62, 66, 69, 73, 99, 167, 168, 185

Quantum corrections, 29, 229, 265, 301, 323, 334, 355

Reflection and transmission coefficients, 79, 82, 101, 128, 183

Regular singularities, 13, 15, 79, 89, 145, 146, 151, 156

Relativistic gas, 268, 349

Riemann zeta function, 34, 315, 317–319, 322, 325, 329, 384, 385

Rotating systems, 130, 135, 163, 172, 218

Sackur-Tetrode formula, 225, 286, 388

Schrödinger representation, 51, 86

Schwartz lemma, 195, 196, 200, 211, 291

Singlet, 128, 157

Specific heat, 25, 196, 201, 204, 205, 251, 265, 291, 340, 342, 365, 368

Spherical harmonics, 9, 115–117, 121, 128, 135, 156, 172, 180, 182, 265

Spherical polar coordinates, 113, 145, 146, 148, 151, 153, 156, 182, 269, 273, 280, 301, 303, 319, 329, 333, 340, 351, 357

Spin, 10, 120, 124, 126, 128, 129, 131, 136, 138, 139, 141, 154, 173, 208, 323
Spin degeneracy, 296, 301, 315–319, 321–323, 325, 327, 329, 330, 332–334, 337, 338, 340, 342, 344–346, 348–351, 353, 355–358, 360, 361, 371, 373, 387, 388
Spring constant, 200, 367
States degeneracy, 39, 128, 129, 135, 157, 163, 164, 167, 168, 174, 204, 212, 227, 229, 235, 236, 246, 261, 265, 271, 279, 282
Stefan-Boltzmann law, 327, 330, 332
Stirling approximation, 33, 203, 204, 208, 212–214, 216, 219, 223, 231, 239, 269, 275, 283, 376, 381
Sun's temperature, 327, 330

Thermal expansion coefficient, 196, 211, 365
Thermal length scale, 234, 239, 284, 297, 318, 342, 383, 388

Thermodynamic potentials, 23, 193, 194, 198, 202, 211, 225, 276, 286, 332, 381, 388
Three dimensional Schrödinger equation, 12, 56, 135, 145, 151, 153, 161
Time evolution, 39, 44, 48, 53, 55, 58, 71, 107, 117, 118, 121, 126, 174
Triplet, 128, 157
Tunnel effect, 186, 189

Ultrarelativistic gas, 216, 223, 319, 355, 385, 387
Universe, 329, 384, 385

Variational method, 6, 93, 96
Virial theorem, 150

Wien law, 327
WKB method, 19, 183, 184, 186, 189

UNITEXT – Collana di Fisica e Astronomia

Edited by:

Michele Cini
Stefano Forte
Massimo Inguscio
Guida Montagna
Oreste Nicrosini
Franco Pacini
Luca Peliti
Alberto Rotondi

Editor at Springer:
Marina Forlizzi
marina.forlizzi@springer.com

Atomi, Molecole e Solidi
Esercizi Risolti
Adalberto Balzarotti, Michele Cini, Massimo Fanfoni
2004, VIII, 304 pp, ISBN 978-88-470-0270-8

Elaborazione dei dati sperimentali
Maurizio Dapor, Monica Ropele
2005, X, 170 pp., ISBN 978-88470-0271-5

**An Introduction to Relativistic Processes and the Standard
Model of Electroweak Interactions**
Carlo M. Becchi, Giovanni Ridolfi
2006, VIII, 139 pp., ISBN 978-88-470-0420-7

Elementi di Fisica Teorica
Michele Cini
2005, ristampa corretta 2006, XIV, 260 pp., ISBN 978-88-470-0424-5

Esercizi di Fisica: Meccanica e Termodinamica
Giuseppe Dalba, Paolo Fornasini
2006, ristampa 2011, X, 361 pp., ISBN 978-88-470-0404-7

Structure of Matter
An Introductory Corse with Problems and Solutions
Attilio Rigamonti, Pietro Carretta
2nd ed. 2009, XVII, 490 pp., ISBN 978-88-470-1128-1

Introduction to the Basic Concepts of Modern Physics
Special Relativity, Quantum and Statistical Physics
Carlo M. Becchi, Massimo D'Elia
2007, 2nd ed. 2010, X, 190 pp., ISBN 978-88-470-1615-6

Introduzione alla Teoria della elasticità
Meccanica dei solidi continui in regime lineare elastico
Luciano Colombo, Stefano Giordano
2007, XII, 292 pp., ISBN 978-88-470-0697-3

Fisica Solare
Egidio Landi Degl'Innocenti
2008, X, 294 pp., ISBN 978-88-470-0677-5

Meccanica quantistica: problemi scelti
100 problemi risolti di meccanica quantistica
Leonardo Angelini
2008, X, 134 pp., ISBN 978-88-470-0744-4

Fenomeni radioattivi
Dai nuclei alle stelle
Giorgio Bendiscioli
2008, XVI, 464 pp., ISBN 978-88-470-0803-8

Problemi di Fisica
Michelangelo Fazio
2008, XII, 212 pp., ISBN 978-88-470-0795-6

Metodi matematici della Fisica
Giampaolo Cicogna
2008, ristampa 2009, X, 242 pp., ISBN 978-88-470-0833-5

Spettroscopia atomica e processi radiativi
Egidio Landi Degl'Innocenti
2009, XII, 496 pp., ISBN 978-88-470-1158-8

Particelle e interazioni fondamentali
Il mondo delle particelle
Sylvie Braibant, Giorgio Giacomelli, Maurizio Spurio
2009, ristampa 2010, XIV, 504 pp., ISBN 978-88-470-1160-1

I capricci del caso
Introduzione alla statistica, al calcolo della probabilità e alla teoria degli errori
Roberto Piazza
2009, XII, 254 pp., ISBN 978-88-470-1115-1

Relatività Generale e Teoria della Gravitazione
Maurizio Gasperini
2010, XVIII, 294 pp., ISBN 978-88-470-1420-6

Manuale di Relatività Ristretta
Maurizio Gasperini
2010, XVI, 158 pp., ISBN 978-88-470-1604-0

Metodi matematici per la teoria dell'evoluzione
Armando Bazzani, Marcello Buiatti, Paolo Freguglia
2011, X, 192 pp., ISBN 978-88-470-0857-1

Esercizi di metodi matematici della fisica
Con complementi di teoria
G. G. N. Angilella
2011, XII, 294 pp., ISBN 978-88-470-1952-2

Il rumore elettrico
Dalla fisica alla progettazione
Giovanni Vittorio Pallottino
2011, XII, 148 pp., ISBN 978-88-470-1985-0

Note di fisica statistica
(con qualche accordo)
Roberto Piazza
2011, XII, 306 pp., ISBN 978-88-470-1964-5

Stelle, galassie e universo
Fondamenti di astrofisica
Attilio Ferrari
2011, XVIII, 558 pp., ISBN 978-88-470-1832-7

Introduzione ai frattali in fisica
Sergio Peppino Ratti
2011, XIV, 306 pp., ISBN 978-88-470-1961-4

From Special Relativity to Feynman Diagrams
A Course of Theoretical Particle Physics for Beginners
Riccardo D'Auria, Mario Trigiante
2011, X, 562 pp., ISBN 978-88-470-1503-6

Problems in Quantum Mechanics with solutions
Emilio d'Emilio, Luigi E. Picasso
2011, X, 354 pp., ISBN 978-88-470-2305-5

Fisica del Plasma
Fondamenti e applicazioni astrofisiche
Claudio Chiuderi, Marco Velli
2011, X, 222 pp., ISBN 978-88-470-1847-1

Solved Problems in Quantum and Statistical Mechanics
Michele Cini, Francesco Fucito, Mauro Sbragaglia
2012, VIII, 396 pp., ISBN 978-88-470-2314-7